T0319283

The Craft and Science of Coffee

Dedication

We would like to dedicate this book to the many millions of coffee farmers around the world, without whom there would be no coffee. The contribution fees of all authors and royalties from this book will be placed in a special fund supporting the *Nespresso* AAA Farmer Future Program, the first ever retirement savings fund created especially for coffee farmers in Colombia. Designed to protect the future well-being of farmers, this platform will also help facilitate the generational transfer of farms from parents to children, motivating young people to carry on coffee production. This program was initiated in collaboration with the Colombian Ministry of Labor, the Aguadas Coffee Growers' Cooperative, Expocafé, and Fairtrade International. More details about this program can be found in the outlook of Chapter 6.

The Craft and Science of Coffee

Edited by

Britta Folmer

Editorial Board

Imre Blank, Adriana Farah, Peter Giuliano, Dean Sanders and Chris Wille

AMSTERDAM • BOSTON • HEIDELBERG • LONDON
NEW YORK • OXFORD • PARIS • SAN DIEGO
SAN FRANCISCO • SINGAPORE • SYDNEY • TOKYO
Academic Press is an imprint of Elsevier

Academic Press is an imprint of Elsevier
125 London Wall, London EC2Y 5AS, United Kingdom
525 B Street, Suite 1800, San Diego, CA 92101-4495, United States
50 Hampshire Street, 5th Floor, Cambridge, MA 02139, United States
The Boulevard, Langford Lane, Kidlington, Oxford OX5 1GB, United Kingdom

Notices
Knowledge and best practice in this field are constantly changing. As new research and experience broaden our understanding, changes in research methods, professional practices, or medical treatment may become necessary.

Practitioners and researchers must always rely on their own experience and knowledge in evaluating and using any information, methods, compounds, or experiments described herein. In using such information or methods they should be mindful of their own safety and the safety of others, including parties for whom they have a professional responsibility.

To the fullest extent of the law, neither the Publisher nor the authors, contributors, or editors, assume any liability for any injury and/or damage to persons or property as a matter of products liability, negligence or otherwise, or from any use or operation of any methods, products, instructions, or ideas contained in the material herein.

Library of Congress Cataloging-in-Publication Data
A catalog record for this book is available from the Library of Congress

British Library Cataloguing-in-Publication Data
A catalogue record for this book is available from the British Library

ISBN: 978-0-12-803520-7

For information on all Academic Press publications
visit our website at https://www.elsevier.com/

Publisher: Nikki Levy
Acquisition Editor: Nancy Maragioglio
Editorial Project Manager: Billie Jean Fernandez
Production Project Manager: Nicky Carter
Designer: Matthew Limbert

Typeset by TNQ Books and Journals

Contents

11. The Roast—Creating the Beans' Signature

Stefan Schenker, Trish Rothgeb

19. We Consumers—Tastes, Rituals, and Waves

Jonathan Morris

20. Human Wellbeing—Sociability, Performance, and Health

Britta Folmer, Adriana Farah, Lawrence Jones, Vincenzo Fogliano

List of Contributors

Juan Carlos Ardila, Cafexport SA, Vevey, Switzerland

Paulo Barone, Nestlé Nespresso SA, Lausanne, Switzerland

Pablo Benavides, Cenicafé FNC, Manizales, Colombia

Benoit Bertrand, CIRAD, Agriculture Research for Development, Montpellier, France

Imre Blank, Nestec Ltd., Nestlé Research Center, Lausanne, Switzerland

Carlos Brando, P&A International Marketing, Espírito Santo do Pinhal, São Paulo, Brasil

David Browning, TechnoServe, Washington, DC, United States

Lee Byers, Fairtrade International (FLO), Bonn, Germany

Michelle Deugd, Rainforest Alliance, San José, Costa Rica

Andreas Dunkel, Technical University of Munich, Freising, Germany

Adriana Farah, Federal University of Rio de Janeiro, Rio de Janeiro, Brazil

Vincenzo Fogliano, Wageningen University, Wageningen, The Netherlands

Britta Folmer, Nestlé Nespresso SA, Lausanne, Switzerland

Alvaro Gaitán, Cenicafé FNC, Manizales, Colombia

Peter Giuliano, Specialty Coffee Association of America, Santa Ana, CA, United States

Arne Glabasnia, Nestec Ltd., Nestlé Product and Technology Center Beverages, Lausanne, Switzerland

Carmenza Góngora, Cenicafé FNC, Manizales, Colombia

Jeffrey Hayward, Rainforest Alliance, Washington, DC, United States

Juan Carlos Herrera, Nestec Ltd., Nestlé Research and Development Center Plant Sciences, Tours, France

Thomas Hofmann, Technical University of Munich, Freising, Germany

Jwanro Husson, Nestec Ltd., Nestlé Research and Development Center Plant Sciences, Tours, France

Lawrence Jones, Huntington Medical Research Institute, Pasadena, CA, United States

Bernard Kilian, INCAE Business School, Alajuela, Costa Rica

Christian Klatt, Mahlkönig GmbH, Hamburg, Germany

Harriet Lamb, Fairtrade International (FLO), Bonn, Germany

Charles Lambot, Nestec Ltd., Nestlé Research and Development Center Plant Sciences, Tours, France

Ted R. Lingle, Coffee Quality Institute, Aliso Viejo, CA, United States

Celia Marsh, Science Writer and Researcher, Geneva, Switzerland

Sunalini N. Menon, Coffeelab Limited, Bangalore, India

Frédéric Mestdagh, Nestec Ltd., Nestlé Product and Technology Center Beverages, Orbe, Switzerland

Shirin Moayyad, Nestlé Nespresso SA, Lausanne, Switzerland

Jonathan Morris, University of Hertfordshire, Hatfield, United Kingdom

Eric Nadelberg, Granite Mountain Market Forecasts, Prescott, AZ, United States

Martin R.A. Noponen, Rainforest Alliance, London, United Kingdom

Carlos Oliveros, Cenicafé FNC, Manizales, Colombia

Juan Pablo Orjuela, CFX Risk Management Ltd., London, United Kingdom

Aida Peñuela, Cenicafé FNC, Manizales, Colombia

Jérôme Perez, Nestlé Nespresso SA, Lausanne, Switzerland

Arne Pietsch, University of Applied Sciences Lübeck, Lübeck, Germany

Luigi Poisson, Nestec Ltd., Nestlé Product and Technology Center Beverages, Orbe, Switzerland

Jaime R. Polit, Be Green Trading SA, Lausanne, Switzerland

Lawrence Pratt, INCAE Business School, Alajuela, Costa Rica

Sabine Puget, Nestlé Nespresso SA, Lausanne, Switzerland

Karsten Ranitzsch, Nestlé Nespresso SA, Lausanne, Switzerland

Alexis Rodriguez, Nestlé Nespresso SA, Lausanne, Switzerland

Trish Rothgeb, Wrecking Ball Coffee Roasters, San Francisco, CA, United States

Siavosh Sadeghian, Cenicafé FNC, Manizales, Colombia

Dean Sanders, GoodBrand, London, United Kingdom

Juan R. Sanz-Uribe, Cenicafé FNC, Manizales, Colombia

Stefan Schenker, Buhler AG, Uzwil, Switzerland

Samo Smrke, Zurich University of Applied Sciences, Wädenswil, Switzerland

Paul Songer, Songer and Associates, Inc., Boulder, CO, United States

Ria Stout, Rainforest Alliance, Antigua, Guatemala

Edouard Thomas, Nestlé Nespresso SA, Lausanne, Switzerland

Dominique Valentin, AgroSup Dijon, INRA, Dijon, France

Martin von Blittersdorff, CAFEA GmbH, Hamburg, Germany

Marco Wellinger, Zurich University of Applied Sciences, Wädenswil, Switzerland

Chris Wille, Sustainable Agriculture Consultant, Portland, OR, United States

Yves Wyser, Nestec Ltd., Nestlé Research Center, Lausanne, Switzerland

Chahan Yeretzian, Zurich University of Applied Sciences, Wädenswil, Switzerland

Yusianto, Indonesian Coffee and Cocoa Research Institute (ICCRI), Jember, Indonesia

Preface

Vigorous debates were important events in coffeehouses of the 17th century. Many of the topics being argued in those days—a time we now think of as "the Enlightenment"—were philosophic, scientific, and political. In those days, coffee was a fuel for engaged, spirited discussions.

This book serves this same purpose for anyone active in the coffee value chain. We have many topics on coffee to discuss, including new scientific advancements, the same persistent social problems, and swiftly changing political landscapes. At the same time, there are many great ideas on how to make progress. For this book, we have carefully sought out coffee leaders to represent their areas of expertise, in an effort to bring to the surface the very best thinking on these topics. Their views, however, are not necessarily reflective of those of the editors or their organizations. We also know that the authors' opinions may differ, one from the other. We embrace these intellectual differences and hope the discourse herein will lead to a more informed debate.

We, therefore, wish that what has been written will engage the reader to delve deeper into the topics, expand their knowledge, and use it, whether as an academic student or as an influential thought leader in the industry, to find innovative approaches and solutions to make our coffee value chain a better one. We believe that connecting current issues in coffee with a more in-depth academic perspective is one important way to achieve this.

We see this book as a broad view combined with an overview of scientific advancements. We offer it in the spirit of discussion and debate, and hope it serves as a small revival of that great coffeehouse tradition.

Britta Folmer, Imre Blank, Adriana Farah, Peter Giuliano, Dean Sanders and Chris Wille

Introduction

Our colorful and enduring relationship with coffee began before recorded history. *Homo sapiens*, *Coffea arabica*, and *Coffea canephora* evolved in the very same place, the forests of eastern Africa. We can only imagine how the prehistorical romance began. Coffee making was not well documented until the 15th century, when it was a spiritual beverage in the Sufi monasteries of Yemen even if the Ethiopians were surely enjoying the fruits of the local *Coffea* trees before then.

The fact is that humans and coffee have traveled together. Humans have helped this plant expand from eastern Africa to Asia and Latin America. Coffee as a beverage stimulated the Ottoman Empire. North African Muslims brought the bean to Italy through Venice's ports during the Renaissance, and from there, the coffeehouse culture began to spread throughout Europe and the Americas. People all over the world celebrated the pleasures of the drink during the Age of Enlightenment and through many cultural mini-epochs such as the Beatnik Era. And it has never been more popular than now, during the Reign of the Millennials.

Writing about coffee has continued over time, and today there is a goldmine of scientific and professional information available from geneticists, agronomists, chemists, engineers, sensory scientists, historians, health professionals, and various other disciplines. Scientific papers are abundant covering every step of the value chain. Coffee professionals write training materials, instructions, guidelines, and blogs equally providing information for other professionals in coffee.

For at least 20 million farmers and millions of others along the coffee value chain, coffee is not just a delicious beverage, it is central to their livelihood. Both for these farmers and for all those who are involved in the coffee value chain, the craft and science of coffee go hand in hand. One of the central themes throughout this book is how we can create new value in coffee, by strengthening the interplay between craft and science. But are we using our knowledge and resources wisely? How can this add value for both producer and consumer? And, how can we build on this knowledge to drive coffee forward, so that it remains central to the lives of successive generations?

As an agricultural crop, coffee is naturally subject to weather conditions and diseases. However, over the years coffee farmers have also had to deal with social unrest, varying market prices, demand for sustainably grown coffee, climate change, and other factors. This has led to various nongovernmental and nonprofit organizations as well as governments and the private sector to focus on farmers, and coffee itself has become a global laboratory for testing models of equitable and sustainable rural development.

The mission has been both to help the growers and to improve coffee production as an economic engine of growth. Various research organizations started focusing on trying to understand how genetics, agriculture, microbiology, and technology could help farmers improve quality and productivity in a cost-efficient way.

Today many farmers, researchers, and different organizations approach coffee growing hand in hand, learning from experience and experimenting with scientific support. At the same time, as they aim to increase quality and productivity, all aspects of sustainability are considered. This includes a cleaner environment, improved social conditions, and a better financial situation for the farmer.

Many different people add value to coffee before it is ready for consumption. The coffee is cupped to identify its character and quality. Contributions of the cupper, grader, master blender, and roaster each add their personal touch to the bean, allowing it to express its aromatic potential.

However, this is not the full story. When great quality is achieved, curious scientists will aim to understand how it was accomplished. They will then further explore the field to find new opportunities to create even better and consistent quality, and perhaps, to expand the flavor dimension. Back in laboratories, researchers gather speed in the complicated chemistry of coffee, decaffeination, ways to extract more flavor out of every bean, the arcane secrets of grinding and roasting, the influence of the water used in brewing, and various other subjects. Engineers play a pivotal role in translating scientific findings in technologically feasible processes.

Although many people working in the coffee industry may never directly observe the scientific developments, the outcome in the form of improved decaffeinated coffee quality, for example, more precise roasting equipment or improved extraction methods have often gone through the hands of many researchers. So the craft is improving, thanks to scientific contributions and technological breakthroughs and science and technology is learning from the craftsmen by carefully studying how value is created and how quality is enhanced.

In the end, whether through craft or science, coffee is produced and processed to please the consumer in different ways and the story of the coffee can further enhance its appreciation. The consumer is the one who attributes the value, by paying the price that is requested. But just like the farmer, there is no single consumer. Consumers' coffee consumption habits differ depending on culture, taste preference, knowledge, moment, and many other factors. But the coffee experience is not limited by habits and traditions. A consumer's knowledge of sustainability, coffee origin, history, processing, and flavor diversity will naturally increase the value of coffee, which can bring better returns to the farmer. Furthermore, when they understand the positive health implications of coffee consumption, it may make them value it even more. And we should not forget that coffee is like a sacred elixir in a cup, bringing people together and fostering communications. Whether at home, the workplace, or a café, we congregate over coffee to chat, pontificate, loiter, and connect with each other.

This book curates corroborative facts and experience-based observations and perspectives from preeminent experts in each subject as we follow the coffee bean from the farm to end-point enjoyment. We wanted to explore the rich and dynamic intersection between coffee related science and craft. The scientists may be passionate about coffee, but they are coolly objective about their research and theories. The craftspeople and connoisseurs want to learn more about their own trade, about the skillsets, art, and magic of others along the coffee value chain, and about the hard science behind the bean. By compiling essays about the science and craft of coffee between two covers, we hope to stimulate cross-learning and silo-hopping explorations among the scientists and craftspeople, provoke curiosity, and answer questions. By providing information we want to stimulate thoughts on how we, together, can help the coffee industry become a better and more sustainable business for all.

Some of the most knowledgeable scientists and eminently experienced tradespeople contributed to this work. We thank everyone of them. We consulted with colleagues and many other experts in the field, who are unnamed but deeply appreciated. Our editorial committee, comprised of two industrial scientists, a professor in nutrition, a leader of the specialty coffee movement, a sage who helps brands learn to create shared value, and a conservationist reviewed every paragraph.

Honore de Balzac, the French novelist and playwright, who is said to have consumed 50 cups of coffee a day, said, "Were it not for coffee one could not write, which is to say one could not live." Most of us contributors and editors could probably survive without writing, but we share Balzac's famous passion for this remarkable drink.

Britta Folmer, Imre Blank, Adriana Farah, Peter Giuliano, Dean Sanders and Chris Wille

Chapter 1

The Coffee Tree—Genetic Diversity and Origin

Juan Carlos Herrera, Charles Lambot
Nestec Ltd., Nestlé Research and Development Center Plant Sciences, Tours, France

The origin, botany, and genetics aspects of coffee have been widely described in many publications and reviews. This chapter intends to provide a general view about those facets of the main coffee species with a special focus on sensory quality. Additional data related to the geographical distribution and biochemical composition of these coffee species are also provided. More detailed information may be found in reviews by Charrier and Bethaud (1985), Davis (2007), Anthony et al. (2011), among others.

1. WHERE DID THE COFFEE COME FROM?

Over the ages, there have been numerous legends about the origin and discovery of coffee. However, it is known that the wild coffee plant (*Coffea arabica*) is an indigenous plant of Ethiopia, where it was discovered in about AD 850. The history of Robusta coffee is more recent. Apparently, the first plant transfer and cultivation took place around 1870 in the Congo basin. Therefore, even if their histories are not comparable, both species are indigenous to the African equatorial forest (Smith, 1985).

Hence, coffee and *Homo sapiens* both began their long evolutionary journeys in Africa. In fact, the highland forests of Ethiopia and South Sudan are considered the cradle of Arabica coffee; but it is also the region where primitive human beings started their long voyage to conquer the world. The current Arabica species is derived from the ancestral trees found in the primary forest of the famous Rift valley, one of the most incredible geological events on Earth. Today, some wild Arabica trees are still growing in some of these forests as evidenced by recent reports of new botanical accessions, which have been described by the botanist Stoffelen et al. (2008).

Humans planted coffee plants in the rest of the world (Fig. 1.1). The history of the Arabica coffee dissemination started in the 8th century when some seeds were transported from Ethiopia to Yemen where they were cultivated till the

The Craft and Science of Coffee. http://dx.doi.org/10.1016/B978-0-12-803520-7.00001-3

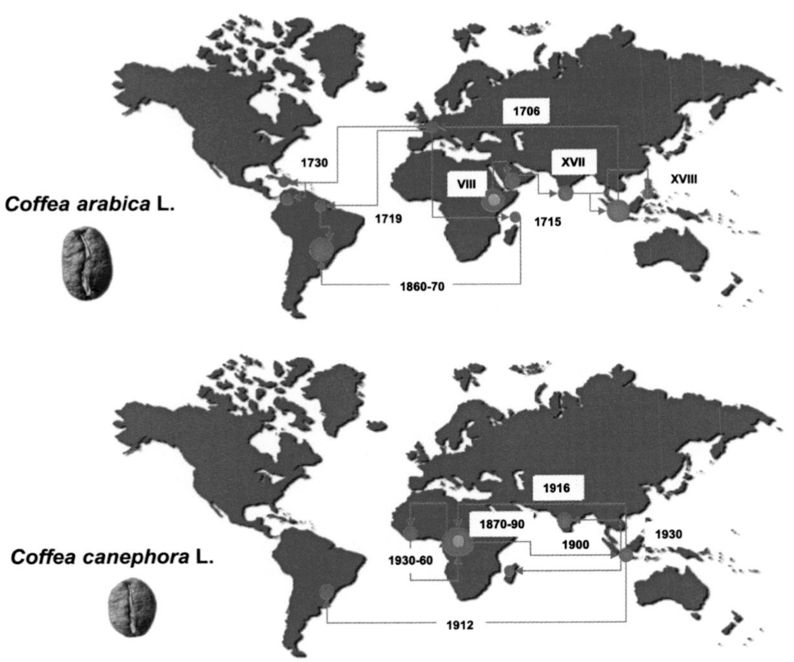

FIGURE 1.1 Progressive dispersion of Arabica and Robusta species from their center of origin (shown as a *green colored point* in the image) to the rest of the world.

end of 14th century by the Arabians, who became the sole providers of coffee for around 100 years. Then, coffee continued its expansion in countries far away, such as India, Ceylon (now Sri Lanka), Java, and Indonesia, where the first commercial plantations were started. Early in the 17th century, coffee arrived in Europe, brought by a Dutch trader in 1616. Several plants were propagated in the Amsterdam Botanical Garden and later taken out to the East Indies to set up new plantations. Other coffee plants also arrived in the *Jardin des Plantes de Paris* as a gift for King Louis XIV. This was the starting point for the future coffee cultivation in French colonies and soon after in other Spanish and Britain colonies. The cultivation of coffee has spread to almost all the intertropical regions around the world.

The Robusta coffee plant dissemination started near to the Lomani River, a tributary of the Congo River in Central Africa. It was through a nursery in Brussels that Robusta coffee moved from the Belgian Congo (Democratic Republic of Congo), where it originated, to Java. After that, early selections were successfully produced and new seeds were used to establish plantations in other countries like India, Uganda, and Ivory Coast. Local natural populations were progressively used for farming in different

African countries and this allowed the dissemination of Robusta coffee on new cultivated areas before they finally reached the American continent in around 1912.

Nowadays, some new wild coffees continue to be discovered in the tropical forests of Africa. Also, new varieties are being developed in different tropical regions around the world. In Central America, Colombia, Brazil, Ivory Coast, or Kenya breeding programs are focusing not only on high productivity but also on high tolerance to pest and disease (mainly for Arabica), better physiological adaptation to new coffee regions or climate changes, and when possible, also differentiating sensory attributes.

2. BOTANICAL ORIGIN AND GEOGRAPHICAL DISTRIBUTION

2.1 The Coffee Plant

For a botanist, any tropical plant of the Rubiaceae family, which produces coffee beans, is considered as a "coffee tree." Over 100 coffee species have been described by botanists since the 16th century, and the number of species has strongly increased since the creation of the genus *Coffea*. The "true coffees" are classified into the Coffeeae tribe, which comprises two genus: *Psilanthus* and *Coffea* (Bridson, 1987). Major differences between these two groups rely on morphological criteria of the flower structure. The *Coffea* genus has a long style and a medium-length corolla tube with protruding anthers. The *Psilanthus* genus is characterized by a short style and a long corolla tube with encased anthers. Phenotype of the coffee plants can vary from small perennial bushes to thick hard wooden trees, whereas their fruits have a particular structure. In fact, each fruit is an indehiscent drupe with two seeds, each of which exhibiting a characteristic deep groove (i.e., invagination) in the ventral part, known as the "coffeanum suture" (Davis et al., 2006).

2.2 Evolutionary History

The African central region appears to be the origin of main *Coffea* species including the two commercially important species *Coffea arabica* (i.e., Arabica) and *Coffea canephora* (i.e., Robusta). Originally, Arabica was a shrub living in the undergrowth of the forests surrounding the southwest of Ethiopia and the north of Kenya, at altitudes between 1300 and 2000 m. Only in recent times has the origin of *C. arabica* been formally recognized after several botanical prospections were carried out throughout the 20th century. In 1999, molecular and cytogenetic analyses finally allowed the elucidation of the origin of this species, which resulted from the natural hybridization between two ecotypes related to *C. eugenioides* and *C. canephora* species (Lashermes et al., 1999). The low genetic divergence found between the two constitutive genomes of *C. arabica* and those of its progenitor species support the hypothesis that *C. arabica* resulted from a very recent speciation event occurring between 10,000 and 50,000 years BC (Cenci et al., 2012).

The *C. canephora* species originated from the humid lowland forests of tropical Africa. Two main genetic types were initially distinguished: Kouillou and Robusta. The Kouillou type was described as a small group with low diversity. A single selection from this type originates from the famous "Conilon" variety widely cultivated in Brazil. The Robusta type was described as the most important among two groups: the Congolese (from Central Africa) and the Guinean (from Côte d'Ivoire and Guinea). Each group was characterized by differences in terms of morphology, growing habits, and adaptation to varied ecological conditions. Subsequent molecular diversity studies allowed the finer differentiation of each group into seven different subgroups: D (Guinean); SG1 or A (Conilon); SG2 or B (Central African Republic); E (Congolese); C (Cameroon); O (Uganda); and R (Democratic Republic of Congo) (Dussert et al., 1999; Cubry, 2008; Gomez et al., 2009).

Despite all the information gathered, the complete elucidation of the evolutionary history of *Coffea* species remains fragmentary. Nevertheless, in recent years, the new genomic technologies applied to the study of coffee plant have given additional insights into the evolution of their genome.

2.3 Geographic Distribution

The original geographic distribution of the *Coffea* genus is restricted to tropical humid regions of Africa and islands in the West Indian Ocean. Comparative studies based on molecular analyses of *Coffea* species demonstrated a strong correspondence between their phylogenetic origin and geographic distribution on four major intertropical forest regions: West and Central Africa, East Africa, and Madagascar (Fig. 1.2), where species originated (Lashermes et al., 1997; Razafinarivo, 2013). The number of *Coffea* species recorded in different countries during the last 15 years reveals the presence of three main hotspots of species diversity located in Madagascar, Cameroon, and Tanzania. Despite increased deforestation in these regions, records of new species in other countries located along the intertropical region of Africa also remain significant (Anthony et al., 2011).

Even if coffee species are found from sea level up to 2300 m above, most species (67%) are adapted to a restricted range of altitude below 1000 m. Some species like *C. canephora*, *Coffea liberica*, *Coffea salvatrix*, *Coffea eugenioides*, or *Coffea brevipes* present a wide distribution in elevated regions, from lowlands (e.g., <500 m up to 1500 m), whereas others are mainly restricted to narrow ranges of variation (e.g., *Coffea heterocalix*, *Coffea kapakata*, *Coffea sessiflora*, or *Coffea stenophylla*). Moreover, *C. arabica* is well adapted to a range between 800 and 2000 m. On the other hand, most of the species having a wide distribution in African mainland (i.e., *C. canephora*, *C. eugenioides*, and *C. liberica*) are commonly found in humid habitats

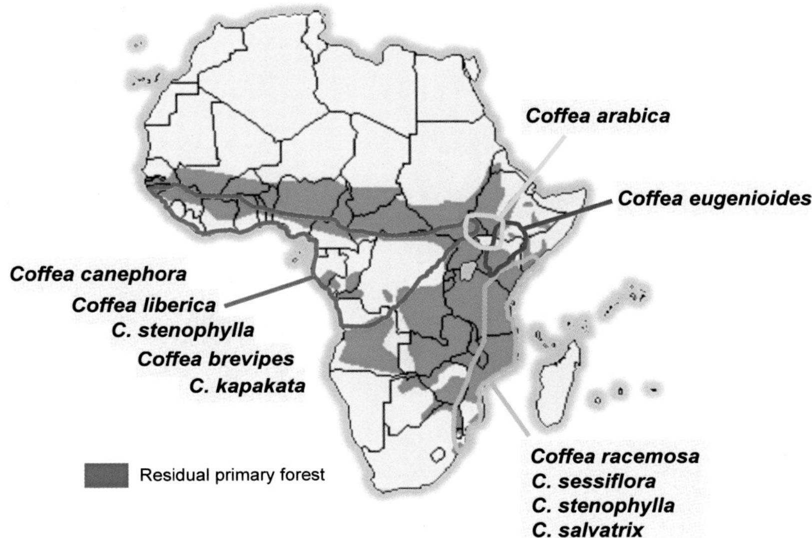

FIGURE 1.2 Natural distribution of some of the most recognized *Coffea* species in Africa.

represented by evergreen or gallery forest. In contrast, other species like *Coffea congensis*, *Coffea racemosa*, or *Coffea pseudozanguebariae* exhibit specific adaptations to habitats with particular soil and climate conditions (Anthony et al., 1987; Davis et al., 2006).

3. BREEDING STRATEGIES AND CULTIVATED VARIETIES

3.1 Arabica Varieties

Around 60% of the total coffee production (84.3 million of 60 kg bags in 2014/2015, ICO statistics) comes from the Arabica species (Fig. 1.3). The most traditional varieties like Typica and Bourbon, derived from Yemen populations, as well as some of their mutants such as Caturra, have been long time considered as highly productive and exhibiting a standard cup quality. Only a few particular selections such as Laurina, Moka, or Blue Mountain have been endorsed by the market as producing premium quality coffees. On the other hand, coffees that originated from the Ethiopian and Sudan regions, such as Geisha and Rume Sudan, are considered as wild or semiwild selections, which are adapted to supply niche markets because of their low productivity.

At the beginning of the 1980s, the necessity to develop resistant Arabica varieties to the coffee leaf rust (CLR) disease caused by the fungus *Hemileia vastatrix* forced the geneticist to search for other sources of resistant genes. Even if the genetic affinity between *C. arabica* and Robusta species may not

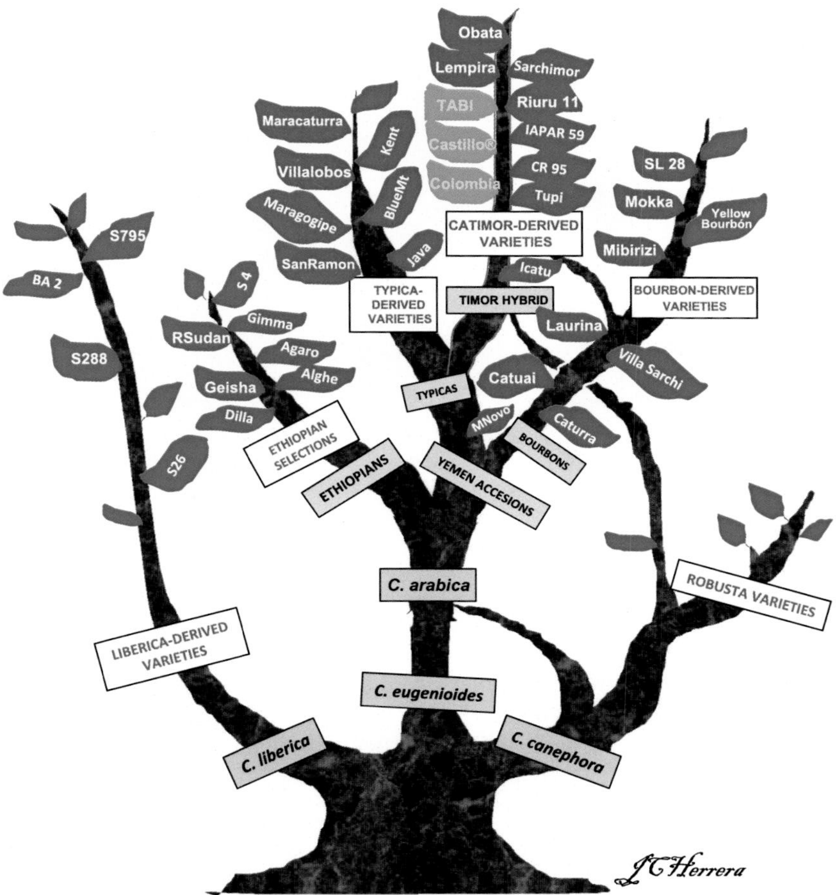

FIGURE 1.3 Tree of the most popular Arabica varieties around the world.

be neglected (Herrera et al., 2002), gene introgression through the way of triploid interspecific hybrids represents a long approach to introduce genetic resistance to cultivated Arabicas. Therefore, most of the breeders decided to use the Timor hybrid (which is a natural hybrid between *C. arabica* and *C. canephora*), highly cross-compatible with Arabica, as the main source of resistant genes. As part of these efforts, different introgressed varieties were developed (e.g., Catimors, Sarchimors) combining high productivity and rust resistance. Even if overall agronomic performance was improved, the cup qualities of new varieties are still in some cases a topic of controversy; this is the case for the variety Costa Rica 95, which has a quality inferior to traditional varieties (Leroy et al., 2006; Van der Vossen, 2009). Nevertheless, breeding experience showed that when the selection for quality is considered as a major criteria, it is possible to obtain introgressed Arabica varieties with an excellent cup quality (Bertrand et al., 2003). Some examples are the Castillo® variety developed in Colombia, the Ruiru 11 variety developed in Kenya, or some resistant varieties developed in Brazil.

In Colombia, the composite varieties Colombia and more recently Castillo® were evaluated for their sensory attributes in different studies (Moreno et al., 1997; Alvarado and Puerta, 2002; Alvarado et al., 2009). Overall, results showed that even if these varieties are composite between several advanced lines (F5 and F6), their sensory profile cannot be differentiated from groups of traditional varieties like Caturra, Typica, or Bourbon. Interestingly, in one study (Alvarado et al., 2009) it was possible to identify particular lines of Castillo® variety exhibiting quality profiles statistically superior to the traditional varieties.

Ruiru 11 is another composite cultivar developed at the Coffee Research Station, Ruiru (Kenya), and released to the growers in 1985. It combines resistance to major diseases of coffee, coffee berry disease (CBD), and CLR with high yield, fine quality, and compact growth adapted to high planting density. In a recent report, Guichuru et al. (2013) pointed to the possibility to select, among the Ruiru 11 components, some hybrid lines combining both high cup quality and CBD resistance.

At IAPAR, the Agronomic Institute of Paraná in Brazil, hundreds of lineages have been selected after 1973. The selection was made from thousands of genotypes derived from germplasms such as Catuaí, Mundo Novo, Icatu, Catimor, Sarchimor, Catucaí, and more recently a number of experimental selections. Overall accumulated experience indicates that quality can strongly fluctuate in progenies obtained even from parents originally exhibiting high cup quality profiles. Many lineages from the Icatu and Sarchimor germplasms, for example, have problems such as a high proportion of shell beans and poor beverage quality, whereas others from the same germplasm show superior quality profiles (Sera, 2001).

3.2 Robusta Varieties

Breeding strategies in Robusta are determined by the strict self-incompatibility of the species and by consequence the objectives will be to select clones vegetatively propagated or clonal hybrids propagated by seeds and obtained under controlled pollination. Two main approaches were developed by coffee Institutes for Robusta breeding: the massal selection, which was mainly applied in Brazil (Gava Ferrão et al., 2007) and in Vietnam, and the selection by hybridization between distant genetic groups developed by the French and Ivorian scientists in Ivory Coast. The breeding by hybridization was extended to a program of recurrent and reciprocal selection in Ivory Coast.

Cultivation of *C. canephora* was initiated early in the 19th century because of significant damages caused by the CLR on *C. arabica* plantations in Asia. According to the first available reports Robusta coffees were introduced to

Java in 1901 from one accession formerly identified in the Belgian Congo (Republic Democratic of Congo). These Robusta landraces were quickly accepted by the first African farmers thanks to their vigor, productivity, and resistance to CLR. Simultaneously, other wild accessions of *C. canephora* species such as Kouillou, Maclaudi & Game, Niaouli, or *Coffea ugandae* were also deployed in different countries such as Ivory Coast, Guinea, Togo, or Uganda, respectively (Charrier and Eskes, 1997).

By comparison, *C. liberica* Hiern, the third most important cultivated species, was planted to a lesser extent than *C. canephora* because of its recognized sensitivity to *Fusarium xylarioides* even if this species showed a clustered fruit maturation, contributing to a better coffee quality and higher seed weight. Despite the close phylogenetic relationship between these two species (Robusta and Liberica), they remain substantially different regarding their morphological characters (N'Diaye et al., 2005).

Only after 1960s, new clonal selections of Robusta were obtained in Uganda, Congo, and then in Ivory Coast. However, most of the polyclonal varieties used by the farmers remained genetically close to wild local accessions. Today, only a reduced number of countries continue with a selection program for commercial clones. Thanks to this work, significant improvements have been obtained, for example, in Ivory Coast where productivity and bean size were dramatically improved (between 30% and 110% gains in yield and 50% in bean size). Polyclonal varieties were also developed in Brazil inside the "Conilon type", which is a genetic subgroup of *C. canephora*. Several Conilon-derived varieties were obtained by massal selection (e.g., Emcapa 8111, Emcapa 8151, Vitória Incaper 8142, Conilon, among others) and are now cultivated on a large scale in the country (Gava Ferrão et al., 2007).

4. GENETIC BASIS OF COFFEE QUALITY

4.1 Biochemical Determinants of Quality

Quality, in the accepted sense of the term, takes into account the physical, chemical, and organoleptic properties of the coffee beverage. This quality is under the influence of a number of factors including not only the genetics and the physiology of the plant but also all agricultural and processing practices related to the harvest and postharvest procedures (see Chapters 2 and 3). Each factor plays a significant role and must not be underestimated. Therefore, the quality potential into the green coffee produced by the plant needs to be preserved during all the chains to deliver the beverage quality expected by the consumers.

Among the overall components of quality, the genetic potential of the species, genotypes, and varieties to produce "quality beans" remains one of the most important. This is true not only in terms of size but also for the content of compounds (i.e., biochemical composition). Furthermore, the quality is also inherent to the level of variability for some genetic factors and their interaction with the environment (Villarreal et al., 2009; Joët et al., 2012).

Studies on different *Coffea* species showed an interesting variation in terms of biochemical compounds currently related to quality attributes (Anthony et al., 1993; Campa et al., 2004, 2005). This biochemical variability is linked to a useful source of genes for coffee breeders; nevertheless, a better understanding of the biochemical and genetic basis of quality remains necessary to use these species for improving of commercial varieties.

On top of the two important commercial species (*C. arabica* and *C. canephora*), there is only a reduced number of other *Coffea* species, which are known to be suitable for human consumption. Among the species that were cultivated, *C. liberica*, *C. racemosa*, *C. stenophylla*, or *Coffea humblotiana* could be mentioned as particular examples of local production and consumption (Anthony, F. and Hamon, P., Personal Communication, November 2015). Some of these species have naturally elevated contents in caffeine, but also in trigonelline and sucrose (Fig. 1.4), which are recognized as two of the main aromas and flavor precursors after roasting (Farah et al., 2005). In the case of Arabica green coffee, higher sucrose and trigonelline contents associated with low caffeine levels were mentioned as potential factors, at least partially, for its better sensory quality (Casal et al., 2000; Ky et al., 2001). However, as already mentioned for the other species, high levels of these three biochemical compounds are not enough to guarantee a high quality coffee.

4.2 Genetics of Quality

In one of the pioneering studies on the genetic variation for quality characters, Walyaro (1983) analyzed 11 coffee varieties (*C. arabica*) and their hybrids for various quality characters over 3 years and under two planting densities. This author came to the conclusion that the most quality characters are highly heritable, especially bean size characters and the overall liquor standards represented by acidity, body, and flavor. Considering this, the genetic improvement for quality appears as an achievable objective. For most of the characters related to quality the observed variation was mainly due to additive genetic effects. In general, the hybrids derived from parents having good coffee quality attributes tend to produce coffee with good quality.

Intense studies have also been organized to understand the genetic mechanisms driving the quality of Robusta coffees (Montagnon et al., 1998; Ky et al., 2000, 2001). Overall, these studies demonstrated that in *C. canephora*, some traits like fat contents, caffeine contents, and bean weight have high heritability values, whereas others like trigonelline, chlorogenic acids, or sucrose contents have high to intermediate values depending on the interspecific or intraspecific origin of the populations under study. In this context, important

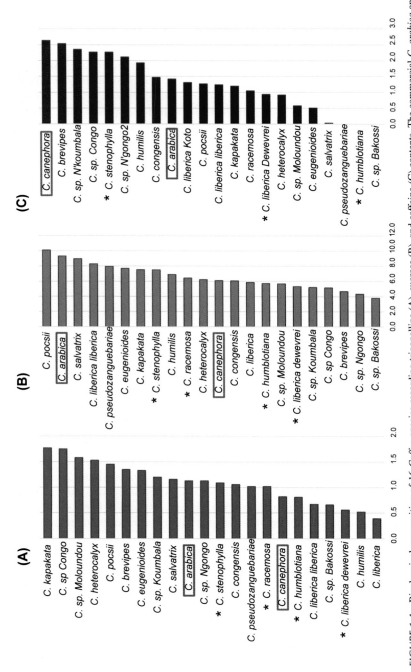

FIGURE 1.4 Biochemical composition of 16 *Coffea* species regarding trigonelline (A), sucrose (B), and caffeine (C) contents. The commercial *C. arabica* and *C. canephora* species are indicated as reference. In addition, some species that are (or have been) cultivated in limited areas only for local consumption (i.e., *C. liberica*, *C. racemosa*, *C. stenophylla*, or *C. humblotiana*) are marked with an *asterisk*. *Data from Anthony et al. (1993) and Campa et al. (2004, 2005).*

genetic gains in Robusta coffees could be obtained by conventional breeding methods on traits like bean size or caffeine content.

Among the hundreds of compounds related to the beverage quality, caffeine is one of the most studied. Caffeine is a purine alkaloid present in a few groups of plant species like tea (*Camellia* spp.), cacao (*Theobroma* spp.), or the exotic plant guaranà (*Paullina cupana*) original from the central Amazonian Basin. Caffeine is accumulated in seeds, cotyledons, and leaves and seems to have a role as allelopathic compound as well as biochemical defense against pest feeding. Recently, the genetic basis of caffeine pathway was also discovered and now well understood. Therefore, thanks to the new biomolecular technologies applicable to the conventional coffee breeding, it could be feasible in the near future to modulate the caffeine contents on a larger spectrum and more efficiently than before the new breeding technologies were available.

4.3 Experimental Approaches

Thanks to the progressive implementation of new analytical methods as well as DNA-based technologies in coffee, the study of the "sensory quality" is more precise and affordable. In this context, the two main complementary approaches are envisaged to assess the genetic basis of the cup quality: (1) the metabolomics approach, which is looking for the identification of "biomarkers," which are storage and volatile compounds widely recognized as important precursors of the coffee aroma and flavor; (2) the transcriptomic approach, which considers the identification of "molecular markers" closely linked to genes, or more recently, to genetic regions, usually involved in the expression of quantitative traits loci (QTLs), associated to the biochemical precursors of quality (Joët et al., 2012). Therefore, it is likely that the increased number of these QTLs recently reported and associated to compounds such as caffeine, carbohydrates, lipids, or phenolic compounds will be the starting point to accelerate modern breeding strategies (Leroy et al., 2006; Mérot-L'Anthoëne et al., 2014).

5. THE FUTURE IN COFFEE BREEDING: STRATEGIES AND PERSPECTIVES

The coffee sector is facing strong challenges such as the negative effects of climatic changes on coffee production and quality. Therefore, the development of varieties of Arabica coffee better adapted to higher temperatures must be the priority for coffee breeders in the coming decades (Ovalle et al., 2015). On the other hand, the world market is also demanding higher volumes of quality coffees, pushing the producing countries to develop a new generation of varieties, showing modification to better adaptability, more consistent productivity, and more stable quality profiles.

For *C. arabica*, which has the finest sensory qualities among the coffee species, the main objective in the future will be to combine a high sensory quality with better adaptation to abiotic stresses induced by the climatic changes. Drought occurrences, as well as the increased incidence of pest and diseases, represent important challenges for the future. Therefore, new varieties should exhibit better tolerance to biotic and abiotic stresses in a context of climatic modifications. In this scenario, hybrids of first generation (F1) become an interesting strategy toward a new "ecologically intensive agriculture" system. These hybrid varieties, first developed in Central America, have proved to be more productive (30–60% more compared to best traditional varieties) and well adapted to different environments, particularly to agroforestry systems (Bertrand et al., 2006). Nevertheless, F1 varieties need vegetative systems of propagation, which are affecting the production cost of coffee plantlets. Some interesting perspectives are under study to reduce this cost by either combination of in vitro propagation methods coupled with horticultural systems or through a seed propagation strategy based on the usage of male sterility.

Robusta varieties have been obtained either by massal selection inside quite narrow genetic groups or by hybridization between genotypes selected in the Guinean and the Congolese groups. Results of these breeding strategies resulted in the significant improvement of yield and vigor, demonstrating the efficiency of these approaches (Leroy et al., 1994). However, the recent molecular studies based on restriction fragment length polymorphisms and simple sequence repeats molecular markers confirmed the existence of an unexploited genetic diversity, with seven groups including those of Guinean and Congolese origins (Dussert et al., 1999; Cubry et al., 2008). At mid-term, it could be expected that breeders would be able to use this available genetic diversity to create mating schemes for taking benefit of the heterotic performance of the hybrids.

Even if significant progress were made in the recent years on the identification of genetic markers not only for sensory quality but also for yield or disease resistance, there are still technical constraints for their efficient usage in coffee breeding programs. In fact, such candidate markers might be transferred and validated on different backgrounds and populations before they can be implemented as a routine tool for the breeders. The recent deciphering of the *C. canephora* genome by an International Consortium (Denoeud et al., 2014) will provide the coffee research community in the coming years with useful as well as powerful tools toward the identification of important genes involved in (1) the natural variability of coffee quality, (2) the biosynthetic pathways linked to their expression, and (3) the identification of genes within this pathway, which will determine the final sensory attributes of a cup of coffee.

6. OUTLOOK

More new coffee species remains to be discovered. Only between 1995 and 2005, a total of 45 new species were identified and about half of the *Coffea*

Chapter 2

Cultivating Coffee Quality—Terroir and Agro-Ecosystem

Charles Lambot[1], Juan Carlos Herrera[1], Benoit Bertrand[2], Siavosh Sadeghian[3], Pablo Benavides[3], Alvaro Gaitán[3]
[1]*Nestec Ltd., Nestlé Research and Development Center Plant Sciences, Tours, France;* [2]*CIRAD, Agriculture Research for Development, Montpellier, France;* [3]*Cenicafé FNC, Manizales, Colombia*

1. INTRODUCTION

Coffee farming is an alchemy of available varieties, knowledge, traditions, technical practices such as pruning, fertilization and pest management, and environmental conditions including soil, topography, altitude, climate, and shade intensity. All the farmers aim to minimize costs and labor and maximize production. With coffee consumers increasingly demanding better aroma and taste, growing high-quality coffee is a key to economic success.

Although outmoded extension programs once insisted that farmers had to choose between, say, environmental conservation and productivity, modern science and experience show a convergence among conservation, community relations, social equality, crop quality, and productivity leading to the concept of sustainability. Farmers working together with scientists, conservation organizations, and coffee companies are demonstrating that following the evolving sustainability guidelines—as defined, for example, by certification programs—provides a path for reaching all of these goals together. Sustainability includes optimizing the environmental, social, and economic conditions on a farm.

Farmers, especially in the main producing countries and regions such as Brazil, Vietnam, Colombia, and Central America, have increasing access to training and new information from national coffee associations, cooperatives, nongovernmental organizations (NGOs), government agencies, agronomic research institutions, and the research and training departments of coffee trading and roasting companies. The standard setting and certification is more advanced in coffee than in any other crop. Trainers and technical assistance providers supporting the NGO-led organic, Fairtrade, and sustainability standard and certification schemes, combined with those from private sector programs, are

The Craft and Science of Coffee. http://dx.doi.org/10.1016/B978-0-12-803520-7.00002-5

reaching many farmers. Even so, many millions of farmers, especially remote small holders, do not benefit from the new science and sustainability guidelines; they must depend on experimentation, the knowledge handed down from preceding generations, and sharing advice and success stories with neighbors. The coffee sector stakeholders recognize this training gap, and there are burgeoning initiatives and coalitions to address it.

Farming is always risky; growing a commodity for export is more so. Coffee farmers are at the mercy of boom and bust price cycles, weather, government policies, changing consumer preferences, social unrest or war, and pests and disease. Uncounted numbers of farmers went broke in the coffee price crisis of the early 1990s. The cyclical outbreak of the coffee leaf rust disease, which may have peaked in 2013, cost Central American coffee farmers 30–70% of their income and an estimated 500,000 coffee jobs.

Producers must make smart decisions in all aspects of farm management, beginning with which seeds to plant in new crop areas. As previously mentioned, farmers do not have to choose between productivity, crop quality, environmental conservation, community relations or ethical treatment of workers, but they can prioritize these and other objectives.

This chapter examines some of the decisions farmers must make with respect to agronomic conditions and practices and primarily consider the impacts of these decisions on crop quality and productivity.

2. CULTIVATING COFFEE QUALITY—TERROIR AND AGROECOSYSTEM

Cup coffee quality is primarily driven by the physical and chemical characteristics of green coffee, which is determined by the combination of three categories: *Environmental factors* × *Genetic factors* × *Agricultural practices.*

All of them are important and also their interaction with each other. This sequence is not neutral and needs to be well understood by coffee producers who aim to produce high-quality coffee. Indeed, environmental factors like the altitude, the climatic conditions, and soil fertility are almost fixed and difficult to modify. To some extent, the agricultural practices could modulate them, with, for example, shade trees, irrigation, and fertilization, but they will never totally compensate for nonappropriate environmental conditions. This will even affect the production cost when strong modifications are required. After properly selecting the production site, the coffee farmer will have to choose the coffee species and variety having the desired quality potential. As mentioned in Chapter 1, there are many studies that demonstrate that all coffee varieties are not equivalent in terms of quality and that some will never produce the finest quality. Then, when sites and varieties are defined the quality potential is determined and will then be valorized according to the agricultural practices applied. If coffee trees are not properly cultivated and harvested, the green coffee quality obtained after processing will not be satisfactory. In the first chapter, the contribution of coffee species and varieties for the quality was elaborated. In the following pages, we will try to demonstrate how the environmental conditions, the propagation techniques, the coffee shading, the fertilization, and the pests and diseases can influence the final coffee quality. Knowledge and scientific evidences have accumulated along centuries of coffee cultivation but many aspects still need further investigations. For example, it will be necessary to better understand how physiological disorders induced by adverse climatic conditions are affecting the quality, plant nutrition, and fertilization, which can be negative or positive.

Scientists should be imaginative and come up with new approaches to overcome this complexity. In the last decades, the development of the near infrared spectroscopy or the metabolomics approach helped scientists to significantly progress in their research. There has been an emphasis on the coffee chemistry, neglecting to some extent the physical components of the coffee grain, which could potentially affect quality. Promising avenues were investigated on the relation between physiological and enzymatic characteristics and the quality measured by the seeds' viability (Selmar et al., 2008). The relation between tissue development in the coffee fruit and the biochemical composition is also an interesting approach (De Castro and Marraccini, 2006). Polysaccharides and, in particular, the cell wall of the endosperm can play a significant role in the sensory quality as they interact with other molecules during the roasting process (Redgwell and Fischer, 2006). All these scientific approaches demonstrate the necessity to combine several domains of expertise to better understand what improves the coffee quality inside the beans.

3. ENVIRONMENTAL CONDITIONS SUITABLE FOR COFFEE GROWING

Climate change represents a major threat for coffee production in the world. According to the Fifth Assessment Report of the Intergovernmental Panel on Climate Change (Stocker et al., 2013), the average surface temperature has increased by 0.85°C from 1880 until now and, depending on the scenarios, would increase from another 1°C (optimistic) to 3.7°C (pessimistic) by 2100. According to these predictions, the average suitability for coffee production of Brazil, which produces 36% of the global Arabica coffee, is expected to drop between 30% and 85% (Bunn et al., 2014).

According to the more pessimistic scenario of IPCC, Mexico will lose 29% of the 27,400 m^2 currently suitable for coffee growing and, Uganda, where coffee represents 20% of the foreign exchange earnings, will also lose the suitable area for coffee production by 25% (Ovalle Rivera et al., 2015). To maintain production levels, Arabica coffee farmers in Uganda and other countries will be required to switch to higher altitude areas where protected mountain forests are located.

Climate change will also bring modifications in the geographical and temporal repartition of precipitations. This will affect Arabica and even more Robusta production, which requires larger volumes of water. According to the more pessimistic scenario of IPCC, Brazil will experience one or two additional dry months per year in 2050 (Ovalle Rivera et al., 2015).

Climate change can also influence coffee rust (*Hemileia vastatrix*) incidence. Observation made by Cénicafé in Colombia during 39 years evidenced a significantly higher occurrence of coffee rust above 1200 meters above the sea level during the warmer El Niño phenomenon when compared to La Niña (Rozo et al., 2012).

To guarantee the sustainability of coffee economy, new solutions will need to be developed to improve coffee resilience to climate change while keeping up with quality market expectations. The combination of resilient varieties selection (Robusta drought tolerant, Arabica pest tolerant, low altitude premium Arabica, premium Robusta, and others) and appropriated agricultural practices (propagation, shade, and others) could help to address this challenging issue.

3.1 Climate in Areas of Origin

Native territories of *Coffea canephora* stretch from Central Africa to the Gulf of Guinea and Uganda. *C. canephora* originates from equatorial rainforests at low to medium altitude between 250 and 1500 m (Davis et al., 2006). The optimum average annual temperatures for the species fluctuate between 22°C and 26°C (DaMatta and Ramalho, 2006), with an annual rainfall pattern between 1200 and 2500 mm and no prolonged dry season. In its areas of origin, *C. canephora* grows mainly on red soils (oxisols, ultisols) that are flat to gently sloping, well-drained, and acidic soils with low native fertility.

When compared to the vast territory covered by *C. canephora*, *C. arabica* appears much more limited. This species originated from a narrow region of southwestern Ethiopia and the Boma Plateau (South Sudan). Wild accessions of *C. arabica* are growing in the mountainous rainforests of Ethiopia, between 1200 and 1950 m (Davis et al., 2006), with average annual temperatures between 18 and 21°C and a rainfall pattern between 1100 and 2000 mm. In the Ethiopian regions, the annual rainfall distribution is unimodal with a minimum from November to February and a maximum between May and September (Liljequist, 1986). In these areas, *C. arabica* grows on deep soils, red or brownish, well drained with a content over 30% with soil acidity varying between pH 4.1 and 6.3. With appropriate processing conditions, the coffee produced in native areas has outstanding sensory quality. It was also well demonstrated that the same beverage quality could be obtained in agricultural production systems if environmental conditions required by those species are met.

3.2 Climatic Conditions in Coffee Cultivation Areas and Coffee Quality

Coffee is produced in more than 50 countries in the intertropical belt. For the two species cultivated, the annual rainfall and the duration of the dry season are the most important water-related parameters, with a significant effect on the yield. The best condition for both species is a high and continuous air humidity.

In a recent study, Bunn et al. (2015) defined the agroecological zones suitable to producing Arabica coffee (latitudinal belt 20°N–25°S). Five clusters were defined by these authors "Hot–Wet," "Constant," "Hot–Dry," "Cool–Variable," and "Cool–Dry" (see Table 2.1).

For *C. canephora* (latitudinal belt 15°N–15°S), the similar clustering is currently not available. However, based on experimental observations, we can propose a description of four agroecological zones suitable to produce *Canephora*. This species is growing well in areas where the minimum temperature is above 17°C and the maximum is below 34°C (Table 2.2).

Specialty Arabica coffees produced under optimum climatic conditions will exhibit distinctive flavors, notably a fine acidity and pleasant aromas (Rinehart, 2009). For Arabica the specialty coffees are mainly produced in the cluster "Constant." Fine Robustas are mainly produced in the cool-wet cluster.

3.3 Plant Symptoms Induced by Climatic Stresses

Arabica is better adapted to cold conditions than Canephora. But under warm conditions Arabica's behavior is quite similar to that of Canephora (Bertrand et al., 2015). When climatic conditions become marginal both species have similar symptoms of stress, which may finally affect the cup quality due to physiological disorders in the plant (Table 2.3).

In most cases, the climate-related stresses reduce the growth or destroy the leaves of coffee. Considering that the beverage quality is correlated to a larger leaf area-to-fruit ratio (Vaast et al., 2006), it can be concluded that all these physiological and mechanical stresses induced by adverse climatic conditions will negatively influence the coffee quality.

3.4 Biochemical Markers Associated to Climatic Conditions and Coffee Quality

As highlighted by DaMatta et al. (2007), a slower ripening process of coffee berries at higher elevations and lower air temperatures, or under shading, induces a better bean filling process (Vaast et al., 2006). In these conditions, coffee beans are denser and far more intense in flavor than those produced at lower altitudes. A longer maturation process appears as one of the key

TABLE 2.1 Bioclimatic Description for Five Clusters Where Arabica Coffee Is Cultivated

Climatic Cluster, World Importance	Main Characteristics of the Climatic Clusters for Growing Arabica	Mainly Found at Elevations (Meters Above Sea Level)
Hot-wet (16%)	High maximum in the warmest month (T° max, 33°C), high minimal temperature in the coldest month (T° min, 16°C), high annual precipitation (>2600 mm), short dry season (<1 month).	946
Constant (26%)	Low temperature seasonality. Maximum in the warmest month (T° max, 25°C), minimal temperature in the coldest month (T° min, 13°C). High annual precipitation (>2600 mm), short dry season (<1 month).	1578
Hot-dry (25%)	Maximum in the warmest month (T° max, 30.5°C), minimal temperature in the coldest month (T° min, 14.5°C). No cold month. Annual total precipitation low (<1500 mm), dry season >3.5 months.	807
Cool-variable (21%)	Maximum in the warmest month (T° max, 29°C), minimal temperature in the coldest month (T° min 11°C). Annual total precipitation moderate (1500–1600 mm), dry season <1.5 months. High seasonality. Mainly in highest latitudes.	825
Cool-dry (12%)	Maximum in the warmest month (T° max, 27°C), minimal temperature in the coldest month (T° min, 9°C). Annual total precipitation low (<1500 mm), dry season >3.5 months.	704

mechanisms leading to high cup quality, possibly by a more complete biochemical mechanism required for the development of the beverage quality (Silva et al., 2005).

Among all the climatic factors, the average air temperature during bean development strongly influences the sensory profile. Positive quality attributes

TABLE 2.2 Bioclimatic Description of Four Clusters Where *C. canephora* Coffee Is Cultivated

Climatic Regions, Examples	Main Characteristics of the Climatic Clusters for Growing Canephora	Mainly Found at Elevations (Meters Above Sea Level)
Wet (Central Africa)	High maximum in the warmest month (T° max 30°C), high minimal temperature in the coldest month (T° min 19°C), high annual precipitation (>2000 mm), short dry season (<1 month).	0–300
Hot-dry (West Africa (Togo), Brazil)	Maximum in the warmest month (T° max, 34°C), minimal temperature in the coldest month (T° min, 19°C). Annual total precipitation low (<1500 mm), dry season >3 months.	0–500
Dry (Vietnam)	Maximum in the warmest month (T° max, 32°C), minimal temperature in the coldest month (T° min, 17°C). Annual total precipitation low and unimodal (<1800 mm), dry season >4 months.	300–900
Cool-wet (Uganda, mountainous regions)	Maximum in the warmest month (T° max, 30°C), minimal temperature in the coldest month (T° min, 14°C). Annual total precipitation (<1600 mm), two short dry seasons.	600–1000

such as acidity, fruity character, and other specific flavors are correlated with air temperature and Arabica coffees are typically produced under cool climates. Two volatile compounds (ethanal and acetone) were identified after roasting as biochemical markers of these cool temperatures on Arabica coffee (Bertrand et al., 2012).

TABLE 2.3 Main Symptoms Induced by Climatic Stresses

Stresses Related to Climatic Conditions	Symptoms on Plant Development	Symptoms on Flowering and Fruit Development	Solutions
Strong wind during several days	Reduce leaf area and internode length	Shedding	Windbreaks, selection of cultivars best adapted, pruning
Low temperatures	Photo bleaching	Reduces the number of fruits per tree	Higher density, shade
Drought	Stop the vegetative growth, high branch die-back, risk of tree mortality.	Reduces the bean size and the number of fruits per tree	Irrigation and/or grafting Arabica on Canephora
Hot wind during the dry season	Stop the vegetative growth	Shedding- poor fruit set, empty fruits increasing	Irrigation and/or shade
Sporadic and low intensity of rains during the dry season		Several blossom periods, flower atrophy, unsynchronized fruit ripening	Irrigation, shade
High temperature and high rainfall	Poor root growth; high branch die back	Flower atrophy, poor fruit set	Drainage Grafting Arabica onto Canephora
Water shortage during the fruit expansion stage		Small fruits	Irrigation, shade, selection of cultivars better adapted

Among detected volatiles, most of the alcohols, aldehydes, hydrocarbons, and ketones appeared to be positively associated to elevated temperatures and high solar radiation, whereas the sensory profiles displayed major defects (i.e., green, earthy flavors). Two alcohols (butan-1,3-diol and butan-2,3-diol) were correlated with a reduction in aromatic quality, acidity, and an increase in earthy and green flavors. High temperatures induce accumulation of these compounds in green coffee and would be detected as off-flavors, even after roasting (Bertrand et al., 2012).

4. PROPAGATION SYSTEMS

Several methods are traditionally used for coffee propagation including seeds, rooted cuttings, grafting, and in vitro methods. They can be classified into two main categories: generative systems (seeds) and vegetative systems (rooted cuttings, grafting, in vitro methods). The choice of the method is guided by the genetic nature of the material to be propagated, the level of technical skills, and the environmental and cost constraints.

4.1 Main Propagation Methods and Their Usage

The decision on the most appropriate system of propagation needs to be taken after a deep understanding of several parameters related to the species, the genetic status of the variety, the technical methods under control, the phytosanitary problems, and the expected production cost of coffee plantlets. Discussions with stakeholders (breeders, agronomists, project sponsors) are necessary to accumulate key information leading to the best method. Main techniques [seeds, grafting, cuttings, and in vitro methods (Georget et al., 2014)] are listed in Fig. 2.1 with some key information about their usage. The different classical propagation methods are extensively detailed in many reference documents (Wintgens and Zamarripa, 2004; Cenicafé FNC, 2013; Almeida da Fonseca et al., 2007) and scientific papers for the in vitro method (Etienne-Barry et al., 1999; Ducos et al., 2007).

4.2 Propagation Methods and Coffee Quality

Vegetative systems of propagation by cuttings, grafting, and possibly in vitro methods are useful techniques to reproduce true to type genotypes and take benefit of their quality potential in terms of sensory, physical, and biochemical characteristics (Fig. 2.2).

With grafting there are risk and opportunities of compatibility between rootstocks and scions. This is particularly the case when interspecific grafting is done for agricultural purposes. Bertrand et al. (2001) studied the compatibility between different *Coffea* species of rootstocks used with Arabica scions. Robusta rootstocks did not modify the coffee composition or quality but significantly improved the yield; nevertheless a negative effect was observed of Robusta rootstock on the plant growth when used at high altitudes due to the poor adaptation of Robusta to low temperatures. Interspecific grafting did not affect the biochemical composition of the green coffee but bean size and aroma were significantly reduced when *C. liberica*

Seeds are used to propagate different type of genetic material (Arabica pure lines or population; Robusta population or clonal hybrids). They are traditionally planted in seedbeds and after two or three months, plantlets are transferred to containers in nurseries. The method is simple and leads to well-developed plantlets at a low production cost. It is the most popular method.

Cuttings are used to propagate clones, mostly on Robusta but recently the technique was also applied on large scale in Central America to propagate Arabica hybrids. Suckers of orthotropic shoots are produced on mother plants and trimmed to produce cuttings. They are placed in appropriate conditions to produce roots and after two or three months transferred to containers in nurseries

Grafting is used to propagate clones for the two species. This technique offers the opportunity to combine the benefit of appropriate combination between the clone (scion) and the rootstock usually propagated by seeds. It is a common practice to control nematode damages by grafting sensitive Arabica varieties on Robusta tolerant population (Nemaya or Apoata). It also give the flexibility to renovate coffee plots by grafting new varieties on suckers or branches of old trees.

In Vitro methods are used to propagate Robusta clones and Arabica hybrids of first or second generation mainly by production of somatic embryos. These methods offer the benefit of velocity in the propagation of unique plants identified in breeding programs. They require high skill levels and appropriate laboratories for the production of In Vitro plantlets. The production cost of plantlets remains a barrier for its broader usage. Results obtained by combining the production of somatic embryos and "mini-cuttings" could be an option to make it more economically viable (Georget et al. 2014).

FIGURE 2.1 Main propagation methods and their usage.

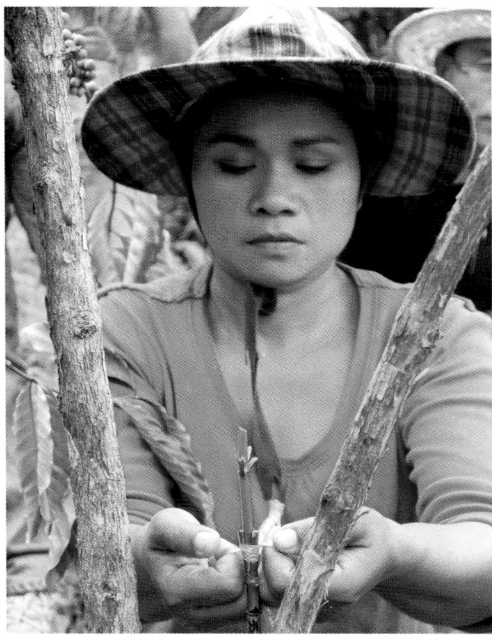

FIGURE 2.2 Asian farmer grafting Robusta coffee with one improved clone.

and *C. dewevrei* were used as rootstock. These two species revealed a low compatibility with Arabica scions. A similar study was reported by Fahl et al. (2001) indicating that grafting Arabica cultivars on *C. canephora* or *C. congensis* resulted in an increase of yield without affecting the sensory quality and biochemical composition of the green coffee. Grafted plants had also a better drought resistance and it was suggested that this benefit was induced by a greater capacity of the root system of *C. canephora* to provide water to the shoot. Trials in Brazil revealed that grafting of one Arabica variety on Robusta Apoata variety did not affect the sensory profile, but the bean size was more homogeneous (Saath et al., 2015). A benefit on bean size could be related to the better capacity of the root system of *C. canephora* to supply water to the plant, as it is generally accepted that water scarcity on coffee has a negative effect on bean size (Charrier and Berthaud, 1988; DaMatta, 2004).

The potential benefit of grafting technique for quality was demonstrated in other crops like tomato. A study made on tomato (*Solanum lycopersicum* Mill) demonstrated that an improvement can be obtained through grafting of a drought-sensitive variety on a drought-resistant rootstock (Sánchez-Rodríguez et al., 2012). In this case, intraspecific grafting was able to improve yield and fruits quality.

Seeds propagation is an easy way to propagate coffee trees and will usually lead to a natural structure of the trees, including the root system. Coffee quality may be impacted when seed production is not properly controlled and may lead to a genetic deviation with the expected phenotype. This is typically the case when open pollinated Robusta seeds are collected without proper control of the parental lines, or when Arabica pure lines are contaminated with foreign pollen during the flowering period. In these situations, progenies will differ from the expected material with potential consequences on the coffee quality.

Some composite varieties like Ruiru 11 selected in Kenya and propagated by seeds may exhibit different cup qualities for the different hybrids, which constitute the composite variety (Gichimu et al., 2012). Any change in the composition of the variety may influence the final quality and it is clear that these composite varieties are under continuous selection. The composite Arabica variety Castillo selected in Colombia has different regional derivatives (Alvarado et al., 2010), which may result in different green coffee qualities.

5. SHADE TREES FOR IMPROVED COFFEE QUALITY

The coffee plant evolved in the undergrowth of tropical forests and is naturally shade tolerant. Wild, uncultivated coffee is still collected in some forests of Ethiopia and other countries where the genus originated. Until the 1970s, most coffee was grown in agronomic (traditional) systems that mimicked the natural forests. To increase production and combat the leaf rust disease, agronomists introduced the “full-sun” monoculture system, where farms were mostly deforested, coffee was planted in dense hedgerows and given heavy doses of chemical fertilizers and pesticides. These high-input (modern) monocultures often significantly boosted yields, and by the 1990s many farms had converted to the shaded or unshaded monoculture systems. A general description of both traditional and modern management systems in Latin America is presented in Fig. 2.3.

At about the same time, countering trends began. Farmers and society began to realize the importance and value of rainforests and to seek ways to conserve them. And coffee consumers began demanding higher quality and sustainably grown coffee. As mentioned earlier, one of the most important environmental factor determining coffee quality is consistent temperatures. The overarching canopy of shade trees creates microclimates, where beans can ripen slowly and naturally, shielded from the harsh tropical sun and rains.

Arabica and Canephora coffees grow naturally in the undergrowth of tropical forests, which are an inhomogeneous habitat (Charrier and Berthaud, 1988). However, the decision to cultivate coffee under the shade needs to be taken after considering many aspects including the environmental conditions, the socioeconomic constraints, the production systems envisaged by the producer, and the final green coffee quality, which is expected. Literature reviews are giving details on the effects and constraints of shaded vs. unshaded production systems (Muschler, 2004; DaMatta, 2004; Steiman, 2008).

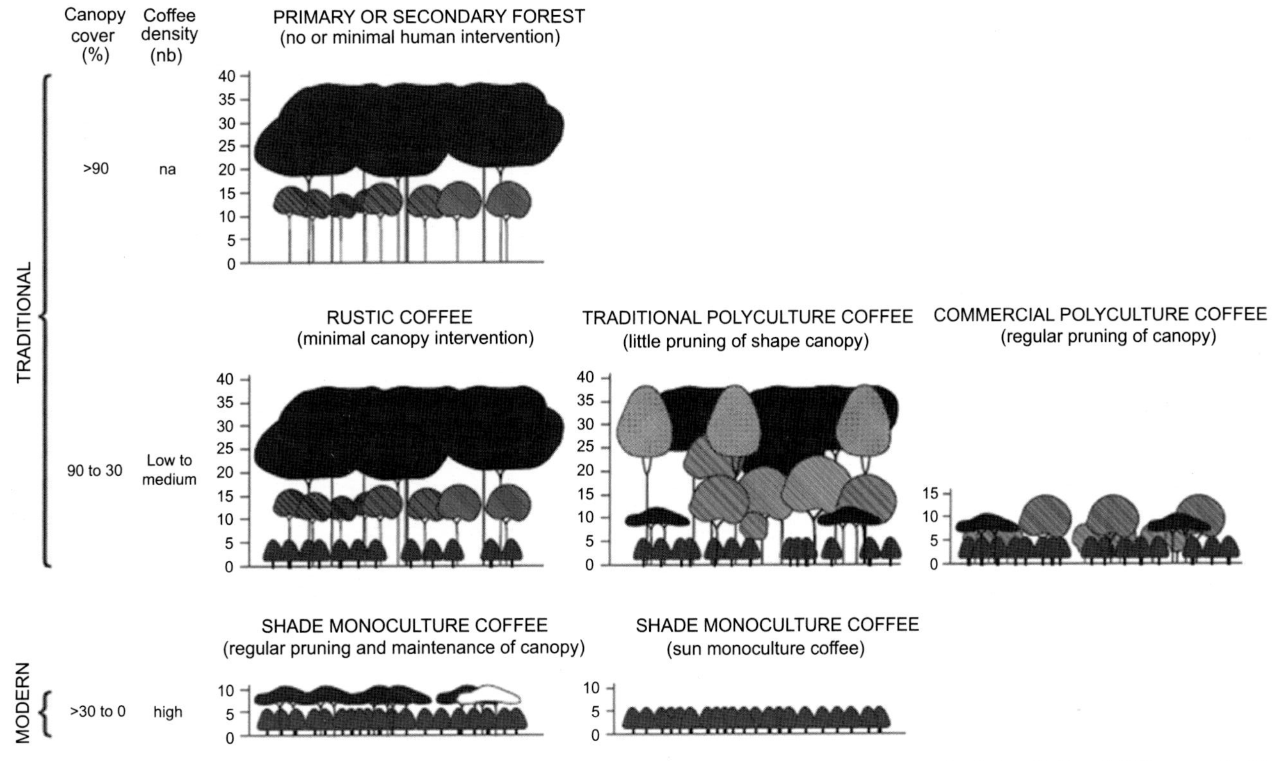

FIGURE 2.3 Classification of forest and coffee management systems in Latin America. *Modified from Moguel and Toledo (1999) and Philpott et al. (2008).*

There are many positive contributions of shade trees on the coffee production system such as the protection against heavy rains, tropical sun, frost, and other unfavorable climatic conditions, the decrease of coffee plant stress to biotic and abiotic factors and enriching of the soil in organic matter and protecting it against erosion. In the case of leguminous shading trees, a significant amount of atmospheric nitrogen is fixed and contributes to the soil fertility. The growth of weeds and their competition with coffee trees is limited under shade.

All these positive contributions make shaded coffee a more sustainable and environmental friendly system compared to the "full-sun" system. However, farmers must adjust the shade to their specific local conditions and farm objectives. Improperly managed shade trees can significantly decrease productivity, worsen pest or disease outbreaks, and compete with coffee plants for water and nutrients.

5.1 Benefits of Trees on Coffee Farms

Conserving and/or planting trees on coffee lands bring benefits to the farmers, farm workers, local communities, consumers, and society at large. The extent of these benefits depends on the selection of tree species, pruning, and other management and the density. The optimal number of trees per hectare has been much debated. Greater density generally means more environmental and social benefits, such as biodiversity, watershed protection, carbon sequestration, fruits and firewood, and other goods. But too many trees will restrict sunlight from reaching the coffee plants, decreasing productivity. In setting the first sustainability standard during the early 1990s, the Sustainable Agriculture Network (SAN), working with farmers and scientists, concluded that in Mesoamerica the optimal density would create 40% shade cover. More than 40% shade frequently inhibited productivity; less than 40% reduced the habitat value for biodiversity. The first standard called for at least 40% density, at least 70 trees per hectare and a mix of at least 12 native species (Jha et al., 2014; Sustainable Agriculture Network standard, 2010).

The SAN standard setters quickly realized that the desirable amount of shade varied from one region to another and even farm by farm. Farmers, agronomists, and certifiers learned to adjust tree density to localized conditions to maximize benefits. For example, Farfan (2013) demonstrated that, in Colombia, an Arabica plot with a shade intensity of 30–40% obtained with *Inga densiflora* planted at 12 m of distance produced 152% more than a system with a shade intensity of 60–70% planted at a distance of 6 m. It is generally considered that the upper limit for acceptable shade is between 40% and 70% (Beer et al., 1998).

Shade tree species selection is as important as planting density; they must meet the requirements of the production system and the environmental constraints (Descroix and Wintgens, 2004; Lambot and Bouharmont, 2004). Ideal shade trees should provide a homogeneous canopy, be deeply rooted and fast growing. Evergreen species that are able to withstand intensive pruning to produce substantial quantities of litter are preferred. Leguminous trees that fix a substantial amount of nitrogen are usually selected and they belong mostly to the genus of *Inga*, *Erythrina*, *Leucaena*, *Gliricidia*, and *Albizzia*.

Forest cover is especially important in areas where environmental conditions are not fully appropriate for coffee cultivation, particularly in areas that are exposed to high temperatures, heavy rainfall and wind, frost or hail, and poor soil fertility. The regulating effects of trees are of tremendous importance considering the decrease of suitable coffee producing areas due to climatic change. Shaded systems have been identified as part of the remedy for confronting harsh new environments in coffee regions that result from climate change (Jha et al., 2014). In addition, shade trees offer a natural protection to excessive temperature variations between day and night. They also protect the soil against erosion and evaporation and reduce coffee plant transpiration. Properly selected and maintained shade trees could help mitigating drought stress.

Shade trees can reduce weed growth, especially of perennial grasses and sedges; the weeds encountered in shaded coffee farms are less harmful and easier to control. Shade trees add leaf litter to the soil acting as a mulch and adding organic matter and minerals in the soil (Mitchell, 1988). The root nodules of leguminous trees fix nitrogen, significantly contributing to soil fertility. The biodiversity of shaded coffee systems is much richer than that found in full-sun monocultures, particularly for birds, insects, and ants. By maintaining higher biodiversity in the coffee farm, there are potential benefits on yield due to more abundant pollinators and on pest control due to the presence of parasitoids and other predators of pests. Carbon sequestration is also significantly better in shaded systems (Jha et al., 2014). The influence of shade trees on the incidence of pests and diseases is complex and can be either positive or negative according to different factors like species, shade intensity, and climatic conditions (Beer et al., 1998).

Inappropriate levels of shade can restrict coffee flowering and fruiting, and consequently reduce yields. Certifiers and other sustainability advocates argue that some reduction in yield is offset by higher prices paid for better quality and the growing demand for "shade grown" coffee. Also, forested farms often require fewer inputs and less rigorous management. Based on a trial organized on Robusta coffee in Madagascar, Snoeck (1988) concluded that yields decreased as shade intensity increased, but without shade it was necessary to apply costly mineral fertilizers. Farmers adopting the full-sun systems must have the capacity to finance the technical package (fertilizers, pesticides, irrigation) required by the intensification. Adopting the full-sun system without the technical package will lead to dramatic failure as coffee trees will suffer from strong biennial production and die back (DaMatta, 2004).

Smallholders, who are often more vulnerable to price fluctuations, have less technical capacity and struggle to finance fertilizer and pesticide inputs, find the traditional shaded, low-input farming systems more secure and sustainable.

Considering the positive effect of associated shade on the environment and the biodiversity, international programs were developed to promote the adoption and conservation of shade cover in coffee plantations. Shade coffee certification programs offer possibilities to economically compensate farmers for the biodiversity service provided by their shaded plantations (Perfecto et al., 2005).

Trees provide a menu of useful goods to farmers and farmworkers, including construction materials (e.g., wood, rattans, and bamboo), fruits, other edibles, oils and resins, and medicinal, decorative, and spiritual plants. Two forest goods are extremely important: watershed protection and firewood. Forested farms have more water and often maintain the watershed for communities down slope. Firewood is the primary energy sources of many coffee growing regions, and it is increasingly scarce. Forest patches and buffers make farms less vulnerable to extreme weather events, flooding, and landslides. By planting multipurpose trees, farmers can diversify their income through the sale of construction materials, fruits, flowers such as orchids, and other products. Coffee farmers enrich their lands with trees like Durian (*Durio zibethinus*) or stink bean (*Parkia speciose*) in Asia, banana, citrus, macadamia nuts, and many others. Multipurpose plots can be more profitable and secure for the farmers (Malézieux et al., 2009).

Shade trees must be properly selected to fit the requirements of the production system and the environmental constraints (Descroix and Wintgens, 2004; Lambot and Bouharmont, 2004). The decision about how much shade to incorporate into a farm should be taken after careful consideration of the same constraints and farm objectives (Muschler, 2004; DaMatta, 2004). The adoption of shade or sun system of production should be taken after deep analyses and understanding of several parameters. Muschler (2004) proposed to consider several parameters to guide the decision:

- Full-sun system could be adopted when environmental conditions are optimal and/or intensive agricultural practices (fertilizers, pesticides, irrigation) can be applied. This will lead to an intensive system of production in which the biodiversity is not a priority.
- Shade system will be encouraged when environmental conditions are not optimal and/or agricultural practices for intensification are not guaranteed. This is leading to a diversified and more environmental friendly system in which the biodiversity is playing a significant role.

5.2 Shade and Coffee Quality

Shade appears to impart its greatest benefit in coffee bean flavor for plants growing in suboptimal and heat-stressed growing regions, where shade can bring environmental conditions closer to ideal levels (Muschler, 2001). This suggests that shade may be particularly important for maintaining coffee quality in the context of climate change, especially in regions with expected temperature increases in future climate scenarios (Jha et al., 2014). It is widely accepted that fruit weight and bean size increase under shade, percentages of increase may differ according to the varieties.

Vaast et al. (2006) observed that the place where the Arabica variety was grown—Costa Rica 95 cultivated under 1180 m of altitude in Costa Rica, under shade—positively affected the bean size and the beverage quality. Negative attributes like bitterness and astringency were higher for sun-grown coffees and positive attributes like acidity and consequently preference were higher for shade-grown beans. The cherries maturity was delayed by until 1 month under shade and the bean composition was significantly modified.

Muschler (2004) compared results obtained by several studies of shade on coffee quality and came to the hypothesis that organoleptic properties could be improved by the delay in ripening due to the shade and its microclimatic effects. This might be the main reason for improved quality of coffee grown in suboptimal environmental conditions.

6. SOILS REQUIREMENTS, FERTILIZATION, AND COFFEE QUALITY

Coffee trees tolerate a wide range of soils, provided that they are deep, porous, well drained, and well balanced for their texture. Volcanic soils are particularly well suited for coffee.

Agricultural practices should preserve the soil fertility, which is the wealth of coffee growers. Efforts should be dedicated to preserve and if possible increase the organic matter in the soil to promote the microbial life and the exchange capacity.

Whenever a soil is depleted it is observed that the coffee quality is affected. By contrast, coffee trees will be healthier on a soil rich in active organic matter, they will have a better leaf area-to-fruit ratio leading to a better quality. Soil is the main reservoir of mineral nutrients for plants. Roots grow and absorb water and nutrients according to the physical, chemical, and biological properties of the soil (Fig. 2.4). For most regions worldwide where coffee is cultivated, the nutritional reservoirs in soil are not sufficient to completely

Physical Properties	Chemical Properties	Biological Properties
• Texture • Structure • Effective depth • Bulk Density • Aeration • Water Infiltration • Water Holding Capacity • Temperature • Consistency	• pH • Salinity • Organic Matter • Nutrients Content • Nutrient Ratios • Toxicity • Cation Exchange Capacity • Base Saturation	• Flora • Fauna • Biological Activity • Enzymes Activity • Organic Matter • Potential Bioindicators

FIGURE 2.4 Soil properties determining its fertility. *Adapted from Malavolta (2006).*

cover the coffee plants' demand. For this reason, it is necessary to continuously supply the soil in a balanced way with sufficient amounts of organic and inorganic fertilizers.

6.1 Fertilization Strategy and Methods

Production can be drastically reduced in poor soils if limiting factors are not properly corrected. In Fig. 2.5 it is illustrated how yield is increased when the soil is fertilized with basic nutrients including the elements nitrogen, phosphorus, and potassium. The effect of terroir and year is illustrated. Clearly, even when overall yields are low due to other factors, fertilization helps to increase the yield.

The production of green coffee leads to the depletion of nutrients in the soil. This depletion needs to be compensated for by an appropriate fertilization. For Robusta and Arabica coffee, the production of 1 ton of marketable coffee leads to the following amount of nutrient depletion (kg of nutrient):

	Nitrogen (N)	Phosphate (P_2O_5)	Potassium (K_2O)	Calcium (CaO)	Magnesium (MgO)
Robusta	33.4	6.1	44.0	5.4	4.2
Arabica	30.9	5.2	44.3	6.0	3.8

The results of long-term research have established the basics of coffee fertilization taking into consideration aspects of soil fertility and

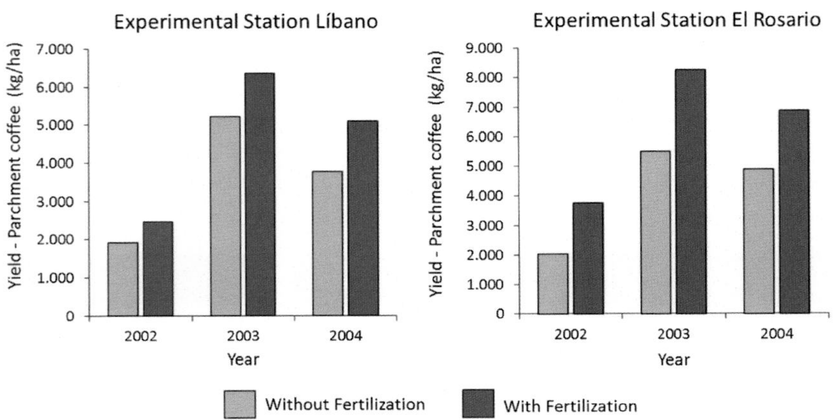

FIGURE 2.5 Yield in two experimental stations showing the impact of fertilization over a period of 3 years. *Adapted from Sadeghian (2010).*

parameters such as the age of the coffee trees, the planting density, and the degree of intensification. The research has allowed the development of analytical software using soil analyses data coupled with multifactorial parameters to establish "tailor-made" fertilization formulas. These systems are applied in different coffee-producing countries where local institutes collaborate with the coffee sector (SIASCAFÉ-Cenicafé, FERTI-UFV) and provide the service to establish fertilization formulas based on a multi-factorial system. It takes into consideration general information of the crops and the environment, the necessity to improve the soil acidity (pH), and finally elaborates the fertilizer formula. A summary of the main principles of these systems is illustrated in Fig. 2.6.

Calculations are based on coffee requirements and experience acquired on the ideal soil conditions leading to the optimal production. In general, coffee requires a soil with a pH between 5.0 and 6.0, an equilibrium between K:Ca:Mg close to 1:12:3 and other characteristics, which are well-documented (Sadeghian, 2008).

All training programs recommend that farmers take soil samples and have them analyzed by a competent laboratory before buying and applying fertilizer, specifically composed for the needs of that soil. Today specific softwares are used to define such compositions. They elaborate on fertilization formula using available fertilizers. They also provide recommendations on how to apply them to the soil and their distribution along the crop cycle trying to optimize the plant feeding, to decrease losses by evaporation and leaching. If necessary, recommendations will include a pH correction by liming or

General information
•Crop characteristics (species, age, planting density, etc..)
•Environmental constraints (Slope percentage, annual rainfall, etc..)
•Financial and technical ressources
Soil acidity management
•Nature of enrichment (lime, gypsum, dolomite) and/or organic matter
•Type of product and dosage determined according soil parameters (pH, Al3+) and content in CaO and MgO
•Modalities of application according the crop development and the fertilization program
Soil fertility and plant nutrition aspects
•Soil fertility diagnostic : texture, content for the main elements and their equilibrium (see table X)
•Nature and doses of fertilizers are determined acccording local availabilities, soil fertility diagnostic, intensification level (expected yield) and compatibility between fertilizers
•Application modalities are determined to maximize the efficiency (where, how and when)

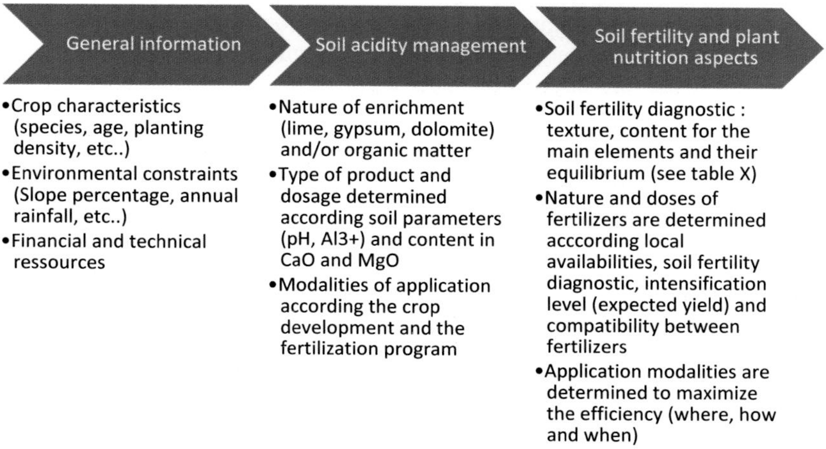

FIGURE 2.6 Main principles for the elaboration of the fertilization program based on soil analyses.

dolomite application. The beneficial role of the organic matter is also considered and is part of the strategy.

For farmers without access to such tools, commercial fertilizers formulated for coffee in most conditions are available. These generally contain between 17% and 25% of nitrogen (N) and potassium (K_2O), between 3% and 6% of phosphorus (P_2O_5), and up to 4% of magnesium (MgO) and sulfur (S), and sometimes micronutrients such as zinc and boron (up to 0.2%) may be included. Another option, when soil analysis is not available, is to prepare fertilizer formulas at the farm by mixing simple fertilizers. For example, four to six volumes of urea can be mixed with one volume of diammonium phosphate and four to five volumes of potassium chloride (KCl).

Coffee plants can also be fertilized through their leaves. Foliar fertilization is often used to compensate for deficiencies in micronutrients like zinc, boron, iron, and manganese. These micronutrient deficiencies are identified by symptoms visible on the leaves or by analysis of the biochemical composition of the leaves. Leaf analysis is an efficient way to monitor the health and nutritional status of coffee plans, especially in intensive systems of production.

6.2 Fertilization and Coffee Quality

The chemical composition of green coffee beans is an important factor in determining the beverage quality (Joët et al., 2010; see also Chapter 12). Fertilization can influence this chemical make-up (dos Reis et al., 2011) and therefore the final coffee quality in the cup. Among the macronutrients, those containing nitrogen and potassium are the most predominant in the bean, usually followed by calcium, magnesium, phosphorus, and sulfur. With respect to micronutrients, the concentrations of iron and manganese tend to be higher than zinc, copper, boron, molybdenum, and chlorine.

Table 2.4 summarizes the main effects observed for the relation between fertilization and coffee quality. This review suggests that the negative and positive influence can be observed and that further research is necessary to better understand the complexity of the relation between coffee quality and plant nutrition.

6.3 Soil Conservation for Sustainable Production

Healthy soils are the foundations of agriculture and, indeed, civilization itself. Typical, traditional practices of tilling the soil increase soil erosion and degradation, undermining agriculture and society. This risk has been known and studied since Plato and Aristotle (Montgomery, 2007). Soil erosion accompanied the declines of civilizations in the Middle East, Greece, Rome, Mesoamerica, and other regions. Healthy top soils that took centuries or even millennia to build can be washed or blown away after hillsides are deforested,

TABLE 2.4 Relationship Between Plant Fertilization and Coffee Quality Attributes

Aspect	Effect on Quality	Sources
Soil fertility	Possible effect on cup quality	Foote (1963)
	Most acidic coffee are produced on volcanic soils	Bertrand et al. (2006)
Fertilization	Reduction in percentage of hollowed fruits	Castillo (1957)
	Chemical fertilization does not affect cup quality	Lazzarini (1961)
	Breaking down fertilization applications does not affect cup quality	Vilella and de Faria (2002)
	Fertilization does not affect cup quality	Graner and Godoy (1970)
Nitrogen fertilization	Fertilization with nitrocalcium and ammonium nitrate produced lower sensory quality. The higher dosage of ammonium sulfate had negative effects on chemical composition and bean quality	Malta et al. (2003)
	Nitrogen fertilization increased bean N content and affected negatively cup quality	de Amorim et al. (1973)
Phosphorus fertilization	Cup quality was negatively affected by the omission of P in the fertilization	de Amorim et al. (1965)
Potassium fertilization	High potassium-K dosage reduced boron and Zinc in the bean. Excessive dosage of K reduced quality in inconsistent manner.	de Amorim et al. (1973)
	Excess of potassium can induce Mg deficiencies and negatively affect coffee quality	Mitchell (1988)
	Bean quality improved with medium dosages (266 kg/ha/year of K_2O)	Silva et al. (2002)
	Potassium sulfate and nitrate sources improved quality when compared with potassium chloride	Malta et al. (2002)
Sulfur fertilization	No effect was found for dose or sulfur source on cup quality nor N, P, and S content in the fruit	Malavolta (1986)

Continued

TABLE 2.4 Relationship Between Plant Fertilization and Coffee Quality Attributes—cont'd

Aspect	Effect on Quality	Sources
Micronutrients fertilization	Zinc supply positively affected bean quality in terms of less percentage of medium and small size beans, lower CBB infestation, lower potassium leaching and electric conductivity, higher contents of zinc and chlorogenic acids, and higher antioxidant activities	Prieto Martinez et al. (2013)
	Cup quality was not affected by using two sources of micronutrients	Jayarama et al. (1992)

or in a few seasons of poor farm management. Many assessments rank soil degradation as one of the greatest challenges—and even threats—to society in this century (Scholes and Scholes, 2013; Food Agricultural Organization – Natural Resource Management and Environment Department; IAASTD, 2009).

Wild coffee plants evolved in the thin soils under the forest canopy in Eastern and Central Africa and most coffee is grown in the place of rainforest. The exuberant vegetation of rainforests suggests rich soils, but most tropical forests keep the nutrients cycling well above ground; fallen leaves and other detritus are quickly broken down and recirculated. As coffee plants were transported around the tropics, they landed on many soil types. Coffee was often planted in agroforest systems where shade trees are maintained or, increasingly, in open and deforested fields, where the coffee plants and the soil are exposed to equatorial sun, monsoon rains, trade winds, and other natural events. Coffee is often planted on steep hillsides, greatly increasing the risk of erosion.

Although there have been significant advances in soil science and countless programs to teach soil conservation methods, erosion outstrips topsoil accumulation by one to two orders of magnitude around the world, including in developed countries with modern, industrial farming technologies. Soil erosion rates are highly variable among coffee farms—even between neighboring farms and different sections of the same farm—but most coffee farms have above optimal rates of soil loss and all farmers must make continuous efforts to maintain sufficient soil fertility.

Erosion is not just a problem on the farm. Sediment, often laced with agrochemicals, contaminates streams, lakes, and wetlands, suffocating aquatic life and polluting drinking water supplies. Miguel Araujo, the former minister of the environment in El Salvador, remembers: "Representatives of the fishing industry thanked the coffee farming association for helping maintain a productive off-shore fishery. Because the farms controlled erosion and agrochemical use, the water flowing from the highlands of El Salvador to the Pacific Ocean was cleaner than water exiting other agricultural or urban areas."

Soil characteristics have been mapped in most countries, including nearly every coffee-growing region. There is abundant knowledge about what soil types are most suitable for coffee, and how to amend and maintain soil for crop productivity and quality. Agronomists with government extension programs; multilateral agencies such as the UN Food and Agricultural Organization (FAO); agricultural research institutions such as Centre de Coopération Internationale en Recherche Agronomique pour le Développement (CIRAD), Centro Agronómico Tropical de Investigación y Enseñanza (CATIE), and Instituto Interamericano de Cooperación para la Agricultura (IICA); civil society organizations such as Solidaridad, Technoserve, and the standard-setting groups; coffee-growing associations and the training programs of coffee roasting and trading companies all know what is required for coffee plant nutrition in each soil type. It is however also important to ensure farmers get this information and can apply it.

The certification, government, and private sector training programs make soil conservation a requirement for participation and access to program benefits and markets. The certification criteria and/or training elements include making a soil conservation plan, soil analysis, organic enrichment, contour planting and terracing, live barriers and cover crops to control erosion, and reforestation where practical.

Soil nutrition is the critical factor limiting productivity and thus profitability for many coffee farmers. Because fertilizer is expensive, only those farmers doing well can afford it. Some governments, such as Rwanda, and some national coffee associations subsidize fertilizer. Fertilizers pose some risks to human health and must be handled with care. Fertilizers can pollute streams, wetlands, and aquifers; using best practices in application can reduce these hazards. Helping farmers use less fertilizer more effectively, finding alternative inputs and adopting organic farm management techniques are keys to the future of coffee farming.

7. PESTS AND DISEASES CONTROL FOR A BETTER COFFEE QUALITY

Viable and profitable coffee production must be supplied with the proper nutrients and protected from pests and diseases. Most farmers who can afford agrochemicals use them to boost productivity and income and protect their investments. Their coffee crop is the only source of income for millions of farmers and can be wiped out by insects or fungus. The stakes for farmers are critical as scientists expect that the traditional cycles of pests and diseases are likely to be aggravated by climate change. The outbreak of rust that started in 2013 and swept across Central America and southern Mexico was a stark reminder of the vulnerability of the coffee farming enterprise. An estimated 4 million people in that region depend on coffee, half a million in Guatemala alone. Farmers reported losses of 30–70% and many went out of business. Guatemala, Honduras, and El Salvador declared states of emergency.

Although there is no evidence that diseases or pests can influence directly the synthesis or accumulation of specific metabolites in the bean, their occurrence and control are critical for coffee quality. Fungi and insects can affect plants and cherries with such intensity that bean development and quality can be affected. In some cases, even small infections or parasitism can affect the physical characteristics of the bean or the beverage quality potential of the green coffee.

A large percentage of defective beans is generated by fungal or insect parasitism during the period of bean maturation. Out of the 16 commonly recognized fungal diseases in coffee worldwide, five involve the attack of coffee cherries: berry blotch or iron spot (*Cercospora coffeicola*), coffee berry disease (*Colletotrichum kahawae*), American leaf spot or "Ojo de Gallo" (*Mycena citricolor*), pink disease (*Corticium salmonicolor*), and oily spot (*Colletotrichum* spp.). For insects, among the list of 19 pests reported in coffee plantations, one is responsible for the main damage of the beans: the coffee berry borer (CBB) (*Hypothenemus hampei*). Nevertheless, some insects can also locally play a significant role in directly damaging the cherries or through microorganism infections leading to off-flavors, such as different species of fruit flies (*Ceratitis capitata*, *Trirhithrum* spp., *Anastrepha ludens*), Antestia bugs (*Antestiopsis orbitalis ghesquierei*), and mealy bugs (*Planocococcoides* spp.). Besides these field damages on the coffee cherries and beans, one insect is also responsible for severe affections on the coffee beans during storage, the coffee weevil (*Araecerus fasciculatus*) that feeds on the beans, making them completely unusable for roasting.

Generally speaking, damages of pests and diseases on the coffee plants will make them weak and consequently affect both yield and coffee quality. Strong defoliation is associated to the coffee leaf rust infection (*Hemileia vastatrix*) or to Lepidoptera attacks (i.e., *Leucoptera caffeina*), sometimes even leading to the death of the trees. There are excellent literature reviews that describe the main pests and diseases affecting Arabica and Robusta coffees in the world, giving information on symptoms, damages, and control methods (Waller et al., 2007; Gaitan et al., 2015).

7.1 Main Fungi Diseases Affecting Coffee Beans and Cherries

In general, the fungal infection on cherries starts with the invasion of the soft pulp tissue of the fruits (the mesocarp). Infection during early stages of cherry development may prompt it to fall off, but at later stages it can remain unnoticeable since fungal invasion may be superficial and pulping will remove the affected tissues, obtaining a clean bean. However, with more aggressive infections, parchment (endocarp) and bean (endosperm) are affected, with the consequence of partial pulping caused by mesocarp sticking to the endosperm. Bean stains occur due to polyphenolic compounds formation or rotten black tissue emerges because of fungal bean consumption. Although fungal growth stops once the bean humidity is under 12%, damages will remain, producing a negative aesthetic effect. The degradation of endosperm constituents such as sugars, or accumulation of fungal metabolites, leads to a lower quality coffee with harsh and woody cup characteristics (De Lima et al., 2012).

Berry blotch or iron spot (*Cercospora coffeicola*) is the most common fruit problem worldwide and its association to malnourishment of coffee plantations is well known, particularly with respect to nitrogen deficiencies. The fungus generates red lesions from induced early ripening that later turns brown and necrotic (Fig. 2.7). Dry periods will make the problem even more important. It means that the first action to take is to improve the nutritional status of the plant, providing fertilization under appropriate soil acidity and water availability conditions. Unfortunately, when malnourishment is the result of poor root development, nematodes, or root-sucking mealy bugs (*Dysmicoccus brevipes* or *Puto barberi*) parasitism, the corrections are harder to make as the plants cannot efficiently absorb the fertilizers. For nematodes and root insects, chemical controls can probably reduce the populations in a plot and improve plant assimilation of nutrients. In general, increasing shade will slow down the plant metabolism, reducing mineral requirements and therefore the incidence of the disease.

Coffee berry disease (*Colletotrichum kahawae*) is a devastating pathogen so far present exclusively in Africa. In its more severe form, the fungus invades the berry during the green stage (4–14 weeks after flowering) producing dark brown spots that end up covering the cherry and affecting bean development and quality. A milder attack results in the formation of a cork tissue (a scab) where the infection remains latent until fruit ripening, affecting mostly the pulp tissue but not the bean. Periodical applications of fungicides are required to keep the disease on check. Bred varieties with genetic resistance have been available for a while and provide a more effective solution.

Oily spot (*Colletotrichum* spp.) is a problem that has a strong inheritance component for susceptibility; therefore, seed selection from trees free of the disease will eliminate its onset in new plantations. Finally, American leaf spot (*Mycena citricolor*) and pink disease (*Corticium salmonicolor*) are problems conditioned to constant high humidity and temperature, and reduced exposure to sunshine, consequently adequate adjustments of shade levels in the farm according to the rain distribution throughout the year, combined with timely preventive fungicide applications that can lead to reduced incidence of these pathogens (Fig. 2.7).

FIGURE 2.7 Fungal diseases affecting the coffee cherry and generating defective coffee beans that affect cup quality. From left to right: iron spot (*red lesions* from early ripening) and oily spot (*gray lesions*) on the same cherries; American leaf spot with formation of needle-like gems protruding from the lesion; pink disease, with a characteristic pink crust growing on the branch.

7.2 Main Insects Affecting Coffee Beans and Cherries

The tiny CBB beetle (*Hypothenemus hampei*) detects coffee berries at a long distance as they represent its only host. It bores a hole at the tip of the cherry (Fig. 2.8) and makes a tunnel into the beans, opening a way to bacterial and fungal infections. It will lay eggs inside the beans, where larvae will feed, damaging the structure and composition of the endosperm. A cup of coffee prepared with up to 50% of grade 2 CBB infested beans (Fig. 2.9) can have an acceptable quality for the coffee market (Castaño-Castrillón and Torres, 1997), but presence of grades 3 and 4 usually causes rejection considering the occurrence of off-flavors like fermented or moldy taste (Montoya, 1999). For premium coffee markets the tolerance to CBB damage is nil. Even in beans for cryoconcentrated freeze-dried soluble coffee purposes, grade 3 CBB infestation will produce very inferior qualities (Castaño-Castrillón and Quintero, 2004). CBB not only affects coffee quality, it also decreases bean weight leading to a reduction in market value. Infested coffee berries lose weight in an average of 10.82%, up to 45.12% (Montoya, 1999), meaning that a farmer could lose almost half of its production under severe attacks. In addition, only 1.5% defective coffee beans are allowed for exportation. CBB can also create food safety issues as it allows the growth of fungi (*Aspergillus ochraceus*, *A. niger*, and *Penicillium viridicatum*) responsible for ochratoxin synthesis, a serious contaminant (Vega and Mercadier, 1998; Velmourougane et al., 2010).

The CBB beetle has spread from its center of origin in Africa to every coffee-producing country worldwide, with the last introduction registered in

FIGURE 2.8 Adult female of CBB infesting a coffee berry in the field.

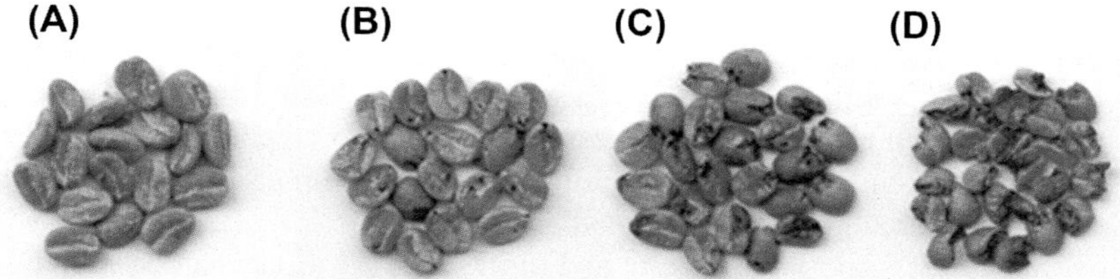

FIGURE 2.9 CBB infestation scale for beans (Castaño-Castrillón and Quintero, 2004). (A) Grade 1: uninfested; (B) Grade 2: 43% infestation; (C) Grade 3: 57% infestation; (D) Grade 4: 100% infestation.

the Kona district of the island of Hawaii in 2010 (Burbano et al., 2011). Since no efficient genetic resistance has been identified for CBB, integrated pest management combining agricultural practices, biological, and chemical controls is required to keep infestation levels below 5% in the field.

7.3 Environmental and Safety Concerns About Plant Protection

As indicated more extensively in Chapter 4, the contamination of coffee with pesticides residues is a growing concern for coffee quality. Rising coffee quality standards and increasing pest and disease outbreak due to climate change could encourage more pesticide use. However, when compared with other crops, the use of pesticide is rather low on coffee crop, meaning that pesticide residues are rarely found in beans prior roasting. It is, therefore, important to pursue research for organic, environmentally sustainable, and low-cost alternatives that can ensure a cleaner and safer coffee production, preserving not only the environment, but also human health and coffee quality. Current developments on natural and biological products to control diseases and pests, combined with self-defense mechanisms present in the coffee plant are encouraging perspectives. They need to be combined with preventive agronomical practices based on weather forecasting and better resources management of the farm. Early and proper identification of pests and diseases in the coffee plantation is necessary to determine appropriate corrective actions. Chapter 4 further describes management of fungi and insects using biological controls.

At the farm level, pesticides pose considerable peril to human health and the environment. Certification programs, government regulations, and extension and training programs seek to reduce these risks through a multifaceted approach that often includes the following elements:

- Prohibit the most toxic and dangerous chemicals. Infamous pesticides such as dichlorodiphenyltrichloroethane are mostly banned in coffee production, and there are campaigns to eliminate the use of other high-risk pesticides such as endosulfan, a product used to control the CBB (*Hypothenemus hampei*). Safer alternative molecules are now available like the active ingredient cyantraniliprole that was recently (2014) registered by the Brazilian authorities. Other insecticides banned or on their way out include chlorpyrifos and the class 1a (extremely hazardous) organophosphates such as methyl parathion and disulfoton.
- Teach farmers to monitor pests and diseases and use mechanical, biological, or cultural remedies as far as possible. Apply chemicals only as a last resort and use the least-toxic product only where and when needed. These are precepts of a system called integrated pest management, which can greatly reduce the reliance on—and the hazards of—pesticides.
- Use best practices in storing, transporting, mixing, and applying agrochemicals to minimize the threats to human health and the environment.
- Do not apply agrochemicals near housing, work areas, or along streams and other bodies of water.
- Any worker touching agrochemicals must have the proper training, personal protection equipment, and access to medical care.
- These and other measures allow farmers to protect their crops and stay within risk perimeters determined by scientists, NGOs, and government agencies.

8. OUTLOOK

Coffee farmers face strong and diversified challenges where climate change is playing a major role. Although there are multiple efforts to provide new information and tools to farmers, it is difficult to reach millions of remote smallholders. There is a need to facilitate the transfer of available solutions to small coffee producers to avoid the crop abandon and social disorders, which will follow. If yield is affected, no doubt that quality will be even more affected as we learned that quality is obtained by optimization of the different factors.

The preservation or restoration of soil fertility will be one of the most serious needs to be solved in the main coffee growing regions in the next decades. Intensive use of pesticides and mineral fertilizers are associated with massive losses of organic matter, leading to soil exhaustion. The consequences include serious soil-borne pests and diseases like nematodes or *Fusarium* sp., forcing hundreds of thousands coffee growers to temporarily abandon the coffee cultivation, trying to restore the soil fertility by fallows of 2–5 years. Soil characteristics have been mapped in most countries, including every coffee-growing region. There is abundant knowledge about what soil types are most suitable for coffee and how to amend and maintain soil for crop productivity and quality. But the challenge is getting this information to farmers in ways that they can accept, absorb, and apply it.

For now, intercropping and/or agroforestry appear as a good solution to preserve the soil fertility. Some alternatives or complementary solutions to mineral fertilizer afford also innovative approaches to stimulate the microorganisms in the soil. New commercial bio fertilizers (e.g., mycorrhizal inoculants) could help the coffee plant to access less mobile nutrients (particularly phosphorus and nitrogen). Other nonmicrobial products such as biochar and vermicompost stimulate the microbial activity and could be used at a large scale. Some bioactive products are now developed by the agrochemical sector and could afford alternatives or complementary solutions to chemical pesticides. These are new avenues that may certainly contribute to making coffee production more sustainable and less damaging for the environment.

Juan Diego Roman, an agronomist and third-generation coffee farmer in Costa Rica, said, "Coffee farming is challenging. We apply hard-earned experience, all the tricks and techniques we've learned over the years. But still there are risks. We need scientists to help reduce the uncertainties. For example, scientists know exactly what nutrients a coffee plant requires to grow high-quality cherries. Agronomists translate the information into practices that farmers can implement, taking some of the guesswork out of fertilization. With challenges such as climate change raising the stakes, we will need more science, more experimentation and better exchange of information throughout the coffee sector."

REFERENCES

Almeida da Fonseca, A., Gava Ferrao, R., Gava Ferrao, M.A., Verdin Filho, A.C., Carvalho Bittencourt, P.S., 2007. Jardins clonais, producao de sementes e mudas. In: Gava Ferrao, R., Almeida da Fonseca, A.F., Bragança, S.M., Gava Ferrao, M.A., Herzog De Muner, L. (Eds.), Café Conilon, pp. 229–255.

Alvarado, G., Moreno, E., Montoya, E., Alarcon, R., 2010. Calidad física y en taza de los componentes de la variedad Castillo R y sus derivadas regionales, vol. 60 no. 3. Cenicafé, pp. 210–228.

de Amorim, H.V., Scoton, L.C., de Castilho, A., Gomes, F.P., Malavolta, E., 1965. Estudos sobre a alimentação mineral do cafeeiro. XVII. Efeito da adubação NPK na composição química do solo do fruto e na qualidade da bebida. Anais da Escola Superior de Agricultura "Luiz de Queiroz" 22, 139–152.

de Amorim, H.V., Teixeira, A.A., Moraes, R.S., dos Reis, A.J., Gomes, F.P., Malavolta, E., 1973. Estudos sobre a alimentação mineral do Cafeeiro XXVII. Efeito da adubação N, P e K no teor de macro e micro nutrientes do fruto e na qualidade da bebida do café. Anais da Escola Superior de Agricultura "Luiz de Queiroz" (Brasil) 30, 323–333.

Beer, J., Muschler, R., Kass, D., Somarriba, E., 1998. Shade management in coffee and cacao plantations. In: Directions in Tropical Agroforestry Research. Springer, Netherlands, pp. 139–164.

Bertrand, B., Bardil, A., Baraille, H., Dussert, S., Doulbeau, S., Dubois, E., Severac, D., Dereeper, A., Etienne, H., 2015. The greater phenotypic homeostasis of the allopolyploid *Coffea arabica* improved the transcriptional homeostasis over that of both diploid parents. Plant and Cell Physiology 56 (10), 2035–2051.

Bertrand, B., Boulanger, R., Dussert, S., Ribeyre, F., Berthiot, L., Descroix, F., Joët, T., 2012. Climatic factors directly impact the volatile organic compound fingerprint in green Arabica coffee bean as well as coffee beverage quality. Food Chemistry 135 (4), 2575–2583.

Bertrand, B., Etienne, H., Eskes, A., 2001. Growth, production, and bean quality of *Coffea arabica* as affected by interspecific grafting: consequences for rootstock breeding. HortScience 36 (2), 269–273.

Bertrand, B., Vaast, P., Alpizar, E., Etienne, H., Davrieux, F., Charmetant, P., 2006. Comparison of bean biochemical composition and beverage quality of Arabica hybrids involving Sudanese-Ethiopian origins with traditional varieties at various elevations in Central America. Tree Physiology 26 (9), 1239–1248.

Bunn, C., Läderach, P., Ovalle Rivera, O., Kirschke, D., December 13, 2014. A Bitter Cup: Climate Change Profile of Global Production of Arabica and Robusta Coffee. Available from: http://rd.springer.com/article/10.1007%2Fs10584-014-1306-x#.

Bunn, C., Läderach, P., Pérez Jimenez, J.G., Montagnon, C., Schilling, T., October 27, 2015. Multiclass Classification of Agro-ecological Zones for Arabica Coffee: An Improved Understanding of the Impacts of Climate Change. Available from: http://journals.plos.org/plosone/article?id=10.1371/journal.pone.0140490.

Burbano, E., Wright, M., Bright, D.E., Vega, F.E., 2011. New record for the coffee berry borer, *Hypothenemus hampei*, in Hawaii. Journal of Insect Science 11 (1), 117.

Castaño-Castrillón, J.J., Quintero, G.P., 2004. Calidad de extractos de café perforado por broca obtenidos por crioconcentración, vol. 55 no. 3. Cenicafé, pp. 183–201.

Castaño-Castrillón, J.J., Torres, M.L.A., 1997. Caracterización de café tostado a partir de café perforado por broca. Cenicafé, p. 51. Available from: http://biblioteca.cenicafe.org/handle/10778/642.

Castillo, Z.J., 1957. Influencia de algunos tratamientos culturales sobre la calidad del grano de café, vol. 8 no. 11. Cenicafé, pp. 333–346.

Cenicafé, F.N.C, 2013. Manual del Cafetero Colombiano, vol. 2, pp. 8–26.

Charrier, A., Berthaud, J., 1988. Principles and methods in coffee plant breeding: *Coffea canephora* Pierre. In: Clarke, R.J., Macrae, R. (Eds.), Coffee: Volume 4—Agronomy, pp. 167–197.

DaMatta, F.M., 2004. Ecophysiological constraints on the production of shaded and unshaded coffee: a review. Field Crops Research 86 (2), 99–114.

DaMatta, F.M., Ramalho, J.D.C., 2006. Impacts of drought and temperature stress on coffee physiology and production: a review. Brazilian Journal of Plant Physiology 18 (1), 55–81.

DaMatta, F.M., Ronchi, C.P., Maestri, M., Barros, R.S., 2007. Ecophysiology of coffee growth and production. Brazilian Journal of Plant Physiology 19 (4), 485–510.

Davis, A.P., Govaerts, R., Bridson, D.M., Stoffelen, P., 2006. An annotated taxonomic conspectus of the genus *Coffea* (Rubiaceae). Botanical Journal of the Linnean Society 152 (4), 465–512.

De Castro, R.D., Marraccini, P., 2006. Cytology, biochemistry and molecular changes during coffee fruit development. Brazilian Journal of Plant Physiology 18 (1), 175–199.

De Lima, L., Pozza, E., Da Silva Santos, F., 2012. Relationship between incidence of brown eye spot of coffee cherries and the chemical composition of coffee beans. Journal of Phytopathology 160 (4), 209–211.

Descroix, F., Wintgens, J.N., 2004. Establishing a coffee plantation. In: Wintgens, J.N. (Ed.), Coffee: Growing, Processing, Sustainable Production, pp. 210–217.

Ducos, J.P., Labbé, G., Lambot, C., Pétiard, V., 2007. Pilot scale process for the production of pre-germinated somatic embryos of selected Robusta clones. In Vitro Cellular and Development Biology Plant 43 (6), 652–659.

Etienne-Barry, D., Bertrand, B., Vasquez, N., Etienne, H., 1999. Direct sowing of *Coffea arabica* somatic embryos mass-produced in a bioreactor and regeneration of plants. Plant Cell Reports 19 (2), 111–117.

Fahl, J.I., Carelli, M.L.C., Menezes, H.C., Gallo, P.B., Trivelin, P.C.O., 2001. Gas exchange, growth, yield and beverage quality of *Coffea arabica* cultivars grafted on to *C. canephora* and *C. congensis*. Experimental Agriculture 37 (2), 241–252.

Farfan, V.F., 2013. Establecimiento de sistemas agroforestales con café. In: Manual del Cafetero Colombiano: Investigación y Tecnología para la sostenibilidad de la caficultura, vol. 2. Cenicafé, Chinchiná, pp. 46–63, 320 p.

Food Agricultural Organization – Natural Resource Management and Environment Department. How soil is destroyed – Erosion Destroyed Civilisations. Available from: http://www.fao.org/docrep/t0389e/T0389E02.htm# Erosion destroyed civilizations.

Foote, H.E., 1963. Factors affecting cup quality in coffee. Coffee and Cacao Journal (Filipinas) 5 (12), 248–249.

Gaitan, A., Cadena, G., Castro, B., Cristancho, M., Rivillas, C. (Eds.), 2015. Compendium of Coffee Diseases. American Phytopathological Society. 88 p. ISBN: 978-0-89054-470-9.

Georget, F., Courtel, P., Malo Garcia, E., Hidalgo, J.M., Alpizar, E., Poncon, C., Bertrand, B., Etienne, H., 2014. A booster for commercial propagation of *Coffea arabica* F1 hybrids : somatic embryo-derived plantlets can be efficiently propagated in nursery via rooted cuttings. In: ASIC, 25th International Conference, Colombia. Available from: http://asic-cafe.org/fr/content/booster-commercial-propagation-coffea-arabica-f1-hybrids-somatic-embryo-derived-plantlets-ca.

Gichimu, B.M., Gichuru, E.K., Mamati, G.E., Nyende, A.B., 2012. Selection within *Coffea arabica* cv. Ruiru 11 for high cup quality. African Journal of Food Science 6 (18), 456–464.

Graner, E.A., Godoy Jr., C., 1970. Adubação do café X. Produção, rendimento, qualidade da bebida e características do fruto e do grao no quarto ano de colheita 1963. Revista de Agricultura 45 (1), 52–57.

IAASTD, 2009. Agriculture at a Crossroads. International Assessment of Agricultural Knowledge, Science and Technology for Development (IAASTD), Global Report. Island Press, Washington, D.C.

Jayarama, R.P., Ananda, A., Naik, C.S.K., D'Souza, M.V., 1992. Effect of micronutrient sprays on arabica coffee. Indian Coffee 8 (56), 3–5.

Jha, S., Bacon, C.M., Philpott, S.M., Méndez, V.E., Läderach, P., Rice, R.A., 2014. Shade coffee: update on a disappearing refuge for biodiversity. BioScience 64 (5), 416–428.

Joët, T., Laffargue, A., Descroix, F., Doulbeau, S., Bertrand, B., Dussert, S., 2010. Influence of environmental factors, wet processing and their interactions on the biochemical composition of green Arabica coffee beans. Food Chemistry 118 (3), 693–701.

Lambot, C., Bouharmont, P., 2004. Soil protection. In: Wintgens, J.N. (Ed.), Coffee: Growing, Processing, Sustainable Production, pp. 278–281.

Lazzarini, W., 1961. Qualidade da bebida do café. O Agronómico (Brasil) 13 (11–12), 6.

Liljequist, G.H., 1986. Some aspects of the climate of Ethiopia. Symbolae Botanicae, Uppsala 26 (2), 19–30.

Malavolta, E., 1986. Efeitos de doses e fontes de enxofre em culturas de interesse econômico. IV – Café. In: Centro de Pesquisa e Promoção do Sulfato de Amônio, vol. 4. Boletim Técnico, São Paulo, 41 p.

Malavolta, E., 2006. Manual de nutrição mineral de plantas. Agronômica Ceres, São Paulo, 631 p.

Malézieux, E., Crozat, Y., Dupraz, C., Laurans, M., Makowski, D., Ozier-Lafontaine, H., Rapidel, B., de Tourdonnet, S., Valantin-Morison, M., 2009. Mixing plant species in cropping systems: concepts, tools and models: a review. In: Sustainable Agriculture. Springer, Netherlands, pp. 329–353.

Malta, M.R., Nogueira, F.D., Guimaraes, P.T.G., 2003. Composição química, produção e qualidade do café fertilizado com diferentes fontes e doses de nitrogênio. Ciencia e Agrotecnología 27 (6), 1246–1252.

Malta, M.R., Nogueira, F.D., Guimaraes, P.T.G., 2002. Avaliação da qualidade do café/*Coffea arabica*/L. fertilizado com diferentes fontes e doses de potassio. Revista Brasileira de Armazenamento 5, 9–14.

Mitchell, H.W., 1988. Cultivation and harvesting of the Arabica coffee tree. In: Clarke, R.J., Macrae, R. (Eds.), Coffee: Volume 4–Agronomy, pp. 50–52.

Moguel, P., Toledo, V., 1999. Biodiversity conservation in traditional coffee systems of Mexico. Conservation Biology 13, 11–21.

Montgomery, D.R., 2007. Soil erosion and agricultural sustainability. Proceedings of the National Academy of Sciences of the United States of America 104 (33), 13268–13272.

Montoya, R.E.C., 1999. Caracterización de la infestación del café por la broca y efecto del daño en la calidad de la bebida, vol. 50 no. 4. Cenicafé, pp. 245–258.

Muschler, R.G., 2001. Shade improves coffee quality in a sub-optimal coffee-zone of Costa Rica. Agroforestry Systems 51 (2), 131–139.

Muschler, R.G., 2004. Shade management and its effect on coffee growth and quality. In: Wintgens, J.N. (Ed.), Coffee: Growing, Processing, Sustainable Production, pp. 391–418.

Ovalle Rivera, O., Läderach, P., Bunn, C., Obersteiner, M., Schroth, G., April 14, 2015. Projected Shifts in *Coffea arabica* Suitability Among Major Global Producing Regions Due to Climate Change. Available from: http://journals.plos.org/plosone/article?id=10.1371/journal.pone.0124155.

Perfecto, I., Vandermeer, J., Mas, A., Pinto, L.S., 2005. Biodiversity, yield, and shade coffee certification. Ecological Economics 54 (4), 435–446.

Philpott, S.M., Arendt, W.J., Armbrecht, I., Bichier, P., Diestch, T.V., et al., 2008. Biodiversity loss in Latin American coffee landscapes: review of the evidence on ants, birds, and trees. Conservation Biology 22 (5), 1093–1105.

Prieto Martinez, H.E., Poltronieri, Y., Farah, A., Perrone, D., 2013. Zinc supplementation production and quality of coffee beans. Revista Ceres 60 (2), 293–299.

Redgwell, R., Fischer, M., 2006. Coffee carbohydrates. Brazilian Journal of Plant Physiology 18 (1), 165–174.

dos Reis, A.R., Favarin, J.L., Gallo, L.A., Moraes, M.F., Tezotto, T., Lavres Junior, J., 2011. Influence of nitrogen fertilization on nickel accumulation and chemical composition of coffee plants during fruit development. Journal of Plant Nutrition 34 (12), 1853–1866.

Rinehart, R., 2009. What Is Specialty Coffee? Article published by the Specialty Coffee Association of America (SCAA). Available from:www.scaa.org/?page=RicArtp1.

that can either enhance quality or give quality consistency, expand flavor potential, or improve environmental aspects. The commonality with each of the methods is that they are mainly driven by the management of one specific parameter: water quantity, oxygen (aerobic or anaerobic), or microbes. The examples given are in some cases specific to a local culture, which lead to a specific flavor of the coffee from that region. The distinct flavor obtained is sometimes accompanied with a beautiful story, making the coffee even more appealing to the specialty coffee industry. By controlled processing one can enhance aromatic diversity, and hence increase the price of the coffee. This in turn can increase revenue for farmers.

2. HARVESTING THE COFFEE

Coffee fruits should be collected from the trees to be taken to the facilities where they are transformed into a stable product. There are two main ways to harvest coffee; by hand picking or mechanically. The selection of the method depends on several factors such as the landscape, slope, coffee variety, labor cost, size of the farm, and distribution of cherry maturity.

2.1 Generalities

Mechanical harvesting is chosen primarily to take advantage of flat and small sloped lands and needs big investment in equipment that may, however, be shared among many growers and which can be depreciated by efficiency and labor cost reduction. A nice example to mention is in the Cerrado area of Minas Gerais, Brazil, where the maturation of beans is synchronized by playing with irrigation; as a consequence the amount of immature cherries to be sorted out before processing is reduced (Borém, 2014). Hand picking is the preferred method where selective harvesting is a requirement, and for steep slopes and difficult landscapes where it is impossible to use large machines. This method depends also on the availability of labor. The scope is the possibility to do selective picking of only ripe cherries. In contrast this implies to do multiple batches of picking to harvest the ripe cherries from the entire farm, which makes the process labor intensive and costly and this must be compensated with the quality and price paid.

There are other harvest methods such as stripping, which is the pulling out all the cherries (ripe and unripe) from each branch. Sorting of ripe and unripe cherries is then needed before processing. The use of vibrating mechanical fingers in hand-held harvesters with a small engine is now widespread in Brazil to perform mechanical stripping with a limited degree of selectivity. These small harvesters are sometimes owned by the coffee pickers themselves. These machines are progressively being tested and used in other countries. Finally, there are some new developments to facilitate the picking of the beans, like using vibration rings applied to the trunk of the coffee plant and using the principle that the ripe coffee falls more easily than the unripe one, or vacuum pumps to select the ripe cherries and pull out from the branch, but these still need to be miniaturized and the cost reduced to be affordable by growers and to create a benefit for them. In addition to these regular harvesting methods where ripe cherries are selected, cherries can also be harvested when overripe, adding novel flavors to the cup.

2.2 Late Harvest

Instead of harvesting the ripe cherries, one can also harvest the coffee cherries at a later stage, i.e., not when they are ripe (*bright red color*) but after leaving them on the tree until they are overripe (*reddish purple color*). This principle is not new in the food industry. The inspiration comes from the wine industry, where in certain regions grapes (*Vitis vinifera*) are left on the vine well beyond optimum ripeness, allowing them to become overripe producing a very sweet wine (Barbeau et al., 2001).

Factors to be taken into account to produce the coffee known as late harvest are terroir and plant variety. In Brazil this process is commonly used in regions that are hot and dry and where it is possible to retain the cherries on the tree until they are dehydrated (moisture close to 30%). Mechanical harvesters pick all the cherries with a very low probability of having immature cherries, which are then dried. These coffees are known as late harvest unwashed coffees. In the Cerrado region of Minas Gerais coffees obtained by this process receive high scores in the international contests such as Cup of Excellence.

In Colombia, this process is not commonly applied but was tested in the region of Tolima in 2010 [unpublished data from the National Federation of Coffee Growers of Colombia (FNC) and Nespresso]. Various existing cultivars were evaluated in different regions of the country that are known to have different soil types and climatic conditions, commonly known as ecotopes (Gómez et al., 2001). The best result was obtained with the Castillo® variety, where the coffee cherries went from a state of ripeness to overripeness without the beans falling from the branches on to the ground.

The maturation of coffee cherries goes through four principal stages (unpublished definition FNC). The first stage lasts from the formation of the bean until approximately the eighth week, the second stage lasts until week 26, the third stage runs to week 32, and the fourth stage known as the overripeness phase lasts approximately until weeks 34 and 35 (Fig. 3.1). After week 32 (more than 225 days), the bean becomes overripe and turns a dark purple color, and then finally dries and loses weight and produces a *dark or black color*. Unique and special flavor characteristics are obtained with sweet winey notes.

Apart from the color of the cherry, the best indication that the cherry has reached optimal overripe stage is by measuring the Brix degrees, which is a

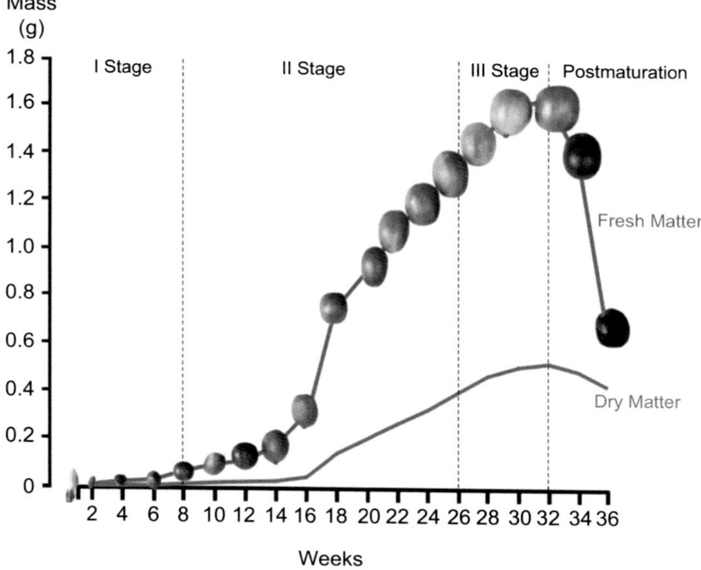

FIGURE 3.1 Ripening stages of the coffee cherry as a function of mass.

measure of soluble solids, including sugar, in the mucilage (1° equals 1 gram of sucrose in 100 mL water, the presence of other sugars will somewhat influence the measure). The Brix degree increases with ripening with a maximum at "overripeness" and then reduces once dry/dehydrated (exact Brix values will depend, e.g., on variety, altitude, humidity). To ensure optimal quality, unripe and dried fruits need to be removed to ensure that only the overripe coffee is processed. Late harvests treated by the wet process require 2 h in addition to normal fermentation time.

3. VARIATIONS IN MUCILAGE REMOVAL TO IMPACT FLAVOR

During fermentation, the mucilage adhered to the coffee parchment naturally degrades and is removed from the coffee beans. This degradation is caused by microorganisms that dissolve the mucilage structure either through enzymatic action or by using it as a carbon source. Enzymatic activity is caused mainly by pectinases (Avallone et al., 2002; Masoud and Kaltoft, 2006). In wet processing, once the mucilage is degraded, it is removed by washing. Fermentation also occurs in semidry and dry processes by decreasing coffee moisture (Evangelista et al., 2013).

Fermentation is a naturally occurring spontaneous process caused by microorganisms growing in the environment. It is influenced by many factors, including the initial condition of the beans, as well as the variety, climate, and fruit maturity. These external factors play an important role in fermentation evolution because they determine the timing of microorganism activity and substrate transformation. During fermentation biochemical reactions act on different compounds, mainly carbohydrates, and cause an increase in temperature and decrease in pH (from ca. 6.5 to 4.1), which may ultimately influence the coffee flavor (Avallone et al., 2001; Peñuela Martínez et al., 2010; Velmourougane, 2013). Microorganism cultures sometimes alter the physical quality and taste of Arabica coffee (Suárez-Quiroz et al., 2004). Microbial populations change during the fermentation process. At the beginning of the process generally bacteria are dominant, followed by different species of yeast in the middle and finally filamentous fungi toward the end (typically occurring coincident with the onset of aerobic conditions) (Avallone et al., 2001; Evangelista et al., 2015; Silva et al., 2008; de Mela Pereira et al., 2014).

In addition to fermentation other processes exist to remove the mucilage from the coffee bean. For example, adding pectinolytic enzymes to the fermentation is a natural way to accelerate the degradation process, with no noticeable effect on quality (Peñuela Martínez et al., 2013). Mucilage can also be removed by mechanical means. Although there is a preference in the market for coffee processed by natural fermentation, mechanical demucilaging results in a more consistent product since all the beans receive always the same treatment. Mechanically demucilaged beans generally have a somewhat more sharp acidity compared to fermentation which delivers more juicy and fine acidity (Borém, 2014). To complete the mechanical removal of mucilage, Pabón Usaquén et al. (2009) discovered that rinsing the demucilaged coffee before drying has positive effects on quality, even if up to 14 h have elapsed. It is also interesting to note that fermentation, as compared to mechanical mucilage removal, causes 2–3% weight loss and even more at higher temperatures.

In the wet process, it is important that ripe coffee cherries are processed immediately to avoid chemical and/or biological damage, leading to loss of flavor and flavor defects. In addition, uncontrolled processing conditions or bad practices may cause medicinal, earthy, musty, moldy, and hidy (tobacco, leather-like) off-flavor (Lingle, 1986).

For instance, during wet processing, fermentation can take place in large tanks, which may include water. This may impact the fermentation conditions and, as a result, affect factors such as coffee mass homogeneity, temperature, oxygen availability, and processing time (Peñuela Martínez et al., 2013; Schwan and Graham, 2014; de Melo Pereira et al., 2015).

For the coffee industry, it has become apparent that fermentation is critical; the factors involved in fermentation, their interactions, and the variation of fermentation methodologies all play an important role. This contributes to coffee composition and formation of both aroma and flavor precursor compounds. And this, in turn, eventually creates different sensory experiences.

Next we will discuss a few variations to conventional fermentation that will increase the value of the coffee either through qualitative or flavor aspects.

3.1 Pulped Naturals and Honey Processed

The Campinas Agronomy Institute, IAC, in Brazil first tried semiwashed processing in the early 1950s. In this process pulp is removed and parchment is dried with some or all remaining mucilage. In the 1980s, a few farmers in the South of Minas Gerais first deployed the practice on a commercial scale with positive quality impacts. This led to its wide adoption after Pinhalense developed the wet milling equipment required, installed pilot mills in different producing areas of Brazil, and presented the new product to domestic and foreign roasters. The final product of the Brazilian pulped natural is also known as Cereja Descascado or CD.

The residual amount of mucilage on the parchment in semiwashed coffees will influence the taste of the coffee. Cup features generally are similar to those of natural coffees, but without the unwanted harsh taste of unripe cherries. In other words, the process enables the production of high-quality natural-like coffees without the typical green, immature taste that may be found, for example, in Brazilian naturals when cherry ripening is not uniform. After a slow start in Brazil in the late 1980s, the new pulped natural process gained weight as the coffees it produced started to be widely used by roasters in the 1990s.

Whereas the pulped natural system was originally developed to cope with a quality problem, the system is now also used to obtain different tastes and qualities and as such expanded from Brazil into other regions such as Central America where the coffee is known as honey coffee and dried in a slower process in a humid environment, leading to a more aromatic richness. Like in Brazil, the cup characteristics vary according to the quantity of mucilage remaining on the bean. In Central America, three types of honey-processed coffee are yellow, red, and black, each described here below (Fig. 3.2).

For Yellow Honey production mucilage is partly removed using mechanical equipment. Coffee can be sundried and the total process can take approximately 8–10 days, leading to a yellow to gold coffee parchment. The flavor is close to the washed coffee, and depending on the coffee variety, the acidity and cereal notes can be reinforced.

For Red Honey production around 50–75% of the mucilage is retained for the drying, which takes 12–15 days and needs regular rotation. Mechanical driers can be used toward the end of the process to guarantee a homogenous moisture of the coffee. The natural and fine acidity of the coffee are expressed while revealing the sweet character.

Black Honey is dried with 100% of the mucilage. Due to the high stickiness of the coffee, frequent rotation and mixing are needed to avoid the formation of coffee blocks. It is the most difficult process and takes a lot of time and care from the grower. It takes approximately 30 days to fully dry the coffee. Mostly

FIGURE 3.2 Honey-processed beans with mucilage remaining—yellow, red, and black honey.

a combination of sun and mechanical drying is applied. The final appearance of the coffee is a dark brownish color. This process attracts interest from the specialty coffee segment due to its flavor which varies from flowery to sweet, and from a mild and gentle acidity to wild juicy notes.

Applying the honey process on special varieties amplifies intensities and their sensory attributes. Honey processed and pulped natural varieties such as Geisha, Bourbon, Catuaí, Typica, Maragogype often obtain the highest scores in programs such as Cup of Excellence, Specialty Coffee Association of the Americas (SCAA), Best of Panama, and many others.

Notwithstanding the options above, it is possible to dry coffee with mucilage attached much faster. Pulped natural coffee with all mucilage attached may require as little as 1 day under full sun with frequent turning-over before it can be dried mechanically with direct feeding into rotary driers.

3.2 Intestinal Fermentation

Some animals eat coffee cherries as part of their regular diet, such as the Indonesian civet cat (*Paradoxurus hermaphroditus*), the Brazilian Jacu bird (*Penelope obscura*), and some African elephants, but do not digest the seeds. All the coffee obtained from the excretions of those animals is available in limited quantities and for that reason the price is high.

Luwak coffee first rose to fame just before the 1950s, when coffee plantations expanded heavily during the reign of the Dutch in the East Indies. However, it only became widely known among gourmet coffee enthusiasts in the 1980s. In 2010, it made the headlines in Australia after the President of Indonesia, Susilo Bambang Yudhoyono, gave Luwak coffee to the Australian Prime Minister, Kevin Rudd, during his visit to Australia.

Luwak coffee is produced in a special way. The civet cat chooses only ripe and sweet cherries. The cat's digestive tract removes the pulp and mucilage of the cherries. The main difference with the wet process is that the microorganisms present in the animal are not the same as those found in fermentation tanks. The digestion process takes approximately between 12 and 24 h (in the case of the Jacu bird the digestion is much shorter). Once the pulp and mucilage are digested, the beans are excreted. Coffee is obtained from the scats of wild civets or from animals grown on farms for this purpose. In both cases the farmer needs to collect the civet feces every day. The beans are soaked and washed, and then the coffee is sundried. The coffee connoisseurs define the flavor of the coffee as less acidic, with more body and with a lingering and strong after taste. There is no impact on the caffeine content (Yusianto et al., 2012).

Malic and citric acid content are used as chemical fingerprint to trace and identify real Luwak coffee from mixed or fake beans (Jumhawan et al., 2013). Kopi Luwak parcel well processed are considered as kosher (Indonesian Ulema Council, Majelis Ulama Indonesia, through fatwa No. 4, dated July 20, 2010) (Figs. 3.3 and 3.4).

FIGURE 3.3 The civet cat chooses to eat the red coffee cherries.

FIGURE 3.4 Coffee parchment from the wild civet.

4. VARIATIONS IN DRYING AND STORAGE

Coffee drying must be closely controlled to avoid loss of flavor potential and the development of fungi. Storing of coffee beans is often considered as a passive activity from drying till roasting. However, it is in reality very active with factors such as humidity, temperature, and time to consider as these can contribute to dynamically controlling, maintaining, or damaging the flavor. Coffee beans undergo many physical, chemical, and (micro)biological changes over time. Standard shelf-life of coffee beans is not the same between countries, regions, and types of coffee. During storage, one type of coffee may undergo changes that still create an acceptable flavor, whereas another type of coffee may become unacceptable.

Stored coffees can be perceived as more acidic due to loss of other flavor attributes, and become milder. However, uncontrolled storage can cause defects. A coffee expert has a simple motto: "If the conditions are good for man, they are also good for coffee." In practical terms, ventilation and insulation are critical during coffee storage as coffee moisture will adapt to the humidity levels of the environments. Therefore there must be enough space and air circulation to maintain constant levels of humidity and temperature. Then even if the temperature rises, good air circulation can prevent the coffee temperature from increasing.

Coffee beans may also experience varying environments further down the value chain. For example, the beans are exposed to different temperatures and humidity levels as they travel to the port. Rapid temperature changes can trigger condensation, a big problem when coffee beans from producer countries in hot, tropical climates are transported to colder temperatures. Risks of unfavorable storage conditions are fungi that can produce toxins—mainly ochratoxin, such as *Aspergillus ochraceus* and *Aspergillus carbonarius*, and *Penicillium verrucosum* (Ismayadi et al., 2005).

When coffee beans are stored for a long time, even under optimal conditions, color of beans changes (Yusianto et al., 2007) and the coffee flavor becomes woody. Studies have suggested that the oxidation of lipids is responsible for such changes. The proximity of proteins and lipids inside the cells may also enable proteins to oxidize together with the lipids. The presence of carbonyl groups in the proteins indicates that reactive oxygen species or oxidation of lips could generate this oxidation. The presence of thiobarbituric acid reactive substances and carbonyl groups at the beginning of the storage suggested that the oxidative process occurred during drying (Rendón et al., 2012).

Although generally drying and storing coffee aims at affecting quality as little as possible, these processes can also be a way to create new flavors and thus create new value. Next we will describe two examples on storing processes that are used in different countries to help create specific flavors.

4.1 Indonesian Labu Coffee

The unique Indonesian Labu processes of microanaerobic fermentation combined with wet hulling are commonly applied in Gayo-Central Aceh, North Sumatra, West Java, Enrekang, Toraja, Manggarai, and some regions in Bali (Kintamani), Bajawa-Ngada, and east Java.

The pulped cherries are fermented in polypropylene plastic bags from which maximum air is evacuated before sealing. Consequently microanaerobic microorganisms take the lead of the fermentation and generate secondary metabolites during the log-phase of growth, which are different from aerobic fermentations (e.g., traditional fermentation tanks). Fermentation time is long and the bags must be turned to accelerate the process. The flavor of the coffee becomes wild red fruits with a sharp acidity.

After fermenting, the coffee is washed and sundried for 1–3 days. Once the parchment skin is dry (but not the bean), it is ready for wet hulling, also called full wash or giling basah (Fig. 3.5). The reason for applying this method is that the raining season arrives in the same period of the harvest, and drying time can become as long as 2–3 weeks. The coffee thus needs to be dried as fast as possible to avoid mold growth. Wet hulling is a complex and difficult process. The risk of bruised beans is increased as a consequence of the pressure submitted to the soft and humid coffee in the hulling chamber. The coffee is sundried semiwet if the weather conditions are favorable. Mechanical drying can also be applied. Labu coffee is highly

FIGURE 3.5 Cracked parchment coffee "ready to hulling."

appreciated by specialty market with high SCAA scores thanks to its characteristic sharp juicy acidity, high body, and complex flavors.

4.2 Indian Monsooned Coffee

India launched the specialty coffee of "Monsooned" in the international market as early as 1972, even before the specialty movement gathered momentum in the world. The story of Monsooned coffee began in the mid-1800s, more than over a century before the word "specialty" was even coined for coffee.

The distinctive coffee of Monsooned was developed by a quirk of fate, a great story in the development of this specialty coffee. Sometime around 1850, a consignment of coffee beans left the shores of the Malabar Coast, bound for a coffee buyer in the Scandinavian region. In those days, the shipping time was very long, with a vessel taking more than 6 months to reach its destination. During the ship's long voyage around the Cape of Good Hope, through the monsoon rains emerging from the Arabian Gulf, the Arabica coffee beans within the jute bags would absorb moisture through the wooden holds of the ship. The seawater and the accompanying humid breeze from the heavy monsoon rains would cause the coffee beans to swell with moisture, undergo chemical changes, change appearance from a blue-gray bean to a creamy-golden yellow bean, become light weight, and double in size. The taste profile was also changing. The Scandinavian buyers and consumers received these golden yellow beans and the customers got accustomed to the particular taste profile of such beans.

Over the years, shipping logistics improved, as well as the packing and containerization of these bags of Arabica. The beans, when they reached their destination in Scandinavia, were no longer large in size or golden yellow in color and had a different taste profile. Quality complaints were lodged by the buyers in these Scandinavian countries and India had to send a special team to ascertain the reasons for the quality complaint.

The officials were taken aback to find that the quality of the delivered Arabica was intact and similar to the quality of the coffee dispatched from India, but that the earlier consignments were different in visual and cup quality. A thorough investigation revealed that the changes in the beans took place during the long sea voyage and acquired the mellow taste profile. Simulation of the sea voyage was made and monsooning trials were carried out on the west coast of India. The monsooning process emerged and with it the specialty coffees of Indian Monsooned Malabar and Basanally.

Today, the monsooning process is well controlled and carried out in a specific location on the west coast of India between June and October, with the latest harvest. During this period, the southwest monsoons, with winds from the Arabian Sea, lash the west coast of the country. High quality natural (cherry) coffees pertaining to Arabica and Robusta are selected for monsooning. The bean size is important, with the cup quality of selected Arabica and Robusta coffees being "clean" and of "fine" cup quality.

After selecting the lot of coffee, the large-sized beans are subjected to the monsooning process, which comprises spreading the beans to a fairly thick level on the concrete or brick floor of a well-ventilated warehouse (Fig. 3.6). The beans are exposed with regular raking for more than a week to the moisture and salt-laden monsoon winds. When the beans have increased in size and changed in color from brownish gray to light yellow, the coffee is gathered and filled in jute bags. They are then stored for more than a week in wind rows and exposed to the monsoon winds, allowing the moist air to pass through the coffee beans in the bags and to absorb even more moisture. The beans are once again spread on the floor and exposed to the monsoon weather after which they are bagged again. This process of spreading and bagging is continued till the coffee obtains a golden yellow color, which is a hallmark of this specialty coffee. The moisture content of the monsooned beans is around 13−14% and even at this high moisture level, mold or fungal damage is not observed and it is presumed, though not scientifically established, that the salt from the seawater could be acting as a preservative.

The taste profile of monsooned Arabica is very interesting with the cup having a smooth, creamy, warm, syrupy and wholesome mouthfeel, dotted with flecks of brightness. Overall, the profile is mellow, with overcurrents of flavors, which could be a potpourri of caramel, dark chocolate, nuts, pipe tobacco, earthy, and a hint of spice enveloping the palate.

FIGURE 3.6 Monsooned Arabica coffee. *Courtesy M/s Aspinwall and Company, Mangalore, India.*

5. PROCESSES AND TECHNOLOGIES FOR IMPROVED QUALITY AND QUALITY CONSISTENCY

If the key variables involved in the process are not well identified and controlled, there is a risk of inducing unreproducible product properties, and thus inconsistent quality. Therefore the raw material should be processed and maintained in a standardized manner. For small growers who postharvest process on their farms, this fact becomes an obstacle of the consistency of the quality. The current trend in the coffee business is that coffee growers, scientists, and engineers are focused on developing new technologies and original ways to process coffee, with the objective of giving special attributes to the beverage. For example, new technologies allow the coffee growers to have access to standardized processes on their farms, or central mills are constructed where rigorous procedures are followed. Environmental, social, and economic aspects of the new technologies and processes should be taken into consideration, as they play important roles in the sustainability of the business.

5.1 Process Precision on Microlot Coffees

The processing steps required for microlots are not different from those required to process larger lots of specialty coffee that in turn are not different from those for commercial coffees.

What changes in the case of specialty coffee and even more in the case of microlots is how one performs some of these steps. In the case of microlots there is the additional challenge posed by the small volumes as it is very difficult, if not impossible, to process in large equipment due to a lack of time to adjust the machines in the brief period it takes for microlots to go through them. In using standard or large machines, quality control will be at risk and precision will be lost in a situation where they are critical. The solution has been to use smaller wet lines, driers with smaller capacity or compartments, and dedicated dry mills that process only microlots. The capacity of wet mills is usually not a problem because they are either small or already divided into multiple lines. The challenge here is rather flexibility, e.g., the ability to separate and to process separately cherries at several degrees of maturation or to accommodate different types of fermentation combined or not with mechanical removal of mucilage.

The most difficult challenge, however, lies on the availability or training of qualified operators who must be able to implement the processing precision and monitoring of quality required with emphasis on the control of incoming cherries and outgoing parchment and the separation of cherries according to ripeness and fermentation. It is also important to control the usual wet milling parameters such as damage to parchment (and green) coffee, percentage of parchment lost with the pulp, percentage of pulp mixed with parchment, and degree of mucilage removal. All these factors interfere with the yields and the economics of processing a much more expensive product and, most importantly, with the final quality that is closely related to the price it will obtain.

Sun drying of microlots does not require special infrastructure but only more labor that must be properly trained to avoid color and quality losses in the process. Mechanical drying does require machines of much smaller static capacity or multiple small compartments. It is important to ensure that the product is not exposed to excess temperature, and that the final moisture content is homogeneous.

The challenge is to be able to process microlots in a structure that has been designed for large capacities that are ever growing vis-à-vis the size of the microlots. The solution that most millers are using is the installation of smaller capacity lines dedicated to microlots which allow enough time to make machine adjustments to ensure quality optimization. Quality control will be tighter, the lines will be richer in processing flexibility (e.g., by-passes, repassing, and the ability to switch lots quickly), and the packing options will be specific for microlots, e.g., regular jute bags, 30 kg plastic or paper bags, etc., and no bulk container loading.

5.2 Move to Central Mills

The vast majority of coffee producers in the world are smallholders, and they process their coffee in different ways that affect quality. This poses a great challenge to meet the growing demand for coffee with consistent quality. One solution lies in wet milling the coffee from many small growers in a central unit that uses the same processing steps to achieve the uniform quality desired. This model is becoming more popular because, besides the quality control, it has also shown environmental, economic, and social advantages.

Although a small grower is usually restricted to the use of a small often manual pulper and natural fermentation in conditions that are hard to control, a central wet mill can use the latest technology available. This may include cherry separation according to degree of maturation, reduced damage to coffee, improved pulp separation, mucilage removal that meets specific quality requirements, reduced water consumption, and proper handling of waste water. The overall result is higher and more consistent quality coming from a group of different growers that could otherwise produce different qualities. Trained personnel monitor and control the process and focus on quality control in a way that individual growers may not.

The positive impact on the environment is related to the centralization of wastewater treatment instead of having tens, or hundreds, of uncontrolled discharges in a region. In addition, biogas generation and composting in a commercial scale are feasible in central mills and of limited scope or use by small growers. The economies of scale of central milling are significant. Small pulpers belonging to individual growers operate only a short time per day, whereas a central mill will run for 6–8 h at least. The price of equipment per kilogram of coffee of capacity is much larger for small pulpers than for larger equipment. Things may be different on the operating side because the cost of labor that goes into individual milling is often hidden—i.e., not computed or accounted for by the grower—but it becomes apparent and has to be paid for in a central mill. In the end this additional cost may be compensated by better quality coffee obtained, which may lead to higher income for the farmer derived from a higher selling price for the produce. Finally, in central mills farmers may receive immediate payment for their crop instead of needing to wait when selling the processed beans.

There are other benefits from central milling that may help offset this extra labor cost and lead to social benefits: facilitation of training (in quality, cost control, growing, etc.) as growers congregate on the central mill site, the ability to purchase inputs together, and, in the longer run, to market coffee together. Overall, the positive social impact is a consequence of a better way of living since coffee growers have more time for personal and familiar projects, instead of spending much time processing his/her crop, but also a healthier environment and in many cases increased income for education and health.

An important drawback in central mills is the loss of traceability to specific regions or even individual farms, as is today often required by specialty microroasters. If small lots have to be processed centrally and kept separately, there will be substantial idle time and also little opportunity to fine-tune the adjustment of the machinery in the short period a small lot is milled, which may in turn cause suboptimal processing and quality. The ideal situation is for small lots to be brought together before milling and to be processed as one large lot with individual owners sharing the outgoing product. An alternative to process small lots is separate lines that central mills should have as explained in the Section 5.1 about process precision on microlot coffees. It is usual for central mills to have separate hoppers and lines; fermentation tanks are usually many in any case. Small driers or larger driers with separate compartments are also addressed in the section about microlots.

5.3 Wet Processing Using Less Water

The only stage of the wet coffee process that requires the use of water is mucilage removal, which is usually performed following two methods: by natural fermentation and washing or by mechanical means (Roa Mejia et al., 1999). The organic load of the wastewaters resulting of this stage depends on the specific water consumption; the lower the specific water consumption the more concentrated the organic load in the wastewaters.

Many machinery makers offer wet milling equipment with minimum water consumption. However, there are important trade-offs between water usage and coffee quality. As less water is used there is a tendency to increase damage to parchment and beans. Likewise there is an increase in both parchment loss with pulp and pulp mixed with parchment with the latter also affecting fermentation and coffee quality. In addition, cutting down water consumption is much more difficult in large than small or micro-wet mills.

Dry transport must be used with elevators and conveyors replacing channels that use water to convey coffee. Dry fermentation combined with dry transport of coffee is also a good alternative to wet fermentation and even mucilage removers that always require some water.

At a time when there is a move toward central milling for quality, consistency, and cost reasons, small to medium machinery with less water consumption should still be pursued because a central mill is often a combination of several different lines that may be mid or even small sizes to also process microlots. These mills also ensure the quality consistency required by the industry. Scientists at Cenicafé developed the ECOMILL® which has reduced water consumption, controlled water contamination, delivering a coffee with high quality standards (Fig. 3.7). The technology has an up-flow mechanical washer in which the wastewaters are expelled out through case perforations through the effect of agitation and/or centrifugal forces (Oliveros Tascón et al., 2013b, 2014). The mechanical washer was designed to obtain the lowest power and water requirements, without damaging the beans.

The ECOMILL® technology comes with cylindrical tanks and a bottom screw conveyor to feed the right amount of coffee to the mechanical washer. Mucilage is removed either by natural fermentation or by adding pectinolytic enzymes. The hoppers have a pronounced angle and a wide exit to allow the coffee with degraded mucilage to flow without using water.

FIGURE 3.7 Ecomill® 3000 technology developed at Cenicafé.

FIGURE 3.8 Dry mucilage obtained from the wastewaters of the ECOMILL® technology.

The thick wastewaters are then dried out with low-cost solar driers (Fig. 3.8)(Oliveros Tascón et al., 2011). The final product has very good fertilizing properties. Another alternative is to mix the thick wastewaters with the pulp (fruit skin) obtained in the previous pulping stage, as was proposed for the Becolsub technology (Roa Mejia et al., 1999).

5.4 Drying Using Solar Energy

The high moisture content of washed coffee, 52.7–53.5% (Puerta Quintero, 2005), its chemical composition, and humid weather conditions are favorable for the development of microorganisms, mainly molds and yeasts, which can affect its quality and innocuousness. The microorganisms can come from the field, trees (total coliforms), contact with the picker hands, and from the water used for washing (Archila, 1985). There is also the possibility of contamination in the facilities of the coffee mill.

There is a physical property that measures the availability of water in foods for biological reactions, known as water activity, which for green coffee is defined by the ratio between the vapor pressure of the grain and the vapor pressure of pure water at the same conditions. When drying agricultural products down to the levels required for food safety, for coffee 10–12%, the water activity is reduced to 0.65–0.68, at which microorganisms growth and metabolic activity are reduced. The minimum water activity for the growth of *Aspergillus ochraceus* varies from 0.77 to 0.83 and for producing ochratoxin A (OTA) it is from 0.83 to 0.87 (Urbano et al., 2001). Pardo et al. (2005) observed no production of OTA at a water activity of 0.80. According to Suárez-Quiroz et al. (2004) at a water activity below 0.80 coffee is protected from *A. ochraceus*.

FIGURE 3.9 Solar tunnel developed for the drying of coffee beans.

Sun drying where the coffee is spread on patios is the most common method to dry coffee. It is cheap and energy efficient. However, weather conditions can influence quality both through rain or too strong sunlight. Therefore, other solutions have been developed which reduces the risk of quality loss during the drying. For example, solar tunnels have been developed in Colombia to obtain a product with high physical and sensorial qualities with an efficient use of solar radiation and air energy (Oliveros Tascón et al., 2013a). They are low-cost structures, with a 40% shade plastic mesh, a transparent plastic cover to take advantage of the greenhouse effect, and to protect the coffee beans against rains. It has side curtains to facilitate operation activities such as loading the wet coffee, to stir the coffee beans, and remove the dried product (Fig. 3.9). In addition, this construction allows a flow of air through the tunnel that removes the humidity and speeds up the drying process.

Another interesting innovation using solar energy is the use of solar panels to operate mechanical dryers. Solar radiation in the area of coffee plantations ranges between 3000 and 4500 Wh/m^2. Transferring this to mechanical dryers can be a very efficient way of using this energy compared to the direct drying method as it has a conversion efficiency of 40–50% compared to only 10–15% in the direct drying method (Mulato et al., 1998). These mechanical dryers can run on solar energy but also on combustion or a combination of both (Fig. 3.10). Benefits compared to traditional mechanical drying are the carbon emissions that can be brought to 0 if fully operated on solar energy. The Center of Indonesian Coffee and Cocoa Research (ICCRI) is one of the institutions that has developed a mechanical coffee dryers operating with solar energy collectors installed on the roof of the building in addition to a wood stove. The drying chamber uses flatbed-type multiplenum, each of which is

FIGURE 3.10 Solar panels and mechanical wood furnace as a heat source for coffee dryer.

equipped with axial fan, but other types of dryers can also be considered. Combustion air flow rate is 100 m^3/h with a maximum air drying temperature of 80°C and the combustion energy output ranges from 50 to 100 kW (Mulato et al., 1998).

5.5 Use of Analytical Methods for Quality Control

In today's coffee industry, there is a growing need for practical and cost-effective analytical methods for large numbers of green coffee samples at different stages along the value chain. First the farmer's main need is to control quality during processing. For farmers, methods need to be simple to apply and with minor cost for investment and operation. When moving along the value chain from buying stations toward roasters, methods can become more sophisticated as large coffee quantities need to be checked. The general goal is to identify the quality of different loads and ensuring that each is used optimally for its character and to ensure value creation is not lost at any stage. In the next section we will describe a few examples of how innovation in quality control methods can lead to very simple instruments that can easily be used by farmers or very sophisticated techniques such as DNA analysis for traceability purposes.

5.5.1 Quality Control Tools for Farmers

Farmers who process coffee on the farm need to reply on experience to optimally process the coffee. Some basic innovative tools have been developed that can help to control the different critical stages of coffee processing; fermentation and drying.

Fermentation is a complex process that depends on many variables (see Section 3 of this chapter) and carelessness can generate physical and cup defects (Almacafé, 2007). Uncontrolled overfermentation, for instance, may lead to off-tastes that devalue the product (Puerta Quintero, 2006). On the other side, an incomplete fermentation leads to physical and cup defects too because the remaining mucilage keeps fermenting during drying, causing stains in the parchment, as well as off-notes in the cup (Roa Mejia et al., 1999).

Peñuela Martínez et al. (2013) developed a novel method, the Fermaestro®, to help control fermentation based on volumetric changes between freshly pulped coffee and coffee with its mucilage completely degraded. The volume reduction is caused by the mucilage layer, which after fermentation is separated from the beans. The method consists of filling a half-liter perforated conic reservoir with the freshly pulped coffee that is in the fermentation tank (Fig. 3.11). Once it is completely full, the cap is tightened and the implement is introduced into the mass of coffee. After the time passes, an empty space on the top of the conic reservoir is to appear, due to mucilage flowing out. When the empty space reaches a predetermined mark it means the fermentation process is completed and the coffee is ready to be washed. In farms, testing has shown that the method can be used for different varieties of the *Coffea arabica* sp., in several altitudes and weather conditions, and different processing ways, including underwater fermentations (Peñuela Martínez et al., 2012).

Another critical process for farmers is the drying of the coffee (see Section 4). To determine the appropriate time to finish the drying process, farmers use traditional methods, such as evaluating the color or hardness of the green coffee and also the sound produced by the parchment coffee

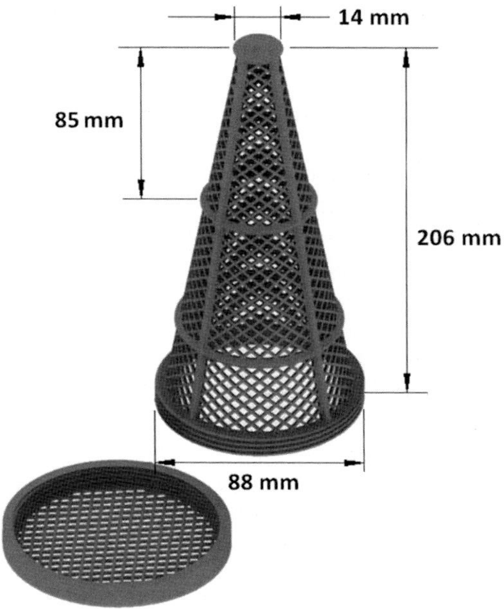

FIGURE 3.11 Implement Fermaestro® to determine the coffee washing time.

FIGURE 3.12 Use of Gravimet method by weighing and placing sample in the coffee mass.

beans when are stirred in the dryer. Sometimes these methods lead to making wrong decisions about the moment when drying is stopped and which can cause quality losses and thus affects the producer income (Roa Mejia et al., 1999).

A gravimetric method for measuring the moisture content of coffee beans during solar drying was developed. The method assumes that during drying only water is removed, i.e., dry matter losses from breathing are negligible (Oliveros Tascón et al., 2013a). The initial moisture of the beans is set, for example, to 53%, in the case of washed Castillo® and the final moisture content of coffee is set to 11% wb, which ensures that no beans have moistures over 12% wb that can cause the rejection in the selling place.

For the application of the gravimetric method, called Gravimet, a small basket made of plastic mesh is used. The basket is filled with a layer of beans equal to the beans in the solar dryers (Fig. 3.12). The sample mass is measured at the beginning of the process and when the weight has been reduced by 47–48% the drying is completed. The mass loss is followed using a digital low-cost scale (US$15) with a range between 0 and 5 kg and a 1 g resolution. The method was validated in controlled conditions with an error between 1.92% and 0.09% (wb) (Jurado et al., 2009) and on farm with an effectiveness of 93% (Oliveros Tascón et al., 2013a). The method can easily be deployed to other varieties and in other countries by taking into account the initial moisture content of the product.

5.5.2 Near Infrared for Chemical Analysis

In research laboratories more sophisticated analytical methods are available that can help to access various coffee quality parameters. In the last two

FIGURE 3.13 NIR hand-held analyzer is taking readings for biochemicals in coffee leaves.

decades, there have been numerous applications using near infrared (NIR) spectroscopy for coffee analysis (Barbin et al., 2014). For example, the determination of the major biochemical compounds of green coffee such as caffeine can now easily be made (Huck et al., 2005; Pizarro et al., 2007; Davrieux et al., 2004). Other NIR application examples on roasted coffee include the discrimination of Arabica and Robusta (Downey and Boussion, 1996; Kemsley et al., 1995; Esteban-Díez et al., 2007), the presence of barley as an adulterant (Ebrahimi-Najafabadi et al., 2012; Wang et al., 2009), and markers linked to sensory attributes such as acidity and bitterness (Ribeiro et al., 2011). NIR spectroscopy combined with chemometrics shows advantages of an easy sample preparation and it allows the prediction of multiple parameters simultaneously.

The different spectroscopy-related fields such as chemometrics, optical components, sensors, and electronics are continuously evolving leading to even more sophisticated methods but also to faster and cheaper ones. Hand-held analytical tools are developed and getting efficient in terms of detection, sensitivity, and easy to use, for example, with application for reading biochemical in coffee leaves (Fig. 3.13).

5.5.3 Nuclear Magnetic Resonance for Chemical Analysis

Nuclear magnetic resonance (NMR) technique is evolving and becoming a control tool for quality-related attributes. Application examples for fruit juices, olive oil, plant extract were reported in literature (Belton et al., 1996; Kim et al., 2012; Fauhl et al., 2000). Eurofins company uses the NMR method to ensure food and beverage authenticity (Jamin, 2010), for example, for detecting the sugars added in honey, the nitrogen compounds that increase the percent of protein in milk or to detect safflower in saffron. Various applications are reported for green coffee quality control, ^{13}C NMR coupled with principal component analysis or orthogonal partial least squares discriminated analysis were used to discriminate the Arabica and Robusta species and samples from six geographic origins (Wei et al., 2012). ^{1}H NMR-based metabolomic analysis was used for grading green coffee beans: specialty or high-grade green coffee beans compared to commercial-grade beans (Kwon et al., 2015). The results from this study demonstrate that high levels of sucrose and low levels of γ-aminobutyric acid, quinic acid, choline, acetic acid, and fatty acids were all markers for coffee quality.

5.5.4 DNA Analysis for Single Variety Traceability

Like in the wine industry where certain terroirs and single varieties are reputed for high quality, specialty coffee is positioned as a premium product, thanks to single origins, and/or single varieties such as Geisha, Maragogype, or Bourbon varieties. Traceability is critical to ensure origin, variety, and processes applied. DNA analysis is already used to ensure food authenticity and traceability of single varieties in a finished product. Nespresso developed a DNA PCR-based method to identify varieties through the value chain, from the field to the finished product (Morel et al., 2012). The method is used as a quality control tool to guarantee the purity and authenticity of the raw material. The DNA test is applied to green coffee bean batches from farms in Southern Brazil, which grow red and yellow bourbon varieties.

6. OUTLOOK

Agroindustry has witnessed great changes in the last decades due to a demand for high-quality agricultural products and differentiated features grown in a sustainable way. Coffee is not the exception. Specialty coffees have risen in popularity in the last years as consumers want to discover new and special flavors. This has become a motivation for coffee growers all over the world to focus on quality and flavor attributes, either from superior varieties or through applying specific postharvest processes. Roasters in turn bring these products to consumers, marketing the coffee variety and the processes applied often combined with the story of the cultural heritage. In addition, attention has been given to improved practices to reduce the environmental impact, with water usage for processing as the major focus.

These different aspects, while considering low investment costs for farmers, has inspired the creativity of cultivators, engineers, and scientists to find original solutions as illustrated by some examples in this chapter. But scientists, agronomists, and farmers do not rest and massive work is ongoing to further achieve even better quality control methods, innovative flavors, and improved environmental practices aiming at creating value for both farmers and consumers.

For example, a better industrial use of by-products is being considered both for environmental aspects and as a high value product that can add income to farmers. Every grower knows that the pulp and the mucilage are rich raw materials and there have been incipient initiatives to transform them as valuable industrial products. Farmers can use nitrogen-rich coffee pulp to fertilize coffee plantations and thus replace artificial fertilizers. But coffee pulp can also be used to produce biofuel (Shenoy et al., 2011). In Japan, a product based on roast and ground coffee and pulp was commercialized (DyDo CoffeeBerry). In addition, the intrinsic goodness of coffee has been shown to have a beneficial impact on human health (Chapter 20). One could thus consider a wide range of cosmetic or pharmaceutical opportunities for creating value using beans that are rejected because of flavor or physical defects. In Colombia, there is an initiative to use wastewaters as described in Section 5.3. Further to the recovery and use as fertilizer, the wastewater is rich in antioxidative agents from the coffee pulp, which can be used to produce molasses and cosmetic products (Naox, see http://naoxantioxidante.blogspot.com.co/). Alejandro Méndez from El Salvador, World Barista Championship 2011, used mucilage to sweeten his winning beverage.

Coffee has still a lot to assimilate from the wine industry where fermentation equally plays a key role. The question is when fermentation tanks will be replaced by reactors with complex controls and instruments in which variables such as temperature, pH, oxygen content, microbiota, and many more are managed to optimize the flavor potential.

More sophisticated spectroscopic methods such as near and mid-infrared can in some cases replace chemical analysis to provide innovative solutions for maximum productivity and cost efficiency (Osborne, 2000; Cozzolino, 2009; Lin and Ying, 2009). A microsatellite high-resolution melting analysis method, allowing high sample throughput with reduced time to results, was developed for grapevine and olive varietal certification (Mackay et al., 2008) and has the potential for coffee origin/variety traceability use.

To conclude, it is the creativity of the researchers, agronomists, and growers together that will continue to investigate original ways to process coffee in a qualitative and sustainable manner. They will on the one hand provide insight into the creation of novel flavors and stories leading to innovative and sustainable products for demanding consumers to discover. On the other hand, a better use of the natural resources can help in the creation of new value for existing and new players along the coffee value chain.

REFERENCES

Almacafé, S.A., 2007. Defectos del café y su incidencia en taza. Almacafé Sede Manizales. Manizales (Colombia).

Archila, G.M., 1985. Análisis bacteriológico de aguas residuales de beneficio de café. Universidad de los Andes. Facultad de Microbiología, Bogotá, p. 40.

Avallone, S., Guyot, B., Brillouet, J.M., Olguin, E., Guiraud, J.P., 2001. Microbiological and biochemical study of coffee fermentation. Current Microbiology 42 (4), 252–256.

Avallone, S., Brillouet, J.M., Guyot, B., Olguin, E., Guiraud, J.P., 2002. Involvement of pectolytic micro-organisms in coffee fermentation. International Journal of Food Science & Technology 37 (2), 191–198.

Barbeau, G., Cadot, Y., Stevez, L., Bouvet, M.H., Cosneau, M., Asselin, C., Mege, A., 2001. Role of soil physical properties, climate and harvest period on must composition, wine type and flavour (*Vitis vinifera* L., cv chenin), Coteaux du Layon, France. In: Proceedings of the 26th World Congress of the OIV Adelaide, Australia, 11–17 October 2001, section viticulture, pp. 105–118.

Barbin, D.F., Felicio, A.L.D.S.M., Sun, D.W., Nixdorf, S.L., Hirooka, E.Y., 2014. Application of infrared spectral techniques on quality and compositional attributes of coffee: an overview. Food Research International 61, 23–32.

Belton, P.S., Delgadillo, I., Holmes, E., Nicholls, A., Nicholson, J.K., Spraul, M., 1996. Use of high-field ^{1}H NMR spectroscopy for the analysis of liquid foods. Journal of Agricultural and Food Chemistry 44 (6), 1483–1487.

Borém, F.M., 2014. Handbook of Coffee Post-Harvest Technology, ISBN 978-0-9915721-0-6.

Cozzolino, D., 2009. Near infrared spectroscopy in natural products analysis. Planta Medica 75 (7), 746–756.

Davrieux, F., Manez, J.C., Durand, N., Guyot, B., 2004. Determination of the content of six major biochemical compounds of green coffee using NIR. In: Near Infrared Spectroscopy: Proceedings of the 11th International Conference, vol. 441.

Downey, G., Boussion, J., 1996. Authentication of coffee bean variety by near-infrared reflectance spectroscopy of dried extract. Journal of the Science of Food and Agriculture 71 (1), 41–49.

Ebrahimi-Najafabadi, H., Leardi, R., Oliveri, P., Casolino, M.C., Jalali-Heravi, M., Lanteri, S., 2012. Detection of addition of barley to coffee using near infrared spectroscopy and chemometric techniques. Talanta 99, 175–179.

Esteban-Diez, I., Gonzalez-Saiz, J.M., Saenz-Gonzalez, C., Pizarro, C., 2007. Coffee varietal differentiation based on near infrared spectroscopy. Talanta 71 (1), 221–229.

Evangelista, S.R., Silva, C.F., Miguel, M.G.P.C., Cordeiro, C.S., Pinheiro, A.C.M., Duarte, W.F., Schwan, R.F., 2013. Improvement of coffee beverage quality by using selected yeasts strains during the fermentation in dry process. Food Research International 61, 183–195.

Evangelista, S.R., Miguel, M.G.D.C.P., Silva, C.F., Pinheiro, A.C.M., Schwan, R.F., 2015. Microbiological diversity associated with the spontaneous wet method of coffee fermentation. International Journal of Food Microbiology 210, 102–112.

Fauhl, C., Reniero, F., Guillou, C., 2000. ^{1}H NMR as a tool for the analysis of mixtures of virgin olive oil with oils of different botanical origin. Magnetic Resonance in Chemistry 38 (6), 436–443.

Gómez, L., Caballero, A., Baldión, J., 2001. Ecotopos cafeteros de Colombia. Federación Nacional de Cafeteros de Colombia.

Huck, C.W., Guggenbichler, W., Bonn, G.K., 2005. Analysis of caffeine, theobromine and theophylline in coffee by near infrared spectroscopy (NIRS) compared to high-performance liquid chromatography (HPLC) coupled to mass spectrometry. Analytica Chimica Acta 538 (1), 195–203.

Ismayadi, C., Marsh, A., Clarke, R., 2005. Influence of storage of wet arabica parchment prior to wet hulling on moulds development, ochratoxin A contamination, and cup quality of mandheling coffee. Pelita Perkebunan (Coffee and Cocoa Research Journal) 21 (2).

Jamin, E., 2010. Profiling des produits alimentaires par RMN à Haute-Résolution. Eurofins Scientific Analytics, pp. 1–4. Eurofins 33.

Jumhawan, U., Putri, S.P., Marwani, E., Bamba, T., Fukusaki, E., 2013. Selection of discriminant markers for authentication of Asian palm civet coffee (Kopi Luwak): a metabolomics approach. Journal of Agricultural and Food Chemistry 61 (33), 7994–8001.

Jurado-Chana, J.M., Montoya-Restrepo, E.C., Oliveros-Tascon, C.E., García-Alzate, J., 2009. Método para medir el contenido de humedad del café pergamino en el secado solar. Cenicafé 60 (2), 135–147.

Kemsley, E.K., Ruault, S., Wilson, R.H., 1995. Discrimination between *Coffea arabica* and *Coffea canephora* variant robusta beans using infrared spectroscopy. Food Chemistry 54 (3), 321–326.

Kim, J., Jung, Y., Bong, Y.S., Lee, K.S., Hwang, G.S., 2012. Determination of the geographical origin of kimchi by ^{1}H NMR-based metabolite profiling. Bioscience, Biotechnology, and Biochemistry 76 (9), 1752–1757.

Kwon, D.J., Jeong, H.J., Moon, H., Kim, H.N., Cho, J.H., Lee, J.E., Hong, K.S., Hong, Y.S., 2015. Assessment of green coffee bean metabolites dependent on coffee quality using a ^{1}H NMR-based metabolomics approach. Food Research International 67, 175–182.

Lingle, T.R., 1986. The Coffee Cupper's Handbook: Systematic Guide to the Sensory Evaluation of Coffee's Flavor. Coffee Development Group.

Lin, H., Ying, Y., 2009. Theory and application of near infrared spectroscopy in assessment of fruit quality: a review. Sensing and Instrumentation for Food Quality and Safety 3 (2), 130–141.

de Melo Pereira, G.V., Soccol, V.T., Pandey, A., Medeiros, A.B.P., Lara, J.M.R.A., Gollo, A.L., Soccol, C.R., 2014. Isolation, selection and evaluation of yeasts for use in fermentation of coffee beans by the wet process. International Journal of Food Microbiology 188, 60–66.

de Melo Pereira, G.V., Neto, E., Soccol, V.T., Medeiros, A.B.P., Woiciechowski, A.L., Soccol, C.R., 2015. Conducting starter culture-controlled fermentations of coffee beans during on-farm wet processing: growth, metabolic analyses and sensorial effects. Food Research International 75, 348–356.

Mackay, J.F., Wright, C.D., Bonfiglioli, R.G., 2008. A new approach to varietal identification in plants by microsatellite high resolution melting analysis: application to the verification of grapevine and olive cultivars. Plant Methods 4 (1), 1.

Masoud, W., Kaltoft, C.H., 2006. The effects of yeasts involved in the fermentation of *Coffea arabica* in East Africa on growth and ochratoxin A (OTA) production by *Aspergillus ochraceus*. International Journal of Food Microbiology 106 (2), 229–234.

Morel, E., Bellanger, L., Lefebvre-Pautigny, F., Lambot, C., Crouzillat, D., 2012. DNA traceability for variety purity in Nespresso product. In: 24th International Conference on Coffee and Science. Association for Science and Information on Coffee (ASIC), Costa Rica.

Mulato, S., Atmawinata, O., Yusianto, Widyotomo, S., dan Handaka, 1998. Kinerja kolektor tenaga matahari pelat datar dan tungku kayu mekanis sebagai sumber panas unit pengering kopi rakyat skala besar. Pelita Perkebunan 14 (2), 108–123.

Oliveros Tascón, C.E., Ramírez Gómez, C.A., Sanz Uribe, J.R., Peñuela Martínez, A.E., Pabón Usaquén, J.P., 2013a. Secado solar y secado mecánico del café. In: Manual del Cafetero Colombiano. Investigación y Tecnología para la Sostenibilidad de la Caficultura, vol. III. FNC. CENICAFE, Chinchiná, pp. 49–80.

Oliveros Tascón, C.E., Sanz Uribe, J.R., Montoya Restrepo, E.C., Ramírez Gómez, C.A., 2011. Equipo para el lavado ecológico del café con mucílago degradado con fermentación natural. Revista de Ingeniería. Universidad de los Andes 33, 61–67.

Oliveros Tascón, C.E., Sanz Uribe, J.R., Ramírez Gómez, C.A., Tibaduiza Vianchá, C.A., 2013b. Ecomill: tecnología de bajo impacto ambiental para el lavado del café, p. 8 (Avances Técnicos N° 432).

Oliveros Tascón, C.E., Tibaduiza Vianchá, C.A., Montoya Restrepo, E.C., Sanz Uribe, J.R., Ramírez Gómez, C.A., 2014. Tecnología de bajo impacto ambiental para el lavado del café en proceso con fermentación natural. Cenicafé 65 (1), 44–56.

Osborne, B.G., 2000. Near-infrared spectroscopy in food analysis. In: Meyers, R.A. (Ed.), Encyclopedia of Analytical Chemistry. John Wiley & Sons Ltd., UK.

Pabón Usaquén, J.P., Sanz Uribe, J.R., Oliveros Tascón, C.E., 2009. Manejo del Café Desmucilaginado Mecánicamente. Avances Técnicos Cenicafé (Colombia) No. 388, 1–8.

Pardo, E., Ramos, A.J., Sanchis, V., Marın, S., 2005. Modelling of effects of water activity and temperature on germination and growth of ochratoxigenic isolates of *Aspergillus ochraceus* on a green coffee-based medium. International Journal of Food Microbiology 98 (1), 1–9.

Peñuela Martínez, A.E., Sanz Uribe, J.R., Pabón Usaquén, J.P., 2012. Método para identificar el momento final de la fermentación de mucílago de café. Cenicafé 63 (1), 120–131.

Peñuela Martínez, A.E., Oliveros Tascón, C.E., Sanz Uribe, J.R., 2010. Remoción del mucílago de café a través de fermentación natural. Cenicafé 61 (2), 159–173.

Peñuela Martínez, A.E., Pabón Usaquén, J.P., Sanz Uribe, J.R., 2013. Método Fermaestro: para determinar la finalización de la Fermentación del mucílago de café. Avances Técnicos 431, 8.

Pizarro, C., Esteban-Díez, I., González-Sáiz, J.M., Forina, M., 2007. Use of near-infrared spectroscopy and feature selection techniques for predicting the caffeine content and roasting color in roasted coffees. Journal of Agricultural and Food Chemistry 55 (18), 7477–7488.

Puerta Quintero, G.I., 2006. Buenas prácticas agrícolas para el café. Avance Técnico 349. Cenicafé, p. 12.

Puerta Quintero, G.I., 2005. Quality and safety of coffee processed by the wet method and dried in solar dryers. Salvador [Brasil]: Workshop Improvement of Coffee Quality Through Prevention of Mould Growth 1.

Rendón, M.Y., Salva, T.J.G., Ribeiro, J.S., Bragagnolo, N., 2012. Oxidation of lipids and proteins in green Arabica coffee during the storage period. In: 24th International Conference on Coffee Science San José (Costa Rica), pp. 294–297.

Ribeiro, J.S., Ferreira, M.M.C., Salva, T.J.G., 2011. Chemometric models for the quantitative descriptive sensory analysis of Arabica coffee beverages using near infrared spectroscopy. Talanta 83 (5), 1352–1358.

Roa Mejia, G., Oliveros Tascón, C.E., Alvarez Gallo, J., Ramírez Gómez, C.A., Sanz Uribe, J.R., Álvarez Hernandez, J.R., Dávila Arias, M.T., Zambrano Franco, D.A., Puerta Quintero, G.I., Rodríguez Valencia, N., 1999. Beneficio ecológico del café. Cenicafé. Federación Nacional de Cafeteros de Colombia, Chinchiná, Colombia, pp. 1–273.

Schwan, R.F., Graham, H.F., 2014. Cocoa and Coffee Fermentations. In: Encyclopedia of Food Microbiology, pp. 466–473.

Shenoy, D., Pai, A., Vikas, R.K., Neeraja, H.S., Deeksha, J.S., Nayak, C., Rao, C.V., 2011. A study on bioethanol production from cashew apple pulp and coffee pulp waste. Biomass and bioenergy 35 (10), 4107–4111.

Silva, C.F., Batista, L.R., Abreu, L.M., Dias, E.S., Schwan, R.F., 2008. Succession of bacterial and fungal communities during natural coffee (*Coffea arabica*) fermentation. Food Microbiology 25 (8), 951–957.

Suárez-Quiroz, M.L., González-Rios, O., Barel, M., Guyot, B., Schorr-Galindo, S., Guiraud, J.P., 2004. Effect of chemical and environmental factors on *Aspergillus ochraceus* growth and toxigenesis in green coffee. Food Microbiology 21 (6), 629–634.

Urbano, G.R., Taniwaki, M.H., Leitao, M.D.F., Vicentini, M.C., 2001. Occurrence of ochratoxin A–producing fungi in raw Brazilian coffee. Journal of Food Protection 64 (8) , 1226–1230.

Velmourougane, K., 2013. Impact of natural fermentation on physicochemical, microbiological and cup quality characteristics of Arabica and Robusta coffee. In: Proceedings of the National Academy of Sciences, India Section B: Biological Sciences, vol. 83 (2), pp. 233–239.

Wang, J., Jun, S., Bittenbender, H.C., Gautz, L., Li, Q.X., 2009. Fourier transform infrared spectroscopy for Kona coffee authentication. Journal of Food Science 74 (5), C385–C391.

Wei, F., Furihata, K., Koda, M., Hu, F., Kato, R., Miyakawa, T., Tanokura, M., 2012. ^{13}C NMR-based metabolomics for the classification of green coffee beans according to variety and origin. Journal of Agricultural and Food Chemistry 60 (40), 10118–10125.

Wintgens, J., 2009. Factors influencing the quality of green coffee. In: Wintgens, J.N. (Ed.), Coffee: Growing, Processing, Sustainable Production. A Guidebook for Growers, Processors, Traders and Researchers. Wiley-VCH, pp. 797–817.

Yusianto, Ismayadi, A., Saryono, D., Nugroho, Mawardi, S., 2012. Characterization of animal preference to Arabica coffee varieties and cup taste profile on domesticated "Luwak" (*Paradoxurus hermaphroditus*). In: Proc. 24th Internatonal Conference on Coffee Science (ASIC) 2010, San Jose-Costa Rica, pp. 136–144.

Yusianto, Hulupi, R., Sulistyowati, Mawardi, S., Ismayadi, C., 2007. Mutu fisik dan citarasa beberapa varietas kopi arabika harapan pada beberapa periode penyimpanan. Pelita Perkebunan 23, 205–230.

Chapter 4

Environmental Sustainability—Farming in the Anthropocene

Martin R.A. Noponen[1], Carmenza Góngora[2], Pablo Benavides[2], Alvaro Gaitán[2], Jeffrey Hayward[3], Celia Marsh[4], Ria Stout[5], Chris Wille[6]

[1]*Rainforest Alliance, London, United Kingdom;* [2]*Cenicafé FNC, Manizales, Colombia;* [3]*Rainforest Alliance, Washington, DC, United States;* [4]*Science Writer and Researcher, Geneva, Switzerland;* [5]*Rainforest Alliance, Antigua, Guatemala;* [6]*Sustainable Agriculture Consultant, Portland, OR, United States*

1. INTRODUCTION

Pointing to his forested coffee farm, El Ciprés, on the slopes of El Salvador's Picacho volcano, Juan Marco Alvarez says, "That's truly my family's bank account, including that beautiful cloud forest you see on top. We try to invest wisely, making it more productive, and avoiding eating up the principle—our natural capital." Juan Marco, the founder of SalvaNatura, a leading environmental organization in El Salvador, is a recognized expert on sustainable agriculture and the values of environmental services, but all successful coffee farmers pay close attention to their ecological balance sheet.

Farms and farmers directly depend on nature—soil, water, biodiversity, climate, and the innumerable interactions that make up ecosystems. Managing this portfolio of living assets has never been easy. It's getting much more difficult. Our actions have so altered the globe—even the climate—that some scientists have declared these times a new geological epoch, the Anthropocene, the era of human dominance (Ackerman, 2014). Farmers are among the first to feel the impacts of earth-shaking phenomena such as climate change. Although feeding the family and growing marketable crops has always been challenging, now agriculturists must adapt to the harsh realities of the Anthropocene.

Earth is in the middle of an extinction crisis, losing species faster than ever before in the planet's history. Rainforests, a thin band around the Earth's equatorial middle, house half of all the world's species of flora and fauna. The destruction of the rainforests is one of the most pressing environmental calamities; recent satellite data show that 10 million hectares of tropical forest

The Craft and Science of Coffee. http://dx.doi.org/10.1016/B978-0-12-803520-7.00004-9

were lost in 2014 (Global Forest Watch, 2015). That is equivalent to an area the size of a football pitch deforested every 3 s. Some of the biggest losses were in coffee-producing countries, including Brazil, Indonesia, Papua New Guinea, Peru, the Philippines, Sri Lanka, and Vietnam.

Coffee farms, cultivated within or in place of tropical forests, cover some 10 million hectares—an area equivalent to that of rubber or oil palm—so their role in helping maintain valuable and species-rich tropical ecosystems is large. Like other agriculture, coffee farming causes and is affected by deforestation, the loss of biodiversity, soil erosion, climate change, water quality and availability, and other environmental factors.

Solutions for environmental challenges should be addressed at different spatial, temporal, and managerial scales. For example, global issues such as climate change and deforestation can directly impact coffee farmers through changes in local weather and water supplies. Likewise, local actions by farmers can cause problems far away, such as when waste waters from a remote coffee mill contributes to municipal water pollution in a different province. Coffee communities, ecosystems, and supply chains—all the way to the consumers—are interconnected and solutions to environmental challenges must be equally integrated.

Coffee is one of a few crops that can be grown in harmony with rainforest conservation, and the concept of sustainable agriculture is well advanced in the coffee sector. Sustainability means understanding the interconnectedness of environmental, social, and economic issues. For example, forest, soil, and water conservation benefit all three of these interlocking spheres of sustainability.

The much discussed tradition of growing coffee under a canopy of native rainforest trees is an example of the synergistic benefits of sustainability. Trees have social, economic, and environmental benefits at the spatial, temporal, and managerial scales. They provide nutrient recycling, thus reducing the need for costly fertilizers; retain soil moisture, thus reducing the need for irrigation; help bind the soil, thus reducing erosion and landslides; supply firewood, fruits, building materials, and other goods; harbor abundant and diverse wildlife, including coffee pollinators, sometimes increasing yield; and moderate the temperature extremes, allowing coffee beans to ripen slowly and naturally, thus improving quality. Trees on farms sequester carbon and protect watersheds—opening the possibilities for some farmers to receive payments for "environmental services."

An integrated synergistic approach toward addressing environmental challenges also means improving conditions and management of the entire coffee supply chain—sound infrastructure and an enabling environment, including proper regulations, technical assistance, financial support, cooperation, and shared value from farmers through to roasters and on to consumers. Linking all the actors in the value chain with standards, training, certification, and seals of approval is more advanced in coffee than in any other commodity (Potts et al., 2014; Panhuysen and Pierrot, 2014).

Multilateral agencies such as the World Bank and United Nations Development Program, government aid agencies, nonprofit conservation and development groups, coffee farming companies and cooperatives, traders and roasters, national coffee associations, agronomic research institutes and universities, standard-setters and certifiers, and many other tireless experts are working with farmers—often in multistakeholder coalitions—to find ways to minimize the environmental impacts and optimize the social and economic benefits of cultivating the beans that make the brew so beloved around the world. This grand campaign is so extensive and diverse, with a dramatic fusion of tradition, science, passion, and innovation, that it merits an entire book. This chapter aims to outline the highlights in three areas: biodiversity, climate change, and natural ways to control pests and disease.

2. COFFEE FARMING AND BIODIVERSITY ARE INTERDEPENDENT

2.1 Coffee Farming During a Global Extinction Crisis

Coffee and the clever primates who drink it both began their long evolutionary journeys in Africa. The highland forests of Ethiopia and South Sudan are the cradle of coffee, and the small *Coffea* tree still grows wild in some remaining forest fragments. Coffee is naturally part of Earth's most diverse ecosystem, the tropical forests. As the ultradiverse rainforests and other tropical forests are destroyed, uncounted species disappear with them. In the late 1980s, scientists and environmentalists began to focus public attention on the ongoing extinction crisis. Species are today disappearing at perhaps the fastest rate in the planet's history. The current extinction crisis is as severe as the one 65 million years ago, when an asteroid crashed into the Yucatan, wiping out the dinosaurs and much of the other flora and fauna. The catastrophe was during the Cretaceous; the new age is called the Anthropocene, as *Homo sapiens* now dominate every aspect of the planet, even its atmosphere and climate. This time, we are the asteroid, causing the sixth great extinction cataclysm (Kolbert, 2014).

No one knows how many species we are losing, since we do not know—even to the nearest order of magnitude, how many species there are. About two million species have been cataloged by scientists, and about 20,000 new plants, animals, fungi, and microbes are described each year (Wilson, 2014). The estimates for the total number of species range from 5 to 30 million (Ecosystems and Human Wellbeing, 2005). With so little knowledge about the total number of species, it is difficult to estimate how fast they are disappearing. Conservation scientists and taxonomists estimate that extinction rates are 100 to 1000 times higher than before the Industrial Revolution and the spread of humans to every corner of Earth (Wilson, 2014).

Edward O. Wilson, professor emeritus, Harvard University, is the best known and most authoritative promoter of biodiversity conservation. He has written more than 20 books, two of which won the Pulitzer Prize (On Human

Nature, 1978, and *The Ants*, with Bert Hölldobler and Wilson, 1990). Wilson and other biologists flock to coffee-producing countries because coffee grows in many of what Conservation International deem "Biodiversity Hotspots" (Possingham and Wilson, 2005). As an example of species richness in rainforests, Wilson and two colleagues found 275 species of ants in just 8 hectares in Peru (Wilson, 1992).

Wilson, other scientists, and nongovernmental organizations (NGOs) emphasize the importance of conserving rainforest, as it holds so much of the planet's life (Wilson, 2014). For example, Ecuador has at least 15,000 and possibly 20,000 plant species. All of Europe—with 31 times greater area—has 13,000 plant species. A single reserve in Peru, the Tambapata, hosts 530 bird species, compared with about 850 in the United States and Canada combined (Myers, 1984 also see Jukofsky, 2002).

Coffee farmers depend on biodiversity in many ways, from pest control to soil fertility. Scientists studying coffee farms in Costa Rica found that a patch of rainforest within 1 km of the coffee plants provided pollinating insects and could increase yields by 20% (Ricketts et al., 2004).

Coffee farmers and consumers should also want to save another kind of biodiversity: genetic. There are 103 species in the genus *Coffea*, but nearly all cultivated coffee is from a few cultivars derived from *Coffea arabica* and *Coffea canephora* (see also Chapter 1).

The original gene pool for Arabica is in the montane forest of southwestern Ethiopia. The living genetic blueprints for Canephora (Robusta) are likely more scattered in central and western sub-Saharan Africa, including South Sudan. The forests containing these ancestral coffee plants are threatened and diminishing. At the same time, locally cultivated coffee varieties are inbreeding with the wild plants. Plant breeders need the genetic variety found in wild relatives—those that have survived for millennia— to make high-yielding hybrids that can better survive drought, disease, pests, and other maladies. Although saving the rainforest is essential to saving biodiversity, conserving the wild coffee forests in Africa—the original genetic library—is key to the industry's future (Fig. 4.1).

2.2 Development and Deforestation Followed Coffee Planting Around the World

Coffee planting spread from its birth place in Africa to the Arabian Peninsula. The Dutch established farms on what is now the island of Java toward the end of the 17th century (Pendergrast, 2010). As the crop took root in the Caribbean and then sailed to the Central American isthmus and down to South America, it brought a wave of development and mostly positive political change, as well as widespread deforestation.

The bean landed in Brazil in 1727; by the mid-1800s, the country was the world's leading producer. In 1865, coffee represented 65% of Brazil's exports

FIGURE 4.1 The coffee plant evolved in African forests, and the traditional way to farm coffee is under a forest canopy. *Photo credit: PUR Projet/Christian Lamontage.*

(Fausto, 1999). In the first century of Brazil's coffee expansion, an estimated 7200 km^2 was cleared in the Mata Atlantica, the Atlantic Forest, now one of the most famously diverse and endangered ecosystems, with only about 5% remaining. Coffee spread from northern Brazil to the southeastern states of Rio de Janeiro and Minas Gerais. Producers began planting in Brazil's once vast savannah, the cerrado, in the 1960s. This shrub and grassland area are almost as rich in biodiversity as the Amazon, and much more suitable for agriculture. About 20% of the cerrado remains, but coffee is a minor factor compared to sugar, soy, and cattle.

Costa Rica's first coffee boom began in 1830 (Molina and Palmer, 2011). The bean may have reached Colombia in 1723, but production there did not really take off until the creation of the Federacion Nacional de Cafeteros in 1927 (Pizano, 2001). Coffee was first planted in Vietnam in 1857, but production did not explode until the 1990s (Doan, 2001). By the time Vietnam was edging out Colombia as the world's second-largest coffee exporter, the rainforest conservation movement had begun. The banks and aid agencies sponsoring Vietnam's coffee ascendancy were criticized for not considering the values of the forest. Although deforestation to create new coffee farms continued apace in Vietnam, scientists, activists, and many consumers were aware of the environmental consequences and knew that there was a better way.

Until the advent of intensified production in the 1970s, most coffee farmers would only cut some trees and clear out the forest understory, replacing it with coffee bushes. Prodded by agronomists and often assisted by banks and multilateral aid agencies such as the US Agency for International

Development, farmers deforested their farms and replaced the traditional varieties with new, dwarf, precocious hybrids. Fully exposed to the tropical sun and given heavy doses of fertilizer and pesticides, the densely planted rows of compact hybrids greatly increased yields. The result was a transformation of the coffee landscape in Northern Latin America with devastating consequences for biodiversity and the environment (Rice, 2000).

2.3 Forested Coffee Farms: a Productive and Natural Environment

In the late 1980s, as scientists and environmentalists were sounding the alarm about the rampant destruction of the world's most bio-rich ecosystem, some were beginning to document a startlingly hopeful proposition: coffee farms could be the salvation for many plant and animal species. Coffee, along with cocoa, vanilla, and a handful of other crops, thrives in the soft, filtered light under the rainforest canopy. This trait, along with the crop's dominance of the middle altitudes in many tropical countries and coffee's energizing role in minds and markets, makes it a central actor—for better or worse—in biodiversity conservation (Schroth et al., 2004).

Farmers, by observation, and biologists, by training, know that traditionally managed coffee farms are nearly naturally functioning ecosystems. Forested farms are almost self-sustaining. For example, during the decade of deadly and disruptive wars in Central America, many farmers had to abandon their fields. When they returned, sometimes years later, they found that the coffee trees were still growing and producing under the forest canopy. With some renovation and management, the farms could once again support families.

In 1996, a landmark paper in BioScience called, "Shade Coffee: A Disappearing Refuge for Biodiversity" showed that, in Latin America, between 1970 and 1990, nearly 50% of shade coffee farms were converted to low-shade systems, ranging from 15% in Mexico to 66% in Colombia. This influential paper provided examples of the higher biodiversity in forested farms and included this conclusion: "These preliminary results suggest that shaded plantations can have a local species diversity within the same order of magnitude as undisturbed forests" (Perfecto et al., 1996).

Scientists and conservationists had been saying the same thing at least since the biological surveys by Griscom in the 1930s and FIIT (the Interamerican Foundation for Tropical Research) in the 1980s (see below). The rich web of life in a forested coffee farm, from detritus-munching micro-organisms to monkeys, birds, and wild cats—and countless other creatures with an almost infinite number of interspecies interactions—rival that of a primary forest. The combination of coffee crop and forest requires few inputs, sustaining natural nutrient recycling as leaves and litter fall from the trees, decompose, and are again taken up by plants. Trees provide many environmental services: for example, some trees are leguminous, fixing nitrogen in the soil, reducing the need to add fertilizers. According to Montenegro (2005) a coffee plantation with abundant *Erythrina* trees can produce up to 144 kg nitrogen/year through pruned branches spread on the soil of the coffee plantation, promoting at the same time improved soil structure.

The forest canopy also protects the soil and coffee plants from pounding tropical rains and desiccating sun, maintaining soil moisture and the watersheds that supply springs and streams. Forested coffee farms are therefore less prone to soil erosion and landslides during heavy rains and more resilient in the face of changing climates.

Some agronomists argue that tree shade increases coffee diseases, but studies and experiments show that properly managed shade has negligible effect on disease rates and can even help coffee plant health (Perfecto et al., 1996; Jha et al., 2014). The forest canopy protects ripening coffee cherries from the elements and improves in-the-cup quality, mainly because the shade cover creates a stable microclimate and fosters the slow ripening of the fruits (Muschler, 2004).

The trees and other indigenous vegetation in a coffee farm provide agronomic benefits in addition to the valuable ecological services. Birds and predatory insects living in the trees can help keep pest levels low and thereby increase productivity. Bees and other pollinating insects service the coffee blossoms. Trees absorb carbon, helping mitigate climate change and opening the possibility for farmers to sell carbon credits, as well as beans. Trees and other natural, native vegetation in coffee farms provide water, fruits, medicine, additional sources of income, and building materials. Forested farms provide firewood, an important source of fuel for most coffee farming families and particularly important in parts of Latin America and Africa where women and children spend hours each day collecting wood.

One of the authors of the "disappearing refuge" paper (Jha et al., 2014), Robert A. Rice at the Migratory Bird Center/Smithsonian Conservation Biology Institute and colleagues updated the study. They found that, as before, coffee farm management decisions are nuanced, localized, and driven by many factors. A few countries had higher percentages of shade production; others showed declines. The researchers had 2010 data from 19 producing countries and calculated that 41% of the coffee area in those countries is full-sun, 35% with sparse shade, and only 24% with traditional diverse shade. The rapid expansion of coffee-growing areas in Vietnam was accompanied by widespread deforestation, which also occurred to a lesser degree in other emerging coffee powers such as Thailand and Indonesia. The paper re-emphasized the values of trees and other natural vegetation to coffee farmers and society, including biodiversity, pollination, pest control, climate regulation, and nutrient cycling.

2.4 Birds Represent Biodiversity on Coffee Farms

As the wave of "technification" washed over the coffee lands, farmers and biologists noticed the dramatic decline in biodiversity. Since birds are highly visible and vocal, their absence was especially evident. John Terborgh is among the scientists seeking solutions for shrinking bird populations. In his book, *Where Have All the Birds Gone*, Terborgh (1989) noted the loss of nesting habitat in North America and wintering habitat in countries to the south. Terborgh called the coffee-growing zone one of the most endangered habitats on the planet, with consequences for all biodiversity, but especially for birds.

The linkage had been documented decades before. In the early 1930s, ornithologist Ludlow Griscom was in Guatemala, collecting birds for the American Museum of Natural History. He noted that coffee growers left much of the natural forest to shade their plants, and that "in such growth, the bird population was little, if any, different from its original condition" (Griscom, 1932). In the 1980s, biologists from the Fundación Interamericana de Investigación Tropical—the first science and conservation NGO in Guatemala and known by its Spanish acronym FIIT—surveyed birds, reptiles, and amphibians in coffee-growing regions of Guatemala. During the period 1987–1991, they retraced Griscom's steps—even going to some of the same coffee farms—and found that many of the migratory species were much less abundant than Griscom had recorded (Vannini, 1994).

Other researchers had surveyed coffee farms for birds (Aguilar-Ortiz F, 1982), ants (Benítez and Perfecto, 1990), beetles, butterflies, and bats (Estrada et al., 1993). But it was the realization that coffee habitat could be the salvation of beloved birds such as warblers and thrushes that connect the continents with their migrations north and south that excited conservationists and scientists. A plethora of studies emerged, giving conservationists more evidence of the value of forested farms (e.g., Greenberg et al., 1995; Wunderle and Latta, 1994; Wunderle and Waide, 1993; Komar, 2006).

Many birds make seasonal migrations—songbirds, shorebirds, waterfowl, raptors, and others. As temperatures drop, days shorten and food supplies dwindle in the north, more than 150 neotropical migrants begin amazing journeys across large distances. Many land in coffee farms. For example, about one-quarter of the birds in coffee farms in southwestern Guatemala are long distance migrants (Vannini, 1994).

The Smithsonian Migratory Bird Center (SMBC, Washington, DC, USA) was founded in 1992, and Russell Greenberg, who had been studying birds in agroforest systems in Mexico, became its director. Greenberg and his SMBC colleagues began to document and publicize the higher levels of biodiversity—especially birds—in forested or "shade" coffee farms. Other scientists were drawing similar conclusions from coffee farms on Caribbean islands (Wunderle and Latta, 1994).

Every scrap of habitat is important. "The North American breeding grounds encompass 40 million square kilometers, but the entire land mass of Mexico, the Bahamas, Cuba, and Hispaniola is only about 6 million square kilometers, and two to five billion more than half of all neotropical migrants funnel into the narrow throat of Mexico and Central America, where they condense into whatever suitable habitat is left; often on coffee farms (Greenberg and Reaser, 1995)."

The biannual migration is one of the greatest—and least understood—natural spectacles on Earth (Greenberg and Reaser, 1995; Wille, 1990). The birds face threats on both ends of the journey—predation, nest parasites, misused pesticides—but loss of habitat is the gravest, most pervasive threat, and the hardest to fix. So, when scientists revealed that forested coffee farms could help save the birds that stitch together the continents with their migrations, conservationists seized on what seemed like the ultimate win–win opportunity. Conserving the coffee-growing traditions could help conserve rainforests and all their teeming biodiversity. Articles with titles such as "The Birds and the Beans" and "Can Coffee Drinkers Save the Rain Forest" began appearing (Hull, 1999; Wille, 1994).

Oliver Komar, formerly the director of conservation science at the NGO SalvaNatura, began studying the avifauna of El Salvador in 1993. Forested coffee farms were especially important in El Salvador, where only 0.5% of natural forest remained but more than 9% of the small nation's landmass was planted in coffee. SalvaNatura, a member of the Sustainable Agriculture Network (SAN), helped develop the SAN standards for sustainable coffee production, infusing them with conservation science. With those standards, SalvaNatura's agronomists and biologists could evaluate farms; those that complied were awarded the Rainforest Alliance Certified seal, a badge of pride for the farmers that had value in the marketplace. In time, Komar and his colleagues at SalvaNatura convinced coffee farmers in El Salvador of the many benefits—economic, social, and environmental—of forested coffee farms. The farms were so richly forested that visiting birders could not easily distinguish farms from parks. SalvaNatura even used certified farms to buffer and link national parks (Langley, 2005) (Fig. 4.2).

2.5 Sustainability, Agroforestry, and Resilience in the Anthropocene

There was a harmonic convergence in the 1990s that set in motion a global movement to incentivize farmers to conserve biodiversity and other natural resources. The converging forces included the rise of the Save the Rainforest movement; the Rainforest Alliance and the Rainforest Action Network which both began in 1986; Friends of the Earth UK was already engaged and the Smithsonian Migratory Bird Center (that developed "Bird Friendly" certified coffee) and other organizations followed. Scientists were calling attention to the extinction crisis. Coffee farmers were suffering from boom and bust price cycles and many were deforesting their farms to accommodate the new, intensive, and monoculture practices. At the same time, the specialty coffee market started to boom; consumers began demanding quality, beans with a backstory, traceability, and sustainability.

Agronomists, NGOs, farmers, and scientists worked together to develop standards for sustainable coffee farming. The premise was that consumers would reward farmers that respected workers, wildlife, and the environment by

FIGURE 4.2 Forested coffee farms abound in biodiversity. Biologists with SalvaNatura identified 103 tree species on a cooperative in El Salvador—more tree species than are native to Europe. Migratory songbirds, such as this yellow warbler (A), nest in North America and winter in countries to the south, often in coffee farms. Hornbills (B) are sometimes seen in coffee farms in India and Indonesia. Butterflies (C) and orchids (D) are among countless other species in sustainably managed farms. *Photo credit: Warbler courtesy of Creative Commons; others by Chris Wille.*

buying certified sustainable coffee. Furthermore, it was assumed that the standards would improve and guide training, providing farmers with precisely the information and assistance they needed, and that the programs would help farmers drive down their cost of production and improve crop quality and yields, thus earning more and bettering their livelihoods.

This seemed like a rare, mutually advantageous scenario. Farmers, their communities, farmworkers, consumers, and the environment should all benefit. Coffee farms became a laboratory for standards and certification, which advanced more in coffee than in any other crop (Potts et al., 2014). Most coffee brands adopted NGO-led certification programs or invented their own, many of which made ecosystem and biodiversity protection and "shade" a priority.

Most farmers that use the standards as guidelines for best management practices are seeing environmental and economic improvements. See Chapter 7 for information on how standard programs are improving impacts. There are abundant challenges. Millions of farmers still are not receiving training or even proper information. Many do not have the resources to make improvements. Too many coffee farmers still struggle to survive (see Chapters 5, 6 and 7 for other aspects of sustainability in coffee farming.

However, most technical assistance and training programs now include the issues covered by the sustainability standards, including integrated farm-management planning; conservation of wildlife habitat protection, soil, and water; pollution control; workers' rights and welfare; waste management; and environmental education and climate adaptation. Deforestation for new coffee farms is now widely seen as unacceptable, and the value of forests and other healthy, functioning ecosystems is increasingly understood. Moreover, standards and training programs now include ways to improve quality and yields and to control costs, giving equal attention to the economic leg of the sustainability triangle.

In 2003, coffee stakeholders, including farmer representatives, NGOs, standard and certification programs, roasters and traders, came together to create a baseline standard, the Common Code for the Coffee Community. The intention was to reach as many farmers as possible with at least basic tools and training. The 4C Association grew with support from major roasters and traders. In 2016, the association and more than 300 coffee-related businesses, NGOs, and others formed the Global Coffee Platform for a Sustainable Coffee World, combining resources and leveraging government support to engage the millions of far-flung small holders (4C Association, 2016).

Other new forces are again converging. Although the standard setters proved that it is not effective to tackle single issues, such as pesticides or deforestation in isolation, there is now growing agreement that is not efficient to address the challenges of a single farm. Farms are not islands. They are part of a socioeconomic and ecological landscape. Modern training and technical assistance programs address the matrix of land uses and issues. Increasingly, we will see integrated, landscape-level programs that mix agroforestry, farm diversification, resilience planning, regenerative agriculture practices, payments for environmental services, microenterprise, and business planning. Farmers, other land-users, rural communities, biodiversity, and the environment will all benefit together (see, for example, Milder et al., 2010; Perfecto and Vandermeer, 2015; Altieri, 1995; COSA, 2014).

3. COFFEE PRODUCTION MUST CHANGE WITH THE CLIMATE

Climate change is having a profound effect on the world's weather patterns. Changes in ecosystem functioning and the loss of biodiversity will affect the provisioning of natural resources and ecosystem services on which we all depend. The effects of climate change, however, will disproportionately affect the most vulnerable regions of the world. People in these countries are often heavily dependent on agriculture and forests as their main livelihood source, yet are increasingly challenged in their ability to maintain these activities.

In the equatorial belt where coffee cultivation takes place, many regions are already experiencing higher temperatures, prolonged droughts, and episodes of intense rainfall. These factors can affect coffee production in many ways: increasing the distribution and occurrence of coffee plant pests and diseases, disrupting pollination, or limiting uptake of necessary nutrients and therefore impacting crop yields and quality. At the same time, coffee production generates greenhouse gas (GHG) emissions, for example when forestland is cleared to cultivate coffee, sometimes to make up for shortfalls in yields due to poor harvests. Agriculture's total direct GHG emissions, including coffee cultivation, generates between 10% and 12% of global GHG emissions, stemming from manure and methane of livestock, management, and inputs such as fertilizers and other agrochemicals. Including the conversion of forests to croplands, agriculture may contribute as much as 25% of total global GHG emissions. Importantly, however, coffee systems can provide a multitude of environmental services, including carbon sequestration, watershed protection, and biodiversity conservation.

A changing climate could lead to a shifting of ecological zones, loss of flora and fauna, an overall reduction in ecological productivity, and changes in the suitability of growing conditions. These impacts can have significant repercussions on livelihoods, food production, and the sustainable development of local communities. Additional pressure on forests and biodiversity will mount, as agricultural production will often be displaced onto previously uncultivated land.

Impacts from climate change will be felt hardest by subsistence farmers in the tropics as they have little access to resources to buffer and adapt to the changing climate (IPCC, 2007). Lack of institutional support and reliance on the natural environment for their livelihoods place the rural poor in the most vulnerable position (Verchot et al., 2007). The distribution and virulence of pests and diseases of crops will change. New equilibria in crop—pest—pesticide interactions will affect crop production. Loss of biodiversity will have impacts, for example, on pollination. Researchers have established a strong link between biodiversity, pollinators, and coffee production (Vergara and Badano, 2009). Climate change will also have impacts on the effectiveness of irrigation, nutritional value of foods, and safety in food storage and distribution.

3.1 Climate Change and the Coffee Industry

Climate change is altering the yields and quality of coffee produced around the world. The economic effects are already being felt across the supply chain, from the farmers all the way to patrons of the corner café. But to develop action plans to adapt coffee farming to climate change, its effects must first be explored and predicted and assessed. One of the challenging aspects regarding climate change is its uncertainty and variability of impacts, which is further being confounded by the heterogeneity of coffee production systems. There is consensus, however, that the severity and frequency of events such as droughts, storms, or pest outbreaks will increase in the years to come.

The most important factors determining coffee yield and quality are temperature and rainfall (Haggar and Schepp, 2012). For the higher quality but more sensitive *Coffea arabica* plant, it is thought that mean temperatures above 23°C may hinder the development and ripening of coffee cherries, and those above 30°C may reduce plant growth and yield (Camargo, 1985). Further, a minimum period of dry weather is necessary to trigger the plant's hormonal response to induce flowering, but prolonged drought can reduce the ability of the plant to photosynthesize and lead to coffee flowers dying off.

Similarly, although Robusta coffee (*Coffea canephora*) is often seen as the more resilient of the two main varieties due to its ability to tolerate higher temperatures, it is less able to cope with intraseasonal variation and lower temperatures than Arabica. As a result of its low frost tolerance, Robusta thrives at lower altitudes; its leaves and fruits are unable to survive temperatures below 5–6°C, or even prolonged periods below 15°C (DaMatta and Ramalho, 2006). To date, the vast majority of coffee–climate research has focused on Arabica varieties, perhaps due to its dominance on the global market. However, research by Bunn et al. (2015), which used climatic models to predict the suitability of global growing areas to coffee production, found that the center of origin of the Robusta plant, the Congo basin, may become unsuitable altogether for its production by 2050.

The challenge for the farmer is that temperature and rainfall patterns are becoming increasingly difficult to predict and the exact effects of climate change can vary greatly between regions. For example, in the coffee growing regions of Colombia where temperatures naturally have shown less variation to date, precipitation has greater potential to adversely affect coffee production: between 2010 and 2011, the excess rainfall (28% more than average) and cloud coverage (16% less sunlight than average) of La Niña (defined as cooler than normal sea-surface temperatures in the central and eastern tropical Pacific ocean that impact global weather patterns) resulted in a reduction of the subsequent year's harvest from an average of 11.7 million bags to just 7.8 and 7.7 million bags in 2011 and 2012, respectively (Café de Colombia, 2014), and therefore a spike in the global price of Arabica coffee—Colombia is the world's second largest Arabica producer in the world. The frequency of La Niña weather events is predicted to increase substantially due to climate change (Cai et al., 2015), meaning that farmers need to be given the resources to better adapt to their growing systems.

Climate change also affects the prevalence of coffee pests and diseases. Globally, the most costly pest to the coffee industry is the coffee berry borer

(CBB), *Hypothenemus hampei*, a tiny beetle that drills into coffee cherries, causing losses of more than US$500 million annually (Vega et al., 2003). Until the 1990s, there were no reports of the CBB at altitudes above 1500 m, which prevented it from becoming a pest in many areas favored for Arabica production. However, rising temperatures have resulted in the pest expanding its range to higher altitudes and spreading to previously unaffected coffee-growing areas. In Tanzania for example, the borer has been found at altitudes 300 m higher than it was found 10 years ago (Mangina et al., 2010). Predictions made using climatic mapping combined with the borer's life history traits suggest that outbreaks are likely to be particularly severe in medium-to high-altitude Arabica-growing areas of East Africa (Jaramillo et al., 2011), not only because they are likely to shift their optimum elevation to higher altitudes, but also because they are predicted to be able to reproduce at a much faster rate: currently, the number of borer generations produced per year ranges from 1 to 4.5, but this is predicted to increase to 5–10 generations per year.

3.2 Impacts on the Coffee Supply Chain

Coffee is a perennial plant, normally taking 3–5 years to produce its first fruit, and 6–8 years until it reaches peak production. Plants can then last for 20–30 years, or longer, depending on how they are managed by the farmer. The often high initial capital investments needed to establish coffee production means that farmers plan and invest in their coffee production systems for the long term, making them particularly vulnerable to short-term climate change-induced events such as severe drought or changing rainfall patterns. Unlike farmers of annual crops, coffee farmers must base their business decisions on long-term market dynamics, but these dynamics are greatly affected by the increasingly unpredictable natural environment. This in turn can have a profound effect on national and local economies; across Mexico and Latin America, for example, over four million people are directly dependent on coffee production for their livelihoods (CEPAL, 2002). With 70% of the world's coffee-growing area being farmed by smallholder farmers who heavily depend on natural resources and ecosystem services as part of their livelihood strategy, climate change has the potential to significantly disrupt the lives of millions of rural and often poor households. One could of course argue that a solution would be for coffee production to migrate to higher latitudes (Zullo et al., 2011) or altitudes (Schroth et al., 2009) to adapt to suitable growing conditions. However, this would not benefit current producers (Baca et al., 2014), but instead displace cultivation from today's traditional and cultural centers of the trade to cooler, mountainous regions far afield, in turn causing migration that could further threaten ecosystems (Laderach et al., 2010) and perpetuate coffee's impact on climate change.

At the global market level, the macroeconomic consequences of climate change impacts will show changes in trade patterns and volumes and increased coffee prices. In an increasingly resource- and agricultural land-restricted world, adaptation strategies for agricultural commodities such as coffee that will depend on the provision of such resources, and an increase in production costs is almost inevitable.

There was a major coffee crisis during 1999–2004, when employment in Central America's coffee sector declined drastically and workers in the sector faced malnutrition and hunger (Tucker et al., 2010). To diminish the chances of such a crisis happening again, assessments are now being carried out to determine the major threats and levels of vulnerability to different environmental and climatic stressors and shocks faced by coffee farmers and the coffee sector as whole. However, vulnerabilities have been found to vary hugely from region to region, meaning that the development of adaptation strategies must be site specific. For example, research by Baca et al. (2014) investigated the exposure, sensitivity, and adaptive capacity (the three elements defining "vulnerability" according to the IPCC) of coffee farmers in four different Latin American countries, and identified areas of high vulnerability that will not be suitable for coffee production by 2050. In areas where coffee production will remain possible but with reduced climatic suitability, improved agronomic practices will enable farmers to continue their production. However, although "exposure" was defined as a region's risk to climatic change impacts, multiple additional factors such as out-migration of the labor work force, access to credit, levels of social organization, and postharvest infrastructure were also all found to have a strong effect on the sensitivity and adaptive capacity of farmers. As the backbone of the coffee industry, farmers must therefore understand the implications of these exposure gradients caused by climate change to successfully develop and adopt adaptation strategies. This in turn will depend strongly on a robust collaboration between coffee supply chain stakeholders to guarantee continued sharing of experiences and knowledge via learning platforms and financial investment into new and innovative approaches.

3.3 Responses and Solutions

To address the many climate-induced challenges outlined above, a new approach called climate smart agriculture (CSA) is being promoted in the agricultural sector. The term was first defined by the Food and Agriculture Organization (FAO) at The Hague Conference on Agriculture, Food Security and Climate Change in 2010, and integrates the three dimensions of sustainable development—economic, social, and environmental concerns—into a strategy that jointly addresses food security and climate challenges. Its three main pillars are focused on,

- Sustainably increasing productivity and resilience, thus helping farmers adapt to climate change
- Enhancing achievement of national food security and development goals, helping to secure sustainable economic livelihoods for farmers

- And reducing/removing greenhouse gas emissions and by doing so contributing to global climate change mitigation efforts where possible.

Example: Teaching Climate Smart Coffee

The Rainforest Alliance and partners are teaching CSA practices to farmers growing coffee and other crops. For example, in Oaxaca, Mexico, the alliance is working with smallholder coffee farmers to advance community-based climate-smart agriculture. Rainforest Alliance Certified farms are reforesting degraded areas and enhancing tree cover. These efforts promote habitat connectivity, soil fertility, and the cultivation of a valuable supply of timber and fruit for the community.

The communities are also working toward meeting the requirements of the Verified Carbon Standard (VCS). Once verified, the project will become one of the first VCS reforestation projects in Mexico, and farmers will be able to sell carbon credits. The projected emissions removal over 30 years for the first group of farmers are estimated at 73,000 t CO_2.

CSA is not a new form of agriculture but rather an approach that combines the different dimensions of sustainable development under a climate change umbrella. The methodology provides the tools and pathways for farmers to build more resilient livelihoods while helping to reduce the impact of farming on climate change through implementation of best-management practices that mitigate the identified climate risks.

Some of the most commonly recognized CSA practices include reducing land degradation and the enhancement of agro ecological systems and functions through appropriate soil management techniques, water conservation practices, and the establishment or improvement of agroforestry systems. At the same time, considering and developing responses to socioeconomic risks and vulnerabilities such as the design and delivery of technical capacity and resources or financial safety nets must be an integral part of any CSA strategy (Initiative for Coffee and Climate, 2014) (Fig. 4.3).

Example: Climate Smart Agroforestry

Nicaragua, the poorest country in Central America, experienced a 40% drop in coffee production during the 2012–13 harvest in part due to leaf rust ("la roya") and other factors aggravated by climate change. An agroforestry company, Nica-France, and a green investor, Moringa Fund, created a project to rejuvenate farms in the traditional coffee-growing area of Matagalpa. Farmers that choose to join the program are provided with modern agronomy practices and new coffee bushes; in exchange, they plant trees.

Trees help improve coffee productivity and quality, provide timber, firewood, and other goods, diversify income streams, increase resilience to climate change, and store carbon. Project designers expect the planted trees to sequester 500,000 tons of carbon. The project aims to protect or create 6000 jobs.

FIGURE 4.3 Climate smart agriculture includes better management of agrochemicals, renovating farms with more adaptable coffee varieties, improved planning, diversification, and agroforestry. *Photo credit: PUR Projet/Christian Lamontage.*

3.4 Adaptation

In response to the environmental and market challenges affecting coffee growers, traders, roasters, and retailers, companies and other stakeholders in the coffee industry are now beginning to identify climate change adaptation as a key element of managing and reducing risk within their supply chain. Utilizing the available predictions of climate change impacts, and building corresponding adaptation strategies suited to local conditions into coffee farming practices can better secure the long-term future and resilience of coffee production, and therefore farmer livelihoods.

Adaptation strategies include increasing productivity and the resilience of agricultural systems to adverse climate change impacts, both from extreme events (short-term adaptation strategy required) and slower-onset changes (long-term adaptation strategy required). It is always essential that the most impacted and at-risk people are empowered to access the proposed services and at the same time are recognized for their knowledge and differences in responding to the identified challenges.

Short-term adaptation strategies can include capacity building and organization among smallholder famers and communities to cope with climate-induced events. Analysis of risks and vulnerabilities, the development of an emergency response plan and improving agronomic practices such as shade and nutrient management would be considered under such a strategy.

Longer-term adaptation, on the other hand, focuses more directly on the process and functions within an enabling environment—availability of improved coffee varieties that are more pest and drought resistant, access to climate data and forecasting information, access to financial services, and the availability of crop or weather insurance products.

To reduce the vulnerability of small-scale subsistence farmers to interannual variability in precipitation and temperature, tree-based systems are often favored as a primary adaptation strategy and promoted by institutions such as the World Agroforestry Center. The benefits of tree-based systems such as agroforestry are multifold. Through diversification and tree inclusion, positive effects on the environment as a whole can be identified (Boye and Albrecht, 2006). For example, beneficial soil properties such as porosity are increased, which lead to higher water infiltration and water retention (Noordwijk et al., 2006) and therefore reduce run-off. Shaded coffee systems create more favorable microclimates, reducing heat stress on plants and, over the long-term, compensating for reduced yields compared with un-shaded crops (Jonsson et al., 1999). According to Verchot et al. (2007), evapotranspiration is increased in comparison to unshaded tree systems, which helps to aerate the soil quicker and reduce water logging, but also shows the potential for shade trees to compete with crops for water. Agroforestry includes rotational systems with tree fallows, which have been demonstrated to significantly improve soil fertility before the next cycle of cropping, especially when they include N-fixing tree species (Sanchez, 1999). In addition, Gallagher et al. (1999) found that short-term improved fallows can significantly improve weed management in tropical environments.

3.5 Standards and Initiatives Helping to Build Resilience in the Coffee Sector

More recently there have been a number of new initiatives and also standards that have acknowledged the challenges posed by climate change and refocused their efforts on providing solutions to achieving greater resilience for smallholder coffee producers and the coffee sector as whole. For example, the revised 2017 Sustainable Agriculture Network (SAN) standard is now explicitly orientated toward the principles of climate-smart agriculture. The standards promote CSA through advancing practices that help improve on-farm resilience, protect native ecosystems, avoid deforestation, maintain healthy soils, reduce on-farm carbon footprints through decreased use of energy, water and agrochemicals, and simultaneously promote gender equity, biodiversity conservation, and other sustainable development goals.

The 2017 SAN standard helps farmers incorporate possible extreme weather events into their cycle of assessment, planning, implementation, and continuous improvement. The carbon footprint of certified coffee farms is reduced through optimized agrochemical use, protection of forests, and other high conservation value ecosystems, carbon sequestration by trees and other natural vegetation on the farm land and soil conservation practices. By improving farming practices, management systems, and coffee farmer knowledge, SAN standards implement the three pillars of CSA—sustainably increasing agricultural productivity and incomes; adapting and building resilience to climate change; and reducing and/or removing greenhouse gas emissions, where possible.

3.6 Coffee's Climate Change Impact

One way of quantifying the contribution of coffee farming to climate change is through carbon footprint assessment. This process quantifies the greenhouse gas emissions that are caused by all inputs and processes associated with the coffee farming system, which include natural resources, energy consumption, and management inputs such as fertilizers or pesticides. With this information in hand, one is able to evaluate the potential to reduce a farm's climate change impact, often by using more energy-efficient technology or equipment or by adjusting agronomic practices such as fertilizer application to better suit the biophysical requirements of a given site (e.g., soil nutrient levels) or changing their timing (e.g., fertilizer application to coincide with certain physiological events or before the onset of a rainy season).

It is universally agreed that at the *farm* level, the application of fertilizers, the degradation and clearing of adjacent forest areas to make way for further coffee expansion, and the management of waste water are the biggest GHG emission sources. If the carbon footprint assessment includes the full value chain, another major emission hotspot is found at the other end of the chain—energy use for the preparation of coffee. This highlights the responsibility of consumers in coffee's contribution to climate change (PCF, 2008).

As a woody biomass-based system, coffee has a great potential to actually mitigate and so reduce or even reverse the negative impacts on climate change and become a so-called "carbon sink" (Noponen et al., 2013). Particularly in shade-grown coffee systems, large stocks of carbon can be found in the tree biomass and in the soil (Albrecht and Kandji, 2003; Nair et al., 2009; Soto-Pinto et al., 2010; Verchot et al., 2007). Conserving and enhancing those stocks are without doubt very important and contribute to not only stabilizing microclimates but also benefitting the global climate by avoiding further emissions through land use change. In addition, there is the opportunity to further increase this sink by converting coffee systems that have little or no shade to shaded systems, or increasing woody biomass elements of farms such as live fencing or windbreaks, through suitable tree-based systems.

Since coffee production systems occupy over 10 million ha globally (FAO, 2011), it is apparent that good management—at the farm level and across the entire industry—can have major implications for our climate as well as for the many livelihoods that depend on it (Fig. 4.4).

FIGURE 4.4 Scientists and farmers agree that planting native trees in coffee farms is one of the most important actions that farmers can take to build resilience to climate change. Trees bring many other benefits, such as sheltering the coffee from the elements; providing fruits, fodder and firewood, and enriching the soil and maintaining soil moisture. A canopy over the coffee helps make a stable microclimate, which can improve crop quality and yields. The PUR Projet, an NGO, helps farmers grow seedlings of appropriate native species and plant them throughout their farms. Here, a father and daughter plant for the future. *Photo credit: PUR Projet/Christian Lamontage.*

4. BIOLOGICAL CONTROL OF COFFEE PESTS AND DISEASES

Forested coffee farms—a perennial crop surrounded by plant, animal and microbial biodiversity—can be quite stable ecosystems. This biodiversity provides the farmers with services in pest and disease control, additional income, as well as healthy surroundings for their families. See Chapter 2 for more about coffee pests and diseases.

What is Integrated Pest Management?

According to the University of California, integrated pest management (IPM) is an ecosystem-based strategy that focuses on long-term prevention of pests or their damage through a combination of techniques such as biological control, habitat manipulation, modification of cultural practices, and use of resistant varieties. Pesticides are used only after monitoring indicates that they are needed according to established guidelines, and treatments are made with the goal of removing only the target organism. Pest control materials are selected and applied in a manner that minimizes risks to human health, beneficial, and nontarget organisms, and the environment. IPM includes the Prevention, Avoidance, Monitoring, and Suppression approach.

Unexplored for many years, the rich biodiversity present in coffee plantations has been revealed after studies in the last three decades, aiding a global trend to decrease reliance upon chemical pesticides in favor of ecofriendly alternatives. The discovery of multiple parasitic relationships on the pests and pathogens affecting coffee has opened the way to including biological controls into every day practices.

In disease management, consistent results have been obtained in controlled environments—such as germination beds and nurseries—where temperature, ultraviolet radiation, water availability, and substrate conditions can be manipulated, therefore, favoring the colonization and activity of fungal products that can be used as controls. These include products based on *Trichoderma harzianum*, *Metarhizium anisopliae*, *Beauveria bassiana*, *Paecilomyces lilacinus*, and Mycorrhizae, to control diseases such as Damping Off (*Rhizoctonia solani*), nematodes (*Meloidogyne* spp.), and Iron Spot (*Cercospora coffeicola*). Under field conditions, *T. harzianum* formulations have been effective in controlling Stem Canker (*Ceratocystis fimbriata*) and Root Rot (*Rosellinia* spp.) (Gaitan et al., 2015).

For pest control, entomopathogenic fungi and nematodes, and a plethora of parasitoids, mostly wasps, and predators such as ants and other insects, have been continuously reported as effective in controlling populations of common problems such as red spider mites (*Oligonychus coffeae*), coffee leaf miners (*Leucoptera coffeellum*), leaf scales, and stem borers.

Considerable research has focused on the CBB due to the lack of genetic resistance among the cultivated coffee varieties and the limitations of chemical controls. Applying the entomopathogenic agent *Beauveria bassiana* (Balsamo) Vuillemin has been the most common biocontrol used in coffee farms since the arrival of CBB (Bustillo, 2006), avoiding expensive economic losses to the coffee industry. Improvements have been required to make this fungus as effective, and competitive with, the chemical sprays, including

- Mixing of highly virulent genetically diverse strains with mortalities of 93% under laboratory conditions (Cruz et al., 2006) and 67% in the field (Cárdenas-Ramírez et al., 2007).
- Application of fungal product to berries on the ground, which caused a 75% decrease in the percentage of infestation of berries in the tree by both killing adults and decreasing the number of laid eggs, affecting subsequent generations of CBB (Vera et al., 2011).
- Mixing of *Beauveria bassiana* and/or *Metarhizium anisopliae* with plant extracts such as garlic (*Allinum sativum*), chili (*Capsicum* sp.), and wormwood (*Artemisia* spp.) that are highly repellant to CBB, causing the CBB to leave their shelters and increasing the exposure and vulnerability to biocontrol agents, increasing CBB mortality up to 87% (Benavides and Góngora, 2015).

African parasitoids of the species *Prorops nasuta* and *Cephalonomia stephanoderis* have been released and wild populations established in different countries in the Americas (Maldonado and Benavides, 2007). Another species, *Phymastichus coffea*, has been tested under laboratory and field conditions (Jaramillo et al., 2002; Aristizabal et al., 2004) and is a promising candidate to effectively reduce CBB numbers. A standardized CBB mass rearing system (Portilla, 1999) was developed that would allow more efficient production of millions of parasitoids.

Some predatory insects native to the CBB-invaded areas are adapting to prey on the beetle. Several predators have been identified in the field (Vera et al., 2007), and three species show the highest potential for an augmentative biological control approach: *Cathartus* spp., *Ahasverus* spp., and *Crematogaster* spp. The first two species are feasibly mass reared on dried corn and can be produced at the farm level. The ant *Crematogaster* requires more research to identify small species that can reach CBB inside the berries, prove its predatory capacity, and assure that this species will not displace other beneficial insects. In all cases, parasitoids with specific hosts are recommended over predators that are generalists to avoid or reduce unintended impacts.

In an agroecological approach, plants such as *Nicotiana tabacum* and *Lantana camara* have been identified as CBB repellents, and *Emilia sonchifolia* as attractant (Castro et al., 2015). Field experiments are under way to test the use of these plants in combination with other strategies to control this pest.

In addition to further research, other key developments are crucial for the continuing success of biological control of pests and disease in coffee cultivation, including better formulation technologies (including UV and dehydration resistance), extended shelf life, implementation of quality control

FIGURE 4.5 Using natural predators to control pests, as described here, is a promising approach. Farmers also use scented traps to catch pests such as the coffee berry borer. *Photo credit: Chris Wille.*

standards, and cost-effective production and competitive prices of biological controls when compared to traditional chemical controls.

Increasing awareness and understanding among farmers of how to apply and what to expect of biological control measures has greatly enabled the practical dissemination of this alternative to the previously unavoidable use of synthesized chemicals in coffee production (Fig. 4.5).

REFERENCES

4C Association, 2016. Over 300 organizations agree to work together with governments on building a more sustainable coffee sector. In: News Bulletin Released March 8, 2016 and Included in the Proceedings of the 4C Association 2016 General Assembly in Addis Ababa, Ethiopia. 4C Association Secretariat, Bonn, Germany.

Ackerman, D., 2014. The Human Age: The World Shaped by Us. WW Norton & Company.

Aguilar-Ortiz, F., 1982. Estudio ecologico de las aves del cafetal. In: Avila-Jimienez, F. (Ed.), Estudio ecológica en el agroecosistemas cafetales. Instituto Nacional de Investigaciones Sobre Recursos Bioticos, Xalapa, Mexico, pp. 103–128.

Albrecht, A., Kandji, S.T., 2003. Carbon sequestration in tropical agroforestry systems. Agriculture, Ecosystems and Environment 99, 15–27.

Altieri, M., 1995. Agroecology: The Science of Sustainable Agriculture, second ed. Westview Press.

Aristizabal, A.L.F., Salazar, E.H.M., Mejía, M.C.G., Bustillo, P.A.E., 2004. Introducción y evaluación de Phymastichus coffea (Hymenoptera: Eulophidae) en fincas de pequeños caficultores, a través de investigación participativa. Revista Colombiana de Entomología (Colombia) 30 (2), 219–224.

Baca, M., Laderach, P., Haggar, J., Schroth, G., Ovalle, O., 2014. An integrated framework for assessing vulnerability to climate change and developing adaptation strategies for coffee growing families in *Mesoamerica*. PLoS One 9 (2).

Benítez, J., Perfecto, I., 1990. Efecto de diferentes tipos de manejo de café sobre las comunidades de hormigas. Agroecologia Neotropical 1, 11–15.

Benavides, M.P., Góngora, B.C.E. Combination of Biological Pesticides. Patent WO2014111764A1. http://www.google.st/patents/WO2014111764A1?cl=pt.

Boye, A., Albrecht, A., 2006. Soil erodibility control and soil-carbon losses under short term tree fallows in western Kenya. In: Roose, E.J., Stewart, B.A.L.R., Feller, C., Barthes, B. (Eds.), Soil Erosion and Carbon Dynamics. CRC Press, Boca Raton, FL, pp. 181–195.

Bunn, C., et al., 2015. A bitter cup: climate change profile of global production of Arabica and Robusta coffee. Climatic Change 129 (1), 89–101.

Bustillo Pardey, A.E., 2006. Una revisión sobre la broca del café, *Hypothenemus hampei* (Coleoptera: Curculionidae: Scolytinae), en Colombia. Revista Colombiana de Entomología (Colombia) 32 (2), 101–116.

Café de Colombia, 2014. http://www.cafedecolombia.com/bb-fnc-en/index.php/comments/how_el_nino_la_nina_affect_production_of_cafe_de_colombia/.

Cai, W., et al., 2015. Increased frequency of extreme La Niña events under greenhouse warming. Nature Climate Change 5, 132–137.

Camargo, A.P., 1985. Florescimento e frutificacão de cafe arabica nas diferentes regiões cafeeiras do Brasil. Pesquisa Agropecuária Brasileira 20, 831–839.

Cárdenas-Ramírez, A.B., Villalba-Guott, D.A., Bustillo-Pardey, A.E., Montoya-Restrepo, E.C., Góngora-Botero, C.E., 2007. Eficacia de mezclas de cepas del hongo *Beauveria bassiana* en el control de la broca del café. Cenicafé 58 (4), 293–303.

Castro, T.A.M., Tapias, J., Ortíz, A., Benavides, M.P., Góngora, B.C.E., 2015. Uso de plantas repelentes y atrayentes en una estrategía de manejo agroecológico de la broca del café en Colombia. In: Congreso de la Sociedad Colombiana de Entomología, Resúmenes, p. 218. Medellín Julio 29 a 31 de.

CEPAL (Comision Economica Para America Latina y el Caribe), 2002. Centroamerica: El Impacto de la Caıda de los Precios del Cafe en el 2001. Available at: http://www.cepal.org/publicaciones/xml/9/9679/l517.pdf.

COSA, 2014. The COSA Measuring Sustainability Report.

Cruz, L.P., Gaitan, A.L., Gongora, C.E., 2006. Exploiting the genetic diversity of *Beauveria bassiana* for improving the biological control of the coffee berry borer through the use of strain mixtures. Applied Microbiology and Biotechnology 71 (6), 918–926.

DaMatta, Ramalho, 2006. Impacts of drought and temperature stress on coffee physiology and production: a review. Brazilian Journal of Plant Physiology 18 (1), 55–81.

Doan, T.N., 2001. Orientations of Vietnam coffee industry. In: Speech at International Coffee Conference May 17–19, 2001, London, UK.

Ecosystems and Human Wellbeing, Current Status and Trends 1, 2005 (From Millenium Ecosystem Assessment).

Estrada, A., Coates-Estrada, R., Merrit Jr., D., 1993. Bat species richness and abundance in tropical rain forest fragments and in agricultural hábitats at Los Tuxtlas, Mexico. Ecography 16, 309–318.

Fausto, B., 1999. A Concise History of Brazil. Cambridge University Press, ISBN 978-0-521-56526-4.

FAO, 2010. Climate-smart agriculture. Policies, Practices and Financing for Food Security, Adaptation and Mitigation. Retrieved from: http://www.fao.org/docrep/013/i1881e/i1881e00.pdf.

FAO, 2011. FAOSTAT: Land-use Statistics. Food and Agricultural Organisation, Rome, Italy. URL. http://faostat.fao.org/site/377/default.aspx#ancor.

Gaitan, A., Cristancho, M., Castro, B., Rivillas, C., Cadena, G., 2015. Compendium of Coffee Diseases and Pests. APS Press, p. 79.

Gallagher, R., Fernandes, E., McCallie, E., 1999. Weed management through short-term improved fallows in tropical agroecosystems. Agroforestry Systems 47, 197–221.

Global Forest Watch, 2015. Citing Research by Hansen et al./University of Maryland/Google/USGS/NASA, High-Resolution Global Maps of 21st-Century Forest Cover Change.

Greenberg, R., Reaser, J., 1995. Bring Back the Birds: What You Can Do to Save Threatened Species. Stackpole Books, Mechanicsburg, PA.

Greenberg, R., Salgado-Ortiz, J., Warkentin, I., Bichier, P., 1995. Managed forest patches and the conservation of migratory birds in Chiapas, Mexico. In: Wilson, M., Sader, S., Santana, E. (Eds.), The Conservation of Migratory Birds in Mexico. Technical Publication, Orono (ME), pp. 178–190 (University of Maine, School of Natural Resources).

Griscom, L., 1932. The distribution of bird-life in Guatemala. Bulletin of the American Museum of Natural History 64.

Haggar, J., Schepp, K., 2012. Coffee and climate change impacts and options for adaption in Brazil, Guatemala, Tanzania and Vietnam. NRI Working Paper Series: Climate Change, Agriculture and Natural Resources.

Hölldobler, B., Wilson, E.O., 1990. The Ants. Harvard University Press.

Hull, J.B., August 1999. Can coffee drinkers save the rain forest? The Atlantic Monthly.

IPCC, 2007. Climate Change 2007: Impacts, Adaptation & Vulnerability. In: Parry, M.L., Canziani, O.F., Palutikof, J.P., van der Linden, P.J., Hanson, C.E. (Eds.), Working Group 2 Contribution to the Fourth Assessment Report of the Intergovernmental Panel on Climate Change. Cambridge University Press, Cambridge, UK, p. 976.

Jaramillo, et al., 2011. Some like it hot: the influence and implications of climate change on coffee berry borer (*Hypothenemus hampei*) and coffee production in East Africa. PLoS One 6 (9), e24528. http://dx.doi.org/10.1371/journal.pone.0024528.

Jaramillo, S.J., Bustillo, P.A.E., Montoya, R.E.C., 2002. Parasitismo de *Phymastichus coffea* sobre poblaciones de *Hypothenemus hampei* en frutos de café de diferentes edades. Cenicafé 53 (4), 317–326.

Jha, S., Bacon, C.M., Philpott, S.M., Méndez, V.E., Läderach, P., Rice, R.A., 2014. Shade coffee: update on a disappearing refuge for biodiversity. BioScience 64 (5), 416–428.

Jonsson, K., Ong, C.K., Odongo, J.C.W., 1999. Influence of scattered nere and karite trees on microclimate, soil fertility and millet yield in Burkina Faso. Experimental Agriculture 35, 39–53.

Jukofsky, D., 2002. Encyclopedia of Rainforests. Oryx Press, Westport, Connecticut, London.

Komar, O., 2006. Ecology and conservation of birds in coffee plantations: a critical overview. Bird Conservation International 16, 1–23.

Kolbert, E., 2014. The Sixth Extinction; an Unnatural History, Henry Holt & Company.

Laderach, P., Lundy, M., Jarvis, A., Ramırez, J., Perez, P.E., et al., 2010. Predicted impact of climate change on coffee-supply chains. In: Leal Filho, W. (Ed.), The Economic, Social and Political Elements of Climate Change. Springer Verlag, Berlin, DE, p. 19.

Langley, N., September 2005. Park or coffee? Shade-grown coffee plantations are one option for linking protected areas to create "biological corridors". World Birdwatch 27 (3).

Maldonado, L.C.E., Benavides, M.P., 2007. Evaluación del establecimiento de *Cephalonomia stephanoderis* y *Prorops nasuta*, controladores de *Hypothenemus hampei*, en Colombia. Cenicafé 58 (4), 333–339.

Mangina, F.L., Makundi, R.H., Maerere, A.P., Maro, G.P., Teri, J.M., 2010. Temporal Variations in the Abundance of Three Important Insect Pests of Coffee in Kilimanjaro Region, Tanzania.

Milder, J.C., DeClerck, F.A., Sanfiorenzo, A., Sánchez, D.M., Tobar, D.E., Zuckerberg, B., 2010. Effects of farm and landscape management on bird and butterfly conservation in western Honduras. Ecosphere 1 (1), 1–22.

Molina, I., Palmer, S., 2011. The History of Costa Rica (reprint). Editorial Universidad de Costa Rica.

Montenegro, G.E.J., 2005. Efecto de la dinámica de la materia de nutrientes de la biomasa de tres tipos de árboles de sombra en sistemas de manejo de café orgánico y convencional (MSc thesis). CATIE, Turrialba, Costa Rica.

Muschler, R.G., 2004. Shade management and its effect on coffee growth and quality. In: Wintgens, J.N. (Ed.), Coffee: Growing, Processing, Sustainable Production, pp. 391–418.

Myers, N., 1984. The Primary Source; Tropical Forests and Our Future. W.W. Norton & Company, New York.

Nair, P.K.R., Mohan Kumar, B., Nair, V.D., 2009. Agroforestry as a strategy for carbon sequestration. Journal of Plant Nutrition and Soil Science 172, 10–23.

Noordwijk, M.V., Saipothong, P., Agus, F., Hairiah, K., Supraygo, D., Verbist, B., 2006. Watershed functions in productive agricultural landscapes with trees. In: Garrity, D.P., Okono, A., Grayson, M., Parrot, S. (Eds.), World Agroforestry into the Future. World Agroforestry Centre, Nairobi, Kenya, pp. 103–117.

Noponen M.R.A., Healey, J.R., Soto G., Haggar J.R., 2013. Sink or Source—The Potential of Coffee Agroforestry Systems to Sequester Atmospheric CO_2 into Soil Organic Carbon, 2013.

Panhuysen, S., Pierrot, J., 2014. Coffee Barometer 2014. HIVOS, IUCN, Oxfam Novib, Solidaridad. WWF.

PCF Pilotprojekt Deutschland, 2008. Case Study Tchibo Privat Kaffee Rarity Machare by Tchibo GmbH, p. 60.

Pendergrast, M., 2010. Uncommon Grounds: The History of Coffee and How It Transformed Our World. Basic Books.

Perfecto, I., Rice, R.A., Greenberg, R., Van der Voort, M.E., 1996. Shade coffee: a disappearing refuge for biodiversity. BioScience 46 (8).

Perfecto, I., Vandermeer, J., 2015. Coffee Agroecology: A New Approach to Understanding Agricultural Biodiversity, Environmental Services and Sustainable Development. Routledge.

Pizano, D., 2001. El café en la encrucijada: evolución y perspectivas. Alfaomega, ISBN 958-682-192-7.

Portilla, M., 1999. Development and evaluation of new artificial diet for mass rearing *Hypothenemus hampei* (Coleoptera: Scolytidae). Revista colombiana de entomología 25 (1/2), 57–66.

Possingham, H.P., Wilson, K.A., 2005. Turning up the heat on hotspots. Nature 436 (7053), 919–920.

Potts, J., Lynch, M., Wilkings, A., Huppé, G.A., Cunningham, M., Vivek Voora, V., 2014. The State of Sustainability Initiatives Review 2014: Standards and the Green Economy.

Rice, R., October 1999. A place unbecoming: the coffee farm of northern Latin America. The Geographical Review 89 (4), 554–579. Copyright 2000 by the American Geographical Society of New York.

Ricketts, C.D., Daily, G.C., Ehrlich, P.R., Michener, T.H., 2004. Economic value of tropical forest to coffee production. Proceedings of the National Academy of Sciences of the United States of America 101 (34).

SAN Standard, 2015. http://san.ag/web/our-standard/our-sustainability-principles/.

Sanchez, P., 1999. Improved fallows come of age in the tropics. Agroforestry Systems 47, 3–12.

Schroth, G., Haggar, J., Hernandez, R., Castillejos, T., 2009. Understanding vulnerability and building resilience to climate change in a high biodiversity mountain landscape in Chiapas, Mexico. In: IOP Conference Series Earth and Environmental Science, vol. 6 (34).

Soto-Pinto, L., Anzueto, M., Mendoza, J., Ferrer, G., de Jong, B., 2010. Carbon sequestration through agroforestry in indigenous communities of Chiapas, Mexico. Agroforestry System 78, 39–51.

Schroth, G.A.B., da Fonseca, G., Harvey, C.A., Gascon, C., Vasconcelos, H.L., Izac, A.N., 2004. Agroforestry and Biodiversity Conservation in Tropical Landscapes. Island Press, Washington, Covelo, London.

Terborgh, J., 1989. Where Have All the Birds Gone? Princeton University Press, Princeton, New Jersey.

Tucker, C.M., Eakin, H., Castellanos, E.J., 2010. Perceptions of risk and adaptation: Coffee producers, market shocks, and extreme weather in Central America and Mexico. Global Environmental Change 20 (1), 23–32.

Vannini, J.P., 1994. Nearctic avian migrants in coffee plantations and forest fragments of south-western Guatemala. Bird Conservation International 4, 209–232. http://dx.doi.org/10.1017/S0959270900002781.

Vega, F.E., et al., 2003. Global project needed to tackle coffee crisis. Nature 435, 343.

Vera, A.J.T., Montoya, R.E.C., Benavides, M.P., Góngora, B.C.E., 2011. Evaluation of *Beauveria bassiana* (Ascomycota: Hypocreales) as a control of the coffee berry borer *Hypothenemus hampei* (Coleoptera: Curculionidae: Scolytinae) emerging from fallen, infested coffee berries on the ground. Biocontrol Science and Technology 21 (1), 1–14.

Vera, M.L.Y., Gil, P.Z.N., Benavides, M.P., 2007. Identificación de enemigos naturales de *Hypothenemus hampei* en la zona cafetera central colombiana. Cenicafé 58 (3), 185–195.

Verchot, L., Van Noordwijk, M., Kandji, S., Tomich, T., Ong, C., Albrecht, A., Mackensen, J., Bantilan, C., Anupama, K., Palm, C., 2007. Climate change: linking adaptation and mitigation through agroforestry. Mitigation and Adaptation Strategies for Global Change 12, 901–918.

Vergara, C.H., Badano, E.I., 2009. Pollinator diversity increases fruit production in Mexican coffee plantations: the importance of rustic management systems. Agriculture, Ecosystems & Environment 129, 117–123.

Wille, C., May 1990. Mystery of the missing migrants. Audubon.

Wille, C., November-December. The birds and the beans. Audubon.

Wilson, E.O., 1978. On Human Nature. Harvard University Press, Cambridge, Massachusetts.

Wilson, E.O., 2014. The Meaning of Human Existence. Liveright Publishing Corporation, New York.

Wilson, E.O., 1992. The Diversity of Life. The Belknap Press of Harvard University Press, Cambridge, Massachusetts.

Wunderle, J.M., Latta, S., 1994. Overwinter turnover of Nearctic migrants wintering in small coffee plantations in Dominican Republic. Journal fuer Ornithologie 135, 477.

Wunderle, J.M., Waide, R.B., 1993. Distribution of overwintering neartic migrants in the Bahamas and Greater Antilles. Condor 95, 904–933.

Zullo Jr., J., Pinto, H.S., Assad, E.D., de Ávila, A.M.H., 2011. Potential for growing Arabica coffee in the extreme south of Brazil in a warmer world. Climatic Change 109 (3), 535–548.

Chapter 5

Social Sustainability—Community, Livelihood, and Tradition

David Browning[1], Shirin Moayyad[2]
[1]*TechnoServe, Washington, DC, United States;* [2]*Nestlé Nespresso SA, Lausanne, Switzerland*

1. INTRODUCTION—A SNAPSHOT OF THE WORLD'S COFFEE FARMS

Global coffee production has tripled over the past 50 years however most of this growth has been secured by two origins; Brazil and Vietnam that have had the most competitive cost structure. The rest of the world's production collectively has remained relatively flat for decades (International Coffee Council, 2014).

In simplistic terms, the world's coffee farms fall into one of three archetypes; large sophisticated agribusiness, family owned estate, or smallholder. Large coffee agribusinesses exist primarily in Brazil (see text box below). They represent less than 1% of the world's coffee farms, but provide approximately 5–10% of global production (TechnoServe unpublished data from client work, industry interviews, and national coffee data sources across 20 major coffee origins, 2016).

Ipanema, One of the World's Largest Coffee Farms
Brazil's Ipanema Agricola is one of the world's largest coffee agribusinesses. With over 10 million coffee trees, its annual average production of 85,000 bags is more than many small coffee-producing nations. Highly technified with deep agricultural capabilities, the company's coffee productivity at 31 bags per hectare is more than double the world average. Ipanema goes beyond just farming and is vertically integrated with sophisticated processing, exporting, and marketing capabilities that are a world away from a more typical smallholder coffee farm.

Family owned coffee estates have many similarities to a family run farm in the Western world and range up to several hundred hectares. The farms are typically multigenerational with highly skilled and educated owners,

The Craft and Science of Coffee. http://dx.doi.org/10.1016/B978-0-12-803520-7.00005-0

employing workers to assist with farm operations such as harvesting. Coffee is not just an important source of income for these communities but a vibrant and vital part of the social fabric stretching back generations. Family run estates can be found in various coffee-producing nations but particularly throughout Central America and Colombia. Collectively they make up less than 5% of all the world's coffee farms but supply approximately 30% of the world's coffee production (TechnoServe unpublished data 2016).

The remaining 95% of the world's coffee farms can be characterized as smallholder with holdings of less than 5 ha. Most of the world's coffee, approximately 60% of total global production, is produced by these smallholder farmers (TechnoServe unpublished data 2016). Productivity rates are well below the global average and as a general trend, these smallholder coffee farms are shrinking in size as farms are subdivided each generation. In Uganda, average smallholder farm is halving in size each generation; approximately from 4 ha in 1960 to 2 ha in 1990 to 1 ha today (TechnoServe and IDH, 2013). This is in contrast to agricultural consolidation that has occurred in the United States over the same period where the average landholding size has doubled (United States Department of Agriculture, 2002).

Of these smallholder farms that comprise most of the world's coffee producers, approximately half are less than 2 ha in size and generate income that falls below globally accepted definitions of extreme poverty.

How did we get here? Coffee is a valuable crop relative to many alternative agricultural options and it provides vital and welcome income for those who grow it. This income is used for sustenance, housing, education, health expenses, and more. And yet from a social perspective most of the world's coffee farmers and farm workers do not enjoy threshold standards of income, nutrition, education, and health.

We will focus our attention here on smallholder farmers who comprise the vast majority of coffee farmers and produce the majority of the world's coffee, to understand the social challenges they face and what can be done to address their situation.

2. AN AGENDA FOR SOCIAL SUSTAINABILITY

In September 2015 world leaders adopted a social and environmental agenda for sustainable development (http://www.un.org/sustainabledevelopment/). This agenda focused on 17 sustainable development goals to end poverty, fight inequality and justice, and tackle climate change by 2030 (United Nations - General Assembly, 2015).

In sequence, these goals are (1) no poverty, (2) zero hunger, (3) good health and well-being, (4) quality education, (5) gender equality, (6) clean water and sanitation, (7) affordable and clean energy, (8) decent work and economic growth, (9) industry, innovation, and infrastructure, (10) reduced inequalities, (11) sustainable cities and communities, (12) responsible consumption and production, (13) climate action, (14) life below water, (15) life on land, (16) peace, justice, and strong institutions, and (17) partnerships for the goals.

This chapter will touch on seven of these goals, which are focused on intractable social issues that are endemic in the developing world, including coffee origins; ending poverty, ending hunger, ensuring healthy lives, providing education, empowering women and girls, ensuring the availability of water and sanitation, and promoting just labor practices. As the ultimate goal is to end poverty, protect the planet, and ensure prosperity for all, the coffee industry is by its very nature uniquely positioned to contribute.

Though we will not discuss in detail here, there is an implicit recognition that there are interlinkages between these different goals. The common threads of poverty, conflict, discrimination, and unsustainable resource management are tightly interwoven. Empowering smallholder farmers is interdependent with tackling climate change, ensuring adequate nutrition, promoting gender equity, and more. We will present examples throughout of instances where the coffee industry addresses these goals. As well, three particular anecdotes will illustrate through real-life examples the many ways in which the coffee industry contributes to solutions.

One further caveat is in order. Although there has been a great deal of work done on the social dimension of the coffee sector at a microlevel (e.g., studying one cooperative or one region) data at a macrolevel (e.g., national or global figures relating to coffee farmers, their lives and incomes) are much more variable. Estimates have been made based on data as far as available.

2.1 Ending Poverty

The first of these goals is to end poverty in all its forms everywhere. Extreme poverty, currently measured as $1.25 per capita per day in purchasing power parity (United Nations - General Assembly, 2015), is estimated to encompass over 800 million people. Based on varying estimates of the number of coffee households globally, approximately 50–100 million people of this total are in families that either own or work on coffee farms (Lewin et al., 2004; TechnoServe unpublished data, 2016).

There are many well-documented and often cited factors that create and perpetuate poverty in coffee communities. These include: *agro-climatic factors* such as natural disasters; *conflict, political institutions (both colonial and postcolonial); governance*; and *social inequities* based on ethnicity, race, gender, or social class.

However other factors also conspire to keep coffee farmers poor, and their influence is less obvious. Latitude is a predictor of poverty with economies in the tropical zone where coffee is grown generally poorer than those outside the tropics (Sachs, 2001). This phenomenon holds true even within economically integrated zones of temperate and subtropical climates; subtropical United

States South lags behind temperate North, Northern Europe is wealthier than Southern Europe. Brazil's tropical northeast is less prosperous than the temperate Southeast and temperate Northeast China has long had higher per-capita incomes than subtropical Southeast China.

Biophysical factors also contribute to poverty in coffee communities. The absence of frost in the tropics inhibits agricultural productivity and exacerbates disease, particularly malaria. Cold winters outside the tropics force insects into a dormant state that inhibits the cycle of insect-borne disease while frost helps the creation of rich fertile topsoil. We typically visualize the tropics as a place of lush fertile biodiversity but tropical rainforest soils are often nutrient poor. Soils in temperate zones build up nitrogen and carbon that remains in the soil as organic matter but in the tropics, insects and microbes break down matter quickly and prevent rich topsoil from developing. Lower agricultural productivity and higher incidence of disease hurt economic growth and exacerbate poverty (Masters and McMillan, 2001).

Landlocked countries including many coffee origins face special challenges that exacerbate poverty. In the 2015 *Human Development Report* five of the six countries with the world's lowest Human Development Index scores are landlocked. Distance to ports, the additional time and cost this creates, and dependencies upon neighbors' infrastructure and stability translate into less access to world markets, and greater poverty (Faye et al., 2004).

Many of these causal forces of poverty converge in the tropical belt where the coffee plant evolved and is grown. Tropical latitude supports cultivation of the coffee tree but inhibits agricultural productivity due to poor soils and exacerbates diseases such as malaria. Warm moist air over tropical oceans creates the engines for cyclones that destroy human, social and economic capital; particularly in the coffee origins of Central America. Landlocked coffee origins such as Uganda, Rwanda, and Burundi lack access to sea-ports which reduce their economic competitiveness. Finally, many if not most of the world's coffee origins have endured major conflict over the past 50 years including coups d'état, revolutions, drug wars, genocide, and more. Against such a backdrop, it should not be a surprise that most of the world's coffee farmers survive on small landholdings with incomes below the poverty line.

Coffee does not create poverty, it coexists with it. Coffee is one of the few commodities where impoverished smallholder farmers in the tropics have a natural competitive advantage over first-world farmers (based on geography). Smallholder farmer production represents the majority of all coffee produced in the world and finds its ways to cafes along the Champs-Élysées and coffee bars on Wall Street. The welcome income smallholder farmers receive from this crop provides the means to cover cash expenses such as additional food, education, and health care.

A clinical discussion of facts and figures tends to shield us from the harsh reality that poverty represents for coffee farmers such as Rwandan Athanasie (see story below), as well as the heroic endurance that is required. Hundreds of millions of dollars have been spent over the past few decades by aid agencies, foundations, nonprofits, coffee companies, and others on efforts to help coffee farmers out of poverty. Athanasie's story is a reminder that with adequate resources and effective interventions, opportunities do exist to transform the status quo and help these farmers overcome poverty.

The Story of Athanasie, a Rwandan Coffee Farmer

Athanasie is a Rwandan coffee farmer. Her day starts with a 1-h walk for water. It will be the first of three daily trips she makes for cooking, washing, and cleaning water. She carries 20 L each trip, similar in weight to a water cooler canister in an office building in the developed world. Athanasie will also collect firewood and prepare breakfast over an open stove. Like hundreds of millions of other women and children around the world, the simple act of securing household water and heat will consume 4 h of Athanasie's day, every day. But today, Athanasie counts herself lucky.

Back in 1994, 2 months pregnant, Athanasie was caught in the maelstrom of a brutal genocide and joined a mass exodus of humanity fleeing the country on foot. She made it to a refugee camp that had within days become the second largest city in Tanzania with a population of over a quarter of a million people. Against all odds, she survived and gave birth to her daughter on the ground in the camp. This was Athanasie's fourth child, all three children before had died. She gave her daughter the name "Nzamwitakuse." The translation; "I'll give you a name if you survive" (Fig. 5.1).

It was coffee that created an important lifeline for Athanasie. Fast forward to her return to Rwanda postgenocide. At 14 years, her daughter Nzamwitakuzi would soon start fourth grade. Like so many children of smallholder coffee

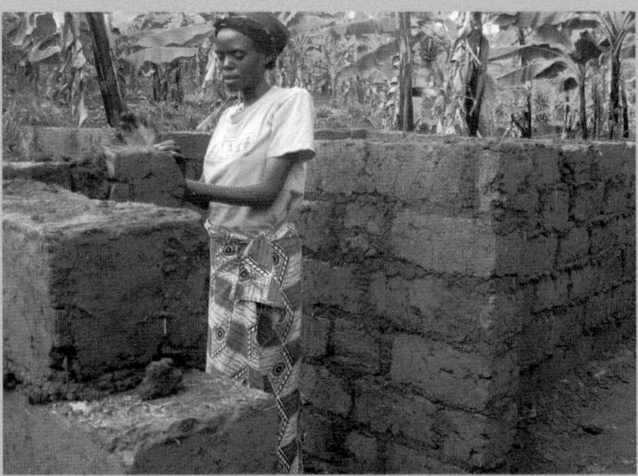

FIGURE 5.1 Athanasie in her coffee plot.

Continued

The Story of Athanasie, a Rwandan Coffee Farmer—cont'd

farmers around the world, a combination of political instability, sickness or other disruptions meant that it would take her more years to complete elementary school education than a child in the Western world. However this did not trouble Athanasie. Her bigger concern was the looming challenge of financing high school education for her daughter. Athanasie knew what a high school education would mean for Nzamwitakuzi. A chance to escape poverty and be successful; the opportunity to go to a doctor when she was sick, to take a bus when she needed to go somewhere, to eat food other than corn, and to have water from a tap rather than to walk 3 h a day to fetch it. But school fees would cost $150 per year, more than twice her typical annual income from coffee of around $70. She owned 500 coffee trees on land less than a quarter of a hectare.

The game changer for Athanasie and her community was a partnership with TechnoServe, aimed at improving the quality of their coffee via postharvest processing. Athanasie was just one of nearly 200,000 coffee families (or 1 million people) whose lives were economically improved by the initiative. It was made possible by a specialty coffee industry which was creating new global markets and higher premiums for high-altitude smallholder coffee farmers worldwide.

The mechanism was straightforward. (1) Identify terroirs where coffee had the potential to be of excellent quality, but lost value either by dint of poor processing, or by being bulked with lesser qualities. (2) Within this terrain, identify underserved farmer groups willing to collaborate to improve their lot. (3) Farmers prepared business plans to secure bank loans to finance small-volume, low-water usage ecopulpers and farmers were taught to process well. The model was predicated on maximizing the use of these pulpers to minimize fixed costs over the highest volume of coffee, and so they were strategically positioned in places where small farmers could easily walk to deliver their daily harvest. Loans were set for 4 years, but some of the more successful co-operatives paid them down in as few as two. Agronomy training was also provided which increased yields by more than 50% on average. In four East African countries the Initiative implemented this model, bringing income, dignity, and self-sufficiency to hundreds of thousands of farmers and putting them on specialty coffee maps.

The coffee income earned gave Athanasie the resources she needed to keep her daughter in school. The addition also allowed her to purchase health insurance—a first. A few months after, Nzumwitakuze contracted malaria. Athanasie had to carry her fever-wracked daughter over the hills to the health clinic, and the health insurance she had purchased permitted treatment, without which Nzumwitakuze may well have died.

Today Athanasie has tripled her income as well as her number of coffee trees, and enjoys a dramatic reduction in labor required to home process her coffee. She has triumphed, and her determination and resilience are a breathtaking inspiration for the rest of us and testimony to the hope that coffee can provide.

2.2 Ending Hunger

The second sustainable development goal calls for an end to hunger and malnutrition by 2030 including access by all people, including infants, to safe nutritious and sufficient food all year round. Adequate nutrition during the critical 1000 days from beginning of pregnancy through a child's second birthday is a particular focus.

Various researchers (Bacon, 2005, 2008; Méndez et al., 2010; Caswell et al., 2012) have analyzed food insecurity in coffee regions and arrived at similar conclusions; many smallholder farmers struggle to meet basic food needs. Often food insecurity is seasonal in nature with families having particular challenges bridging approximately 3 months each year before the next harvest. Families resolve this food gap by strategies such as reducing caloric intake, going into debt, or reducing cash expenses (such as school fees, propane gas) to purchase food.

Smallholder coffee farmers rarely rely only on coffee for their livelihood. A typical coffee farm may have coffee for cash income, staple crops such as maize for subsistence, livestock as an additional source of income (e.g., milk), off-farm labor such as seasonal harvesting or a family member in the city, as well as remittances from a family member in the United States or Europe.

Each of these income sources have risk; a drought may decimate rain-reliant staple crops, a family member in the United States may get injured or laid off reducing remittances, coffee prices may fall or a pathogen such as leaf-rust may reduce yield, and seasonal work may be difficult to secure.

Importantly, these various risks are frequently uncorrelated; the price of coffee on the New York Exchange is decoupled from localized rainfall of Nicaragua or employment prospects of a family member. This lack of correlation is important for allowing families to weather the volatile uncertainties of their lives. If one income source is reduced or eliminated, another source can provide a safety net.

Collectively these income sources form a diversified portfolio. For this reason coffee income, and subsistence crops are best understood not in isolation or in competition but rather as complementary components of smallholder farmer income.

Although every family makes different choices, a common choice for smallholder coffee farmers is to allocate approximately one-quarter of land holdings to staple crop production (TechnoServe unpublished data based on interviews with field staff across Africa and Latin America, 2016). For a 2-ha farm this means approximately half a hectare for maize, beans, or some other staple. This provides an important minimum safety net for families (perhaps 700 kg of food per family per year)[1] (World Bank, 2015) but falls short of both

1. Based on cereal yield in Sub-Saharan Africa of 1381 kg\ha in 2013.

the minimum recommended caloric intake to prevent malnutrition and also the diversity of nutrition required for a healthy diet including adequate protein, calcium, and vitamin A and C and other micronutrients (Tulchinsky, 2010).

2.3 Ensuring Healthy Lives

The third sustainable development goal establishes a range of health related goals including reducing maternal and child mortality, ending the epidemics of AIDS, tuberculosis, and malaria; combating water-borne diseases; reducing the number of deaths and illnesses from hazardous chemicals and more.

Coffee farmers contend with many of the challenges documented including infectious diseases, occupational health and safety, and sexual violence. The severity and incidence of different health challenges vary greatly by geography and income level. For larger farms in Brazil, occupational health and safety issues related to injury from machinery, harmful pesticides, and access to safety equipment may take precedence whereas for smallholder coffee farmers in Africa with little or no access to mechanization or inputs, "diseases of poverty" including malaria, tuberculosis, and HIV pose the greatest threat.

Each year malaria infects 300–500 million people, and kills over a million children. Malaria is endemic in many coffee origins although smallholder Arabica coffee famers living above 1500 m receive some respite. The higher altitudes result in lower temperatures that inhibit development of both the malaria parasite itself as well as the population of anopheles mosquito that serves as the vector for malaria (Centers for Disease Control and Prevention, 2015). Over time, climate change may result in greater incidence of malaria in higher altitude coffee communities (Siraj et al., 2014).

Tuberculosis is an infectious disease caused by bacteria that usually attacks the lungs. It used to be called "consumption" and is spread when infected people cough, sneeze, or spit. The World Health Organization estimates that one third of the world population (including many of the world's coffee farmers) are infected, resulting in 1.5 million deaths in 2014 (World Health Organization, 2015).

The HIV causes HIV infection and AIDS (Douek et al., 2009; Weiss, 1993). More than 30 million people have HIV worldwide resulting in over a million deaths each year. Women account for more than half the total number of people living with HIV in eastern and southern Africa (Institute for Health Metrics and Evaluation, 2016; UNAIDS 2016).

These major health challenges are exacerbated by contributing factors including inadequate sanitation and contaminated water, which contribute to malaria and parasitic diseases; lack of access to health care and health information, malnutrition, and open cooking fires which contribute to acute respiratory infection. Many of these conditions act in a vicious reinforcing cycle; malnutrition weakens the human immune system leading to an increased risk of contracting infectious disease and the onset of disease reduces the body's ability to take in and retain adequate nutrition to ward off further infectious disease.

Behind these grim statistics there is cause for hope. Mortality rates for tuberculosis malaria, and HIV have experienced dramatic decline in recent decades (World Health Organization, 2015) due to advances in the efficacy and availability of medicine and expanded health interventions. This does not mean that the battle has been won; prevalence rates remain stubbornly high and emerging drug resistance poses a serious challenge, but it does indicate that progress can be made.

Coffee farming is an occupation that comes with certain hazards related to, for example, the use of agrochemicals, or accidents from the use of farm equipment. Again, these are seen more on large farms with bigger labor pools and more employees and the certification movement has helped encode best practices against such risks.

Pesticides, for example, pose a health threat for farmers. The issue has been the focus of attention by a wide range of organizations over the past two decades including governments, nonprofits, aid agencies, coffee associations at origin, certification programs and more that have worked tirelessly to make improvements in this area. Endosulfan is an example of an extremely toxic insecticide that was historically used in the coffee industry but is now being phased out. Over the past 5 years many countries have banned its use. Much progress has been made although Endosulfan is not yet entirely eliminated. It can still be found in use in some coffee origins. A recent article by the Sustainable Coffee Partnership detailed farmer testimony regarding the social and environmental hazards of endosulfan and also the challenges of safe disposal and compliance with pesticide storage (Pesticide Action Network UK and 4C Association, 2014).

As with poverty, greater health issues such as malaria or tuberculosis coexist with growing communities because of the regions coffee grows in, not because coffee causes them. There are insufficient existing industry measures that protect smallholder farmers against endemic tropical and other diseases, but certification schemes have put in place standards to give workers—particularly on large farms with a hired labor pool—access to basic health care. Over the past few decades, certification bodies and coffee programs have contributed to providing a safer environment for coffee growers and coffee workers. Efforts include providing frameworks for safe pesticide use, health care, accident prevention codes, and good hygiene, to name a few. Above and beyond this, there are specific organizations like Grounds for Health that deliver better solutions to specific issues.

The nonprofit, Grounds for Health, developed from the observation that cervical cancer rates were particularly high amongst women in small Central American coffee-growing communities, where they had no access to health care.

The organization used extension workers from coffee growing cooperatives to spread the message to women in their communities, recruiting them for simple screening and effective field treatment of precancerous cells where detected. Over the past 16 years, 60,000 women have been screened and 4000 have been treated for a positive screening result (Grounds for Health, 2016). Born in the coffee industry, Grounds For Health today is the only nongovernmental organization (NGO) devoted exclusively to health issues in coffee communities. The screening and treatment are recognized by the World Health Organization as a "best buy," meaning as a cost-effective preventative service.

The organization leverages both the reach and the trust of coffee cooperatives' extension workers and network to recruit women from the communities to be screened. While there is no connection between coffee and cervical cancer, there is a definite correlation between rural areas where coffee is grown and cervical cancer rates. Typically, women in these areas do not have access to adequate screening and treatment services that could save their lives.

2.4 Providing Education

The sustainable development goal for education calls for all children, boys and girls, to complete primary and secondary education by 2030.

Education and poverty levels are highly correlated. Lower income countries, including many coffee origins, have lower education enrollment rates than wealthier countries. Within coffee origins, a similar relationship tends to exist between lower income rural communities and wealthier urban communities; secondary school enrollment rates are typically lower in rural than in urban areas.

Primary school education and basic literacy have been a policy priority for developing countries, including coffee producing countries, for decades and this is reflected in very high levels of primary school attendance. World Bank data (World Bank, 2015) suggests primary enrollment rates in most coffee-producing countries of 90% or higher. Supporting childhood education has long been a focus of efforts in the coffee industry, including minimum standards embedded in certification programs regarding access to school and prohibition of school age children working during school hours, initiatives by coffee companies to build schools, to name just two examples. Due to illness, financial shocks, or other challenges, it is not uncommon for students to be delayed in their completion of primary education by several years and to graduate as teenagers. Although there are often associated costs with primary school education (such as school fees, uniforms, books, etc.) these are typically modest and within the means of impoverished smallholder farmers.

Secondary school education for coffee growing communities presents a very different story. These costs are subsidized by the government to a much lesser degree and the costs very quickly can become prohibitive for smallholder farmers. As a result, the data show a precipitous drop-off for rural children attending secondary school. Logistical arrangements can also become far more challenging. Less students mean less schools in a given area that often demands large distances to be covered to attend secondary school. This can incur ancillary costs of transport and boarding, and/or imposition on friends and family. As children grow older, both boys and girls represent important economic earning potential for families. This can be in the form of manual labor on the farm, off-farm earning opportunities, or taking over responsibilities from other adult family members to free up adult earning potential. Coffee smallholder families are keenly aware of the advantages of education for their children, but if financially stretched, may not be able to afford the long-term investment.

Enrollment rates do not tell the whole story. Western observers frequently treat education as a generic good, as evidenced by the many well-intentioned school building initiatives that occur in coffee communities. In one instance for example, a well-meaning roaster donor built a school on a coffee plantation in Papua New Guinea, to educate workers' children. In a uniquely New Guinean dilemma, the workers came from a tribe hostile to the indigenous one and were only there because they had no land of their own. Therefore, working parents were unwilling to leave their children at risk of tribal retribution in the schoolhouse, preferring to take them into the field. School attendance was duly sporadic.

Moreover, education can be measured not just in quantity but also in quality. Coffee farmers are highly attuned to differences in educational quality, and the implications for future success of their children. Coffee farming households with adequate financial resources will often look beyond local secondary school offerings and ask themselves not "where will my child go to school" but rather "where can my child receive the best possible education" even if that may represent a school several days travel away from home.

One homegrown coffee industry NGO, Coffee Kids, began in 1988 and has a mission to "collaborate with the next generation of coffee farmers to realize their full potential as part of the global coffee community." Coffee Kids provides training, mentoring, and financial support to young coffee farmers so that they can implement creative solutions to the challenges they face in the changing coffee industry (www.coffeekids.org).

2.5 Adult Education

Adult education for coffee farmers has been implemented by a wide range of organizations for at least half a century including local government agencies or coffee organizations, bilateral and multilateral institutions, philanthropic foundations, nonprofit organizations, and more recently participants within the coffee supply chain (including coffee roasters and retailers.) Historically the resources available have been modest relative to the scale of the need and as a result, most of the world's coffee farmers could benefit from additional technical and business skill training (see Chapter 7).

2.6 Empowering Women

The fifth goal calls for gender equality and the empowerment of all women and girls. This includes an aspiration to end discrimination and violence against women and girls everywhere, ensure equal access to education, and equal rights to economic resources.

Women do much of the work on smallholder coffee farms, including key cultivation and processing activities that affect coffee yields and quality. Typically, however, women have less rights to coffee income, fewer land rights, and land tenure, which flows through to reduced access to other economic resources such as credit, training, or inputs. Social biases favoring men are deeply rooted in cultures and traditions throughout the coffee growing world.

Numerous initiatives have been undertaken in the coffee sector to advance gender equity. The Coffee Quality Institute (CQI) launched an initiative in 2014 entitled the Partnership for Gender Equity (PGE) which draws attention to the coffee industry specific status quo (Coffee Quality Institute's Partnership for Gender Equity, 2015). Further, in a published report of the initiative's first findings, it maps a path forward for the industry. Another NGO, the International Womens Coffee Alliance was founded in 2003 with the goal of empowering women in the international coffee community to achieve meaningful and sustainable lives and to encourage and recognize their participation in all aspects of the industry. All certifications include the call for gender equity, whereas the CQI PGE's next stage aims to draft an industry-wide policy. In short, there is widespread recognition that women need their fundamental human right of equality to be respected.

The coffee arena presents an interesting scenario insofar as many coffee growing regions have hosted wars and conflicts that took men away from their traditional farming roles, leaving women in their stead. With peace and the return of the menfolk, traditional roles were re-evaluated. One such example is Nicaragua, where the civil war between the Sandinistas and Contras continued throughout the 1980s.

The SOPPEXCCA Gender Story

Fatima Ishmael, the leader of the SOPPEXCCA cooperative (Union de Cooperativas Agropeguarias) of Jinotega, Nicaragua has shepherded it through a transformational journey of quality enhancement, which ultimately benefitted an entire community. Remarkably, in a way that allowed women their fair share of the pie. By focusing on coffee quality, Fatima's vision contributed to life's enrichment for a group of women formerly cut-off from direct trade opportunities and, hence, control over their destinies and livelihoods. This transformation has meant that they no longer live with insecurity. Fatima herself began her career as an

The SOPPEXCCA Gender Story—cont'd

agronomist, a job choice opened up to women because of the absence of men, who were off fighting wars.

The story began when Fatima noticed that coffees delivered by female cooperative members tended to taste better than average. She supposed that as natural caregivers, women were simply more attuned to the needs of the earth. Their instinct to nurture extended to the trees under their care.

This led to a more systematic separation and marketing of the women's coffee. With time, the cooperative put in place actions to level the playing field for women and allow them equal access to coffee funding (both credit to buy more land and prefinance for coffee cultivation). First though, women were given assistance in registering land title, a prerequisite to being a full co-op member. Technical trainings and field workshops were formed on subjects from basic agricultural best practices through to hedging: the financial instrument that protects farmers against commodities exchange fluctuations. A café was built in the cooperative headquarters to teach youth barista skills. The coffee was branded under the name of "Las Hermanas," the sisters. What is good for the women, the cooperative realized, will be good for their families and, hence, their communities.

Today, the framework that supported these women continues to strengthen. As coffees are delivered during the harvest season, farmers are taught the basics of defect cupping and good processing practices. They learn to distinguish an excellent preparation, a commercial preparation, and a low quality coffee both visually and in the cup. They taste their own coffees at least three times during harvest: at the start, peak, and end of the season, getting feedback each time from the cooperative. In this way, they understand how cultivation practices affect their own coffee. And what they can do to improve quality and ultimately earn better prices (Fig. 5.2).

FIGURE 5.2 Coffee deliveries checked and tracked by lot.

Continued

The SOPPEXCCA Gender Story—cont'd

By taking control into their hands and caring for quality, they earn a better price for their coffee, improve the quality of their lives, send their children to school, build homes for themselves, gain access to better health care, and greater influence over financial decision making.

SOPPEXCCA cooperative had always wanted to emphasize children's projects, since they see their progeny as the hope for the future. The youth and children's groups were integrated in the extension training services of the cooperative, particularly in the fields of environmentalism and gender equality education. Moreover, each member farmer now signs an agreement on joining the cooperative, undertaking to school their children. The cooperative is intent on eliminating the belief among farmers that just because they *are* farmers, they should lack education. An educated farmer, it says, understands better quality and thus education is what guarantees sustainability. Supported by the NGO, Coffee Kids, the cooperative built three community schools which are now being administered by the Ministry of Education.

The gender relations and humanitarian principles of SOPPEXCCA incorporate progressive measures. Domestic violence results in mandatory expulsion and no one practicing discrimination is allowed to join. The cooperative prioritizes women and their need to progress. It codifies the need to prioritize women in a Gender Policy which states that cooperative members strive to establish equality between female and male members. The policy was designed to promote justice, respect, equality, solidarity, integration, and participation to improve the lives of all members and their families. Cooperative members are exhorted to act as examples in their communities, maintaining good relations with their families and neighbors with special emphasis on honesty, responsibility, and decency in their everyday lives.

Membership in the cooperative empowered these women to move beyond subsistence and into self-sustenance: from impoverished landless laborers to smallholders free of crushing debt. They now own their own land, have access to cooperative financing, agronomical training, and social programs.

In mid-2015 SOPPEXCCA received a formal acknowledgment from HIVOS, a Women's Certification. As the certification verifies internal processes, it took 3 years to obtain. The wording: "Gender equality is the key for successful cooperative development."

2.7 Ensuring Availability of Water and Sanitation

The sixth sustainable development goal calls for universal and equitable access to safe and affordable drinking water for all, and adequate and equitable sanitation and hygiene for all.

Water scarcity and sanitation issues represents one of the most pressing challenges facing humanity. Over a billion people, one-fifth of the world's population live in areas of physical water scarcity (United Nations - Water, 2007; United Nations - General Assembly, 2015). The latter report states that 2.5 billion people do not have access to "safe" water (for human consumption). Women and girls in Africa and Asia walk on average 6 kilometres to fetch water (United Nations - World Water Development Report, 2016). Over 2 billion people lack sufficient water for proper sanitation (UNICEF and World Health Organization, 2015). Nearly 1000 children die each day due to preventable water and sanitation-related diarrheal diseases. By 2050 more than one in four people are predicted to live in a country affected by chronic or recurring shortages of fresh water.

Agriculture is vital to our sustenance but heavily tied to the water challenges we face. Agriculture is responsible for 70% of the freshwater consumed by human activity (Food Agriculture Organization, 2012). When well-managed, agriculture can protect the watersheds on which communities and ecosystems depend; maintain healthy soils and ecosystems, protect aquifers, and guard against deforestation and erosion. When agriculture is managed poorly, it can damage or destroy these same watersheds and biospheres beyond repair. Aquifers can be depleted in a "tragedy of the commons," watersheds stripped of vegetation, and streams contaminated with agrochemicals or other pollutants. In addition to the environmental challenges this creates, poor water stewardship also creates social challenges including water scarcity and sanitation issues, particularly for downstream communities.

Fortunately, relative to many forms of agriculture, coffee production in general has a light environmental footprint (Van Rikvoort, 2014). Most coffee farms are rain-fed; although irrigation is used by approximately 5–10% of the world's coffee farms, particularly in Brazil and Vietnam (TechnoServe unpublished analysis from client work, industry interviews and national coffee data sources across 20 coffee origins comprising over 90% of global production, 2016). In many coffee origins, coffee production has played a role protecting the biosphere and aquifers by giving an economic reason for rainforest to be protected rather than cut down. Coffee is a tree crop, often grown in combination with additional rainforest tree canopy, and the deep root systems serve to protect the soil from erosion compared to agricultural land-use such as cattle grazing.

However there are challenges. In Vietnam, over-irrigation in the coffee industry is widespread contributing to water shortages particularly during droughts. Mono-cropping requires far greater stewardship if watersheds and biospheres are to be protected. Where coffee is processed using the "fully washed" method, process water is not always managed effectively, resulting in pollution of adjacent water-streams.

Treaties, national laws, and certification and verification programs have an important role to play by codifying best practices including conserving natural ecosystems, protecting watersheds and wetlands from erosion and pollution, and developing water conservation programs. These measures and others are making a difference.

Wet mill water pollution is a good example where multifaceted initiatives are increasingly addressing the problem in a range of coffee origins. A combination of new technologies to reduce water use and prevent pollution, new laws and regulations, certification standards and growing awareness at all levels of the coffee supply chain from farmer to roaster are slowly transforming the way coffee is processed around the world. In many ways, the coffee industry has been the pioneer in this regard, creating templates for water preservation and protection that have been replicated in other industries.

Innovative solutions exist. In Costa Rica, some municipalities are paying land-owners for water and other environmental services. In Ethiopia, Mother Parkers Coffee and Tea is championing an innovative program called "Water Wise" using vetiver grass to filter water from wet mills to prevent pollution to adjacent rivers. One trial at a Jardín mill in Colombia has demonstrated the potential for centralized eco-milling to provide considerable social benefits including reduced water consumption, improved processing water treatment, and time savings (Nespresso, 2013).

A specific example can be seen in traditional washing mills, which use much water to process coffee from the receiving and sorting, to the pulping, fermenting, washing, grading, and transport of the raw product. Newer equipment—so-called "eco-pulpers"—can operate efficiently on as little as 1/10th of the water usage offering potential for drought-prone regions like Brazil's Cerrado, that nevertheless wish to wash their coffee to maximize quality potential (see Section 5.3 in Chapter 3 for greater detail on water efficient milling).

Despite the progress that has been made, we cannot be complacent. Population pressures are likely to add additional strain to scarce water resources. Climate change is likely to bring new water-related challenges including increased weather pattern volatility, greater challenges protecting soil moisture, preventing erosion, and protecting the integrity of water resources.

The Yemen Water Story

In one country particularly beset by water shortages, we find a powerful example of a coffee community's activities positively impacting the issue of supply.

Yemen is a very dry place. Here coffee grows, one tree to a terrace, each owned by a different farmer, mulched not by organic matter, but by rocks to retain precious moisture. The trees are planted on stony hillside terraces where the surrounding rocks prevent water outflows and evaporation (Figs. 5.3 and 5.4).

The Yemen Water Story—cont'd

FIGURE 5.3 An oasis of coffee gardens.

FIGURE 5.4 One farmer, one tree.

Arguably, the biggest natural threat facing Yemen today is water. The country has extremely limited natural freshwater, it suffers from desertification and, controversially, a problem with qat cultivation. This mildly narcotic plant provides an alternative cash crop to farmers, but draws heavily on Yemen's already scant water resources (anecdotally, it uses 70% of all water supply).

Continued

The Yemen Water Story—cont'd

In the district of Haraaz, the heart of Yemen's coffee-growing mountains, an extraordinary project exists, led by the Al-Ezzi Industries group and founded by the Dawoodi Bohra community. In a country where allegedly only 2.2% of land is considered arable, this project is a light in the darkness. The settlement sits at 2366 m above sea level and is surrounded on all sides by defunct coffee terraces—formerly productive plots that Al-Ezzi is reviving. From this base, the Dawoodi Bohra community is engaged in infrastructure improvement: building immaculate smooth stretches of road. In medical services: their doctors built a clinic with minor operation facilities and a maternity delivery unit to minimize child mortality. In education: they established a small school directly next to the coffee nursery, to educate boys and girls alike. Most significantly of all, however: in coffee revival and water conservation (Fig. 5.5).

FIGURE 5.5 Traditional coffee terraces.

Al Ezzi purchase farmers' coffee, additionally giving vouchers for every 10 kilos of dried coffee cherry they deliver. Farmers can redeem these vouchers against medical services, school fees, or agricultural inputs such as new coffee seedlings. Of critical value to them however, farmers can redeem vouchers for a barrel of free irrigation water.

In response to the danger that irresponsible agricultural irrigation will deplete groundwater reserves, as well as to separate potable resources from irrigation water, the Dawoodis built a water plant with a 250-m deep well that pumps *drinking water*. This they distribute daily to all surrounding villagers. For *irrigation* purposes, on the other hand, they built a completely modern and fully automatic sewerage treatment plant. The cleansed water goes out to irrigate grass and is distributed to farmers for coffee.

The community's extraordinary insights into water issues and what coffee could contribute to redressing the problems led them to action. Seeing water as the

The Yemen Water Story—cont'd

single largest catastrophe facing Yemen, they estimate that the water table level on the Yemen plateau drops 3 m every year and if it continues thus, within 30 years people will be killing each other for a glass of water. The pull of the big city spells the migration of labor to towns and the ancient coffee terraces are not maintained as they should be. This is significant because, a unique benefit to coffee growing with the Yemeni terracing system is that it "percolates" the water, thus retaining it. For more in-depth reading on the water crisis in Yemen, much exists on the Internet, but the documents in reference provide more insight (Glass, 2010; Al-Asbahi, 2005; Handley 2001; Heffez 2013). The community maintains that today many farmers use groundwater to irrigate qat, whereas the Dawoodi community forbids the use of groundwater for farming: only rain water can be used. In general (whether on terraces or in households), farmers no longer harvest rain water sufficiently. Whereas coffee farmed the traditional way, in carefully protected stone terraces, uses the age-old rain harvest methodology, and thus protects the future of both crop and water.

Wherever the truth may lie, this much is certain: water there is at risk, and encouraging coffee growth on traditional stone terraces is the best mitigating strategy available.

2.8 Decent Work for All

The eighth sustainable development goal focuses on a wide ranging set of issues related to sustainable economic development, full and productive employment, and decent work for all. This includes an aspiration to eradicate forced labor and end child labor in all its forms by 2025.

Many actors including coffee roasters and retailers, the International Labor Organization, governments, civil activists, nonprofits, and certification programs have worked to develop and implement minimum standards including Freedom of Association, readable contracts, limited working hours, as well as efforts to ban or minimize forced labor and child labor. However, there is still more work to be done in this area.

Minimum wage compliance is also an issue in the coffee industry, particularly for harvest labor. Harvest workers are typically paid based on productivity (pounds of coffee picked rather than hours worked). Although labor laws differ from country to country, a common interpretation is that workers can be paid using a productivity measure as long as actual wages achieve the minimum daily wage rate. A variety of reports suggest that the nationally mandated minimum wage is not being universally paid for coffee workers (United States Department of Labor, 2013; SCAA, 2016).

A great deal of effort has been made by certification bodies, nonprofits, governments, coffee companies, and others to provide a framework for codifying standards for worker rights and child labor. These include codification of practices around forced labor, child labor, freedom of association, readable contracts, limitations of working hours, and limitations on hazardous work. The coffee sector has for several decades been at the forefront of developing sustainable supply chain standards and has provided a template for other industries seeking a sustainable sourcing solution. ISEAL Alliance (the global membership association for sustainability standards) members including Fairtrade, Rainforest Alliance/sustainable agriculture network (SAN), and UTZ (a sustainable farming certification) are now working together to go beyond a "minimum wage" baseline to a "living wage" standard.

2.9 Aging of Farmers

Many coffee-producing origins are enjoying sustained economic growth above global averages (World Bank, 2015). Each year over the past 5 years, Colombia has outpaced the United States in economic growth, and Ethiopia has grown faster than China. This growth is often accompanied by a rising minimum wage, strengthening currency, and expanding work opportunities outside traditional fields.

These are encouraging developments worth celebrating, however, they pose new challenges for traditional export industries such as coffee by increasing the cost of production relative to competing coffee origins and increasing the challenge of finding labor market interest in coffee farming or laboring. The Sustainable Trade Initiative's report on Colombia analyzed that farming a 2 ha coffee farm was considerably less remunerative than a variety of other jobs including truck driving, work in a textile factory, or even the wages as an informal agricultural worker (TechnoServe and IDH, 2014).

These trends, combined with increasing education levels lead to frequently cited concerns regarding the aging of coffee farmers. The same IDH (Dutch sustainable trade initiative actively involved in the coffee sector) study reported an average age of a Colombian coffee farmer as 56 years. These figures reflect genuine concerns regarding the long-term viability of a thriving coffee sector in growing economies where actual costs and opportunity costs are rising. However the Farm Credit Council in the United States offers an interesting perspective on this issue. In the United States the median age of a farmer is even older than 58 years and has also been steadily climbing over at least the past 40 years (Duncanson & Hays n.d; Johr, 2012). The authors show that this increase is to a large degree driven by the overall aging of the United States workforce over the same time period, rather than any specific disaffection for farming. In other words, one factor contributing to the observed aging of coffee farmers is the general demographic aging occurring in coffee-producing countries as fertility rates decline and life expectancy rises. In Colombia the median age of the population has nearly doubled over the past half century (United Nations - General Assembly, 2015). This does not diminish the very real challenges faced by an industry that cannot easily respond by either mechanizing harvesting (due to topographic and quality constraints) or raising wages to attract labor from other more attractive opportunities due to a globally competitive market place. It does suggest that more analysis may be warranted to understand the true dynamics that are occurring.

3. OUTLOOK

Coffee is a progressive industry in the world of consumerism. If you can "get coffee right," it can certainly do good. As our understanding of sustainability has grown, one learning is that there is tremendous social diversity within the coffee industry and individual needs have to be considered based on geography, national context, and farm size to name a few. The advent of certification and other standards has drawn attention to social issues in the coffee industry that has heightened consumer awareness and creates a framework for improvement over time. As a daily consumable, coffee also weighs on the palates of humankind. And where there are social issues that have been associated with a consumable, it becomes incumbent upon the consumer to demonstrate concern. At the same time, as a tropical cash crop that can co-exist with subsistence crops—one that is loved in the wealthier consuming world—coffee is unique in its ability to offer the poor a means of improving their lot. Juxtapose these two aspects—the daily reminder that the beverage represents to its drinker and its ability to generate cash for the otherwise disenfranchised—and we can have hope that coffee offers a solution. As a commodity that exists with this tension, coffee is a beacon of hope.

We will leave the last word with Nelson Mandela;

There is no passion to be found playing small, in settling for a life that is less than the one you are capable of living.

It always seems impossible, until it is done.

After climbing a great hill, one only finds that there are many more hills to climb.

REFERENCES

Al-Asbahi, Q.Y.A.M., 2005. Water resources information in Yemen. IWG-Env, International Work Session on Water Statistics. Available from: http://unstats.un.org/unsd/environment/envpdf/pap_wasess3a3yemen.pdf.

Bacon, C., 2005. Confronting the coffee crisis: can fair trade, organic and specialty coffees reduce small-scale farmer vulnerability in Northern Nicaraguan. World Development 33 (3), 497–511.

Bacon, C.M., 2008. Chapter 7 "Confronting the coffee crisis: can fair trade, organic, and specialty coffees reduce the vulnerability of small-scale farmers in northern Nicaragua?". In: Bacon, C.M., Mendez, V.E., Gliessman, S.R., Goodman, D., Fox, J. (Eds.), Confronting the Coffee Crisis: Fair Trade, Sustainable Livelihoods and Ecosystems in Mexico and Central America. MIT Press.

Caswell, M., Méndez, V.E., Bacon, C.M., 2012. Food Security and Smallholder Coffee Production: Current Issues and Future Directions. University of Vermont.

Centers for Disease Control and Prevention, 2015. About Malaria. Available from: http://www.cdc.gov/malaria/about/biology/mosquitoes/.

Coffee Quality Institute's Partnership for Gender Equity, 2015. The Way Forward: Accelerating Gender Equity in Coffee Value Chains. Available from: http://www.coffeeinstitute.org/our-work/partnership-for-gender-equity/.

Douek, D.C., Roederer, M., Koup, R.A., 2009. Emerging concepts in the immunopathogenesis of AIDS. Annual Review of Medicine 60, 471.

Duncanson, B., Hays, J., n.d. Different ways to look at the ageing of U.S. Farmers. Farm Credit Council, pp. 1–6. Available from: http://www.fccouncil.com/files/Different%20Ways%20to%20Look%20at%20the%20Aging%20of%20U%20S%20%20Farmers.pdf.

Faye, M.L., McArthur, J.W., Sachs, J.D., Snow, T., 2004. The challenges facing landlocked developing countries. Journal of Human Development 5 (1), 31–68.

Food Agricultural Organization of the United Nations, 2012. Coping with water scarcity: an action framework for agriculture and food security. FAO Water Reports 38, ix. Available from: http://www.fao.org/docrep/016/i3015e/i3015e.pdf.

Glass, N., 2010. The water crisis in Yemen: causes, consequences and solutions. Global Majority E-Journal 1 (1), 17–30.

Grounds for Health, 2016. Impact – Key Performance Indicators. Available from: http://www.groundsforhealth.org/impact/.

Handley, C.D., 2001. Water Stress: Some Symptoms and Causes. A Case Study of Ta'iz, Yemen. Ashgate.

Heffez, A., 2013. How Yemen Chewed Itself Dry, Farming Qat, Wasting Water, Foreign Affairs Magazine (July).

Institute for Health Metrics and Evaluation, 2016. IHME, University of Washington, Seattle, WA. Available from: http://www.healthdata.org/research-article/estimates-global-regional-and-national-incidence-prevalence-and-mortality-hiv-1980.

International Coffee Council, 2014. World Coffee Trade (1963 – 201): A Review of the Markets, Challenges and Opportunities Facing the Sector. International Coffee Organization, pp. 4–9. Available from: http://www.ico.org/news/icc-111-5-r1e-world-coffee-outlook.pdf.

Jöhr, H., 2012. Where are the future farmers to grow our food. International Food and Agribusiness Management Review 15, 9–11.

Lewin, B., Giovannucci, D., Varangis, P., 2004. Coffee Markets: New Paradigms in Global Supply and Demand. World Bank – Agriculture and Rural Development Department. World Bank Agriculture and Rural Development Discussion Paper 3.

Masters, W.A., McMillan, M.S., 2001. Climate and scale in economic growth. Journal of Economic Growth 6 (3), 167–186.

Méndez, V.E., Bacon, C.M., Olson, M., Petchers, S., Herrador, D., Carranza, C., Trujillo, L., Guadarrama-Zugasti, C., Cordon, A., Mendoza, A., 2010. Effects of fair trade and organic certifications on small-scale coffee farmer households in Central America and Mexico. Renewable Agriculture and Food Systems 25 (3), 236–251.

Nespresso, 2013. Helping Farmers to Share Their Workload in Jardín, Colombia. Factsheet 2.2. Available from: http://www.nestle-nespresso.com/asset-library/documents/nespresso%20-%20aaa%20program%20-%20helping%20farmers%20share%20their%20workload%20jardin%20colombia%20-%20factsheet.pdf.

Pesticide Action Network (PAN) UK and 4C Association, 2014. Growing Coffee without Endosulfan: Key Findings, Lessons and Recommended Next Steps, pp. 1–4. Available from: http://www.4c-coffeeassociation.org/assets/files/Documents/Reports-Brochures/Key_lessons_from_Growing_Coffee_without_Endosulfan_project_and_recommended_next_steps_01.pdf.

Sachs, J.D., 2001. Tropical Underdevelopment (No. W8119). National Bureau of Economic Research.

SCAA, 2016. Understanding the Situation of Workers in Corporate and Family Coffee Farms. http://www.scaa.org/chronicle/wp-content/uploads/2016/07/Coffee-Workers-report-ING-V81.pdf.

Siraj, A.S., Santos-Vega, M., Bouma, M.J., Yadeta, D., Carrascal, D.R., Pascual, M., 2014. Altitudinal changes in malaria incidence in highlands of Ethiopia and Colombia. Science 343 (6175), 1154–1158.

TechnoServe and IDH, 2013. Uganda: A Business Case for Sustainable Coffee Production, p. 6. Available from: https://issuu.com/idhsustainabletradeinitiative/docs/131206_uganda/1?e=10013468/6785051.

TechnoServe and IDH, 2014. Colombia: A Business Case for Sustainable Coffee Production, p. 7. Available from: https://issuu.com/idhsustainabletradeinitiative/docs/131206_uganda/1?e=10013468/6785051.

Tulchinsky, T.H., 2010. Micronutrient deficiency conditions: global health issues. Public Health Reviews 32 (1), 243. Download via. http://www.publichealthreviews.eu/upload/pdf_files/7/13_Micronutrient.pdf.

UNAIDS, 2016. factsheet. Available from: http://www.healthdata.org/research-article/estimates-global-regional-and-national-incidence-prevalence-and-mortality-hiv-1980.

UNICEF and World Health Organization, 2015. Progress on Sanitation and Drinking Water – 2015 Update and MDG Assessment. UNICEF and World Health Organization, p. 13. Available from: http://apps.who.int/iris/bitstream/10665/177752/1/9789241509145_eng.pdf?ua=1.

United Nations - Water, 2007. Coping with Water Scarcity. UN Water, p. 4. Available from: http://www.unwater.org/downloads/escarcity.pdf.

United Nations - World Water Development Report, 2016. Water and Jobs. Available from: http://unesdoc.unesco.org/images/0024/002439/243938e.pdf. Earlier reports are available from: http://www.unwater.org/publications/world-water-development-report/en/.

United Nations, Department of Economic and Social Affairs, Population Division, 2015. World Population Prospects: The 2015 Revision, New York, 2015. Available from: https://esa.un.org/unpd/wpp/Download/Standard/Population/.

United Nations - General Assembly, 2015. Transforming Our World: The 2030 Agenda for Sustainable Development, A/Res/70/1. Available from: https://sustainabledevelopment.un.org/content/documents/21252030%20Agenda%20for%20Sustainable%20Development%20web.pdf.

United States Department of Agriculture, Office of Communications, 2002. Agriculture Fact Book 2001-2002, Chapter 3. Available from: http://www.usda.gov/factbook/chapter3.pdf.

United States Department of Labor, 2013. U.S. Labor Department Sues Yauco Coffee Grower for Minimum Wage Violations. U.S. Department of Labor, Wage and Hour Division. Release Number 13-1870-NEW/BOS 2013-170. Available from: http://www.dol.gov/whd/media/press/whdpressVB3.asp?pressdoc=Northeast/20131021.xml.

van Rikxoort, H., Schroth, G., Läderach, P., Rodríguez-Sánchez, B., 2014. Carbon footprints and carbon stocks reveal climate-friendly coffee production. Agronomy for Sustainable Development 34 (4), 887–897.

Weiss, R.A., 1993. How does HIV cause AIDS? Science 260 (5112), 1273–1279.

World Bank, 2015. World Development Indicators. The World Bank. Available from: http://data.worldbank.org/data-catalog/world-development-indicators.

World Health Organization, 2015. Global Tuberculosis Report, twentieth ed. Available from: http://www.who.int/tb/publications/global_report/gtbr2015_executive_summary.pdf?ua=1.

Chapter 6

Economic Sustainability—Price, Cost, and Value

Jérôme Perez[1], Bernard Kilian[2], Lawrence Pratt[2], Juan Carlos Ardila[3], Harriet Lamb[4], Lee Byers[4], Dean Sanders[5]

[1]*Nestlé Nespresso SA, Lausanne, Switzerland;* [2]*INCAE Business School, Alajuela, Costa Rica;* [3]*Cafexport SA, Vevey, Switzerland;* [4]*Fairtrade International (FLO), Bonn, Germany;* [5]*GoodBrand, London, United Kingdom*

1. INTRODUCTION

Oscar Wilde's definition of a cynic as someone who knows the price of everything and the value of nothing presents useful insight when considering the topic of economics and sustainability in coffee. In fact, it would suggest that the entire coffee sector is characterized by cynicism and managed by cynics. Moreover it is certainly the case that there are widely differing perspectives on the best solutions to address some of the economic issues at the heart of the industry.

For decades, all of the major coffee stakeholder groups [producers, intermediaries and traders, governments, roasters, and even nongovernmental organizations (NGOs)] have been obsessed with price as the main determinant of economic sustainability. This is understandable, given the commoditized nature of the raw material; however, this thinking has not necessarily generated improvements in the quality of life of smallholder farmers. Neither has it secured an economically vibrant future for the coffee sector, in which quality, productivity, climate change, farmer profitability, and other factors threaten long-term supply.

Some industry experts suggest that the focus on price is counterproductive as it does not take into account other critical factors like coffee quality, yields, and productivity. It also fails to account for farmers' production costs, such as labor, fertilizers, disease management, processing costs, and the resulting impact on farmers' profitability. Neither does it take into account the need to demonstrate clear value propositions for consumers. Finally, it disregards the influences of government intervention and entrepreneurialism.

The Craft and Science of Coffee. http://dx.doi.org/10.1016/B978-0-12-803520-7.00006-2

This chapter aims to cover these diverse perspectives and consider the potential solutions for creating a more sustainable coffee value chain by asking experts with a unique perspective to share their views and experiences. We start with the son of a Colombian coffee farmer, who now represents groups of cooperatives exporting coffee to Europe. He outlines his views on the economic realities most coffee farmers face and the kind of support they need to grow coffee as a sustainable enterprise. We then move on to senior managers at the Fairtrade Labeling Organisation, who share perspectives on the vexing issue of price. The team at the Center for Intelligence on Sustainable Markets at INCAE Business School, Costa Rica, then discuss the microeconomics of coffee and how to optimize costs to drive net profits at farm level. We also hear from a consultant in corporate social innovation who presents additional perspectives on the importance of and opportunities for greater value creation. Then a Head of Sustainability at a leading coffee roaster shares thoughts and reflections on the economic outlook for the industry and what is needed to ensure a dynamic and attractive sector for the next generation.

By combining these different perspectives, we aim to shed some light on the very challenging economic situation along the coffee value chain. We also aim to offer readers a cross-section of views on coffee economics without identifying any one philosophical, political, or economic solution as the most promising. And although there are many initiatives already in place to help farmers create more sustainable coffee production, there is no consensus on the best way to achieve this. It is our intention that this chapter will stimulate thinking, while provoking and inspiring people and organizations along the coffee value chain to make their own unique contributions to ensuring the resilience and strength of the industry.

2. ENVISIONING COFFEE GROWING AS A SUSTAINABLE ENTERPRISE: PERSPECTIVES OF A FARMER'S SON

Juan Carlos Ardila

"Coffee: the further from the tree, the better."

-Popular saying in the coffee lands-

After working a lifetime in the coffee plantation, my father had two conflicting dreams: that I would take over the family business and follow his passion for growing coffee, and at the same time, that I would leave the farm to get the best formal education possible. Parents across the countryside struggle with this dilemma of wanting to pass on the family farm, yet hoping for a better future for their children.

I fulfilled my father's second goal and took off for a new life far from the coffee-carpeted mountains. What I left behind was the vivid cultural heritage of coffee growers who share fundamental family values, a deep sense of duty, strong spiritual beliefs, and a tenacious industriousness to work the land.

In Colombia, coffee culture is more than just a business; it unites the society, drives the impulse toward peace and progress, and employs a large portion of the population. Among Colombians growing coffee is a cause for pride and hope, a means to resist resignation in the face of hardship and conflict, and an opportunity to work together as shown by the strong coffee institutions created to unite against adversity.

Paradoxically, coffee cultivation is one way to constrain poverty, but certainly not the path to prosperity; even so, coffee farmers embraced their livelihood with such fervor and vibrancy that it shaped the character of the nation.

When coffee hit its lowest price in a century, the drama it unleashed in my homeland awoke something deep in me. After years away, I returned to the coffee business and realized that there lay my future. Like my father, I recognized a fundamental viability problem, and as he did, I am determined to help future farmers overcome it.

Small farmers have spent entire seasons producing coffee beans at prices far below the break-even cost. They have neglected to account for their land's value or their family's labor, and failed to factor-in the necessary environmental conservation efforts or infrastructure upgrades.

Calculating the cost of producing a kilogram of coffee constitutes an academic challenge, and often an ideological debate. This is because it is linked to the realities of rural economics where the laws of finance operate differently, and where there is no firm bottom-line, since so many people live in different levels of poverty.

For a long time, pricing was a race to the bottom, with an industry that claimed it had no ability to distance itself from market forces. The fact that buyers consistently attribute prices to simple market forces reinforces farmers' resignation to accept whatever price is offered.

When earnings barely cover production expenses (on top of the cost of living), small-scale farmers absorb the gap by sacrificing their family's living standards and disinvesting on the farm. This only perpetuates the vicious cycle of low productivity, declining profitability, and future instability. As a result, smallholders have almost no control over the profitability of their own business, and by extension, their own quality of life.

Coffee prices are always highly volatile due to fluctuations in currency exchange rates and the futures market price. In addition, the cost of inputs can also vary, further destabilizing coffee production expenses. The cost of fertilizers, for example, oscillates along with oil prices. Additionally, production volumes are severely affected by two weather phenomena: El Niño and La Niña; so the smallholder's income stream consequently swings erratically from profit to loss.

This instability makes producers highly vulnerable whenever revenues fall below production costs. In a myopic search for higher sales prices—a variable that they cannot influence—farmers tend to neglect in-the-field factors that they could directly control, like increased farm efficiency, decreased production costs, and improved tree productivity and quality. These are also factors that equally affect coffee profitability.

As long as coffee yields and quality remain low, farmers will stay poor, regardless of the price. One solution is to stop focusing on top-line revenue driven solely by price and move to assess bottom-line profit.

Currently, farmers focus on one financial tool: cash flow, while ignoring the profit and loss statement and the balance sheet. To improve their economic conditions, farmers need to learn more about basic financial accounting and small business skills, which is no simple task for thousands of illiterate farmers in developing countries. They also need to consider developing more producer cooperatives, which can help provide this training and education, while also negotiating prices for the group and offering greater economies of scale.

If growing coffee is not economically viable, both current and potential coffee growers will see no appeal in such a turbulent and risky business. Growing coffee is not lucrative, does not afford social security, and is hard work. The average age of an African coffee farmer is 60 (ICO, 2015), with the average Colombian producer at 56 years old (SCP, 2014b). With no generational succession in sight, who will be left to grow future crops? Many youth are drawn to urban professions that seem less risky and—like I did—abandon their rural roots, breaking the family farming tradition. The extreme volatility in prices and climate, the lack of succession planning, plus rural violence due to internal conflicts, paint a dark panorama for the future of coffee in some countries of origin.

At the same time, the industry further down the value chain is looking for solutions to these growing problems. One potential solution is certification schemes. Such schemes incorporate both "voluntary standards" and technical solutions, such as implementing Good Agricultural Practices, to help ensure farmers produce high quality, sustainable coffee that will help them obtain a higher price for their beans. This has the added benefit of helping the buyers ensure the long-term stability of their coffee supply (Opitz, 1995).

During the past decade, coffee roasters' mission-driven procurement strategies have emerged based on multistakeholder cooperation (including trading organizations, certification organizations, governmental, and technical assistance organizations) and approaches to real shared-value creation. These strategies are based on the understanding that producing sustainable coffee requires not only compliant but stable and prosperous farmers.

Such programs rely strongly on technical assistance teams to transfer know-how; however, these teams should also go beyond agronomy and teach farmers how to run their farms as small businesses. They also need to encourage communal work to achieve efficiency through economies of scale and promote long-term thinking to increase resilience to off-farm challenges.

Sustainability programs need to pragmatically focus on reducing labor costs through improved technology, productivity, and efficiency, whereas applying a deeply rooted entrepreneurial approach. They should also help farmers better understand and manage the external risks they might face, and facilitate their adaptation to climate change through intelligent precision farming.[1]

Once producers feel confident that their farms are financially stable, the educated youth of the next generation will see coffee farming as a lucrative venture. They will then become the new "*agripreneurs*" who run the agricultural operation as a small business, transforming coffee farmers into viable and sustainable drivers of rural prosperity.

Small farmers will certainly need a lot of support to achieve sustainability, and more ambitiously, to achieve prosperity. However, this is not optional. We need to accelerate our efforts to ensure coffee can continue providing pleasure to people for centuries, that farming families can continue their traditions while providing for their children, and that coffee farmers worldwide can cultivate coffee to create "a second opportunity on earth" (Garcia Marquez, 1967).

3. PERSPECTIVES ON PRICE: FAIRTRADE AND BEYOND

Harriet Lamb and Lee Byers

Farmers are often born speculators. They wait to sell, hoping that the price will go up and often end up selling when the price has dropped. As outlined in the previous section on the perspectives of the son of a coffee farmer, for many coffee farmers, the coffee price almost feels like an act of God over which the farmers have no control. They live in hope that it will rise; in fear that it will plummet.

From 1962 until 1989 the International Coffee Organization (ICO) used a quota system to regulate supply and demand. This achieved a stable price, typically around US$1.20—40 per pound for most of the 1980s. However, the quota system led to overproduction, the United States pulled out and in July 1989 it collapsed. Key coffee-producing nations could no longer afford to smooth prices or manage coffee stocks. Thereafter, the industry followed the global rush to liberalization. In most countries, stocks were released quickly and prices collapsed to 40—50 cents, well below what it cost farmers to grow and harvest coffee (further details can be found in Chapter 9).

At the height of the coffee crisis in the mid-1990s (Osorio, 2004), in all coffee-producing countries many farmers abandoned their jobs and lost their livelihoods. Poverty increased while export earnings went down. Shade trees were cut down for timber and illicit crop production increased.

1. Precision farming or site-specific crop management is a farming management concept based on observing, measuring, and responding to inter- and intrafield variability in crops.

Prices have never really recovered, staying low until 2005, then rising and falling again to new lows as recently as September 2013, when the composite price reached US$1 lb, once again way below the costs of production, for nearly all arabica farmers. In fact, market prices for raw coffee have not kept up with increases in the cost of living and they have not kept pace with the high cost of inputs such as fertilizer or labor. All these factors in turn have reduced farmers' ability to invest for the future.

According to the ICO composite price indicator, prices in real terms (adjusted for inflation) for coffee during the 1990s and 2000s were still lower than the prices paid during the 1980s before the collapse of the coffee agreement system. This is why the issue of pricing is absolutely central to any debate about the future of coffee—for companies, producers, and whole countries.

3.1 Living With Price Volatility

When market prices fall, producers bear the brunt of the impact as they are the "price takers"—who usually need to accept whatever the market will offer, even when market prices do not cover their production costs.

In addition to volatile and unpredictable price fluctuations, coffee cooperatives may also need to deal with the challenge of obtaining prefinancing, while managing other risks such as weather and disease. Prefinancing enables cooperatives to obtain enough money to buy large volumes of coffee from their members, which they can then turn around and sell to their buyers. This typically means that the need to agree on a sale and a price before the crop is actually ready, which brings in new risks if the crop cannot be delivered due to problems with yields, bad weather, diseases, or other factors.

There are a wide range of market-based price risk management instruments[2] but these hedging tools are only accessible to those with enough money to pay for them. For most producers and exporters, the only feasible risk management instrument is to agree on contracts ahead of harvest. With forward contracts, they know what price they are going to receive, and they can use them as the main collateral for financing their working capital needs. These open price contracts can be fixed against the coffee futures market price of New York (or any other futures market for coffee) when they purchase the coffee, reducing the risk (see also Chapter 9).

When farmers' cooperatives get this wrong, it can prove damaging. For example, if they sign contracts with buyers at a set price and then the market prices rise, their members are tempted to sell to the passing middlemen coming to their door with ready money. The result is that the cooperatives may default on their contracts, unable to get enough coffee from their members. Such defaults are rare, but nevertheless a real issue for those concerned.

2. Traded on organized futures and options exchanges or the over-the-counter market, incorporated into the pricing formulas of physical trade transactions, or encapsulated in financing deals.

3.2 Speculation and Differentials

In recent years, the coffee trade has sometimes hedged its risks through derivatives to create more price stability. At the same time, derivatives have opened up new markets for speculative investments, which have fueled market volatility. Although coffee companies want price stability, speculators can only make money if prices fluctuate.

Speculative investments in coffee have increased steeply since 1989. In the 1990s the average daily contracts were 8500; by 2015, this had increased to more than 32,000. In New York alone, the contracts that changed hands on the futures market amounted to the equivalent of 15 times the total world production of coffee (Intercontinental Exchange, 2015).

This already volatile pricing panorama in coffee has generally become even more unstable with new financial instruments and trading strategies such as derivatives for hedging risks, Index Trader speculators, and "flash traders" with automatized trading programs. There is therefore growing concern that short-term speculation is influencing international coffee prices. With the physical trade now detached from speculation, this is creating even more unpredictability for producers.

The second component of a forward contract is the differential, the difference between the New York price and a particular origin and quality of coffee. Differentials are not published as a price reference. Instead they contain a blend of quality, origin, supply, and demand in the domestic market. The sum of the differential is simply a negotiation. Most of the time, this is a buyer's market with very few organizations that buy in large quantities dominating trade.

Differentials are usually less volatile than the international reference price—but not always. In 2008–09 the differentials for Colombia UGQ (Usual Good Quality) increased from +10 to +100 in just a few weeks, dragging behind it the differentials of other origins. Imagine what this means for an export manager conducting business as usual, selling a substantial percentage of expected sales volume forward, to secure financing and manage risk. The farmers in the cooperative involved would be up in arms, believing that they could have made much more money.

3.3 The Significance of Fairtrade

Fairtrade can be seen as an enormous pilot project, testing how the whole supply chain can seek to manage price volatility more effectively. Fairtrade Labeling (as part of the wider Fair Trade movement) arose from the human and economic crisis of the coffee price collapse in 1985.

Mexican coffee farmers teamed up with Fairtrade pioneers Max Havelaar, in the Netherlands. In close cooperation with coffee producers and traders, the first Fairtrade Minimum Price for arabica coffee was set (initially with a 10%

premium, which became a five cent Fairtrade premium in 1995) based on the previous ICO reference price at 126 USD cents per pound (Zehner, 2002).

The idea is that the Minimum Price acts as a safety net to protect coffee growers from falling prices. The premium is a separate sum to invest in their future. If market prices go above the minimum, then the traders must pay this market price, always with the premium on top.

The liberalization of the coffee market gave producer organizations the opportunity to access the world market directly and they had no experience in this. So the new Fairtrade market also provided a useful incubator in which producer organizations could benefit from some price protection and access to finance as they built their organizations and businesses and learned about international trade.

It gave them the stability needed to plan and invest in the long term, learn organizational skills, and develop the expertise that comes with managing a cooperative.

The Fairtrade Minimum Price and Premium have evolved over time and are based on the so-called Free on Board price or FOB, the price of delivering goods to the nearest port for export. There are many categories of pricing based on the type of bean. For example, prices for organic washed and unwashed arabica are different. However, prices are based on the costs of production; following a wide consultation process involving the whole Fairtrade coffee supply chain, with production data from around the world.

As of 2016, Fairtrade certified cooperatives could count on at least the Minimum Price of US$1.40 per pound for washed arabica coffee sold on Fairtrade terms (30 cents more if organic). They also receive an extra 20 cents per pound Fairtrade Premium to invest as they wish, with 5 cents dedicated to productivity and quality investments.

In 2014, cooperatives invested almost half the 49 million euro premiums earned in their infrastructure, facilities, and processes to improve productivity and quality. They invested an additional 43% in direct services for members such as quality training (Fairtrade, 2015).

As Fatima Ismael, General Manager at Nicaraguan Fairtrade cooperative Soppexcca says, “In a price free-falling system Fairtrade generates a base price that really helps us. It gives us stability for our families.” (Personal interview with Fairtrade).

Premium funds are also used to cover core social needs. For example, the Gumutindo cooperative on Mount Elgon, Uganda, has contributed to build a school and a clinic, protect water sources, and provide working capital to reduce the need for expensive loans (Frank and Penrose-Buckley, 2012).

Research studies also show Fairtrade is making a difference. The University of Göttingen, for example, found Fairtrade created a 30% increase in disposable income among coffee farmers in Uganda, as a result of higher prices and because the farmers then delivered higher quality beans (Chiputwa et al., 2015).

Strong, well-managed cooperatives can help empower farmers to take their own decisions, engage with the market, and gain individual incentives for quality. Fairtrade is one route for them to achieve this and there are others: working toward optimizing costs for the farmers, creating more value, and ensuring there is as much income security as possible.

Increasingly sustainable solutions will likely be a blend of public, private, and nongovernmental organization investment and interventions, to deliver impact at scale, with farmers at the heart of such initiatives.

4. COSTS AND INCOME: RESULTS OF AN INVESTIGATION BY THE CENTER FOR INTELLIGENCE ON MARKETS AND SUSTAINABILITY AT INCAE BUSINESS SCHOOL

Lawrence Pratt and Bernard Kilian

For several decades, the role of science in coffee farming has increased steadily. Farmers, and more recently research institutes, have sought to improve the favorable aspects of genetic quality (productivity, quality, and disease resistance) for nearly 200 years, through selection and cross-breeding (for more information see Chapters 1 and 2).

As modern plant genetics evolved, agricultural research in coffee developed new varieties and attributes that are favorable to farmers. In parallel, the "Green Revolution's" advances in chemical use proved beneficial to many farmers in increasing their yields and protecting against some pests. Interestingly, despite these scientific advances, the application of scientific principles in on-farm agronomic and business practices could be further developed.

CIMS research in the late 2000s showed quite convincingly that there was a particularly worrisome gap in farmer understanding of which agronomic practices relate to key business variables—particularly their own income. A better understanding of revenue and cost drivers appeared to be a critically important variable limiting farmers' ability to be more successful in their businesses.

As mentioned previously in this chapter, farmers have generally focused on coffee price as the key variable that determines their profitability. This tendency appears to derive from several possible perceptions: (1) farmers tend to believe they are at a production optimum and cannot do anything else to increase income, (2) they understand they are not at optimum production but look to price spikes to finance their investment in production techniques and inputs when conditions permit, and (3) farmers are not aware of their own production costs and yields.

Most discussion about coffee farmer's well-being has also tended to focus on price, based on the obvious observation that farmers are generally better off when prices are high. Although this overly simplistic view has dominated the study of coffee economics, both international coffee prices and the costs of production have always been volatile as described previously in this chapter.

There are two possible strategies to generate more economic stability for farmers: (1) ensure that the price farmers receive is above their production costs or (2) help farmers lower their production costs through better economic management of their costs and productivity, along with related investments.

Strategy 1 is complex because in any given coffee region, the most efficient coffee farmer may have costs that are less than 20% of the least efficient one (Kilian et al., 2006). At the same time, coffee buyers do not set prices based on how efficient individual farms are, and they are not going to pay more than the going market price for coffee unless their competitors do the same. In the next sections we will focus on Strategy 2 and measures to lower production costs for farmers.

4.1 Coffee Farm Microeconomics: Knowledge and Behavior

Investigating the Real Farmer Income can help to define key drivers behind farm income. A general question that one can ask is what factors are the main contributors to farm income? To answer this question, one needs to consider how coffee farmers address the business side of their farming. For example, it is important to understand how they manage their accounting. Without a system (even something as simple as a notebook), farmers will not be able to identify their production costs over time, and therefore will not understand their profitability.

Although large differences will exist between continents and countries, considering the low education levels of most farmers one can suspect that many farmers rely only on their memory for their bookkeeping. Booklets are also commonly used, which of course, if used well, can be a good tool. Computers are rare. The above implies that many farmers may not manage their production costs and their knowledge about on-farm economics and managerial issues is usually very limited. If farmers do not know their production costs they may be focusing on the management variables that will not help improve their economic situations, such as the price of the coffee.

4.2 Drivers of Real Farmer Income

Productivity: As productivity increases, the quantity of coffee available to be sold increases which in turn may lower costs per unit. This in turn may lead to higher net profits for the farmer. Although the price may, in certain cases, influence the income for the farmer, other factors such as productivity, cost of production, and quality may be even more important factors determining the net income on the farm.

Quality: Many of the variables that determine quality of a specific farmer's crop are predetermined based on the altitude, soils, rainfall patterns, and other variables outside the farmer's immediate control. Others, such as varieties of coffee and planting strategies, are possible to change with sufficient time and investment. But there are many other variables, such as pruning, shade management, harvest strategy and timing, and particularly postharvest management that are very important determinants of coffee quality and can have very large effects on the quality of the coffee and the price premium it receives.

Small Farm versus Large Farm Efficiency and Profitability: Larger farms can reasonably be assumed to be more likely to have direct access to support on technical issues and accounting. Small farmers, however, may be just as profitable as large farms on a profits per hectare basis, the prerequisite is that they are well managed. For companies working to advance farmer income and long-term profitability the focus should thus be on small-holder farmers who benefit more from additional technical and management support.

Technical support and certifications: Technical support given through support programs and certifications is likely to help farmers increase quality, yield, and sometimes price premiums. A direct comparison between farms is, however, difficult to make as many factors need to be considered; the number of years that technical support has been given (most impact is observed in the first years), farm size, and the type of support given. Hence, when looking at the differences between farms that are certified or part of a technical support program and those that are not, it is important to consider the results in the local context. Generally, farms that have support from a sustainability program, certification, and training are often better off than those who do not (Garcia, 2014; TechnoServe, 2010).

This cluster-specific economic landscape also demonstrates the importance of real farmer income strategies that are tailored to the conditions in each cluster, driven by area-specific data and a good understanding of local market dynamics. There is no "one size fits all" strategy that works for everyone.

4.3 Orienting Farm Income Stability Through Farmer—Buyer Relationship Mechanisms

It is strategically important for coffee buying companies to understand how and where they could best orient their farmer assistance programs (or certification schemes) to most effectively help advance farmer profitability. Although it is clear that increasing productivity is always a key desired variable, the end result on farmer income depends on other variables that vary based on market conditions, farmers' preferences, and over time. For example, a great debate in the coffee industry has moved beyond the need for a minimum price to agreeing on an appropriate price level (usually through price premia) that should be paid in order to help farmers increase their income.

Generally, to increase farmers livelihood by focusing on productivity and quality will likely yield more benefit over a longer time horizon. Some guiding principles to increase farmer income are listed below:

- Farmers need to be seen as, and helped to evolve into, entrepreneurs who can manage their farms more effectively and successfully for the long term. This will mean looking at a broader range of variables than those typically looked at by capital-constrained farmers. It is quite likely that access to different types of capital (particular medium to long term), and understanding of all the different levers that could make coffee a successful business could help evolve many farmers toward a more entrepreneurial approach to business.

- Every coffee growing region has its own economic and social history, competitive dynamic, challenges, and opportunities. "One size fits all" strategies are unlikely to lead to optimum results for farmers or coffee buying companies. Entrepreneurial strategies, business sector strategies, and national development strategies will all need to be articulated and coordinated according to the needs of very specific zones.
- Maximizing farm income requires a more complete perspective of farm income drivers, and the complex interaction among productivity, quality, farm management (ranging from short-term practices to long-term investments), and price volatility. Broadening understanding offers the possibility of more creative mechanisms to assure quality coffee availability for the long term through better designed and targeted programs. This also implies that the social and environmental aspects as discussed in Chapters 4 and 5 need to be incorporated.

5. THE VALUE PROPOSITION: REFLECTIONS ON THE NATURE OF VALUE IN COFFEE

Dean Sanders

In this section, we consider how value is attributed to coffee. We will not just consider the product and the people who consume it, but will also imagine potential solutions to the systemic challenges the global coffee industry faces.

A consumer wishing to purchase a coffee in a more advanced coffee consuming country may simply walk into a coffee shop and order the beverage of their choice, often as part of a routine. For a good quality cup, they may expect to pay around $5 per cup, or perhaps up to $10.

No matter what they buy and what price they pay, behavioral economics teaches us that when making a purchase decision, consumers must feel that they are receiving a higher value from their purchase than the money or time they invest in the transaction. So how does this increase in value occur, and what makes this interaction seem so appealing to consumers? The answer lies in the multiple benefits added as the coffee makes its way from farmer to consumer—along the so-called value chain.

5.1 Understanding the Value Chain

This value chain consists of all the people and organizations that help carry a product from its inception to completion. In the case of coffee this includes producers (farmers), suppliers, processors, and servers.

Importantly, the total amount of value created is ultimately decided when the end consumer makes a purchase decision based on their perceived sense of value and their own level of satisfaction from the transaction(s).

This discussion of value is and always will be contextualized by the two factors explored elsewhere in this chapter—cost and price. This is true at the farmer end of the value chain as well as the consumer end when walking into a coffee shop. When we treat coffee purely as a commodity, it creates a fixation on price that greatly limits value—if all coffee is the same, then why would anyone pay more? It is only the consumer value proposition and the "premiumization" of coffee that will allow us to move beyond this pure commodity conception to achieve profitability for producers and others along the value chain.

When we educate consumers and show them the true value of coffee, and they become willing to pay more for coffee products and experiences, we create a new system with greater total value.

Brands constantly strive to create new value propositions by creating new product experiences and telling stories and providing information about the people who grow, produce, and consume their products—in this case, coffee.

Professor Jonathan Gutman's "Means End Chain" model (Gutman, 1982) is extremely useful for understanding why consumers make certain choices in daily transactions. This theory helps explain why consumers' priorities have shifted from basic desires for stimulation and flavor into more highly evolved concepts such as self-image and social responsibility.

Gutman argues that consumers make choices that produce desired consequences and minimize undesired consequences. In doing so, they look to spend their money on goods and services that will bring them satisfaction. However, as time passes, the factors that contribute to their happiness become more sophisticated.

5.2 Understanding Added Value

To assess the value of coffee, we need to start from the ground up, with the most basic properties of the coffee bean—stimulation and flavor. We can then consider the importance of origin and terroir, and look at how innovative brands and producers use value-based thinking to persuade consumers to pay higher prices for more differentiated, better quality, "higher value" coffee. Social responsibility and environmental concerns, as well as coffee's potential to create value in the form of self-expression, are also important variables that offer the possibility of even greater value creation.

Psychoactivity: Coffee's most fundamental value attribute is its properties as a stimulant. For some consumers, the thought process may begin and end here. Indeed, the hugely important rise of the coffee houses of the 17th century was propelled by the power of caffeine.

However, this stimulating effect of coffee, although initially providing a new and innovative beverage, had become somewhat commoditized. Early traders undoubtedly discovered that their buyers wanted caffeine delivery, so other attributes (such as unique flavors and aromas) were less important. Anyone traveling to the United States before the rise of the specialty coffee sector will recall cheap coffee—human jet fuel to kick start the working day. In contrast, decaf has become a value proposition that actually removes what was once considered the key property of coffee—supporting Gutman's hypothesis that consumers' desires can vary significantly based on context.

Flavor: Coffee's distinctive flavors are today an essential part of its consumer value proposition. Specialty coffee drinkers talk about coffee flavors in the same language they use for wine or craft beer. The final "in-cup" flavor is a product of factors that collectively contributed at every stage of the value chain. This creates an intriguing value proposition that allows consumers to choose the coffee that best suits their palate.

For coffee drinkers who do not like the taste of pure coffee, retailers and baristas have created new coffee beverages. These range from the iced and sugared Vietnamese *cà phê sữa đá* to the hugely popular Starbucks Pumpkin Spiced Latte, which began its successful market life as a limited time beverage associated with autumn.

Origin and *Terroir*: Specialty coffee brands have worked hard to educate consumers about flavor, and in doing so have naturally linked flavor to *terroir* and "origin centric value." Like wine, coffee takes on the characteristics of the places and people who create it. Therefore, information on the coffee origin provides consumers with intuitive knowledge about many aspects of its value, such as flavor, aroma, and rarity. Certain origins are already associated with significant premiums—for example, Jamaica Blue Mountain may cost an impressive $58 for 16 oz (an example from www.coffeeam.com in spring 2016).

Environmental Value: Coffee cultivation has been historically linked to deforestation as much coffee land was formerly tropical forest that was partly or largely deforested to plant the coffee. However, as a tree-crop native to the tropics, coffee can coexist well with virgin tropical forest. Biologists agree that shade-grown coffee farms are systems with "high biodiversity (Perfecto et al., 1996), but consumers need to be educated and encouraged to ascribe value to this. In fact, in El Salvador, the remaining coffee farms represent nearly all of the standing forest in the country following massive deforestation and decades of civil war" (FUSADES, 1996; Quezada Díaz, 2012). Coffee has been associated with ecosystem degradation, particularly soil erosion and loss of shade trees (and their related biological diversity) in efforts to maximize coffee yields under input-intensive growing systems (see also Chapter 4).

Many brands have sought to leverage responses to these concerns by marketing coffee as bird friendly, certified organic, certified sustainable, and carbon neutral. Typically environmentally certified coffees command a consumer premium of $1.1 per kg compared to those without certification when there is sufficient demand for those coffees and certifications (Carlson, 2010).

Portioned and Convenience: The rapid expansion of portioned coffee permits nonexperts to create high quality and consistent espresso coffee in their own homes. Portioned coffee has created excellent opportunities for companies who are able to extract more value from the same volume of raw material input. It has also empowered consumers to involve themselves in the process of creation, while choosing from a wider range of flavors and origins. Globally, the value share of this segment of the total coffee retail market is almost twice its share of volume.

Location and Cultural Context: Consumers have more choice than ever in choosing the location of their consumption. Coffee shops are socially loaded institutions—think of the differences between an American diner and a traditional French *café*.

The choice of serving method and setting empowers consumers to create their own identities, as well as to work in a social space where their cup of coffee acts as their letter of invitation.

Across the globe, including in the developing world, coffee shops are once again the centers of conversation, creativity, and self-expression they were in Europe 300 years ago. A cup of coffee, personally selected, expertly prepared, and responsibly sourced, is a powerful tool for self-image for many consumers.

Mass Luxury and the Era of the Aspirational Consumer: Economists and marketing experts refer to specialty coffee as "mass luxury." Essentially, this means that coffee can deliver both high value and high volume sales. This is because the absolute differential between high end and mainstream in the sector is marginal—unlike, for example, the automobile sector. In specialty coffee, the extra fraction of a dollar per cup to have a more luxury coffee experience is far more attainable for average consumers in wealthy country markets.

An article in the Harvard Business Review that shows the evidence of America's "middle market consumers … trading up to higher levels of quality" (Silverstein and Fiske, 2003) uses coffee to illustrate the concept. Customers will pay a 40% price markup or more, because paying a premium makes implicit statements about the consumer's value priorities. In this context, marketing has helped stretch the differential and create added value concepts that command higher consumer premiums.

A 2014 global consumer segmentation conducted by BBMG (a brand and innovation consultance company) has defined a significant group of consumers, known as "aspirationals." These two billion individuals "unite style, social status and sustainability values to redefine consumption". Aspirational consumers can be described as "representing a shift in sustainable consumption from obligation to desire," and argues that this consumer segment will "define what it means to find value in coffee in the coming decades." In the case of coffee, it is clear that such "aspirational" consumers seek not only highest quality coffee experiences but also sustainability in the value chain that brings an additional dimension of meaning to the consumption.

A Case Study in Third Wave Consumer Education

Intelligentsia, as just one example, understands how to show consumers the value of its coffees through a series of layered value propositions. Like other third-wave brands, they create unique flavor profiles linked to direct farmer relationships to help drive the viability of coffee farming as a stimulating and rewarding livelihood.

On their Website, a potential customer can access information about the farmers, biological information about the plant itself, photographs of the stunning *terroir*, and a detailed description of the flavor.

Intelligentsia's success is proof positive that when potential buyers are shown the substantive and interesting facts about their coffee, its place of origin, the people that grow it, and then how these factors lead to a complex finished flavor profile, their understanding of the product changes. Now seeing past price, they see value—and are happy to pay a premium.

5.3 Meaningful Consumption and Cultural Performance

If good marketing can predict or pre-empt consumers' needs and desires, then correctly marketed coffee has the potential to create a truly satisfied consumer. When they see a direct connection to the people responsible for creating their coffee, consumption increasingly becomes a social act. This fulfills value propositions such as "impressing others" while also reassuring consumers that they are caring and conscientious people, as well as likely greater self-satisfaction with a given consumption choice.

We then arrive at what marketers define as "meaningful consumption." For example, many people believe that brands could do more to support good causes by collaborating with them, and in addition make it easier for them to make a positive difference. Moreover, consumers worldwide believe that business needs to place equal weight on society's interests and business interests.

The result of increasingly targeted segmentation of customers is that it encourages coffee consumption as a form of cultural performance, as explored in an article by anthropologist William Roseberry (Roseberry, 1996), who draws comparisons with the complex issue of class in the United States.

5.4 Future Value Creation

As the coffee industry looks for solutions to the economic problems that continue to threaten it, the same tools that have helped to de-commoditize coffee in the past may prove invaluable. The difference this time is that farmers, buyers, and others earlier in the supply chain will need to drive more innovation to make themselves key parts of the value proposition.

Coffee's future must involve a commitment to de-commoditize and to revalorize the product and to treat it as the physical heart of a wider consumer experience that links the entire value chain positively and constructively—from farm to cup. We must ask ourselves how the coffee production process can increase value at the consumer level, through the creation of new forms of relevant and meaningful consumer value higher upstream.

This will include closer links between discerning and aspirational consumers and entrepreneurial farmers. It will also mean not only educating producers on how to differentiate their product and profit from terroir and farm management practices but also educating consumers on how to appreciate the outcomes of this agricultural effort.

If coffee's value exists in the mind of consumers, it is their perception that will shape the future of its global trade. The value innovations we have explored all prove the effectiveness of creative vision supported by an understanding of the consumer experience. The challenge, as ever, will be to ensure the appropriate allocation of this added value along the entirety of the coffee value chain. But without added value innovation in consumer propositions, and in the current volatile market price-driven context, we can be certain that there will not be sufficient value created in consuming markets to ensure the economic dynamism required in the coffee-producing regions.

6. OUTLOOK: CREATING SHARED VALUE FOR THE NEXT GENERATION

Jerome Perez

The previous sections of this chapter set out the central themes that need to be considered in addressing the economic challenges and issues inherent in coffee supply chains. These intrinsic challenges are a function of wider systemic shortcomings, such as the adverse impacts of price and currency volatility; the tension between the forces that drive toward commoditization, those that support de-commoditization, the looming threats related to climate change, and future production patterns. With so much uncertainty, how does the future look and how can we begin to design a blueprint for action and intervention that can mitigate the worst effects and start to provide more structural support for an industry that is essential to the economic development of communities, regions, and nations? One thing is for certain—there is no panacea to such complex and interconnected problems.

Effective intervention requires three levels of action. First, at farm level stakeholders need to work together to support farmers to continuously improve quality, productivity, and efficiency, requiring transfer of agronomic and business skills through effective training programs. Second, groups of farmers or "clusters" must work collaboratively to address inefficiencies or inadequacies such as landscape resilience and improved collective processing of the crop as well as increased collective capacity in purchasing, marketing, and management. Finally, all stakeholders and actors in the industry must step up to some of the bigger challenges, adopting more precompetitive and cross-sectorial approaches to innovate where old ways of doing things are not working and innovation is required to pilot and scale new projects.

In terms of economic theory, this three-tier intervention brings together the ideas of thinkers who, in different ways, sought to optimize the capitalist model, where they saw it failing. John Maynard Keynes and Joseph Schumpeter were both born in 1883. The two men, rivals who always held "a distaste for each other's work (McCraw, 2009)," were united in their rejection of classical economic theory. Both men saw that "the diffusion of money induces a radical modification into the way in which an economy works" (Bertocco, 2006). Both men would dedicate their lives to a common aim of improving upon the prevalent model of classical economics, both insisting that they understood the necessary steps to build a better, fairer, and more sustainable economy.

At the heart of Keynes's philosophy was his conviction that institutional or state intervention will always be necessary to manage the inherent and potentially harmful cycles of a capitalist economy. Keynes focused his attention on the short-term regulation of capitalist contracycles, famously saying "the long run is a misleading guide to current affairs. In the long run we are all dead." On the other side, Schumpeter was preoccupied by the fear that short-term state interventions would actually destroy the entrepreneurial dynamic which he saw as the core strength of capitalism. He believed "the fundamental impulse that sets and keeps a capitalist economy in motion comes from … innovation initiated by the entrepreneur." Poon on the Mannkal Economic Education Foundation[4] calls this innovative process of the recreation of obsolete business models "creative destruction." This powerful oxymoron, even if unintended, was prescient as we consider the need to reject outdated models that do not serve the pressing needs to protect the planet and secure the livelihoods of communities at the base of the economic pyramid. It is sobering to consider that, at the beginning of the 21st century, we have not resolved the "intervention" debate on issues as wide ranging as commodity price speculation to global warming. In the case of coffee, we see, as a consequence, the vulnerability of millions of coffee producers and the threat to the stable future supply of high quality raw materials despite growing and healthy demand for these high quality materials in finished products.

The "Price, Cost, and Value" sections above have given the reader an insight into the complex dynamics of the global coffee trade and the

4. http://www.mannkal.org/downloads/scholars/schumpeter-keynes.pdf.

microeconomics at farm level, as well as some of the potential solutions to improving the situation. Reflecting on the points of view laid out in this chapter, we are confronted with what for many may seem an odd dilemma—to choose between innovative entrepreneurship, market regulation, or some set of price agreements, when in all likelihood these three will all be the necessary tools to help us resolve the issues at stake.

6.1 Examining the Key Challenges

Coffee is "one of the most important commodities traded globally…playing a crucial role in the livelihoods of millions of rural households across the developing world." But it took the "coffee crisis" (the collapse of commodity coffee prices from \$1.50 per pound to \$0.49 between 1989 and 1992) to "bring the economic situation of coffee producers to the forefront of media and policy discussions" (UN Conference on Trade & Development and International Institute for Sustainable Development, 2015). There is a range of challenges that lead to the current instability and insecurity that characterizes the industry.

Supply and Demand: At the heart of the problem is a structural imbalance between supply and demand, one that is so significant, it threatens the coffee supply chain from bottom to top. Demand for coffee will increase by nearly 25% over the next 5 years (ICO, 2016), with a disproportionate percentage of that being exceptional quality coffees, such as those currently desired by most specialty roasters. Coffee is exploding in popularity in emerging economies such as China, Russia, South Korea, and Turkey (Euromonitor, 2016). These nations are rapidly acquiring the same taste for beans as their more experienced and coffee-savvy counterparts. It is arabica coffee, grown in isolated, low yield farms, that places the greatest strain on demand, and often permits the smallest margins for local smallholder farmers. Meanwhile, supply is stagnating. 2.3% less coffee was produced in the 2014–15 harvests than the year before (both robusta and arabica beans). Market analyst group Allegra estimates that this situation could lead to a shortage of high quality coffee in as little as 5 years (Allegra, 2015).

Although it may seem obvious that constrained supply of high quality arabica coffee, required by the specialty sector, coupled with increasing demand should lead to increasing prices, this has not been the case historically. Robusta varieties of coffee, considered of lower quality (and much lower price) in general compared to arabica coffee represent over 40% of all coffee grown, with significant annual growth (notably in Vietnam and Indonesia in recent years). For discerning coffee drinkers, robusta is not a good substitute for arabica. However, for the many coffee drinkers who drinks robusta containing commodity coffee blends, there is potential for significant substitutability. When arabica prices rise substantially compared to robusta prices, some mainstream roasters may begin to substitute more of the cheaper robustas for the now more expensive arabicas. This lowers the cost to the roaster, and is not that noticeable to drinkers of those brands of coffee. When many actors serving the mass market segment with blended coffee change their recipes to maintain profits (or hold price), the aggregate effect is enormous, reducing arabica demand substantially, and dragging the price down with it (Reuters, 2013).

Robusta production volumes are growing faster than overall coffee consumption, meaning "over supply" and downward pressure on prices of both. So, the incentives for substitution will be ever greater as robusta becomes even more plentiful (and cheap) and arabica becomes scarcer. The substitution will continue then, until massively fewer mainstream coffee drinkers accept the declining quantity of quality beans in their coffee.

Consequently, the only way prices are going to increase for high quality arabica growers is through making their coffee "un-substitutable"—pricing themselves specifically out of the substitution game. This will not be easy and is coupled with some of the ideas set out in the preceding "value" section of this chapter.

Related to the above dynamic price volatility is an additional complicating factor. The New York "C" market price for arabica coffee averaged over the long term represents a price that is above the average cost of production and equivalent to other estimates of the appropriate market price. However, this does not help smallholders who are trying to secure the best prices for their harvest in the context of extreme short-term market volatility. And as noted in the earlier section on cost, the notion of an "average" price can be misleading, as 50% of all coffee producers, by definition, have production costs higher than the median—with many more than double this level. As such there is a clear need for forms of intervention that might amortize high price peaks against the inevitable downward troughs.

Real Farmer Income: On top of this, farmers are earning less and less. For many farmers there is little economic viability in growing coffee, so alternative crops have become more lucrative and widespread where governmental institutions are weak or nonexistent. For example, during the international coffee crisis of 2001–03 and again, albeit to a much lesser extent in the 2012–13, many tens of thousands of Colombian farmers are believed to have abandoned coffee, and many switched to (often illicit) crops with a quicker cash cycle, although still more simply left their fields to wait for future opportunities (Bischler and Parra-Peña, 2015).

Environmental Challenges: The elephant in the room is climate change, which threatens a quarter of coffee output in Brazil (the world's largest coffee producer). The *El Niño/La Niña* weather-related incidents have exacerbated existing uncertainties and encouraged more of the aforementioned speculative market price activity. Another consequence is the higher incidence of disease and pest epidemics such as coffee leaf rust, which wreak havoc with quality and productivity. In an attempt to ensure consistent harvests, many desperate farmers overuse agrochemicals, causing environmental damage and reducing the long-term fertility of soils. With the predicted growth in coffee consumption in the coming years, further deforestation of tropical rainforest to increase the area of coffee cultivation is beginning to appear on the agenda of concerned environmental agencies. As always, environmental concerns exist in a delicate balance with economic ones. Many coffee-producing countries already hold a deep-seated respect for the environment, but more needs to be done to ensure a consistent integration of the value of ecosystem services and natural capital.

Initiatives are already being formulated to engage the private sector in developing possible strategies to reconcile environmental challenges with farmer and community needs, but a more focused take up by coffee enterprises is required. Otherwise, the challenge will be left to farmers. With production costs high, environmental measures that further increase cost will be met with resistance at ground level. Farmers may not be able to afford the costs of compliance associated with restrictions on land use, or water management, or other relevant areas.

As indicated in the Sustainable Coffee Program "New environmental laws ... are likely to make compliance more difficult and costly for individual small farms in the future. Improving the profitability of farming would make it easier for farmers to absorb added sustainability costs and thus improve the business case for them to invest in sustainability" (Sustainable Coffee Program, 2014a).

What both environmental and economic challenges have in common is simple—they make the business of coffee production risky, volatile, and unappealing. Coffee's strange commodity status, the fluctuations of its prices, and the threat of environmental destruction naturally lead to an exodus of potential producers.

6.2 The Hopeful Science: Solutions for a Better Future

Thomas Carlyle branded economics "the dismal science" as early as the 19th century (Carlyle, 1849), and the gloomy moniker has stuck around. Carlyle's contemporaries predicted a future of unchecked population growth, leading to hunger and global misery. But, speaking broadly, along with technology and science, economics deserves credit for improving the lives of millions around the world. In the words of writer Derek Thompson "students of economics should be proud: their 'science' was then (as it can be, today) a force for a more just and, crucially, less dismal world" (Thompson, 2013).

So, if the primary challenge in coffee is declining real prices and volatility, set against increased input costs and the desire to increase farmers' income the question is how, at every step of the supply chain, we can act to make the economics of coffee production more stable, reliable, and sustainable. This means not only understanding the range of risk factors in detail but also creating game-changing solutions to address them.

As mentioned above, this will require a consolidation of microeconomic enterprise innovation, work at local community, and landscape level and greater collaboration in designing new system-level solutions to some of the macroeconomic issues. Sometimes these solutions will be initiated by the private sector, often from international bodies, nongovernmental organizations, and national governments; increasingly they will emerge from multi-sector collaborations and public private partnerships. For maximum effect, the theories and approaches of both Schumpeter and Keynes are inevitably present, forming the basis for new plans of action, at once creatively destroying redundant models of coffee farming while designing and building new ones that are fit for the future. These efforts need, in turn, to be supported by a favorable policy and regulatory context for the industry as a whole, at the national and international levels, in some cases intervening where market dynamics are not working.

Profitable farms are successful farms, and create self-sustaining systems where profits can be reinvested into better production, equipment, safeguards, and other long-term investments. The road to profitability is different for every farm. For some, the solution may be to diversify farm enterprises beyond just coffee. For others, specialization is the answer, with single-origin craft coffees curated for the palates of premium consumers. In this the challenge is in line with Schumpeter's call for innovation and enterprise.

In other cases something more closely resembling Keynesian intervention may be required. Addressing systemic problems such as generational succession planning in coffee farming may require a blend of regulatory engagement and cultural change. As an example, to create opportunities for parent-to-child farm transfers Nespresso, Fairtrade, the Colombian government, and local cooperatives and exporters in the region of Caldas have piloted the Farmer Future Program. This aims to win the long-term commitment of smallholders to retirement planning by arranging for savers to receive up to 20% matched funding from the Colombian government. Early results have been encouraging and there is evidence to suggest that young people are taking a fresh look at the coffee sector as a livelihood. This program increases the likelihood of a successful generational succession by ensuring that the retiring parent does not need much income from the farm, leaving the child the cash flow necessary to develop the coffee business and support him or herself from the farm's income. Other areas of smallholder risk exposure that need equally innovative approaches are insurance instruments for health, accident, and crop risk (Nespresso, 2014) as well as new tools to manage currency and price volatility. This bundle of services may create the right conditions (access to cash flows and lower risk profile) needed for successful generational change.

The key strategies to encourage the next generation of farmers into the industry will be better financial incentives, more productive farms, driving "premiumization," and increasing the attractiveness of the sector in terms of other social and economic value for famers and customers beyond just price.

The spread of technology and the benefits of science may offer a lifeline for struggling farmers, with the promise of higher yields, better agricultural practice, and disease-resistant, higher quality varietals. Enabling farmers to apply more scientific methods to farm management will drive cost optimization and profitability. These will include a commitment to precision agriculture, with more efficient and better-diagnosed farms becoming the new standard. High yield and disease resistant coffee plant hybrids could double or treble productivity on the same area of land in line with the Word Business Council's Vision 2050 Agricultural Pathway (World Business Council for Sustainable Development, 2010).

The industry needs to collaborate around a shared intention to improve the level of skills of coffee farmers, both on topics related to agronomy and how to approach the management of their farms as small professional businesses. Farmers can be helped with improvements such as optimized fertilization, integrated pest management, control of soil erosion, proper tree establishment, and improved labor productivity. However, it is clear that encouraging some of the current generation of older farmers to employ new best practices might be difficult at best. The coffee industry must be presented to young people as a modern industry where they can use business and marketing skills to gain a competitive edge.

6.3 Creating Shared Value for a Shared Future

In his 2011 article in the Harvard Business Review, Professor Michael Porter (Porter and Kramer, 2011) described how Nespresso is seeking to go beyond current frameworks of Corporate Social Responsibility and Sustainability: "A traditional approach of deconstructing the value chain into its respective sections might not be the right way to progress. Instead, at every stage, we should aim for the concept of shared value."

How can macroeconomic theory help us to better understand the nature of shared value creation? Both Keynes and Schumpeter acknowledged that capitalism was not a self-regulating system, but that it contained the potential for self-destruction. The key message of Keynesian economics is that constructive intervention by organizations such as governments creates a freer and fairer system. Past experience has shown that often, libertarian noninterventionist policies have failed to address smallholder poverty, local middlemen exploiting vulnerable producers, and an exodus of farmers from the sector, to say nothing of the failures in accounting for wider externalities such as natural capital depletion. Schumpeter believed that such intervention would ultimately suppress the entrepreneurial genius that drives capitalism and improves living standards. We can imagine his frustration if he were to see the issues affecting the coffee trade today.

However, a complementary integration of their thinking may light the way to a new and better model. We offer the argument that through a context-driven synthesis of the principles of Schumpeter and Keynes, we may find a framework that successfully combines economic efficiency, social justice, individual liberty, and ecological sustainability. Our challenge as an industry is to commit to a collaborative approach to create shared value for the next generation of coffee consumers and producers.

Producers, buyers, millers, roasters, branded coffee companies, civil society, and governments alike should review where the market is failing to create the conditions needed for a healthy sector that is viable in the long term and where constructive intervention is required through policy, practices, and perhaps even market restructuring. But this commitment to an integrated systemic approach should not eliminate the need for innovation and action at farm and community level.

In some cases, companies can innovate to improve their own business model in ways that farmers and others in the value chain can also benefit (see for example programs such as Starbucks CAFÉ Practices and Nespresso AAA), whereas elsewhere precompetitive and cross-industry platforms will need to be established to address landscape level and institutional policy gaps. The successful track record of public—private partnerships and the growth in impact investing may lead to more capital flowing into the coffee industry, partly financing the necessary changes, while delivering economic returns alongside a positive social and environmental impact.

As spelt out before, the consumer has a critical role to play. Unless value is created for him or her, fulfilling existing needs and desires and stimulating new ones, then no meaningful shared value can be created.

In summary, the path to a better set of economic outcomes for the coffee sector and more created shared value for all parties in the value chain is twofold:

First, solutions will need to be designed that work at farm, community, and wider systems level. Second, we need to drive and support innovation and entrepreneurship at farm level and throughout the value chain and be open to macroeconomic interventions when we see the market failing to solve issues such as price speculation and the incorporation of external costs such as climate change into commodity prices. In short for the coffee sector to flourish we should ask ourselves the question "Is it time to reconcile Schumpeter and Keynes and, in so doing, are we able to create shared value for a sustainable future for the coffee industry?"

REFERENCES

Allegra Group, July 2015. The Future of Coffee — Specialty Coffee in the UK. Available at: http://www.worldcoffeeportal.com/LatestReports/The-Future-of-Coffee.aspx.

Bertocco, G., 2006. Are banks special? A note on Tobin's theory of financial intermediaries. Economics and Quantitative Methods. Available from: http://eco.uninsubria.it/dipeco/quaderni/files/QF2006_5.pdf.

Bischler, J., Parra-Peña, R.I., October 10, 2015. How Violence Affects Farmers in Colombia and Beyond. Available at: http://reliefweb.int/report/colombia/how-violence-affects-farmers-colombia-and-beyond.

Carlson, A.P., 2010. Are Consumers Willing to Pay More for Fair Trade Certified™ Coffee? Department of Economics, University of Notre Dame, Paris. Available from: https://economics.nd.edu/assets/31977/carlson_bernoulli.pdf.

Carlyle, T., 1849. Occasional discourse on the Negro question. Fraser's Magazine for Town and Country XL, 670–679.

Chiputwa, B., Spielman, D.J., Qaim, M., 2015. Food standards, certification, and poverty among coffee farmers in Uganda. World Development 66, 400–412.

Coffeeam.com, 2016. Market Information.

Conservation International, 2016. Sustainable Coffee Challenge. Available from: http://www.conservation.org/stories/Pages/Sustainable-Coffee-Challenge.aspx.

Euromonitor, 2016. Coffee in China. Available from: http://www.euromonitor.com/coffee-in-china/report.

Fairtrade, 2015. Monitoring the Scope and Benefits of Fairtrade, seventh ed., pp. 77–78. Available from: http://www.fairtrade.net/fileadmin/user_upload/content/2009/resources/2015-Fairtrade-Monitoring-Scope-Benefits_web.pdf.

Frank, J., Penrose-Buckley, C., 2012. International Institute for Environmental Development with Twin 2012 'How Can Farmer Organisations and Fairtrade Build the Adaptive Capacity of Smallholders?'.

FUSADES and Harvard Institute for International Development, 1996. From Peace to Sustainable Development. FUSADES, El Salvador.

Garcia, C., Ochoa, G., Garcia, J., Mora, J., Castellanos, J., 2014. 'Use of polychoric indexes to measure the impact of seven sustainability programs on coffee growers' livelihood in Colombia. In: Proceedings of the 25th International Conference on Coffee ASIC, Armenia.

Garcia Marquez, G., 1967. One Hundred Years of Solitude. Harper & Row, US.

Gutman, J., 1982. A means-end chain model based on consumer categorization processes. The Journal of Marketing 60–72.

ICO, 2015. Sustainability of the coffee sector in Africa. In: Presented at the ICO Workshop, March 2–5, 2015, London.

ICO, 2016. Coffee Consumption Expected to Jump. Available from: http://www.wsj.com/articles/coffee-consumption-expected-to-jump-1424119985?mod=e2tw.

International Comunicaffe, 2016. CEO of the FNC Advocates for Higher Prices at 4th World Coffee Conference. Available from: http://www.comunicaffe.com/ceo-of-the-fnc-advocates-for-higher-prices-at-4th-world-coffee-conference/.

Kilian, B., Jones, C., Pratt, L., Villalobos, A., 2006. Is sustainable agriculture a viable strategy to improve farm income in Central America? A case study on coffee. Journal of Business Research 59 (3), 322–330.

Intercontinental Exchange, 2015. Market Information. Available from: http://www.theice.com.

McCraw, T.K., 2009. Prophet of Innovation. Harvard University Press.

Nespresso, 2014. AAA Farmer Future Program: Guarding Farmers Against the Risks of Instability in the Coffee Growing Sector Through Innovative Solutions to Farmer Welfare. Available from: http://www.nestle-nespresso.com/asset-library/documents/nespresso%20project%20backgrounder%20-%20aaa%20farmer%20future%20program%20-%202014.pdf.

Opitz, M., 1995. Reducing Risks for Small Coffee Farmers, GTZ GATE: 1995.

Osorio, N., June 2004. Lessons from the World Coffee Crisis: A Serious Problem for Sustainable Development. Submission to UNCTAD XI, Sao Paulo, Brazil. International Coffee Organization (ICO), London.

Parizat, R., Van Hilten, H.J., Tressler, E.G., Wheeler, M., Nsibiwra, R.A., Morahan, R., Modelo Ruiz, J.M., De Smet, J., Pineda, P., David, F., 2015. Risk and Finance in the Coffee Sector: A Compendium of Case Studies Related to Improving Risk Management and Access to Finance in the Coffee Sector. Agriculture Global Practice Discussion Paper No. 2. World Bank Group, Washington, DC. Available from: http://documents.worldbank.org/curated/en/2015/02/24051885/risk-finance-coffee-sector-compendium-case-studies-related-improving-risk-management-access-finance-coffee-sector.

Perfecto, I., Rice, R.A., Greenberg, R., Van der Voort, M.E., 1996. Shade coffee: a disappearing refuge for biodiversity. BioScience 46 (8), 598–608.

Porter, M.E., Kramer, M.R., 2011. Creating shared value. Harvard Business Review 89 (1/2), 62–77.

Quezada Díaz, J.E., Nieto Cárcamo, S.E., 2012. El Salvador Formal R-PP Presentation October 21, 2012, Brazzaville, Republic of Congo, Ministerio de Medio Ambiente y Recursos Naturales El Salvador (MARN). Available at: http://www.forestcarbonpartnership.org/sites/fcp/files/Documents/tagged/El%20Salvador%20FCPF%20PC13%2021%2010%2012%2012%20JEQD%20SN.pdf.

Reuters, McFarlane, S., October 30, 2013. Coffee Drinkers Treated to More Arabica as Prices Sink. Available at: http://www.reuters.com/article/us-coffee-arabica-idUSBRE99T14420131030.

Roseberry, W., 1996. The rise of yuppie coffees and the reimagination of class in the United States. American Anthropologist 98 (4), 762–775.

Silverstein, M.J., Fiske, N., 2003. Luxury for the masses. Harvard Business Review 81 (4), 48–57.

Sustainable Coffee Program (SCP), 2014a. Colombia, a Business Case for Sustainable Coffee Production. An Industry Study by Technoserve for the Sustainable Coffee Program. Powered by IDH. Available from: www.sustainablecoffeeprogram.com/site/getfile.php?id=377.

Sustainable Coffee Program (SCP), 2014b. Colombia a Business Case for Sustainable Coffee Production, IDH Industry Report by Technoserve, p. 9.

TechnoServe, 2010. Brewing Prosperity in East Africa. Coffee Initiative Final Report. Available from: http://www.technoserve.org/files/downloads/Coffee-Initiative-Final-Report.pdf.

Thompson, D., 2013. Why Economics Is Really Called the Dismal Science. The Atlantic. Available from: http://www.theatlantic.com/business/archive/2013/12/why-economics-is-really-called-the-dismal-science/282454/.

UN Conference on Trade & Development and International Institute for Sustainable Development, 2015. Sustainability in the Coffee Sector: Exploring Opportunities for International Cooperation. Available from: https://www.iisd.org/pdf/2003/sci_coffee_background.pdf.

Vision2020, 2016. Market Information. Available from: http://www.vision2020.coffee.

Wallengren, M., 2013. Paper Presented at the 10th AFCA Convention. Uganda, Kampala.

World Business Council for Sustainable Development, 2010. Vision 2050: The New Agenda for Business. Available from: http://www.wbcsd.org/WEB/PROJECTS/BZROLE/VISION2050-FULLREPORT_FINAL.PDF.

Zehner, D.C., 2002. An economic assessment of ‘fair trade’in coffee. Chazen Web Journal of International Business 21.

Chapter 7

Experience and Experimentation: From Survive to Thrive

Paulo Barone[1], Michelle Deugd[2], Chris Wille[3]
[1]*Nestlé Nespresso SA, Lausanne, Switzerland;* [2]*Rainforest Alliance, San José, Costa Rica;* [3]*Sustainable Agriculture Consultant, Portland, OR, United States*

1. INTRODUCTION

Coffee farmers live in a world of constant flux. Global economics, consumer needs and preferences, and agricultural science are all changing. In this increasingly complex context, farmers must juggle both emerging issues such as climate change and perennial questions such as how much fertilizer to apply and how to control soil erosion. Farmers must be agile and adaptable, students, and teachers. Successful farmers are observant, and learn both from their own mistakes and successes and those of their peers. They constantly seek new knowledge.

Since the coffee crisis of the early 1990s, when crop disease and plunging prices broke many farmers, there has been an explosion of information about coffee farming and a revolution in training, technical assistance, and innovation. Coffee cooperatives and associations, national extension services, agricultural research institutions, multilateral aid agencies, and companies all along the value chain—especially, civil society groups [or nongovernmental organizations (NGOs)]—have developed and launched multiple initiatives to train coffee farmers and help them share information among themselves. The growth of certification standards generated a boom of training programs to support farmers on their journey toward true sustainability.

Although training programs abound, a majority of smallholder farmers still rely on traditional sources of information: parents, other relatives, peers, and neighbors. This inherited knowledge, handed down through generations, is

The Craft and Science of Coffee. http://dx.doi.org/10.1016/B978-0-12-803520-7.00007-4

based on accumulated experience and cultural practices. In places where coffee has been grown for decades, sometimes centuries, the accumulated wisdom is rich in detailed practices and techniques to adapt coffee farming to local environmental and cultural conditions. But to succeed, farmers must infuse tradition with modern science. Through human ingenuity and the interrelationship between science and craft, farmers are learning from trainers, and trainers are learning from farmers as they collaborate to adapt to multiplying global challenges.

2. THE ADVENT OF SUSTAINABILITY STANDARDS AND CERTIFICATION CREATED A REVOLUTION IN TECHNICAL ASSISTANCE AND TRAINING

For decades, government extension programs and national coffee institutes provided the only sources of training for coffee farmers, and there were international pacts to control prices. However, even with these efforts, farmers, the ecosystems that supported them, and rural communities continued to suffer. Coffee prices plunged to an all-time low from 1990 to 1992. At the same time, the "save the rainforest" movement was gaining momentum and the connections between coffee and deforestation were coming to light. Social activists were beginning to pressure coffee companies and raise awareness among consumers about the plight of farmers as well as the coffee industry's complicated role in politics, conservation, and national economies.

Multilateral development banks and international aid agencies took notice of the crisis and began investing in improving coffee farming. The World Bank, International Development Bank, US Agency for International Development, Deutsche Gesellschaft für Internationale Zusammenarbeit GmbH, and other institutions converged on coffee. They sponsored research, conferences, and training. NGOs also accepted the challenge, responding in various ways but most importantly with tools to guide, implement, and incentivize sustainability: standards and certification.

The concept was simple yet revolutionary: Set criteria, train farmers to meet the criteria, train auditors to evaluate progress and confirm compliance, urge coffee roasters and retailers to get involved, and use a green seal of approval on product labeling to engage consumers. The organic movement had been doing this for two decades, before it began working in coffee in the 1980s.

The coffee price crisis of the 1990s coincided with an environmental emergency: the rampant destruction of the rainforest, including the deforestation and "technification" of coffee farms (coffee grown with high levels of agrochemical inputs and in full sun). Some scientists, environmentalists, farmers, and coffee company representatives studying these interlinked developments proposed a new, holistic standard that would define "conservation coffee."

The farmers and researchers in the Interamerican Foundation for Tropical Research (FIIT in its Spanish acronym), based in Guatemala knew that the social, economic, and environmental challenges (the "triple bottom line") on coffee farms were all intertwined, that a program would only be effective by tackling them all at once. Using the Sustainable Agriculture Network's (SAN) comprehensive standard, which was already beginning to change banana farming practices in the region, FIIT developed a standard for coffee labeled ECO-OK, which for the first time gave equal consideration to all three pillars of sustainability—social justice, economic viability, and environmental conservation. The first ECO-OK coffee farm was certified in Guatemala in 1994.

As the concept of sustainable coffee took root, the Rainforest Alliance and SAN connected with other organizations on similar quests. Scientists at the US Smithsonian Institution's Migratory Bird Center (SMBC) were also studying the importance of coffee farms to birds. The SMBC became a leading advocate of forested coffee farm habitat and developed a standard and certification program called Bird-Friendly coffee. Additionally, the Natural Resources Defense Council, Conservation International, and other conservationists joined together with the Consumer Choice Council to find common ground on what farm-management practices would best benefit farmers, the environment, and consumers (Rice and Ward, 1996; Rice and McLean, 1999).

These were all important developments, as at that time most mainstream coffee buyers did not even consider the origins of the beans they traded, let alone the environmental or social impacts. Today the coffee industry fully embraces sustainability, and nearly every coffee brand—specialty or mainstream—talks about its sustainability credentials. Other major standards were launched by the late 1990s, and in the new millennium private company standards began to emerge.

The certification seals gave farmers a voice in the marketplace, a way to prove to buyers that they were reforesting instead of deforesting, conserving natural resources, and treating workers with respect. Although standards and certification transformed coffee marketing by engaging brands and consumers in sustainability, their most significant impact was facilitating information flow from scientists and agronomists to farmers. For the first time, farmers and trainers had comprehensive guidelines—the standards—and tools for assessing farms to determine what kind of information or technical assistance was most needed on each individual farm. The standards set performance targets and allowed farmers to compare their progress against a norm and to their peers. Certification provided incentives for continuous improvement. NGOs created their own training programs, and the standards influenced the training efforts of companies and governments.

3. WHAT DO SUSTAINABILITY STANDARDS AIM TO ACHIEVE?

The organic movement promotes farming without the use of most synthetic fertilizers and pesticides. The Fairtrade organizations began by guaranteeing small-holder farmers organized into cooperatives a fair price for their crops and later broadened their standards to include plantations and environmental issues. Sustainability standards include most of the elements defined by the SAN outlined below, but with differing degrees of importance and rigor:

1. Develop a farm management plan that includes social and environmental needs;
2. Conserve natural ecosystems; no deforestation, reforest where practical;
3. Protect wildlife;
4. Conserve water and protect water quality;
5. Ensure fair treatment and good working conditions for farm workers;
6. Ensure occupational health and safety;
7. Contribute to the local community; be a good neighbor;
8. Minimize the threats of agrochemicals to human health and the environment; employ biological, mechanical or cultural controls to pests and disease;
9. Stop erosion and build healthy, fertile soils;
10. Properly manage all wastes.

To meet the SAN standard and earn Rainforest Alliance certification farmers had to stop deforesting, begin reforesting, protect streams with buffer zones, strictly control agrochemical use, and use other practices that were in some cases contrary to the farm modernization practices that were in the 1990s heavily promoted by governments, banks, and aid agencies.

FIIT and the other conservation and rural development groups in the SAN began convincing farmers that implementing the sustainability standards on their farms would maintain productivity, conserve the natural resources on which they and their neighbors depended, and give them access to premium markets, thus compensating for the presumed opportunity costs of not deforesting their lands and switching to the new, technified, full-sun monoculture system that often was more productive, but also more expensive to manage.

Conrado Guinea, formerly with FIIT and now policy manager for the SAN recalls, "It was a tough sell at first. Farmers asked the logical question, 'what's in it for me?' But as they began changing practices, they began seeing the benefits. Coffee farmers have to be smart and cautious to survive; with accurate information, they make good decisions."

Defining Sustainable Agriculture

The SAN defines sustainable agriculture as "Farming that is economically viable, environmentally sound and socially equitable. It improves on-farm habitat for wildlife and safeguards the ecosystems and environmental services on which agriculture and all life depend. Sustainable farms have minimal environmental footprints, are good neighbors to human and wild communities and are integral pieces of regional conservation initiatives. They are resilient and regenerative. Beneficiaries of sustainable agriculture include present and future generations of farmers, farmworkers, consumers and wildlife."

4. MAJOR SUSTAINABILITY STANDARDS SERVING THE COFFEE SECTOR

The coffee sector has proven fertile ground for voluntary sustainability standards with some of the earliest multisector sustainability initiatives such as Fairtrade and Rainforest Alliance using coffee certification as a basis for early innovation and growth. Over time, various other initiatives have formed. According to the State of Sustainability Initiatives Review (2014), coffee produced in conformity with a voluntary sustainability standard represented 40% of global coffee production in 2012. This number included some double counting, as several of the standards and certification programs overlap. The amount of standard compliant coffee grew an estimated 26% per annum during the period 2008–12. Compliance was most advanced in Brazil, Vietnam, Colombia, Central America, and Peru. About one-quarter of the certified coffee actually sold as such from the certificate holder to the first buyer (Potts et al., 2014).

Below is a high level description of some of the most important initiatives at the international level. The information is taken from the State of Sustainability Initiatives Report (2014) with updates derived from personal correspondence with the program managers (Potts et al., 2014).

4.1 Major Multistakeholder Standards for Coffee Farming

4C ASSOCIATION *for a better coffee world*

The 4C Association: The 4C Association was the product of a public–private partnership between the German government and the German Coffee Association and became an independent, member-based organization in 2006. The 4C Association provided a baseline sustainability standard as a stepping stone to full certification with any one of the established labeling initiatives. In 2016, the 4C Association merged with the IDH-Sustainable Trade Initiative's Sustainable Coffee Platform to create the Global Coffee Platform. The 300-plus members of the Global Coffee Platform (GCP) include farmers, traders, industry, civil society, and many other organizations. The GCP unites both public and private sector

actors to create a common vision on the most critical sustainability challenges at a producing country level and feeding these national priorities into a global agenda. The GCP also supports the use of the Baseline Common Code, a set of globally referenced baseline principles and practices for coffee production and processing.

The first Fairtrade labeling initiative was formed under the name of Max Havelaar in the Netherlands in 1988 with a goal of ensuring improved terms of trade for producers and supporting the development of democratic producer organizations. The internationally agreed Fairtrade Standards cover economic, social, and environmental criteria that are applied to producers and traders. Fairtrade International now comprises 29 labeling initiatives and marketing organizations, and three producer networks, and represents more than 1.65 million farmers and workers. Fairtrade farmers and workers have representation on the Board of Directors and Standards Committee, and as of 2013 have 50% of the vote in the Fairtrade General Assembly, Fairtrade's highest decision-making body. In coffee, certification is only open to smallholder co-operatives and associations.

Utz Certified: Launched in 1999 by the Ahold Foundation, Utz Certified was later founded under the name of Utz Kapeh (meaning "good coffee" in Mayan). The organization changed its name to Utz Certified and, in 2016, to simply Utz. The first Utz coffee standard was modeled on the retailer-developed EurepGAP food-safety standard with the addition of specific social and price reporting requirements.

Rainforest Alliance/SAN: Founded in 1986 with a mission of "protecting ecosystems and the people and wildlife that depend on them by transforming land-use practices, business practices and consumer behavior," Rainforest Alliance was instrumental in starting or facilitating certification across a number of sectors including coffee. The sustainable agriculture standard used by the Rainforest Alliance standard is set and managed by the SAN a coalition of local and international conservation organizations. The first Rainforest Alliance Certified™ coffee was marketed under the ECO-OK label in 1994. The Rainforest Alliance training and certification program has grown significantly through a number of corporate partnerships.

IFOAM—Organics International: Although standards vary at the national level, IFOAM-Organics International has provided the reference for international organic standards since 1995, through its IFOAM Basic Standards prior to 2010, and after that through its IFOAM Standard and its IFOAM Family of Standards program. Although organic certification

is best known for its restrictions on the use of synthetic inputs, prohibition of genetically modified organism (GMO) use, and attention to soil health, the IFOAM Standard also addresses other issues, such as social justice or preservation of primary ecosystems.

Major Brand-Managed Standards and Technical Assistance Programs:

***Nespresso* AAA Sustainable Quality Program**: Developed by Nespresso in collaboration with the Rainforest Alliance in 2003, this sourcing program demonstrated that sustainable farm management practices can improve crop quality. The guidelines in the AAA TASQ™ Tool for the Assessment of Sustainable Quality were modeled on the Sustainable Agriculture Network standards and included physical quality practices designed to help farmers meet Nespresso's rigorous quality requirements. Maintaining direct links with farmers, the program helps them to organize into groups and provides a permanent technical assistance. AAA also collaborates with farmers in a continuous improvement path, and rewards them by paying premiums for the coffee.

To address the increasingly important exogenous risks faced by farmers like climate change, lack of social stability, the AAA program is collaborating with other private, and public stakeholder to design solutions at the community and landscape levels. In 2015, *Nespresso* reported that 85% of its supply was from farms involved in the *Nespresso* AAA Sustainable Quality™ Program.

Starbucks C.A.F.E. Practices: Starbucks began working with Conservation International in 2001 to develop socially responsible coffee buying guidelines. The Coffee and Farmer Equity Practices were an effort to develop a system of sustainable practice that was integrated within the corporate business plan and decision-making structure. C.A.F.E. practices combine social and environmental standards with a number of quality-based parameters. Scientific Certification Systems helped develop the evaluation and auditing scheme in 2005. Producers must score certain levels against the requirements to maintain preferred buyer status. In 2014, Starbucks reported that 96% of its coffee was sourced through C.A.F.E. practices or another externally verified system.

Nescafé Plan: Building on nearly 50 years working with coffee farmers, in 2010 Nescafé - the world's largest coffee brand - launched the Nescafé Plan with partners like the Rainforest Alliance and the 4C Association. The program aims to enhance livelihoods of present and future farmers, help their communities and landscapes. The company has been significantly increasing Responsibly Sourced (4C-verified or compliant to other voluntary sustainability standards) coffee supplies and the distribution of disease resistant coffee plantlets. Independent research began to show meaningful gains in economics, particularly productivity increases, and decreases in losses to pests and disease. The focus of the program will remain on generating positive impacts all along the value chain. By 2015, 55% of the coffee was responsibly sourced.

5. DO VOLUNTARY STANDARDS DEVELOPED BY NGOs AND COMPANIES BRING POSITIVE CHANGE TO FARMERS?

The NGO certification and company managed supply chain assurance programs suggest to farmers and farm workers that following the standards and achieving certification will improve their livelihoods, as well as conserve the environment and make their lands more vigorous and resilient. The scheme managers, farmers, coffee companies and the donors, foundations, and multilateral aid agencies supporting standard setting and certification all need data to demonstrate that the programs are indeed having these intended effects.

At the turn of the millennium, there were only a few independent studies of the impacts of standards. This is changing as the standards makers refine their theories of change and tune the criteria to be less prescriptive and more outcome based. In addition, the International Social and Environmental Accreditation and Labeling (ISEAL) Alliance has made quantifying impacts a cornerstone of credibility. ISEAL is the global membership association for voluntary sustainability standards; its aim is to improve the credibility and effectiveness of standards, as well as increase their uptake.

As the number of peer-reviewed studies increased, scientists began comparing methodologies, as well as results. The harmonization of research terms and methodologies will help improve comparability and consistency of ongoing studies (Milder et al., 2014; Seville et al., 2016).

The studies that corporate and civil society standard setters and scheme managers are conducting are important for adaptive management and continuous improvement. However, independent research is required for full credibility. Therefore, numerous independent researchers are examining the performance of the standards (Giovannucci and Koekoek, 2003; Panhuysen and Pierrot, 2014; Romanoff, 2008; Ruerd and Guillermo 2010; Voluntary Sustainability Standards, 2015). The Committee on Sustainable Assessment (COSA), a nonprofit global consortium, conducts or coordinates many of these studies. COSA advances practical and science-based tools to help collect, assess, and interpret reliable field data so standard setters can make the most effective agricultural interventions (Fig. 7.1).

In 2012, COSA scientists presented to the Rio+20 United Nations Conference on Sustainable Development summaries of findings on 5193 farms in six countries. They reported wide variations among countries and certification schemes, but all certification programs improved yields, net income, and the implementation of conservation measures as compared to farms not part of a certification program. Other studies had similar conclusions (Rueda and Lambin, 2013; Hughell and Newsom, 2013; Haggar et al., 2012; Barham and Weber, 2012; Znajda, 2009; Romanoff, 2010; Garcia et al., 2014; Milder and Newsome, 2015; Tuinstra and Deugd, 2011; Deugd, 2003; Kuit, 2010).

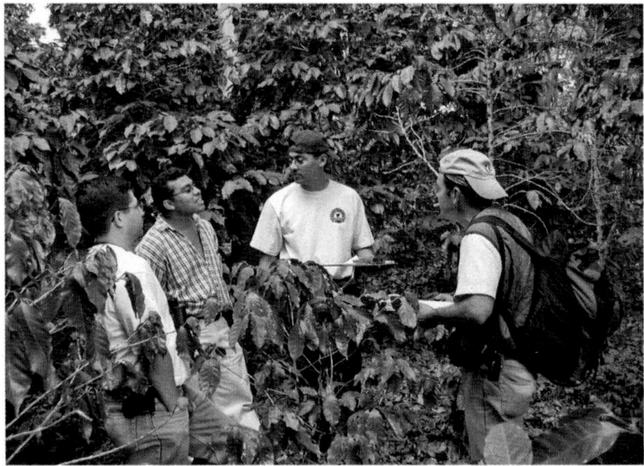

FIGURE 7.1 Assessing farms according to sustainability standards, which occurs directly in the field, is an efficient way to identify the training and technical assistance needs of each individual farmer. Assessing and auditing is done by teams of skilled and experienced scientists, agronomists, and other specialists. *Published with permission from SalvaNatura, El Salvador.*

5.1 Standards and Certification Continue to Improve

Despite the generally positive results of the impact studies, questions have been raised regarding the efficacy of the standards, auditing, and certification model. The programs have not yet reached many smallholder farmers in remote areas. There is some duplication of effort where farmers need to certify to several standards, requiring multiple audits, increasing the cost without necessarily assured increases in benefits. Even though many farmers widely acknowledge the benefits of certification (Tuinstra and Deugd, 2011), it often takes between 2–4 years before the positive effects of the implemented practices to be perceived.

These concerns are being addressed by the standards. For example, most standards have a parallel system for smallholders, which is simpler and less demanding. Costs are reduced when farmers organize into groups, co-operatives, or producer associations, all of which have played a major role in creating an enabling environment for the incorporation of small farmers in sustainable production initiatives. Focusing on the development of these farmer organizations has made it possible for standard organizations and other assistance groups to reach millions of small-holder farmers, thereby reducing the costs of training, certification, and audits, as these can then be shared.

Although they are not a panacea, many farmers, brands, and consumers consider standards, certification, and labeling to be the best way to credibly communicate sustainability claims. Equally important, they guide the development and delivery of training programs and incentivize farmers to adopt improved practices.

6. TRAINING COURSES, CURRICULA, TOPICS, AND THEMES

The voluntary standards have affected modern training curricula, which include essential farm-management practices such as farm renovation, pruning, keeping shade trees, worker health and safety, pollution control, water and soil conservation, and disease and pest control. Increasingly, training includes subjects related to planning, productivity, crop quality, business skills, and adaptation to climate change.

In Kenya and Tanzania, for instance, the Sustainable Commodity Assistance Market (SCAN), a group of international organizations providing technical assistance on sustainable production and business management, offers a Financial Literacy Toolbox. Designed to increase access to finance for sustainable small and medium-sized coffee enterprises, the toolbox is presented to trainers who then work directly with the coffee farmers to strengthen their financial capacities.

Several organizations are offering training to help farmers adapt to changing climatic conditions. For example, the Rainforest Alliance and SAN collaborated with scientists and farmers to make the SAN standard a sharper tool for improving farm resilience to climate change. They and other groups developed training materials for adapting to climate change. For example, with support from a foundation of the coffee trading company, Neumann Kaffee Gruppe, the Initiative for Coffee and Climate developed a step-by-step guide for adapting to climate change (Coffee and Climate Initiative, 2015).

To address the gender and age imbalance in training, many initiatives now offer courses such as management and decision making skills for female farmers and training programs that interest young people in taking over their parents' farms. In Rwanda, Sustainable Harvest, an importer of high quality specialty coffee; along with the NGO, Women for Women International; and the Government of Rwanda, run a special project to help train female coffee growers to deliver a higher quality product, increasing its market value. As of early 2015, some 890 participants had graduated from the program, which taught them how to grow more from small plots of land, as well as to enhance their environmental sustainability, and reduce their financial risk. In addition, they now have better access to international buyers and can get higher prices for their crops (Griswold, 2016).

After attending a training event, David Griswold, CEO of Sustainable Harvest, said, "It's hard to describe what a touching experience this was for me—two years ago these women were making very little money from their coffee and in many cases, felt hopeless about their future. Now they are successful businesspeople engaged in transparent supply chains; they know who buys their coffee and how to produce what they demand" (Griswold, 2016).

Advancing Together: An Example From Guatemala

Coffee farming cooperative, ADESC (Asociación de Desarrollo Social Los Chujes) in Huehuetenango, Guatemala, enables farmer partners to share their knowledge and expertise and negotiate together for a better outcome for all—including higher coffee prices and access to credit.

Founded in 1998 with just 15 farmers, the association boasted 83 active members at the end of 2015, along with a number of achievements. Not only was ADESC the first coffee farmer association in the region to become Rainforest Alliance Certified, it was also the first to become Climate Friendly Verified, and was one of the first that was able to access its own financing.

"Coffee producers become ADESC partners to improve their life situation," says Servando Del Valle, President and Legal Representative of ADESC. "After several years trying to sell coffee together, it was when a foreign company recognized the quality of our coffee, and started buying every year higher quantities, that we started to become stronger and grow our association."

Coffee farmers benefit from joining the cooperative in many ways, explains Arnoldo Cifuentes, a small coffee producer and the ex-Manager of ADESC. This includes technical assistance, many different types of training, access to credit, higher coffee prices, and ultimately, "a better harvest and better income for families."

ADESC partners have also enjoyed opportunities to visit other countries to learn and share ideas with other farmers, to host foreign visitors, and to incorporate new technologies into their work. For example, ADESC farmers used digital tablets to access a variety of information, including coffee prices, facts about modern coffee consumers, and coffee farming methods in other countries.

What began as a small association of friends and colleagues, is now a way of life for some of the cooperative's partners. "ADESC means a lot in my life, and I carry it in my heart," says Servando Del Valle.

7. TYPES OF TRAINING METHODOLOGIES

Government extension agencies and other training providers once offered standardized courses. The certification standards helped modernize training by demonstrating the importance of listening to farmers and understanding that every farm, and every farmer, is different. By first conducting assessments, trainers learn the priorities for each farmer's needs. At the same time, the training can be tuned to the local language, culture, literacy level, and learning styles.

"Farmers usually have their own perception of what's missing and where they want to go," says Reiko Enomoto, Senior Training Manager at the Rainforest Alliance. "For example, most are very keen on learning how to combat pests and diseases or improve fertilization, because these activities affect productivity and thus income. We listen to them and also conduct an assessment of the whole farm—all the challenges—to make sure that the training covers all the critical issues. Talking together, we find the links among the conservation, social, yield and quality aspects."

Whether online or in printed format, training material for farmers must be practical and demonstrate clear socioeconomic or environmental benefits to farmers. Training materials should go beyond knowledge transfer to encourage changes in behavior. One way to do this is to build a network of farmers who have good farming practices and technical knowledge and train them to assist others in a "train the trainer" approach (Fig. 7.2).

Since the late 1980s, Farmer Field Schools (FFS) around the world have used "discovery-based" group learning, involving experiments, field observations, and participatory analysis. FFS typically cover broad topics designed

FIGURE 7.2 Trainers are often farmers themselves, sharing information they have gained from their own experience and learned from agronomists and scientists.

to benefit both farmers and their communities, such as pruning, waste-water management, and integrated pest management. UN agricultural organizations like the Food and Agriculture Organization and the International Fund for Agricultural Development have promoted participatory training, where farmers can learn by doing and from each other.

In some rural communities, FFS and model farms employ local youth as ambassadors to discuss and share best practices with neighboring communities. This serves the dual purpose of empowering other farmers with more knowledge and information, while also encouraging young people to stay in their local communities and get involved in farming.

There are many different perspectives on training methodologies. Niels van Heeren, a consultant for Co-Crear, a sustainable agriculture organization, advocates a "sushi approach," where you roll up a bit of everything in a quick and easily digestible way on a certain topic. For example, in 1 h, a trainer can train a small group on wet-milling practices, using demonstrations as the main training mechanism.

"It doesn't take too much time or budget, and the demonstration is actually done by the farmers themselves. Everyone is working together and enjoying the process," says Niels.

When demonstrations are not possible or practical, trainers can use visuals to help explain new topics or tools. For example, if trainers are talking about plant diseases, they can use photos or drawings to show farmers how to identify the disease. Or if they want to demonstrate a new pruning technique, they can show a video of someone applying that technique.

"We develop training materials that visually describe what the farmers need to learn," says Reiko Enomoto. "And, we design our training program in a way that is most practical for trainers, taking into account the local context. Some villages prefer videos and facilitated training sessions. Some do not have electricity and prefer to use flipcharts or distribute posters to the farmers. And in some cases when farmers are very literate, we provide them with a complete summary of the training, such as an illustrated guide."

Many coffee farmers can participate in face-to-face training programs and workshops in their local communities. However, due to the high cost of maintaining training staff, NGOs and companies have also developed libraries of information online and online training modules in the most used languages. For example, the Rainforest Alliance's online resource, www.SustainableAgricultureTraining.org, offers online training in various languages on topics such as Protecting Rainforests and Preventing the Drift of Agrochemicals to Ecosystems. Although this platform is designed primarily for trainers, farmers can also access it (Fig. 7.3).

Similarly, farmers and stakeholders involved in sustainable agriculture can access the online portal www.sustainabilityxchange.info, from the Global Coffee Platform. The portal includes a library of material on sustainable supply chains, as well as a community discussion forum. Other organizations including

FIGURE 7.3 An excerpt from a training poster about soil conservation in Bahasa Indonesian. This is a typical use of photographs with minimal text to covey complicated ideas in an understandable way. Posters help remind farmers of the practices learned in training programs. *Photo published with permission from Rainforest Alliance.*

Technoserve, Federación Nacional de Cafeteros de Colombia (Fedecafé), FarmerConnect, the European Coffee Federation, and IDH, the Sustainable Trade Initiative, also provide training courses and materials for farmers.

Successful training is a continuous and circular process. Farmers learn from trainers, and trainers learn from farmers. Training processes, methodology, and materials continue to evolve and change over time in ways that blend traditional knowledge with new science, tools, and innovations.

The Growing Role of Technology

Computer-based and mobile technologies such as satellites for monitoring and data collection, mobile phones and other digital devices for reporting and communications, and e-learning, webinars, and online training videos are taking on an increasingly important role. Technology can help cut costs, reinforce previous face-to-face training, and remind farmers about best practice in a wide variety of areas. But how effective is it?

"Technology needs to be used judiciously," says Daniele Giovannucci of COSA. "It does not replace live training. However, someone can go to a village maybe only once a month, but you can send interesting messages or a small video to reinforce bilateral training more frequently. In this way, technology supports the training process." For instance, COSA surveyors used tablets and smartphones to conduct surveys in the field. They also use geospatial mapping to improve the targeting of interventions and to help contextualize the situation in a specific region and correlate important factors.

The commodity trading company ECOM has a division—Sustainable Management Services—dedicated to providing technical assistance to coffee farmers. It has created a mobile APP to feed information about farms to a central database which benefits farmers as well as coffee buyers. This helps guiding the best technical assistance to each farm as it gives ECOM customers a lens through which they can view the performance of the farms in their supply chain.

Sustainable Harvest uses CheckMark, an iPad-based audit program that streamlines the key data from different certification and code-of-conduct assessments. The information is maintained in a cloud-based system so that producers and buyers can verify good practices.

Although there have been enormous advances in bringing technology to farms, many farmers do not have access to devices or the infrastructure. High speed Internet, and the tools to access it, will enable farmers to take ownership of their learning tools and processes. It can also excite young people about agriculture and help farmers see the value of their work. Social media can help farmers connect with other farmers and trainers as well as access information about coffee consumers and the marketplace.

8. OUTLOOK

From long-standing problems such as improving quality and yields, to new issues around climate change and cultivating the next generation of farmers, coffee farmers must continuously cope with a wide array of challenges to improve their resilience and maintain their livelihoods. Even farmers with decades of experience must now be able to adapt to a constantly changing situation. And this in turn is creating a new era of continuous learning and adaptation.

Perhaps the best known and most widely discussed new challenge is climate change. Learning "climate smart" farming will help producers deal with rising temperatures, irregular rainfall patterns, drought, flooding, and robust pests and plant diseases all of which reduce coffee quality and yields.

Although large, successful commercial farms are business that are attractive to young entrepreneurs and can support several or many families, most small farms earn barely enough to feed one family. Farming small plots is a hardscrabble existence and young people are fleeing to the cities. This accelerating migration raises a serious concern about the future supply of coffee and other farm goods. The median age of coffee farmers is rising. For example, in Colombia it is 56 years (TechnoServe and IDH, 2014). Where will the next generation of farmers come from?

Thus the goals of the training programs include making farming more businesslike and profitable, to build economies and communities, to restore dignity and even pride in rural areas. Some programs are going deeper into the socioeconomic challenges, tackling, for example, the lack of succession planning in rural areas. Few farmers have a transition plan to pass their lands on to their heirs. Also, very few farmers have any kind of pension plan or savings program that will enable them to quit working and pass on their farms to younger generations.

To resolve some of these challenges, different parties are striking innovative partnerships. In 2014, Nestlé Nespresso, the Colombian Ministry of Labor, the Aguadas Coffee Growers Cooperative and Fairtrade International, created the first multistakeholder retirement savings plan for coffee farmers. A pilot project with around 850 Colombian coffee farmers, who are part of the Nespresso AAA Sustainable Quality™ Program, provides a flexible, government subsidized retirement savings scheme for workers who are not traditionally covered by an official pension scheme (more details are described in Chapter 6).

Successful farmers gather information from all available sources—traditional and modern—and use it to experiment and innovate. Clement Ponçon, who manages a productive farm in the Matagalpa region of Nicaragua, says that, "coffee farming was never easy, but every day it gets more challenging. We can succeed by learning from experience, applying best agronomic practices, respecting the environment, being good neighbors, and not being afraid to try new things, to experiment and innovate." Ponçon's farm, La Cumplida, is a matrix of high-quality Arabica coffee, commercial hardwood stands and primary forest guarded in conservation areas. Other crops include ferns, which are exported to the floral market. The coffee is Rainforest Alliance and Utz certified. The managed forest is FSC certified. La Cumplida also generates hydroelectric energy, and Ponçon hopes to sell carbon credits.

When asked about the future of coffee farming, Ponçon says he prefers to try to set a good example rather than give opinions or advice. But he avidly promotes planting trees. La Cumplida has generated three million tree seedlings, he estimates. "I planted many of them myself. I plant a tree every day. Trees are the good for the coffee farm and the best investment in our future."

Monitoring, Evaluation, and Continuous Improvement and Learning

It is difficult to evaluate the effectiveness of training, because the "cause" (e.g. learning to apply fertilizer more efficiently) and the desired "effect" (increased crop yield) are so distant, and there are many intervening factors. Trainers begin measuring effectiveness simply by counting the participants. What percentage of the local farmers came to the session? Was there a proper age and gender balance? Did they return for subsequent sessions?

Then there are studies to evaluate the rate of adoption of better practices. These show that farmers learned from the training and changed their behavior. These changes are important, as they suggest that farmers are working with more knowledge and confidence, that they are eager to learn and try new techniques. But what trainers, standard setters, and researchers ultimately want to know is if their efforts resulted in the desired, long-term outcomes. Are yields and crop quality improving? Are the livelihoods of farmers and farmworkers demonstrably improved? Are soils healthy and rivers running clean? The physical changes are difficult to measure; the social and socioeconomic changes are even more complicated. Many factors affect coffee yields and quality, livelihoods, watersheds, and wildlife. But carefully constructed, long-term and replicated studies are beginning to show that—in general and with abundant exceptions based on local conditions—standards, training, and certification bring the intended positive benefits. See J.C. Milder et al. (2014) for a full discussion.

Sometimes training does not produce intuitive outcomes, according Daniele Giovannucci of COSA. "If you train a farmer who has a limited set of skills, we would expect to see a great difference soon afterward. However, with farming you simply can't apply training in the same way you would train people in an industrial process where a clear causal chain may show training yielding immediate results. Unfortunately, it doesn't work this way."

"One of the most important uses of data is to stimulate and encourage farmers to keep implementing the training by consistently and persistently measuring and monitoring their results in collaboration with them," says Daniele. "The data has to be straightforward; it has to be accurate and realistic, and it has to be participatory."

COSA has developed systems that collect information on aspects such as crop quality and quickly feed it back to farmers. The systems work by gathering simple survey data on key performance indicators from traders or local staff during normal field operations. The data are analyzed and presented in easy-to-understand dashboards. When a deeper understanding of the situation is needed, the systems can also link back to baseline and impact assessments.

"Measuring what matters and measuring it well is key," says Daniele. "This is how you can understand impact and manage for results while continuously improving performance."

REFERENCES

Barham, B.L., Weber, J.G., 2012. The economic sustainability of certified coffee: recent evidence from Mexico and Peru. World Development 40 (6), 1269–1279. http://dx.doi.org/10.1016/j.worlddev.2011.11.005.

COSA, 2014. The COSA Measuring Sustainability Report.

Deugd, M., 2003. Crisis del café: Nuevas estrategias y oportunidades. RUTA - FIDA, Costa Rica.

Garcia, C., Ochoa, G., Garcia, J., Mora, J., Castellanos, J., 2014. Use of polychoric indexes to measure the impact of seven sustainability programs on coffee growers' livelihood in Colombia. In: Proceedings of the 25th International Conference on Coffee ASIC, Armenia.

Giovannucci, D., Koekoek, F.J., 2003. The State of Sustainable Coffee: A Study of Twelve Major Markets. International Coffee Organization, London; International Institute of Sustainable Development, Winnipeg; United Nations Conference on Trade and Development, Geneva.

Griswold, D., 2016. Sustainable Harvest. USA Interview by the Authors, Portland, Oregon.

Haggar, J., Jerez, R., Cuadra, L., Alvarado, U., Soto, G., 2012. Environmental and economic costs and benefits from sustainable certification of coffee in Nicaragua. Food Chain 2 (1), 24–41. http://dx.doi.org/10.3362/2046-1887.2012.004.

Hughell, D., Newsom, D., 2013. Impacts of Rainforest Alliance Certification on Coffee Farms in Colombia. Rainforest Alliance, New York.

Initiative for Coffee & Climate, 2015. Climate Change Adaptation in Coffee Production: A Step-by-Step Guide to Supporting Coffee Farmers in Adapting to Climate Change. Embden Drieshaus & Epping Consulting Gmbh. http://www.coffeeandclimate.org/.

Kuit, M., van Rijn, F., Jansen, D., 2010. Assessing 4C Implementation Among Small-scale Producers. 4C Association, Bonn, Germany.

Milder, J.C., et al., 2014. An agenda for assessing and improving conservation impacts of sustainability standards in tropical agriculture. Conservation Biology: Society for Conservation Biology, 1–12. http://dx.doi.org/10.1111/cobi.12411.

Milder, J.C., Newsome, D., 2015. SAN/Rainforest Alliance Impacts Report: Evaluating the Effects of the SAN/RA Certification System on Farms, People and the Environment. Rainforest Alliance, New York, NY, USA.

Panhuysen, S., Pierrot, J., 2014. Coffee Barometer 2014. HIVOS, IUCN, Oxfarm Novib, Solidaridad, WWF.

Potts, J., Lynch, M., Wilkings, A., Huppé, G.A., Cunningham, M., Vivek Voora, V., 2014. The State of Sustainability Initiatives Review 2014: Standards and the Green Economy. International Institute for Sustainable Development and the International Institute for Environment and Development, Winnipeg, Manitoba, Canada, ISBN 978-1-894784-45-0.

Rice, R.A., Ward, J.R., 1996. Coffee, Conservation and Commerce in the Western Hemisphere: How Individuals Can Promote Ecologically Sound Farming and Forest Management in Northern Latin America. Migratory Bird Center, Smithsonian Institution, Natural Resources Defence Council, Washington, DC, USA.

Rice, P., McLean, J., 1999. Sustainable Coffee at the Crossroads. The Consumer Choice Council, Washington, DC.

Romanoff, S., 2008. Shade Coffee in Biological Corridors: Potential Results at the Landscape Level in El Salvador. USAID.

Romanoff, S., 2010. Shade coffee in biological corridors: potential results at the landscape level in El Salvador. Culture & Agriculture 32 (1), 27–41.

Ruerd, R., Guillermo, Z., 2010. How Standards Compete: Comparative Impact of Coffee Certification in Northern Nicaragua. Radboud University Nijmegen, Centre for International Development Issues, The Netherlands.

Rueda, X., Lambin, E.F., 2013. Responding to globalization: impacts of certification on Colombian small-scale coffee growers. Ecology and Society 18 (3), 21.

Seville, D., Shipman, E., Daniels, S., 2016. Towards a Shared Approach for Smallholder Performance Measurement: Common Indicators and Metrics. Sustainable Food Lab, Hartland, Vermont, USA. http://www.sustainablefoodlab.org/performance-measurement/tools-resources/deep-dive/.

TechnoServe and IDH, 2014. Colombia: A Business Case for Sustainable Coffee Production, p. 7. Available from: https://issuu.com/idhsustainabletradeinitiative/docs/131206_uganda/1?e=10013468/6785051.

Tuinstra, A., Deugd, M., 2011. Rainforest Alliance Certification in Coffee Production: An Analysis of Costs and Revenues in Latin America 2010–11. Rainforest Alliance, Costa Rica.

Voluntary Sustainability Standards: Market Report 2015, 2015. ITC.

Znajda, S.Z., 2009. Examining the Impacts of the Rainforest Alliance/SAN Coffee Certification Program: A Summary of Local Perspectives from San Juan del Rio Coco, Nicaragua. Dalhousie University, Canada bib entry 0092.

Chapter 8

Cupping and Grading—Discovering Character and Quality

Ted R. Lingle[1], Sunalini N. Menon[2]
[1]*Coffee Quality Institute, Aliso Viejo, CA, United States;* [2]*Coffeelab Limited, Bangalore, India*

1. WHY DO ROASTERS CUP AND GRADE THE COFFEES THEY BUY?

Coffee is not a true commodity. Coffees grown in Colombia are not one-for-one substitutes for coffees grown in Brazil. A Central American Arabica coffee is not a substitute for Arabica coffees grown in East Africa. Robusta coffee grown in Uganda is not a substitute for a Robusta coffee grown in Indonesia. Consequently, roasters continuously cup and grade the coffees they buy to determine how best to use them in their blends, or in the case of specialty coffee, to discover which coffees are of sufficiently high quality to be sold as "single origin" coffees.

2. TRADITIONAL COFFEE CUPPING

Traditional coffee cupping, weighing out beans individually into small cups, grinding the beans, then pouring boiling water over them, and after 4–5 min slurping the brew on to the palate with a large spoon, is said to have originated at Hills Bros. Coffee in San Francisco. This was an essential practice for coffee roasters. Although coffee may have a good, general visual appearance, as indicated by its color, bean uniformity, and lack of "defective beans," it can have an "awful" taste because it was contaminated either in the processing, storing, or transporting from the coffee farm to the roaster's warehouse. The only way to tell is by actually "tasting" the coffee. This is why everyone in the value chain, from roasters, importers in the consuming countries, and exporters in the producing countries, "cups" the coffee. Traditionally, the only ones in the value chain not cupping the coffees are the small producers, who now need to be educated on the cup nuances of their coffees, to help in effective marketing and cost-effective price realization for their produce. Today, cupping protocols are available for both arabica and robusta coffees. Training in cupping is provided not only by the

The Craft and Science of Coffee. http://dx.doi.org/10.1016/B978-0-12-803520-7.00008-6

Speciality Coffee Associations of America and Europe but also by organizations, roasters guilds, and cupping labs in producing and consuming countries. Thus, at present, the small coffee farmers have an opportunity to understand the taste profile of their coffees, the steps to be taken to improve the quality of their produce and to better understand market requirements, which are undergoing dynamic changes from time to time.

For the better part of 400 years, coffee cupping was an informal art, passed on through the generations by word of mouth. It was a skill set relegated to the largest roasters, importers, and exporters and thought of as a very specialized skill, taking years to acquire and belonging to a very few select individuals. Up until 1984, there was no printed text on cupping, other than a general description of the process reported by William H. Ukers in his classic work, *All About Coffee*, first published in 1922 (Ukers, 1935). In 1984, the Specialty Coffee Association of America (SCAA) published the first edition of the *Coffee Cuppers' Handbook*, written by Ted R. Lingle (1984).

The *Coffee Cuppers' Handbook* helped transform the "craft" of cupping, based on experience and practice, into the "science" of cupping, based on coffee's physical chemistry. Physical chemistry developed the framework for the scientific separation of coffee's primary flavor attributes. First, there is the "fragrance" that comes from the freshly roasted and ground coffee beans. Second, there is the "aroma" from the extracted coffee when nearly boiling water is poured over the grounds. The composition of the aroma molecules from the roasted and ground coffee will be significantly different from the ones of the liquid coffee. One reason for this is the kinetic equilibrium between the two phases (the solid–gas phases and the liquid–gas phases respectively). The partition coefficient, i.e., the ratio of aroma molecules in the two phases, changes with temperature. With increasing temperature, more molecules will move from the solid or liquid phase toward the gas phase. There is also a transport effect based on different physical phenomena, when CO_2 is released and water evaporates, taking other volatiles with it. Finally, various transformation reactions are taking place in both the roasted and ground coffee and in the liquid coffee that influence the composition of the aroma in the two situations. Third, there is the "flavor" of the coffee brew, which is the marvelous combination of taste sensations on the tongue, caused by the dissolved molecules in the brew, and the retro-nasal sensations in the nose, from the gaseous molecules initially present in the brew but which are liberated, thanks to a liquid–air surface area increase when the coffee is vigorously sipped into the mouth. Fourth, there is the "aftertaste" of the coffee brew, which comes from any residual taste compounds on the back of the tongue, usually the result of any less water-soluble compounds in the fluid, as well as the heavier gaseous molecules that may still be trapped in the brew. Fifth, there is the "acidity" of the brew, which is a measure of the amount and type of organic acids in the brew that is linked to the titratable acidity of the fluid, which is sensed by the tongue. And sixth, there is the "body," or mouthfeel, of the brew, which provides "texture" or feel. Exact correlations between body and coffee components are not known, but factors such as soluble and insoluble fibers, melanoidins, lipids (oils, fats, and waxes) in the roasted and ground beans, as well as microfine particles of the bean may play a role.

3. EVOLUTION OF THE SCAA ARABICA CUPPING FORM AND PROTOCOL

The formal SCAA protocol for cupping and grading arabica coffee grew out of a specialty coffee promotion program of the International Coffee Organization that began in 1999. Although there were five countries involved in this program, the cupping protocol was originally developed for Brazilian arabica coffees. The promotion of the Brazil coffees hinged on creating a cupping competition that would be immediately followed by an Internet auction. To conduct the cupping competition, a standardized cupping form was needed as well as a standardized format for roasting and preparing the coffee. Through trial and error, over a 5-year period, the SCAA cupping form evolved into one that arrayed 10 important quality attributes, each worth 10 points, so that evaluations would be based on a 100-point scale. A 100-point scale was determined to be the one most easily understood by people both in and outside of the coffee industry. The final version of the cupping form included 10 quality attributes: (1) fragrance/aroma; (2) flavor; (3) aftertaste; (4) acidity; (5) body; (6) uniformity; (7) balance; (8) clean cup; (9) sweetness; and (10) overall. The first five quality attributes were based on the physical chemistry of coffee's flavor that were originally developed in the *Coffee Cuppers' Handbook* (Lingle, 1984). *Uniformity, Clean Cup, and Sweetness* were added because they represented the quality of the green coffee's preparation during the harvesting and processing of the coffee beans, as had been traditionally evaluated in the traditional coffee cuppings. "Uniformity" is measured by weighing out into five individual cups the *roasted* coffee beans before the coffee is ground, thus isolating off-tasting beans in one or more individual cups, instead of dispersing them through the 250 beans under evaluation, which would happen if all the beans were ground before weighing. Off-tasting beans will stand out in a 50 bean sample, making one or more of the five cups taste different from the others.

"Clean Cup" means that there are no "noncoffee" smells or tastes in any of the five cups. "Noncoffee" smells and taste are generally caused by lack of cleanliness at the wet or dry mills where the coffee is processed, by storing or transporting the green coffee in areas that are contaminated by foreign odorous materials, or by manufacturing the hemp or jute bags with nonfood grade oils that are used for storing and shipping the coffee beans, or even by storing in poorly ventilated warehouses for a length of time in jute bags. "Sweetness" means a slight perception of sweetness, like very dilute solutions of sugar, but which can be caused by low levels of both acids and salts. The hallmark of a great coffee is a significant sweet taste when the coffee is at room temperature.

"Balance" is a concept introduced to SCAA by the Arthur D. Little Company, a major food testing company in the United States. As they also do work for major coffee roasters, they took an interest in the SCAA cupping project. Through a series of meetings and discussions, key members of Arthur D. Little's staff convinced the Technical Standards Committee of SCAA that not only was "balance" an important quality attribute, it was actually an essential quality attribute. SCAA incorporated "balance" into the cupping form, training cuppers that the primary attributes of flavor, aftertaste, acidity, and body must be in balance, meaning present in equal proportions, to receive a favorable score.

"Overall" was added in the beginning because it is often found in wine judging forms as judge's points. In the SCAA protocol, this is the only attribute for which the cupper is encouraged to render a personal appraisal on the coffee's quality. For every other attribute, cuppers are trained to rate individually, not subject to the cupper's personal preference or opinion. In this manner, the protocol and form are designed to make as objective an evaluation of the coffee's quality attributes as humanly possible. Vast empirical evidence in hundreds of cupping competitions using the SCAA protocol and cupping form has demonstrated that the system works, with the best coffees always rising to the top.

4. SCAA ARABICA CUPPING PROTOCOL

Samples should first be visually inspected for roast color. This is marked on the sheet and may be used as a reference during the rating of specific flavor attributes. The sequence of rating each attribute is based on the flavor perception changes caused by decreasing temperature of the coffee as it cools:

Step #1—Fragrance/Aroma: Within 15 min after samples have been ground, the dry fragrance of the samples should be evaluated by lifting the lid and sniffing the dry grounds. After infusing with water, the crust is left unbroken for at least 3 min but not more than 5 min. Breaking of the crust is done by stirring three times, then allowing the foam to run down the back of the spoon while gently sniffing. The Fragrance/Aroma score is then marked on the basis of dry and wet evaluation.

Step #2—Flavor, Aftertaste, Acidity, Body, and Balance: When the sample has cooled to around 70°C (approximately 160°F), 8–10 min from infusion, evaluation of the liquor should begin. The liquor is aspirated into the mouth in such a way as to cover as much area as possible, especially the tongue and upper palate. Because the retronasal vapors are at their maximum intensity at these elevated temperatures, Flavor and Aftertaste are rated at this point. As the coffee continues to cool (70–60°C; 160–140°F), the Acidity, Body, and Balance are rated next. Balance is the cupper's assessment of how well the Flavor, Aftertaste, Acidity, and Body fit together in a synergistic combination.

Step #3—Sweetness, Uniformity, and Cleanliness: As the brew approaches room temperature (below 37°C; 100°F) Sweetness, Uniformity, and Clean Cup are evaluated. For these attributes, the cupper makes a judgment on each individual cup, awarding two points per cup per attribute (10-point maximum score).

Step #4—Overall Score: Evaluation of the liquor should cease when the sample reaches 21°C (70°F) and the Overall score is determined by the cupper and given to the sample as Cupper's Points based on ALL of the combined attributes.

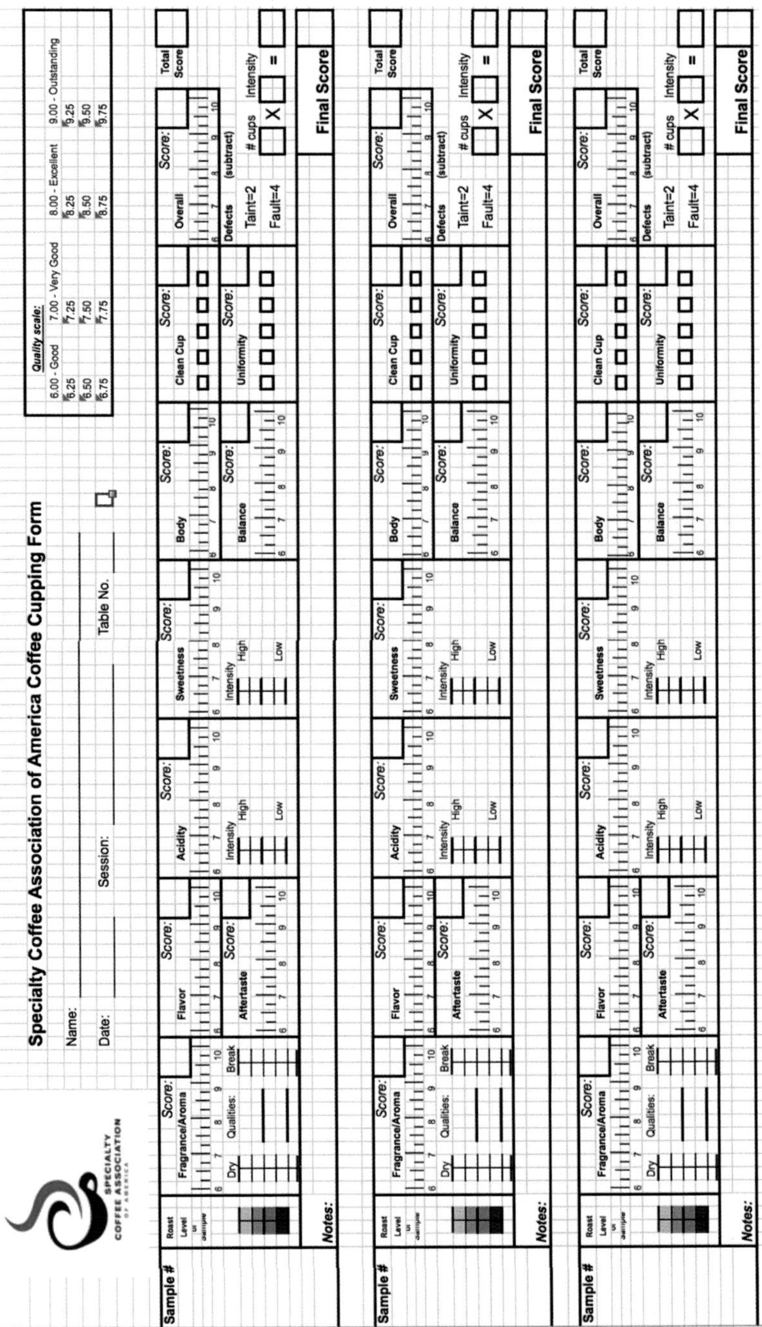

SPECIALTY COFFEE ASSOCIATION OF AMERICA

Specialty Coffee Association of America Coffee Cupping Form

Name: ______

Date: ______ Session: ______ Table No. ______

Quality scale:

6.00 - Good	7.00 - Very Good	8.00 - Excellent	9.00 - Outstanding
6.25	7.25	8.25	9.25
6.50	7.50	8.50	9.50
6.75	7.75	8.75	9.75

Sample # | Roast Level of Sample | Fragrance/Aroma Score: 6 7 8 9 10 — Dry, Qualities:, Break | Flavor Score: 6 7 8 9 10 | Aftertaste Score: 6 7 8 9 10 | Acidity Score: 6 7 8 9 10 — Intensity High, Low | Sweetness Score: 6 7 8 9 10 — Intensity High, Low | Body Score: 6 7 8 9 10 | Balance Score: 6 7 8 9 10 | Clean Cup Score: ☐ ☐ ☐ ☐ ☐ | Uniformity Score: ☐ ☐ ☐ ☐ ☐ | Overall Score: 6 7 8 9 10 | Defects (subtract) Taint=2 Fault=4 # cups ☐ X Intensity ☐ = ☐ | Total Score | Final Score

Notes:

Sample # | Roast Level of Sample | Fragrance/Aroma Score: 6 7 8 9 10 — Dry, Qualities:, Break | Flavor Score: 6 7 8 9 10 | Aftertaste Score: 6 7 8 9 10 | Acidity Score: 6 7 8 9 10 — Intensity High, Low | Sweetness Score: 6 7 8 9 10 — Intensity High, Low | Body Score: 6 7 8 9 10 | Balance Score: 6 7 8 9 10 | Clean Cup Score: ☐ ☐ ☐ ☐ ☐ | Uniformity Score: ☐ ☐ ☐ ☐ ☐ | Overall Score: 6 7 8 9 10 | Defects (subtract) Taint=2 Fault=4 # cups ☐ X Intensity ☐ = ☐ | Total Score | Final Score

Notes:

Sample # | Roast Level of Sample | Fragrance/Aroma Score: 6 7 8 9 10 — Dry, Qualities:, Break | Flavor Score: 6 7 8 9 10 | Aftertaste Score: 6 7 8 9 10 | Acidity Score: 6 7 8 9 10 — Intensity High, Low | Sweetness Score: 6 7 8 9 10 — Intensity High, Low | Body Score: 6 7 8 9 10 | Balance Score: 6 7 8 9 10 | Clean Cup Score: ☐ ☐ ☐ ☐ ☐ | Uniformity Score: ☐ ☐ ☐ ☐ ☐ | Overall Score: 6 7 8 9 10 | Defects (subtract) Taint=2 Fault=4 # cups ☐ X Intensity ☐ = ☐ | Total Score | Final Score

Notes:

5. ADVENT OF THE "Q" COFFEE SYSTEM

In 1995, SCAA established the Specialty Coffee Institute (SCI) as an educational foundation inside the parent organization of SCAA. The purpose was to create a scientific center for the evaluation of specialty coffees that could be funded by tax-deductible contributions from members and other organizations and not part of SCAA's funding from member dues and conference revenues. When the "coffee crisis" hit in 1998 and coffee prices plummeted, bringing the price for washed arabica coffee to levels well below their production costs, SCI changed its mission and its name. Becoming the Coffee Quality Institute (CQI) and taking on the mission of "Working internationally to improve the quality of coffee and the livelihoods of those who produce," CQI began working with coffee producers in Central and South America to assist them in getting better prices through improved quality.

The cornerstone of this program was the development of the Q Coffee System, which was a formalized method of cupping and grading coffee, based on the SCAA Cupping and Grading protocol. The initial pilot program was funded through a grant from the United States Agency for International Development (USAID), and the first classes for a group of students from Colombia were presented in Spanish at SCAA's headquarters in Long Beach, CA, USA. In addition to teaching the Cupping and Grading protocol, the students were put through a battery of sensory tests to measure their ability to taste and smell, as well as their sensory acuity in differentiating coffees based on their origins, and their consistency in actually rating different quality levels using the SCAA cupping form. In all, the students took 22 individual tests, with the requirement that they pass each test to earn the title of Q Grader.

This was the first time the coffee industry had ever created a "formal" cupping training program; open to everyone willing to take the week-long course, and it was a huge success. USAID continued to fund training programs in Central and South America, then later in East Africa, whereas the private sector in the consuming countries provided funding for training cuppers in the roasting and importing community. By the end of 2015, CQI had trained over 6000 cuppers and certified more than 3500 Q Arabica Graders in more than 60 countries, establishing the Q Coffee System as the international standard for defining "specialty coffee," as coffees with a cupping score of 80 or more points on a 100 scale and with no primary defects. "Great taste—no defects" became the simple definition of "specialty coffee."

6. EXPANSION OF THE Q SYSTEM TO INCLUDE ROBUSTA COFFEES

By 2009, arabica coffee prices had fully recovered and its producers were beginning to prosper. However, robusta coffee prices stubbornly remained at

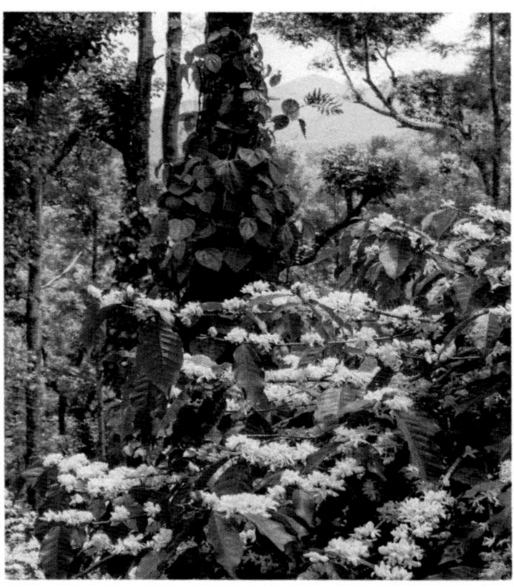

FIGURE 8.1 Robusta coffee plants. *Courtesy: Photograph by Mr. Vivek Muthuramalingam.*

critically low levels, with many coffee farmers still struggling to survive. In part, this was due to the incredibly and inexcusably low standards set by the robusta price futures trading center in London—London International Finance Futures and Options Exchange (LIFFE). Although arabica standards for futures trading set in New York—International Coffee Exchange required a cupping evaluation approval (coffee like flavor) and no more than 15 defects per 350 g sample, LIFFE had no cupping evaluation and permitted up to 450 defects per 500 g sample for robusta lots tendered against the LIFFE futures contracts. As a result, robusta commercial prices tended to be just half of arabica commercial prices (Fig. 8.1).

Realizing that low robusta prices also indirectly held down arabica prices, CQI took steps to develop a Q program for robusta coffees based on the highly successful Q arabica model that clearly demonstrated higher qualities that led to higher prices. Although physical grading of robusta coffees was very similar to that of arabica coffees because the defects are also very similar, the real challenge was developing a cupping protocol suitable for robusta coffees. This was particularly difficult for a variety of reasons. Robusta coffees are substantially different in their chemical makeup than arabica coffees due to plant genetics, as robusta coffees have 22 chromosomes, whereas arabica coffees have 44. Also, there were no official protocols for robusta cupping.

7. EVOLUTION OF THE UGANDA COFFEE DEVELOPMENT AUTHORITY CUPPING FORM AND PROTOCOL

Thanks to a grant from USAID, CQI was able to start a pilot program in Uganda, the birthplace of robusta coffees. Over a 2-year period, 53 coffee professionals from 18 different countries took part in four "workshops." Robusta samples from Uganda and Tanzania were cupped alongside of robusta samples from Brazil, Central America, India, Indonesia, and Vietnam. The outcome from the 3500 man-hours of work was a Robusta Cupping Form and Cupping Protocol along with a *Green Robusta Coffee Grading Handbook*. Kenneth Davids and Sunalini Menon provided invaluable assistance and expertise due to their prior experience in cupping robusta coffees. The workshops proved conclusively that robusta coffees can be differentiated by origin and cup quality, and they can be evaluated on a 100-point scale, with the same standard of 80 points or higher used to separate "fine robusta" coffees from the commercial grades. The terminology of "fine robusta" was selected to differentiate these coffees from "specialty arabica" coffees.

8. UGANDA COFFEE DEVELOPMENT AUTHORITY ROBUSTA CUPPING FORM

The Cupping Form provides a systematic means of recording 10 important attributes for robusta coffee: (1) fragrance/aroma, (2) flavor, (3) aftertaste, (4) bitter/sweet aspect ratio, (5) mouthfeel, (6) balance, (7) salt/acid aspect ratio, (8) uniform cups, (9) clean cups, and (10) overall. Defects, both taints and faults, can also be recorded on the form. The specific flavor attributes are positive scores of quality reflecting a judgment rating of the cupper; the defects are negative scores denoting unpleasant flavor sensations; the overall score is based on the flavor experience of the individual cupper as a personal appraisal.

The aromatic aspects include "*dry* fragrance" (defined as the smell of the ground coffee when still dry) and "*wet* aroma" (the smell of the coffee when infused with hot water). One can evaluate this at three distinct steps in the cupping process: (1) sniffing the grounds placed into the cup before pouring water onto the coffee; (2) sniffing the aromas released while breaking the crust; and (3) sniffing the aromas released as the coffee steeps.

"Flavor" represents the coffee's principal character, the mid-range notes, in between the first impressions given by the coffee's initial aroma and taste to its final aftertaste. It is a combined impression of all the gustatory (taste bud) sensations and retronasal aromas that go from the mouth to nose. "Aftertaste" is defined as the length of positive flavor (taste and aroma) qualities emanating from the back of the palate and remaining after the coffee is expectorated or swallowed. In robusta coffees it is often driven by the potassium level found in the coffee, with high levels resulting in brackish (high saltiness and displeasing aromas) aftertastes and with low levels resulting in savory (low saltiness and pleasing aromas) aftertastes.

Both "Bitter" and "Sweet" taste sensations are present in robusta coffees. The bitter component stems principally from the chlorogenic acid and potassium levels present in the coffee, whereas the sweet component is derived from the fruit acids, and sugar levels in the coffee. Fine robusta coffees have a low bitter and high sweet aspect in their taste, whereas commercial robusta coffees have a high bitter and low sweet aspect ratio in their taste. In determining the "Bitter/Sweet Aspect Ratio Score," the cupper rates the relative bitterness, giving the higher score to the lower perceived bitterness, whereas at the same time, the cupper rates the relative sweetness, giving the higher score to the higher perceived sweetness.

The quality of "mouthfeel" is based upon the tactile feeling of the liquid in the mouth, especially as perceived between the tongue and roof of the mouth. Most samples with heavy mouthfeel may also be perceived as high quality due to the presence of brew colloids. Brew colloids are formed as the oils extracted from the ground coffee coagulate around the microfine bean fibers suspended in the brew. Mouthfeel has two distinct aspects: weight and texture.

How all the various aspects of flavor, aftertaste, bitter/sweet aspect ratio, and mouthfeel of the sample work together and complement or contrast to each other is "balance." As the intensity of each of these attributes increases, it is more difficult for the sample to remain in balance. If each attribute increases equally in intensity, then the balance score is high. If the sample is lacking in one or more attributes or if some attributes are overpowering, the "balance" score would be reduced.

Salt/acid aspect ratio, or "Softness," refers to a pleasing and delicate taste that is derived from distinguishable acidity and sweetness in the sample, stemming from the presence of fruit acids and sugars. Lower levels of potassium and chlorogenic acid also contribute to this character. It is comparable to the "strictly soft/strictly hard" categorization of Brazilian coffees. It is one of the striking taste differences between fine robusta and commercial robusta coffees.

"Uniform cups" refer to consistency of flavor of the different cups of the sample tasted. If a single sour, ferment, phenolic, or other off-tasting bean is present in any of the cups, one or more of the cups will exhibit a different taste. This inconsistency in the flavor of the coffee is a very negative attribute. This type of inconsistency should be so distinct that the cupper can easily identify the off-cup in a triangulation with the other cups in the sample set. The rating of this attribute is calculated on a cup-by-cup basis, with two-point award for each cup that is uniform.

"Clean cups" refers to a lack of interfering negative impressions from first ingestion to final aftertaste, of cup. In evaluating this attribute, notice the total flavor experience from the time of the initial ingestion to final swallowing or expectoration. If a single moldy, dirty, and baggy, or other off-tasting bean, is present in any of the cups, one or more of the cups will exhibit a non-coffee taste. Any non-coffee like tastes or aromas will disqualify an individual cup.

The "overall" score attribute is meant to reflect the holistically integrated rating of the sample as perceived by the individual cupper. A sample with many highly pleasant attributes, but not quite "measuring up" to the cupper's expectation would receive a lower rating. A coffee that met expectations as to its character and reflected particular origin flavor qualities would receive a high score. An exemplary example of preferred characteristics not fully reflected in the individual score of the individual attributes might receive an even higher score. This is the step where the cuppers make their personal appraisal of the coffee. Good cuppers do not allow their personal preference for a coffee to interfere with the rating of the other flavor attributes of the sample.

9. UGANDA COFFEE DEVELOPMENT AUTHORITY ROBUSTA CUPPING PROTOCOL

9.1 How to Identify "Fine" Robusta Coffees

To identify "fine" robustas in the cup, the Coffee Quality Institute (CQI) of the Specialty Coffee Association of America (SCAA), along with the Uganda Coffee Development Authority (UCDA) conducted a number of workshops in various robusta coffee growing countries, to arrive at a protocol, including a scoring system, that needs to be followed to identify "fine" robustas, from "commercial" ones.

For carrying out the sensorial evaluation of robusta coffee, the samples of green coffee are first subjected to roasting, with the roast color being 58 on the Agtron. Thereafter, the roasted beans are allowed to rest for at least 8 h, as robusta beans are denser than arabica beans and present greater resistance to heat. In layman's language, robusta whole bean roast color should be medium to medium dark, unlike the arabica beans wherein the roast color is light to medium light. While preparing the robusta brew for cupping, the ground coffee to water ratio is 8.75 g for 150 mL of water, with the coffee beans being ground immediately prior to cupping and positively not more than 15 min before infusion with water.

When sensory evaluation or "cupping" of robustas is carried out, the most important factor to be kept in mind is the reason and purpose of the evaluation and how the results will be utilized thereafter.

During cupping, the important taste attributes that are evaluated are fragrance/aroma, flavor, aftertaste, bitter/sweet aspect ratio, mouthfeel, balance, salt/acid aspect ratio, uniform cups, clean cups, with the overall rating of the coffee being carried out by the person evaluating the cup based on her or his personal appraisal of the quality of the coffee sample.

Step #1—Fragrance/Aroma: Within 15 min after samples have been ground, the dry fragrance of the samples should be evaluated by lifting the lid and sniffing the dry grounds.

Both the type and intensity of the dry fragrance are rated. The type of dry fragrance will range from flowery to fruity to herbal. After infusing with water, the crust is left unbroken for at least 3 min but not more than 5 min. Breaking of the crust is done by stirring three times, then allowing the foam to run down the back of the spoon while gently sniffing. Both the type and intensity of the wet aroma is rated. The cupper should also note the type of wet aroma on the small horizontal line. The type of dry fragrance will range from fruity to herbal to nut like. In addition caramel and/or cocoa may be detected in the wet aroma.

Step #2—Flavor, Aftertaste, Bitter/Sweet, Mouthfeel, and Balance: When the sample has cooled to about 70°C (160°F), 8−10 min from infusion, evaluation of the liquor should begin. The liquor is aspirated into the mouth in such a way as to cover as much area as possible, especially the tongue and upper palate. Because the retro nasal vapors are at their maximum intensity at these elevated temperatures, flavor and aftertaste are rated at this point. As the coffee continues to cool [70−60°C (160−140°F)], the mouthfeel, bitter/sweet aspect ratio, and balance are rated next. Mouthfeel is a combination of weight and texture. The weight comes from microfine fiber particles swept off the ground-up beans and the texture comes from the oils extracted from the coffee particles and suspended in the brew. Both the weight (heft on the tongue compared to pure water) and texture (slipperiness compared to pure water) are rated. Bitter/sweet aspect ratio is the relative balance between the bitter and sweet taste sensations, with the optimum result coming from a low bitterness and high sweet combination. Balance is the cupper's assessment of how well the flavor, aftertaste, mouthfeel, and bitter/sweet aspect ratio fit together in a synergistic combination. All four attributes should be present in equal intensities to achieve "balance" in the cup. The greater the intensity, while still maintaining balance in the cup, the higher the rating.

Step #3—Salt/Acid Aspect Ratio, Uniform Cups, and Clean Cups: As the brew approaches room temperature (below 37°C; 100°F) salt/acid aspect ratio, uniform cups, and clean cups are evaluated. Salt/acid aspect ratio, or "softness," is perception of the sample's acidity and sweetness that is not diminished by a high salty-bitter perception. Uniform cups and clean cups are rated on a cup-by-cup basis. For these attributes, the cupper makes a judgment on each individual cup.

Step #4—Overall: Evaluation of the liquor should cease when the sample reaches 21°C **(70°F)** and the overall score is determined by the cupper and given to the sample as "cupper's points" based on all of the combined attributes.

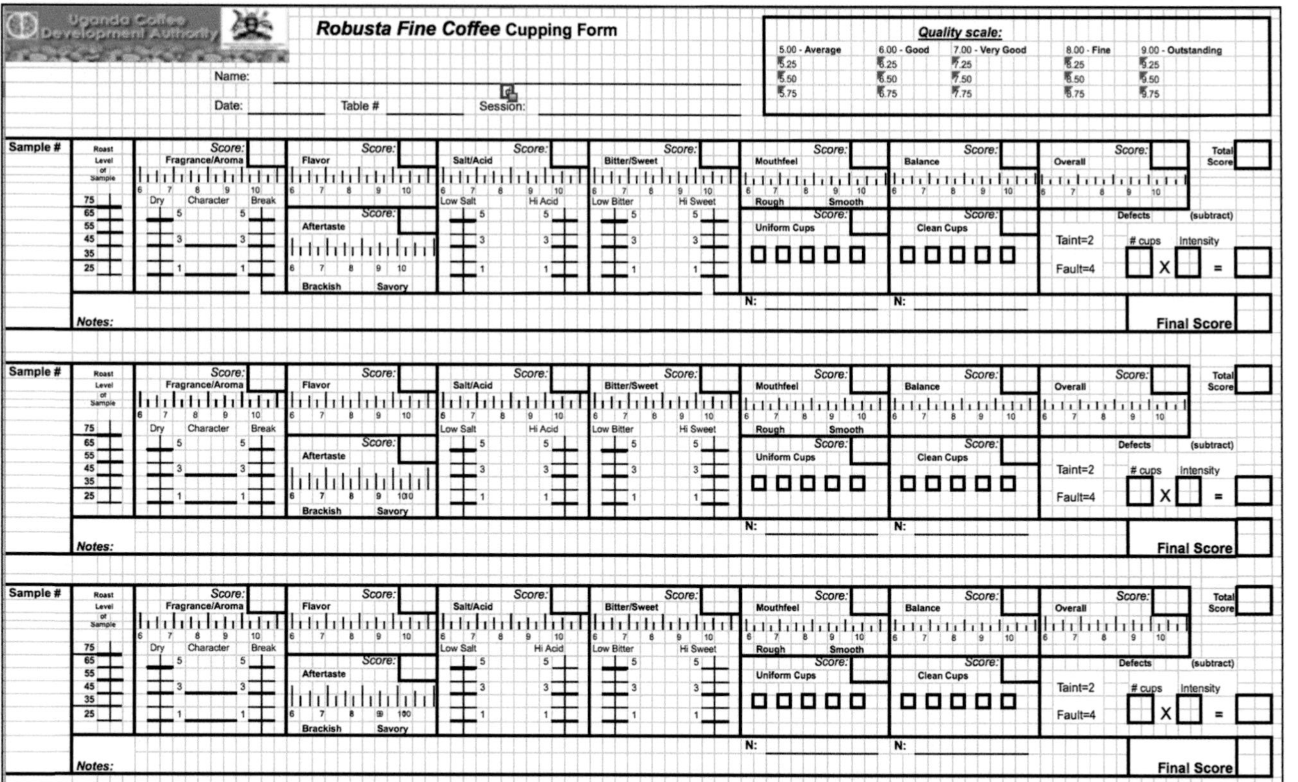

Uganda Coffee Development Authority

Robusta Fine Coffee Cupping Form

Name:

Date: Table # Session:

Quality scale:

5.00 - Average	6.00 - Good	7.00 - Very Good	8.00 - Fine	9.00 - Outstanding
5.25	6.25	7.25	8.25	9.25
5.50	6.50	7.50	8.50	9.50
5.75	6.75	7.75	8.75	9.75

Sample #
Roast Level of Sample: 75, 65, 55, 45, 35, 25
Fragrance/Aroma Score: 6 7 8 9 10; Dry 5 3 1; Character; Break 5 3 1
Flavor Score: 6 7 8 9 10
Aftertaste Score: 6 7 8 9 10; Brackish Savory
Salt/Acid Score: 6 7 8 9 10; Low Salt 5 3 1; Hi Acid 5 3 1
Bitter/Sweet Score: 6 7 8 9 10; Low Bitter 5 3 1; Hi Sweet 5 3 1
Mouthfeel Score: 6 7 8 9 10; Rough Smooth
Uniform Cups Score:
Balance Score: 6 7 8 9 10
Clean Cups Score:
Overall Score: 6 7 8 9 10
Total Score
Defects (subtract)
Taint=2 # cups Intensity
Fault=4 X =
N: N:
Notes:
Final Score

Sample #
Roast Level of Sample: 75, 65, 55, 45, 35, 25
Fragrance/Aroma Score: 6 7 8 9 10; Dry 5 3 1; Character; Break 5 3 1
Flavor Score: 6 7 8 9 10
Aftertaste Score: 6 7 8 9 10; Brackish Savory
Salt/Acid Score: 6 7 8 9 10; Low Salt 5 3 1; Hi Acid 5 3 1
Bitter/Sweet Score: 6 7 8 9 10; Low Bitter 5 3 1; Hi Sweet 5 3 1
Mouthfeel Score: 6 7 8 9 10; Rough Smooth
Uniform Cups Score:
Balance Score: 6 7 8 9 10
Clean Cups Score:
Overall Score: 6 7 8 9 10
Total Score
Defects (subtract)
Taint=2 # cups Intensity
Fault=4 X =
N: N:
Notes:
Final Score

Sample #
Roast Level of Sample: 75, 65, 55, 45, 35, 25
Fragrance/Aroma Score: 6 7 8 9 10; Dry 5 3 1; Character; Break 5 3 1
Flavor Score: 6 7 8 9 10
Aftertaste Score: 6 7 8 9 10; Brackish Savory
Salt/Acid Score: 6 7 8 9 10; Low Salt 5 3 1; Hi Acid 5 3 1
Bitter/Sweet Score: 6 7 8 9 10; Low Bitter 5 3 1; Hi Sweet 5 3 1
Mouthfeel Score: 6 7 8 9 10; Rough Smooth
Uniform Cups Score:
Balance Score: 6 7 8 9 10
Clean Cups Score:
Overall Score: 6 7 8 9 10
Total Score
Defects (subtract)
Taint=2 # cups Intensity
Fault=4 X =
N: N:
Notes:
Final Score

Coffees, which secure a score of 80 and above on the cupping score sheet, will be rated as "fine" robusta coffees, not only securing a better price in the market, but also enabling its usage in a variety of ways, commencing from being sold as a pure robusta blend to robusta beans that would be utilized in a blend with arabica, providing a stronger coffee base to the brew and enhancing the flavors of the arabica beans.

Currently, the UCDA Robusta Cupping Protocol is followed internationally to identify and evaluate "fine" robusta coffees.

10. FACTORS INFLUENCING ROBUSTA FLAVOR

10.1 Plant Strain

In the Rubiaceae family *Coffea canephora* species or Robusta species occurs in many forms in the wild, especially in the Congo basin. *C. canephora var. Pierre*, C. *canephora* var. quillouensis, *C. stenophylla, Congensis*, and *C. bukobensis* are forms of different species, but popularly known as robusta. It is now becoming apparent that each of these "forms" of robusta could have their own distinct and unique cupping characteristics.

With respect to the arabica species, research has been carried out on the cup quality of different arabica varietals such as Caturra, Catuai, Mundo Novo, Villa Sarchi, Sarchimor, Colombia VCR, Castillo, etc., with the cupping characteristics known to the coffee farmer, the coffee buyer, and the trader, resulting in not only the preparation of particular specialty coffees, but also single origin branded coffees for the market. Unfortunately, with respect to the robusta species and its various forms, research has not been carried out on their individual sensory attributes, and also on the best practices for cultivation in different microclimates, which perhaps are the most important requirements to understand on how to improve the robusta species. Understanding the factors that create "fine robustas," if not "boutique robustas," is essential to increase the incomes of smallholder robusta farmers worldwide.

India has made the most advanced studies of robusta coffees over the past two decades. In India, there are three very important robusta forms, which are being cultivated on a commercial basis, namely Old Robusta/Peradeniya, S.274, and C×R. Although the Old Robusta or Peradeniya was introduced into India from Sri Lanka (formerly known as Ceylon) during the early 19th century, S.274 was the first robusta selection that was released by the Indian Coffee Research Station in the late 1940s.

Although the Old Robusta strain has interesting flavor nuances of chocolate and malt, lined with bright notes of citrus, S.274 has nuances of chocolate, caramel, and nuts, with flecks of spices brightening the cup. On the other hand, C×R is a hybrid cultivar, which has been developed through interspecific hybridization involving *Coffea congensis* and *Coffea canephora*. The Indian Central Coffee Research Station has also developed this cultivar, and its salient features are large and bold beans, with the liquor being soft, smooth, and buttery, with flavor notes of fruit and hardly any bitterness. Thus, the inherent quality characteristics of the plant strain are revealed in the cup and could be a major factor in determining the quality of the robusta as being "fine" or "commercial."

10.2 Altitude

The altitude at which robusta grows has an effect on the cup quality. Altitudes above 1000 m produce hard beans, and the cup has clear flavor, besides brightness. A great deal of research on the cup quality and classification of robustas into "Fine" and "Commercial" has been carried out by the Coffee Quality Institute (CQI) of the Specialty Coffee Association of America (SCAA) as early as August 2009, when work commenced on differentiating robustas into the categories of "Fine" and "Commercial." The effect of altitude was seen in the first workshop that was held in Uganda in August 2009, when it was observed that a natural or cherry coffee of Tanzania grown at 1500 m upwards and in volcanic soil, had delicious flavor notes of fruit, with smooth texture and sweet acidic hues, comparable to a quality arabica coffee.

Indian robusta coffees grown at altitudes above 1000 m tend to have clarity of flavors, with sweet acidic hues, enabling the branding of such coffees. Robusta coffees, which are grown under shade at altitudes above 1000 m are soft in the cup, with brightness and varied flavors of lemon and dry fig, layered with caramel and cocoa depending on the plant strain. Although the varied flavors could be an intrinsic attribute of the plant strain, the cultivation at high altitude, which results in slow growth and development, highlights and intensifies these inherent flavors, in addition to ensuring their clarity. It was observed that, the same strain grown at lower altitudes does not have this pronounced clarity of flavors and exhibits subdued flavor nuances.

10.3 Shade Trees

An interesting observation in India has been that, just like for arabica coffees, the type of shade trees, under which robusta cultivation is being carried out, could have an effect on the cup quality. Further details about the impact of shade on quality can be found in Chapters 2 and 4. It has been observed that, when robusta is grown under the shade of fruit trees, the cup profile changes for the better. Robusta coffees, which are grown under the shade of oranges, bananas, and sapodilla fruit, possess cup quality of decreased bitterness with brightness, flavors of fruits, nuts and chocolate, besides the texture becoming smooth and silky. A second finding on certain Indian farms is that, robusta coffees, which are grown in close proximity to pepper, with pepper vines even climbing up the stem of the plants, have fairly distinctive "spice" notes in the cup. All these are only organoleptic findings and there is no scientific study, which has been carried out, to confirm these findings. At present, there is no scientific data on the fruit sugars and the organic acid content of such coffees, to confirm that growing robusta coffees under fruit trees and/or spices could bring about enhancement and/or development of distinctive flavors in the cup.

10.4 Processing

There is a correlation between the cup quality of robusta and the processing steps to which the coffee is subjected (further information can be found in Chapter 3). In India, it was observed that, the washed robusta has a tastier profile than an unwashed robusta, though there are exceptions to this finding, with unwashed or natural robustas also being distinct, but complex in the cup. By and large, it was observed that the processing techniques followed on the farm have a bearing on its cup quality. Wet processing of robusta helps not only to mute and mellow the sharp notes of toasted corn and bitterness, which are often seen at the core of the robusta cup, but also helps in developing soft, buttery mouthfeel, and bright acidic nuances, which play a major role in softening the cup, besides highlighting the intrinsic flavors of the robusta strain.

Generally, a commercial robusta has very thick husk and toasted corn nuances and striking bitterness in the cup, besides unfavorable notes such as woodiness, staleness or rancidity, with the mouthfeel being harsh or coarse. These attributes are due to improper care taken during processing, which could be the result of low prices being offered in the market for the Robusta species. In India, experimentation with processing of robusta coffee has helped to produce beans with varying taste profiles. It has been observed that during harvesting, the cherries need to be well-developed and blackish red in color to ensure the development of their intrinsic flavors and to prevent astringency in the cup. This is a marked difference from the harvesting of arabica coffee, where the cherries have to be picked in a ripe red condition, as otherwise there could be an off note of "fermented" when picked in a darkish red condition.

During pulping, it should be ensured that pulpers are working correctly and adjusted to prevent the robusta cherries from getting "cut," thus resulting in "off notes" and lowered cup quality. After pulping, the cherries are often subject to fermentation to remove the sticky mucilage enveloping the parchment cover enclosing the coffee bean, after which it is washed through an aqua washer. Just like pulping, care needs to be taken to ensure that there are no "cuts" during the washing process. In India, it is observed that natural fermentation could be the best for removal of mucilage. However, this step has to be verified with every coffee season, as the nutritional status of the plants, the quantum of mucilage within the cherries, and the temperature on the farm could vary from year to year.

During fermentation, the coffee is constantly mixed to bring about uniform breakdown of the mucilage. Thereafter, the coffee is washed through an aqua washer taking care to ensure that there are no "cuts" during the washing process.

After fermentation and washing, the coffee is dried carefully, either under the sun or with a combination of sunshine and mechanical dryers.

During drying, care should be taken to prevent under drying or over drying, as these could also affect the "cleanliness" of the robusta cup. Thereafter, the beans are bagged in jute bags, which are manufactured with vegetable oil to preserve the intrinsic quality of the beans. Coffee is stored in well-ventilated warehouses to protect its flavor until further processing at the dry milling factory.

Robustas, which have been prepared by the washed or unwashed methods, not only contribute to provide crema for the much sought after espresso, but have also helped in highlighting the flavor nuances of the blend. Well-washed robusta beans provide clean notes of strength and mild yet subtle flavors to the blend, thus providing wholesomeness to the coffee brewed either as espresso or as a filter drip coffee.

It should be noted, however, that applying the wet processing method to robusta is a lot more difficult than preparing washed arabica, as the mucilage content in this species is much thicker and stickier than in arabica coffee. In some countries robusta fermentation may not be complete even after 72 hours, and considering that robusta is cultivated in lower altitudes, the high temperatures in these areas could make the process riskier, requiring extremely careful monitoring to avoid over fermentation. Also, the long fermentation time and the thick robusta mucilage would require good infrastructure at the pulping station, especially greater tank space. Additional water may also be required for the preparation of washed robusta, and the effluents from the pulping station would need to be carefully monitored and treated to preserve the water quality environment.

Today mobile and motorized processing units are available, combining the two steps of depulping and mucilage removal as one step, with minimum usage of water, helping even small farmers to prepare wet processed robustas that have greater demand in the market and also bring better returns to the coffee farmer. Some of these mobile processing units are also fitted with small mechanical dryers to help in uniform drying after washing, and thus help in the preservation of quality without development of off-flavors, such as woody, stale, rancidity etc., which would affect the quality of the cup (International Trade Centre, 2002)[1].

In addition to the washed method of processing, today, the "honey sun-dried"/"pulped natural" method of processing is also being utilized for the robusta species. This type of preparation could also be an excellent way to obtain and present a high quality, delicious tasting "fine robustas" to consumers. Processing of robusta coffee is much more difficult than arabica coffee, especially when prepared by the washed or the honey sundried methods of preparation. It is important that meticulous, organized, and careful processing steps are followed to preserve the intrinsic quality of the robusta beans

1. If the drying is not carried out carefully, especially in the mechanical dryer, with the coffee beans being subjected to high temperatures of 40°C and above, there is every likelihood of off-flavors such as woody/stale/rancidity being developed, which could affect quality in the cup.

and to avoid the off-tastes of commercial robustas, which could have woodiness, aged, stale, or rancid off-notes in the cup.

Although in India it is firmly believed that fermentation is required to highlight those very delicate flavor notes that are present deep down in arabica and robusta coffee beans, the farmer first experiments with the coffee beans on his farm, whether arabica or robusta, to examine if fermentation is required, and if so, the time of fermentation and the number of days required for sun drying.

In the past, India has experienced adequate sunshine for patio drying, although recently the impact of climate change is challenging the industry with the occurrence of unwanted and untimely rains. Mechanical dryers are now being studied, and experimentations so far have shown that a mechanical dryer could only be used to finish the drying process, with the major part of drying being carried out under direct sunshine to avoid any off notes in the cup.

During drying, absolute care is taken to prevent under drying, or over drying with the optimal temperature in a mechanical drier not exceeding 40°C, as these could again affect the "cleanliness" of the cup. The coffees are thereafter bagged in jute bags, which are manufactured with vegetable oil, to preserve the intrinsic quality of the beans. The coffees are stored in well-ventilated warehouses to protect the flavor of the coffee beans, and when required for marketing, the beans are processed at the dry milling factory.

Thus, "cleanliness" of the robusta cup can be affected at every stage of processing. It is for this reason, whether it is arabica or robusta, the coffee farmer in India is trained and educated to take meticulous care at the various steps in processing to ensure the clarity of flavors and to prevent any off odors or off notes from developing in the coffee cup.

11. TRADITIONAL COFFEE GRADING

Nature does not provide roasters with perfect coffee beans. Every coffee tree, regardless of species or origin, produces defective coffee beans due to a variety of factors. In addition, poor practices in picking and processing also produce beans that are "defective" in either their taste or appearance or both. For the most part these physical defects can be identified by the naked eye, and in well run mills, they are removed either by machine or by hand. Depending on the percentage of defective beans removed, coffee lots are then classified by comparing the number of remaining defective beans to a given sample weight, usually 350 g.

Classification systems vary with country, but all of the systems divide defects into two groups: (1) primary defects that negatively impact cup quality, such as black or sour beans; and (2) secondary defects that negatively impact the visual appearance and/or cup quality of the lot, such as broken, chipped, or cut beans. All classification systems also establish a standard for converting secondary defects into the "equivalent" of one "full" defect. For example, five "chipped" beans would be considered the equivalent of one "black" bean. Thus, in grading a particular lot of coffee, the physical grade could be described as 12 defects, of which five would be primary defects (five black beans) and seven would be secondary defects (35 chipped beans). Table 8.1.

Concurrent with the development of the SCAA Cupping Form, SCAA's Technical Standards Committee also developed a Green Arabica Coffee Classification System based on the following standards: Table 8.2 and Fig. 8.2.

In respect to robusta, the CQI and UCDA have drawn up a fine robusta coffee classification system based on the following standards: Tables 8.3 and 8.4 and Fig. 8.3.

TABLE 8.1 Table of Defect Equivalents in 350 g of Green Arabica Coffee

Category 1 Defects	Full Defect Equivalents
Full black	1
Full sour	1
Dried cherry/pod	1
Fungus damaged	1
Foreign matter	1
Severe insect damage	5

TABLE 8.2 Table of Defect Equivalents in 350 g of Green Arabica Coffee

Category 2 Defects	Full Defect Equivalents
Partial black	3
Partial sour	3
Parchment/pergamino	5
Floater	5
Immature/unripe	5
Withered	5
Shell	5
Broken/chipped/cut	5
Hull/husk	5
Slight insect damage	10

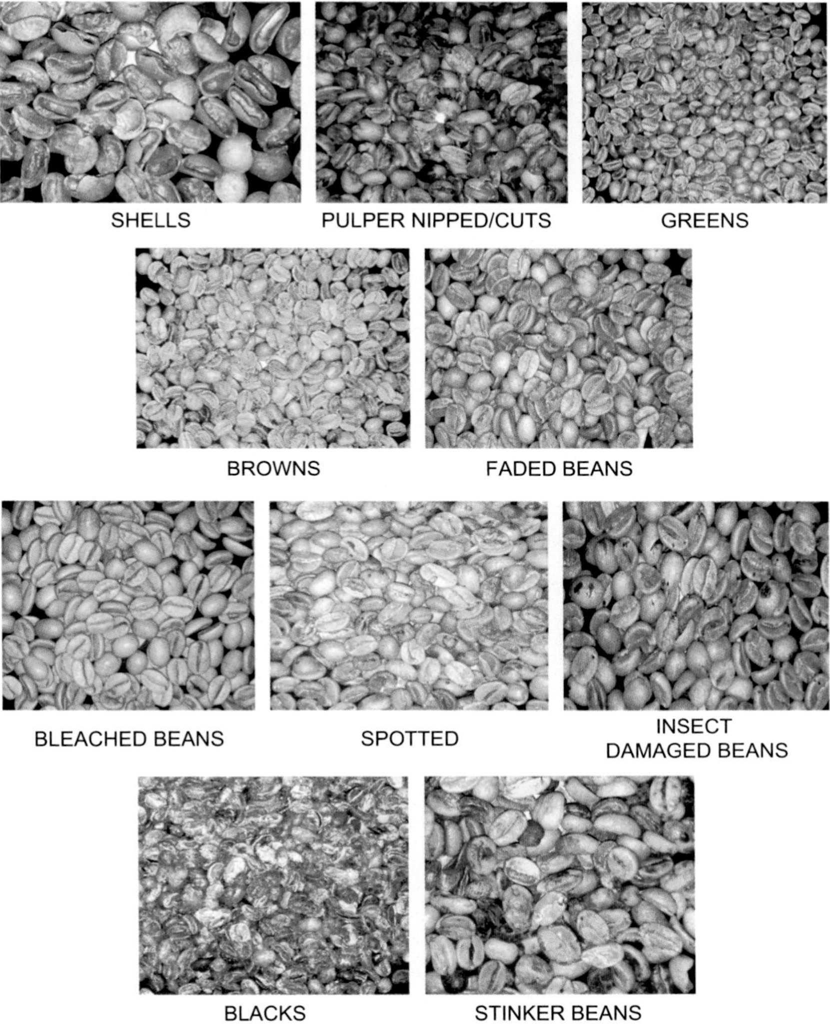

FIGURE 8.2 Defectives in arabica.

As part of the physical grading standards for SCAA, there is also a physical grading of the roasted coffee sample. Under this test, 100 g of a roasted sample is inspected for "quaker" beans. These are beans that do not develop properly during the roasting process and can be identified by a light, pale brown color distinctly different than the dark brown color of the other fully roasted beans. This condition is caused by the coffee cherry being picked before it is fully ripe, resulting in an incomplete chemical development of the two coffee seeds

TABLE 8.3 Table of Defects Equivalents in 350 g of Green Robusta Coffee

Category 1 Defects	Full Defect Equivalents
Full black	1
Full sour	1
Dried cherry	1
Fungus damaged	1
Foreign matter	1
Severe insect damage	3

TABLE 8.4 Table of Defects Equivalents in 350 g of Green Robusta Coffee

Category 2 Defects	Full Defect Equivalents
Partial black	3
Partial sour	3
Immature/unripe/green	5
Withered/shriveled	5
Floater/spongy	5
Chalky white/bleached	5
Broken/chipped/cut	5
Parchment	5
Shell	5
Hull/husk	5
Slight insect damage	10

inside the cherry. This is why high quality coffees are handpicked, with pickers trained only to pick the ripe, red cherries.

It is apparent that grading systems play an important role in determining the quality of a product and the returns to the farmer in particular and to the coffee producing origin as a whole.

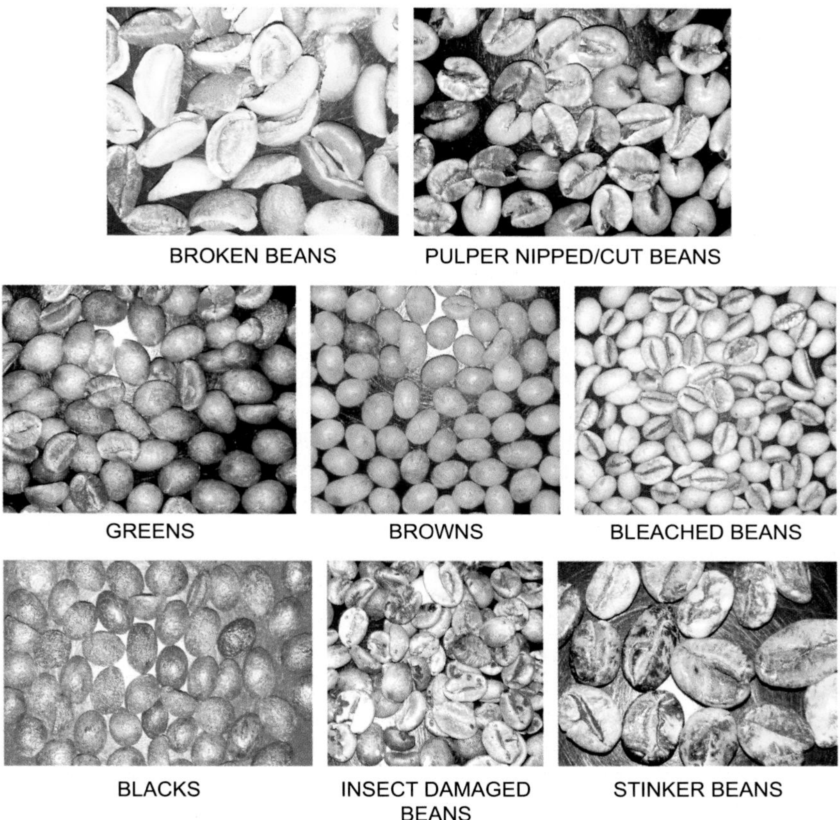

FIGURE 8.3 Defectives in robusta.

12. OUTLOOK

With the growth of coffee consumption in Asia, particularly the current rapid growth in China, the long-term prospects for all coffee producers is quite promising. Although Japan has led the way, South Korea, Taiwan, China, and the other Southeast Asian countries are developing a healthy coffee drinking culture. Of particular importance, it is the younger generation of Asians that is developing a coffee palate. Japan, South Korea, and now China, have the largest total number of Q Graders of any consuming country. Virtually all of these Q Graders are under the age of 30 years old, and the vast majority of these young coffee professionals are extremely proud of their coffee cupping and grading skills.

In addition, both SCAA and SCAE, through their professional development programs and the World Coffee Events competitions, have done a great job in promoting a wide range of coffee skills to young professionals in the area of brewing, espresso beverage preparation, and roasting. These critically important activities mean that the coffee drinking trends in the Asian market will follow those of the specialty coffee markets in North America and Europe.

For arabica coffee, part of this trend in the specialty coffee markets is the search for great tasting varietals. This quest was ignited by the now famous Geisha varietal from the Esmeralda farm in Panama that commanded record high prices in an Internet auction in 2009. It prompted coffee growers worldwide to try this particular varietal on their farms to see if they could match the incredible cup quality found in Panama. This trend is the forerunner of the coffee industry moving closer to the wine model in which regions became famous for producing a particularly high quality wine from a specific grape varietal.

In addition, arabica producers in Central America began experimenting with new types of processing, going beyond the traditional washed or natural processing methods. Aqua pulping, variances in pulped naturals, and honey sundried are being experimented with around the globe to see if a specific type of processing method is best for either a varietal or microclimate. The market place is responding, albeit slowly, to the importance of providing consumers with better tasting coffees, not lower cost coffees.

Although it may take another generation of consumers, robusta producers will ultimately benefit from the changes now taking place in the arabica market. Besides giving a great yield and being less costly to produce than arabicas, robustas are cross-pollinating, which gives them the potential to produce a wide range of cup qualities. One of the great lessons learned from the Uganda workshops was the realization that robusta coffees actually show greater differences in their cupping quality than arabica coffees. This comes from the fact that robusta coffees, like most plants, require cross-pollination to reproduce, whereas arabica coffees are self-pollinating. This means the robusta coffee beans take on the cup character of other coffee plants located nearby (even beans from the same tree can have different cup characteristics), leading to distinctive "land-race" varieties, which was found to be particularly true of robusta coffees grown in Indonesia. This was demonstrated by the relative ease of triangulating the difference in robusta origins as compared with the difficulty in separating arabica origins using the same method of triangulation testing.

Evaluating the quality of a robusta cup is a lot more challenging than the arabica cup due to the complexity of the taste profile of the robusta bean. However, "fine robustas" have special, acceptable, and distinct taste profile, which are unique to these beans, providing a basket of flavors, and thus, varied cups to the consumer. "Fine robustas" can be used not only in Espresso making, but also for drip filter preparations, to fortify milk-based drinks such as cappuccino, latte, besides being a component of high caffeine blends. All the mentioned special attributes provide an opportunity to develop and promote "fine robustas" and expand its use not only among specialty coffee consumers but also across the general population of all coffee consumers.

The next step in developing the full potential for "fine robustas" is to increase roaster awareness of the special features of fine quality robustas, which in turn would help in promoting the species of robusta, besides offering diversity in taste profiles to the consumer. Most "fine robusta" coffees from India, Indonesia, and Uganda are now securing a premium based on their quality as they are being used in high-end blends to enhance the flavor and market acceptability of the brands that use them. "Fine robustas" add a balance to any blend of coffee that cannot be achieved by the use of any other type of coffee. They create a flavor profile that is highly valued by all consumers around the world.

An **"R Certified"** robusta coffee from India in 2014 sold at five times the current commercial price. The market for "fine robusta" coffees is small, and it is mostly out of sight, as the small group of traders for this category often keeps sources, buyers, and prices secret. But "fine robusta" from Uganda is a significant part of the high quality espresso blends marketed in Europe; "fine robusta" coffees from India are being roasted and sold as pure robusta blends under a brand name; "fine robusta" coffees from Indonesia end up in the high quality blends sold in Japan; but "fine robusta" coffees from Brazil are lost in the Brazilian soluble industry.

In summary, the specialty arabicas will continue their journey in an upward spiral, ultimately reaching the stature of the wine industry and sold by varietals from specific growing regions. Although the specialty market for arabicas will grow, the market for fine robustas will also develop and grow in parallel, with cupping playing a major role in distinguishing and identifying unique and distinctive coffees. It is important for exporters, importers, roasters, and consumers to be educated on the nuances of "specialty arabicas," as also on the nuances of "fine robustas," to be able to use these coffees such that their uniqueness and distinctiveness can be highlighted in the cup, enabling the consumer to understand the difference between commercial and "specialty arabica" coffees, as well as commercial and "fine robusta" coffees.

REFERENCES

International Trade Centre - Product and Market Development - Coffee An Exporter's Guide, 2002. Coffee Quality - Robusta, pp. 266–271.

Lingle, T.R., 1984. Coffee Cuppers' Handbook, the Specialty Coffee Association of America (SCAA).

Uganda Coffee Development Authority, 2010. Robusta Cupping Protocols, pp. 1–19.

Ukers William, H., 1935. All About Coffee, Chapter XXII, the Botany of the Coffee Plant. New York The Tea and Coffee Trade Journal Company, pp. 268–285.

Chapter 9

Trading and Transaction—Market and Finance Dynamics

Eric Nadelberg[1], Jaime R. Polit[2], Juan Pablo Orjuela[3], Karsten Ranitzsch[4]
[1]*Granite Mountain Market Forecasts, Prescott, AZ, United States;* [2]*Be Green Trading SA, Lausanne, Switzerland;* [3]*CFX Risk Management Ltd., London, United Kingdom;* [4]*Nestlé Nespresso SA, Lausanne, Switzerland*

1. INTRODUCTION

The way coffee is bought by the roaster has gone through changes in recent years and very different philosophies are being applied. Some of these buying models aim to eliminate some of the many steps and participants in the value chain, which are between them and the farmer, and also almost disconnected from the futures market. Also even though we will not aim to come with the ultimate right suggestion, we will still suggest some ideas about the potential future of buying models, which could lead to a more sustainable and coherent valuation of green coffee.

This chapter addresses the utilization of derivatives in coffee and their relevance for the different market participants who are involved in the coffee value chain. The main market places that are used as hedging vehicle will be described as well as its historical evolution. This chapter builds, in particular, on Chapter 6.

2. TRADING AND ITS HISTORY: SETTING THE CONTEXT

2.1 How Trading Started and Why

The history of futures trading goes back to 1750 BC when the first code (code of Hamurabi)[1] allowed the sale of goods and assets to be delivered at a future date at an agreed price. Trading at the time took place in temples.

1. The Code of Hammurabi—refers to a set of rules or laws enacted by the Babylonian King Hammurabi (reign 1792–50 BC.).

The Craft and Science of Coffee. http://dx.doi.org/10.1016/B978-0-12-803520-7.00009-8

The reasons for having futures contracts were quite straightforward. Agricultural crops around the world are in most cases harvested once a year, sometimes twice. This is the time during which many harvest activities require upfront financing by the farmer to pay labor for the harvest or to rent/operate harvesting equipment. Many other costs are also due during the harvest time, such as the cost for schooling the kids or to pay back the loan for inputs.

In other words, during harvest time cash flow is crucial for the farmer. The farmer's harvested goods need to generate income to cover many expenses. If there is no buyer at the time, the farmer's goods are collected or if the bank wants to know how much the harvest is worth before it is collected, there could be a problem.

We can assume that coffee consumption is constant and linear throughout the world. For example, we want to drink coffee every day, and our favorite coffee shop or coffee company will expect to be able to provide roasted coffee throughout the year and not just once or twice. But how do we ensure that there is somebody buying coffee when the farmer needs to sell it, and not just months later when the roasters are ready to provide freshly roasted coffee to consumers?

This is why there needs to be a place where coffee can go if there is no roaster in sight, a place where it can be physically stored and a place where the roaster can go to buy the coffee. Futures contracts provided the perfect solution.

Modern day futures were originally conceived by the Chicago Board of Trade in 1864 as a way to help mid-western farmers lay-off their harvest risk by creating standardized "exchange traded" forward contracts, which were called futures. These helped farmers avoid having to sell their crops at prevailing cash market prices and if they so choose, they could hedge their upcoming November harvest in May or July and avoid the crush of harvest-time availability that can act to depress prices. With these new instruments, credit and counterparty risk were also negated. Therefore price risk could be passed along to opposite side hedgers (e.g., bakers and millers) or to speculators looking to profit from future price changes.

On March 7, 1882, coffee futures first started trading on the New York Coffee exchange. The exchange was founded in response to a disastrous market collapse in 1880. The Coffee Exchange of New York provided an arena that set standards for different grades of coffee and market where various commercial players could hedge against losses in the cash market. It also established an arbitration system to settle disputes and recorded and disseminated current market information to its members (ITC, 2011).

Transactions on the exchange were organized based upon the rules defining nine grades of deliverable coffee. No coffee with a grade below No. 8 was allowed into the United States, and Grade No. 7 was the basis for price quotations. All other grades were looked at in relation to the basis grade. In contrast to today where coffee is graded and categorized by origin, there was no geographical or processing type referenced. The exchange accepted coffee from North, Central, and South America as well as coffee from the West and

East Indies. Natural Robusta Coffee was not deliverable into the contract (Daviron and Ponte, 2005).

In 1928, the New York Coffee Exchange created a new contract to deal exclusively in Brazilian coffee and this used Santos 4 grade coffee as its benchmark. In 1976, this Brazilian or "B" contract was inactivated and the Coffee "C" contract was initiated. This is today's coffee futures contract and currently deals in 20 different origins with Central American mild coffees as the basis grade. Washed and semiwashed Brazilian coffee are the lowest types of deliverable coffee holding a six cents a pound discount to the usual market price (Daviron and Ponte, 2005).

With the changes in the contract and the increasing use of speedy electronic communications, the exchange also has taken on a more globalized perspective and expanded its delivery capacities. Physical coffee can now be delivered into exchange designated warehouses located in the following ports: the Ports of Antwerp, Hamburg/Bremen, Barcelona, New York District, Houston, New Orleans, Miami, and Virginia (International Commodity Exchange (ICE) available from https://www.theice.com/products/15/Coffee-C-Futures).

Notwithstanding the occasional challenge to its supremacy, the ICE Coffee "C" contract is the world's most active coffee pricing vehicle (see ICE website above). In 2014, more than 7.0 million coffee contracts were traded and 2015 saw 8,108,135 coffee contracts traded (Historical Monthly Volumes available from ICE website listed above). To put this into a more interesting perspective, in 2015/16 the US Department of Agriculture's (USDA) Foreign Agricultural Service estimates total world Arabica production at 84.986 M 60 kg bags. Since one contract of coffee is equal to 37,500 lbs of green coffee, or 284 60 kg bags, the slightly more than eight million contracts is therefore equivalent to just over 2.00 B 60 kg bags of green coffee, 27 times the amount of arabica coffee produced globally in a single crop year.

2.2 The Three Waves of Coffee

An important part of coffee history is the way the coffee industry has evolved through "three waves" (see also Chapter 19). This, in turn, has affected the industry's relationship with the futures exchanges, the commoditization of coffee, and the relationship between the reference price and what consumers are demanding to drink.

During the first wave, which can be traced back to the 1800s, entrepreneurs saw the opportunity to supply affordable and ready to make coffee to a wide audience. The industry saw the emergence of vacuum packed coffee and soluble coffee, which at the time pleased the majority of the consumers. In addition, coffee was fully commoditized in the sense that consumers were not demanding any quality differentiation and roasters focused on supplying good volumes with consistent quality.

In fact, both producers and roasters of coffee found the exchange a very efficient way to offset risk, and use it as a reference price. In general coffee was a commodity and quality differentiation was minimal; all that growers and roasters sought were consistent volumes and quality.

As coffee consumption became more common, the second wave began when roasters started to differentiate their product, and consumers demanded better quality. Specialty beans became the way to market the better coffee and although the grand majority of consumption still remained in the first wave, the second wave started the process of decommoditizing the coffee. Consumers now started to demand different grades of coffee, some traceability of the coffee they drank, and a better understanding of the different roasting styles.

As in every product life cycle consumers and producers continue to specialize and innovate as the industry matures. The third wave of coffee, which is a recent phenomenon, is one in which consumers have become more interested in the way the high quality coffee they drink is being produced. Often consumers want to know the exact location the coffee comes from and how it was grown and processed. This means that the process of decommoditization that started in wave two has advanced further in wave three. When the commodity has a reference price based on coffee of the most basic quality, quality differentiation is normally achieved by setting premiums on top of the reference price. For the first and second waves this was a suitable way to operate. During the third wave, new business models by roasters have challenged the way coffee is referenced to the exchange price and the way premiums are used to differentiate quality.

2.3 The Physical Flow of Coffee

Chapter 6 elaborated on the economic sustainability of coffee, and we could read in some detail about the issue of permanent exposure to high volatility in both currency as well as in the futures market. The New York market dictates the price of the commodity over which the farmers have no control. But who is controlling the market where the coffee sellers meet the buyers? Coffee prices are no longer in the hands of those whom we consider as fundamental traders. In fact, futures and options trading is now dominated by many other market participants who have no real interest in using the physical goods. Before looking at the coffee futures market from a trader's perspective, let's first look at coffee trading via an exchange versus the physical flow of coffee.

The physical flow of coffee varies from country to country and from region to region. It depends on both the situation and the preferences of participants in the flow. Both physical traceability and financial traceability are becoming increasing concerns and this is one of the reasons why we consider different buying models. The flow of arrows in Fig. 9.1 shows the different pathways coffee follows along the supply chain, and it illustrates that there are some

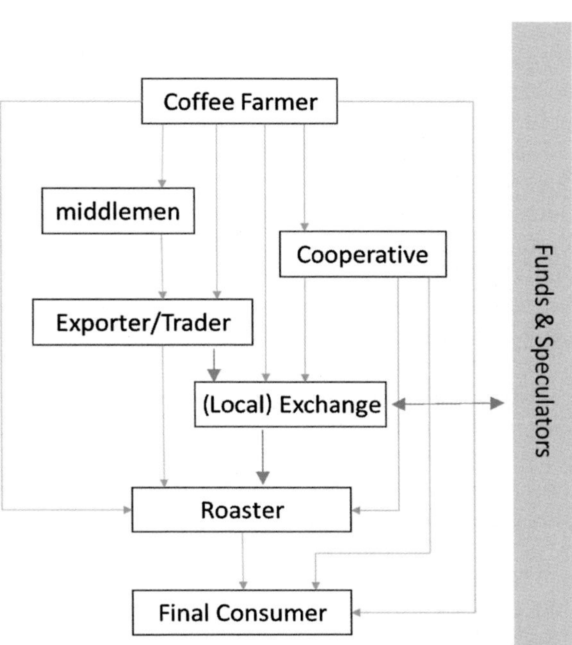

FIGURE 9.1 The physical flow of coffee.

participants who will actually not get anywhere close to the coffee itself. We therefore differentiate between hedgers and speculators (further described in Section 3.2).

3. THE COFFEE FUTURES MARKET

For many casual observers futures are synonymous with uncontrollable risk and confusion. This Trading Places derived popular image of trader chaos remains fixed in the public mind as the face of futures. But times have changed in that respect. Today, those colorfully garbed and apparently out of control pit traders are mostly gone. Trading has evolved into a silent flickering ballet of changing prices reflecting back at the trader from ubiquitous computer screens. The world's trading floors have mostly gone all dark and electronic.

But the frenzied open outcry trading portrayed by Hollywood always had an underlying order among the incomprehensible pit activity. Real contracts with legal rights and obligations were being traded by floor traders attempting to keep up with the order flow and emotional tides that would occasionally sweep over the trading ring. The popular image of an out of control environment created the perception that futures trading is akin to gambling, obscuring the bedrock principle for which these instruments were created, that of hedging, or price risk insurance.

Of late, however, this foundational function of the futures markets has been partially eclipsed by the giant shadow cast by the emergence of commodity index funds. These long only funds invest billions of dollars to create "exposure to the asset class" of commodities as a way of increasing net returns and reducing total portfolio volatility. In addition there are the more active and speculative commodity hedge funds looking to profit from commodity price movements. Activity of either group can temporarily obscure the impact of market fundamentals, but these distortions are mostly short lived, although they can occasionally be disruptive to commercial hedge programs.

For most people, the cash market is more familiar than the futures market because we are used to doing business that way. When we buy a cup of coffee at the shop and pay for it at the counter, we are acting in the spot or cash market. The price may change every now and again, but we have no real way of avoiding those changes. By contrast futures markets are "paper" markets used for hedging price risks or for speculation rather than for negotiating the actual delivery of physical goods (ITC, 2011).

As explained in Section 2, a grower delivering his coffee to a local mill is a familiar type of cash or forward market transaction. Another is a roaster buying coffee on the spot from a merchant. A futures contract on the other hand is a legally binding agreement to buy or sell a specified quantity of a particular commodity for delivery in a specified time in the future. A trader who buys or sells a futures contract makes a commitment to deliver or receive the contracted amount of the specific commodity of a specific grade (deliverable quality) to an exchange designated warehouse (specific location) by a specified time (delivery period). It is a process.

Futures market prices bear economically important relationships to the spot or cash price of the commodity. This is because prices in both markets tend to move together as traders in futures contracts are entitled to demand, or make delivery of physical coffee against their futures contracts (ITC, 2011).

The relationship between cash and the futures price is called the "differential" or the "basis". Also, the futures price in 6 months will be related to the futures price of 3 months. This link is called the spread value, and this can be an economically important variable to market participants as well (Kolb and Overdahl, 2006, Chapter 3).

3.1 The Benefits of Futures to the Trader

The main advantage of futures contracts over their cash forward or over the counter counterparts is that they are standardized and can easily be closed out by offsetting transactions. This means that any contract traded in that marketplace, coffee futures for example, can be substituted for any other coffee contract. This flexibility allows for the free flow of trading activity as it is not necessary for the parties of one futures transaction to close out that trade with one another. It is through the mechanism of the clearinghouse that all trades are offset, and the obligation of delivery is reserved for those who want to use that function of the exchange.

Although there are risks involved in futures trading, such as excessive leverage that can lead to large margin calls, when trading takes place solely in the spot or cash market, there is less ability to plan for future needs and availabilities. In addition, oftentimes the ability to exit a bad market position is not possible through an illiquid cash market.

3.2 Market Participants and Hedging and Speculating

Futures market traders' break down into two categories, hedgers and speculators. Each uses the market for different purposes—the hedger to fix prices or mitigate other risks, the speculator to make money by assuming directional risk that the hedger wants to reduce or eliminate entirely. If one or the other were not in the market, there would be less liquidity, and less commercial usefulness for the prices created through the auction system.

Hedging is the temporary purchase or sale of futures as a substitute for a transaction in the cash market (Kolb and Overdahl, 2006, Chapter 1). Hedgers in coffee are made up of growers, exporters, importers, and roasters. All use the futures market as a means to reduce their exposure to future price risks. Growers and importers tend to be short hedged (sellers) while roasters are long hedgers (buyers). Exporters tend to the long side of the market (buyers) as they usually have not yet bought and fixed all of the coffee they will have to ship, i.e., short physicals.

In comparison to dealing solely in the cash market, hedgers in the futures market can be more creative in generating a profit in the trade. For instance, using a futures hedge, the commercial trader can exercise on his discretion when to lift the hedge. If there is a strong trending environment hedges can be taken off, and market action can improve the final price. However, this flexibility is not available to users of forward contracts that must be liquidated through the delivery of the physical goods.

3.3 A Practical Example of Hedging

How does a hedge actually work and why is it not speculation? Let's look at an example with three stakeholders: A farmer who produces a crop and has different costs to cover, a roaster who needs coffee to roast to sell his product to consumers, and a coffee cooperative has members and roaster clients who expect to have coffee available. In most cases there would be at least one additional party involved which is the Green Coffee Trader. The real notion of the complexity is illustrated in Fig. 9.1 showing the number of participants who can be involved in the trading of coffee.

The roaster in this example has an established business that allows him to plan forward as his clients like to order their coffee supply 6–8 months in advance. After signing an agreement in early February with a customer in which the sales prices were fixed, the roaster wants to minimize his price risk. He, therefore, hedges the equivalent of the sales volume the very same day: He does this by buying one September Futures Contract @ 123 c/lb at the ICE Exchange in New York (NY)—and he is now one lot long.

It is not until late April that the roaster sees more and more new coffee from the Brazilian coffee crop on offer. He needs to buy unwashed arabica of NY2 fine cup quality, which is the commercial grade of the coffee he agreed to roast for his client and does not want to risk not having the right green coffee secured. He contacts the cooperative in Brazil from which he traditionally buys his coffee. The commercial manager is willing to offer 300 bags of NY2 fine cup as he knows that the members are expecting a good crop and that the timing will be normal. However, the cooperative does not have the coffee yet and the agreement is therefore only for the quantity (300 bags), the quality (NY2 fine cup), and the time of delivery (August shipment).

The final contract price remains open except for the differential of the premium above the New York exchange for this particular quality which is plus 12 c/lb in our example. Therefore, the roaster knows his cost (in contrast to the contract price with the cooperative) for the coffee will be 135 c/lb (123 + 12). The cooperative manager knows that there is a margin of 6 c/lb which is needed to cover the cost of the cooperative, including the cost for service to members, overheads, admin fees, and other expenses.

In May, one of the cooperative members, a large Brazilian farmer who produces 2000 bags of coffee per year, looks at the market price that is trading at 158 c/lb. His harvest should start in June and he knows that there will be some payments due. The market level is above his cost of production, allowing for a profit. Therefore, he decides to presell some coffee immediately. He contacts the cooperative and he agrees on a price of 164 c/lb for 300 bags of NY2 fine cup quality. He needs to deliver his coffee to the cooperative warehouse in July. The commercial manager now has a commitment to take coffee at 158 c/lb and he only knows that his customer will buy it at a differential of plus 12 c/lb. Therefore, he sells one futures contract[2] for September immediately at 152 c/lb at the ICE Exchange in NY—he is now one lot short.

July arrives and the farmer delivers his coffee, and the roaster is informed that his coffee is ready to be shipped. The farmer expects to receive his

2. One futures contract or lot equals 37,500 lbs, which in practice used to fix 300 60 kg bags of coffee.

payment of 158 c/lb as agreed, however, the market has lost value over the previous weeks and at the time of delivery, NY September came down to 137 c/lb. So who is covering the difference? How does the hedge secure the margin of the roaster who has an agreed price with its client? Is the cooperative covering its 6 c/lb margins? Does the farmer get his 164 c/lb? We can do this calculation once profits and losses are realized.

The roaster who is long one future at 123 c/lb now sells one future to the cooperative as an "exchange of futures for physicals" (EFP)[3] transaction at market level, which is at 137 c/lb. He then realizes a profit at the exchange of 14 c/lb and his position is even, i.e., the long (previously bought) future was offset by selling one future.

The cooperative buys one lot from the roaster as an EFP via the exchange at 137 c/lb and offsets it against the earlier short of 158 c/lb. It then realizes a profit of 11 c/lb, and the position in NY is now as well even as it is for the roaster—The financial details for all three are as follows.

Consequence for farmer:	Payment received from cooperative	158 c/lb
Consequence for cooperative:	Payment received from roaster	149 c/lb
	Payment to farmer	158 c/lb
	Loss on sales	9 c/lb
	Close of result from exchange	15 c/lb
	Net margin	6 c/lb
Consequence for roaster:	Payment to cooperative	149 c/lb
	Close out result from exchange	14 c/lb
	Cost of coffee	135 c/lb

The example above shows that the theoretical perfection of the exchange is that it provides the means to protect the business of each market participant, with a hedge securing the different business objectives. A reality check will first point out that there are not many farmers who can sell 300 bags of coffee at the same time, and who would be able to rely on a counterpart willing to take the risk and accept a presale of the crop, which

3. EFP is a transaction between two parties in which a futures contract on a commodity is exchanged for the actual physical good. Such transaction requires a bilaterally agreed exchange of a futures position for a corresponding position in an underlying physical.

during that time will still be exposed to all kinds of risk. So it is not a perfect solution for every situation.

3.4 The Monetizing of Coffee Hedges

Other benefits of having futures market hedges are that creditors are more inclined to lend against hedged inventories as the hedge creates a reference price for delivery or receipt of the physical coffee. In a further refinement of the practice, physical traders can choose to have their coffee certified by the exchange and that receipt can be used as a negotiable instrument. Here a trader who holds the warehouse receipt and the hedge against that coffee can get financing through banks and nontraditional sources. For example, brokerage firms will lend at favorable rates for what is called a cash and carry transaction. Some financial entities offer commodity repo agreements where they lend money against the security of the warehouse receipt, which they receive in return as collateral.

Cash and carry, and repo transactions are traditionally done at a discount to the hedge price, and margin financing is frequently a part of these transactions. Businesses traditionally perform these transactions as a way of monetizing warehoused inventories, and reducing costs as working capital is received against the receipts. The lender assumes inventory management during the time the financing is in place as ownership shifts away from the merchant during the time the repo is in place.

3.5 The Changing Influence of Market Sectors

Commodity futures at the start were almost entirely for those who were involved in the commercial aspects of the specific market, such as the coffee exporters, importers, and roasters. But the markets require more than hedgers to operate; they need speculators to provide liquidity in order for them to be successful. For the most part, these mostly small and very individual commodity markets follow the rhythm of their own supply and demand fundamentals. However, due to the increasing globalization and financialization of commodity assets, today's futures markets have a closer relationship than previously seen.

Even so, we can see from the movements of fund positions along with those of the hedgers and speculators as viewed through the lens of the weekly Commodity Futures Trading Commission Commitments of Traders Report (available from http://www.cftc.gov/MarketReports/CommitmentsofTraders/HistoricalViewable/index.htm), core supply and demand basics specific to each market, over time, still drive long-term trends. Therefore, the markets continue to fulfill their underlying practical role. They provide price insurance to industry and agriculture, while shifting risk from that sector onto the shoulders of the speculators.

A speculator is a trader who enters the market in search of profit and, by so doing accepts increased risk (Kolb and Overdahl, 2006, Chapter 4). Although the word speculation carries with it the connotation of gambling, the difference between speculation and gambling is that while gambling serves no economic purpose, speculation does. Speculators provide liquidity to the market place and enable the hedging process to proceed more smoothly, with buyers and sellers active at all price levels.

3.6 Futures Margins, the Exchanges and the Clearinghouse

The possibility of earning large profits with a small cash outlay, or margin deposit, is one of the most enticing aspects of futures trading. In that vein, the coffee contract that is large and volatile is seen as one of the most challenging and rewarding of all the agricultural markets to trade.

Coffee futures require about a 5% margin (which can be changed by the clearinghouse if market conditions are overly volatile or have lapsed into a quiet period). Margin represents a security deposit, or performance bond, which says the trader will fulfill his contract obligations. So for $2500 in initial margin a trader can enter into a contract for 37,500 lbs of coffee at 1.25 a lb the total value of which is $46,875.00. However, less than 1% of all contracts ever go through the delivery process (in late March 2016 there were 1,428,449 bags, or just 5029 contracts of coffee, sitting in the exchange certified warehouses, whereas futures volume to that point already totaled 1.65 M contracts.) although at times in coffee futures it can be larger depending on conditions in the physical market (Kolb and Overdahl, 2006, Chapter 1). This is because futures markets are designed mainly as tools for pricing and speculation. Although commercial interests can use futures markets as outlets for making or taking delivery of the physical goods, normal commercial channels are better equipped and more efficient for day-to-day trade. The one key condition that the trader must meet in futures is the margin call. If a loss on a particular trading position exceeds the acceptable margin, called the maintenance margin, the trader holding this position has to pay the clearinghouse the amount of money needed to bring his account up to the full margin level. Margin calls, however, depend on the amount of money in the trader's entire account, and are not just related to the specific market position that is losing money. Coming back to the example we had earlier in which the roaster took a position early February for September delivery @ 123 c/lb. If we assume that the market prices dropped to for example 110 c/lb the next day he would have received a margin call of 13 c/lb. On the other hand, if the market would have gone up, he would have received the credit accordingly. The simple idea being that any market participant has the liquidity to settle its position at any point in time.

Commodity exchanges are the forums in which the buying and selling of futures takes place. Once these were all physical spaces where traders transacted business in direct negotiation with one another. These trading areas were called trading rings or pits, based on their physical shape and layout. In 2016, however, all coffee and virtually all futures trades are transacted electronically. The notable exception is the Chicago Mercantile Exchange and its subsidiary Chicago Board of Trade where a small pit trading community is still operational, but shrinking.

At last count, there are 110 commodity exchanges around the world from where traders can receive quotes for more than 1000 separate instruments (Commodity System Inc., 2016). A trader can achieve all of this from the comfort of his or her desk, wherever it is located, as they all trade with varying degree of liquidity over the internet. There are an extraordinary number of trading choices, and each of these thousands of instruments has individual contract months. Some, such as Eurodollar Bonds and Crude Oil trade as far out as 10 years. Also, as many of these instruments have options available to trade, the number of trading possibilities is large and so complex that expert knowledge across the entire financial spectrum becomes virtually impossible for one person to claim.

All of these instruments trade on a recognized exchange, and many are available via the Globex trading platform that ties national exchanges into a global marketplace. All recognized exchanges have a clearinghouse where membership is mandatory for any firm that wishes to be a clearing member of that exchange. These members are then legally able to receive and pay out client funds directly.

The exchange itself, for instance the ICE where coffee futures are traded, does not engage in the buying or selling of commodities. That is the business of the exchange's customers. The exchange only provides the structure for trades to occur within a well-defined set of rules and arbitration processes. It also communicates price information and physical delivery rules, and oversees warehouses, graders, and weighers. To ensure that the trading function, as opposed to the trade, is risk free to the participants on either side, each futures exchange has a clearinghouse. The clearinghouse is either part of the exchange, as an affiliated but separate corporation, or is an overarching clearing corporation that clears a number of exchanges. The main function of the clearing house is to act as the buyer to every seller and the seller to every buyer (Kolb and Overdahl, 2006, Chapter 1). The clearinghouse guarantees every trade completed on the exchange, and contract holders have a financial relationship with the clearinghouse and not with the trader on the other side of the trade. Consequently, the trading parties do not need to know the identity of their counterparty, or trust the counterparty. The two parties simply need to trust the clearinghouse and settle their daily margin calls as described before.

The clearinghouse conducts all futures business, including the delivery of physical products under the terms of the futures contract. Through its system of financial safeguards and transaction guarantees, the clearing house protects the interests of the trading public, members of the exchanges and the clearing members of the clearing corporation.

In addition, the US futures markets, exchanges, clearing houses, clearing firms, hedge funds, brokerages, brokers, and traders are regulated by a constellation of regulatory bodies, chief among them is the Commodity Futures Trading Commission. Established in 1975, this is an independent Federal agency that acts as a governmental licensing and regulatory body, it superseded the USDA's Commodity Exchange Agency, and has the power to fine and suspend anyone who does not follow the rules. Futures brokers and Futures Commission Merchants also come under the authority of the National Futures Association (NFA), the self-regulatory body of the futures industry. Membership in the NFA is mandatory for all futures commission merchants, brokers, swap dealers, Commodity Trading Advisors, and Commodity Pool Operators.

4. THE THREE MAJOR COFFEE TERMINAL MARKETS

There are three active coffee futures contracts in existence, although there are a number of inactive contracts (mostly in Asia) as well. The most active is the "Coffee C" contract traded on the ICE and the robusta coffee contract traded on the London-based London International Financial Futures and Options Exchange (LIFFE) (which is owned by the ICE), and an unwashed arabica contract traded on the Bolsa de Mercadorias & Futuros (BM&F) in Brazil.

4.1 The ICE Exchange

As mentioned previously, the ICE Exchange in New York trades the Coffee C contract and deals in 20 different origins with Central American mild coffees as the basis grade (https://www.theice.com/products/15/Coffee-C-Futures). For more information on grading please see Chapter 8.

4.2 The LIFFE Coffee Exchange

Robusta coffee is traded on the LIFFE a subsidiary of the ICE (available from https://www.theice.com/products/37089079/Robusta-Coffee-Futures).
Activated in 1958 by the Coffee Terminal Market Association of London, the robusta contract originally called for the delivery of "Uganda unwashed," native grown robusta coffee. Over the years this market has expanded its acceptable delivery types, and become the reference market for all robusta coffee.

The London coffee market as it is commonly called accepts for delivery uniform lots of 10 metric tons of robusta coffee originating out of Africa, Asia, and Brazil, delivered into exchange certified warehouses in the following Ports/Delivery Areas: Amsterdam, Antwerp, Barcelona, Bremen, Felixstowe, Genoa-Savona, Hamburg, Humberside (including Hull), Le Havre, Liverpool, London & Home Counties, Marseille, New Orleans, New York, Rotterdam, Teesside, and Trieste (see https://www.theice.com/products/37089079/Robusta-Coffee-Futures).

4.3 The BM&F

The São Paulo Futures Exchange was founded in 1917. The present BM&F was established by members of the Bolsa de Valores de São Paulo (BOVESPA) stock exchange in 1985. A long awaited consolidation took place in 1991 with the merger of the BM&F and the original Sao Paulo Futures Exchange. In 1997, a further merger with the Brazilian Futures Exchange of Rio de Janeiro cemented the BM&F's position as the leading derivatives trading center in the Mercosur free trade area (Arabica exchange profile and history available from http://www.bmfbovespa.com.br/en_us/products/listed-equities-and-derivatives/commodities/4-5-arabica-coffee-futures.htm).

Coffee contracts on the BM&F are traded in 100 bags of 60 kg each, making the exchange accessible to smaller growers, which is important given that the BM&F operates in a producing country. The arabica futures contract trades seven positions: March, May, July, September, and December plus the next two positions of the following year.

The contract basis is the commercial grade 4-25 (4/5) or better, good cup or better, classified by BM&F, with prices quoted in US$ per 60 kg bag. Delivery may be made in BM&F licensed warehouses in 29 locations in the states of São Paulo, Paraná, Minas Gerias, and Bahia (deliveries outside the city of São Paulo incur a deduction for freight costs). The contract is dollar-based, and for nonresident firms is banked outside of Brazil which facilitates linkage with the global export market (Regulation/Non-resident investment available from http://www.bmfbovespa.com.br/en_us/products/listed-equities-and-derivatives/commodities/4-5-arabica-coffee-futures.htm).

Through the GLOBEX electronic quotation and trading platform the BM&F is linked to the Chicago Merchantile Exchange Group, which is a minority shareholder in the BM&F. Therefore its coffee contracts are accessible to non-residents of Brazil. This enables foreign traders (upon opening a local brokerage account) to speculate or hedge purchases of Brazilian physicals against Brazilian futures and avoid steep local taxes as well as any currency risk that might be an aspect of a physicals transaction.

5. DIFFERENT BUSINESS MODELS—DIFFERENT BUYING

As we turn our focus from futures markets to the coffee buying community, we can see that the coffee market's very real and unmanageable volatility has generated a segmentation of coffee buyers.

As this book goes to print we have a clear definition of three separate physical coffee buying programs. Each of these programs has its own history and in some cases a cult-like following. In many instances there is an obvious overlap between strategies.

It becomes somewhat obvious that the way coffee is bought relates to the business model of the roaster. What came first, the buying behavior which defined the value proposition or the value proposition which dictates the way coffee is bought? There is no one answer to it and discussing three different approaches hereafter does suggest that the profitability of those is based on different criteria. If we would have to choose three for each we would list:

- Efficiency savings, flexibility, and interchangeability for commodity buying
- Value creation, innovation and differentiation for specialty, second and third wave buying
- Consolidation, synergy leverage, and cash flow optimization for the investors buying

In all there is a clear evolution from one buying program to another as new ownership structures with nontraditional owners changed as quickly as the consumers desire to have transparency and traceability in their morning coffee. The irony is that the two latter phenomena in the world of coffee procurement, head into opposite directions.

5.1 The Commodity Buying Model

To clearly understand the nuances of today's buying programs we need to step back just 30 years to see how the segmentation originated. For the most part, roasters referred to themselves as large, medium, or small depending on the volume of coffee they roasted. The more you roasted, the bigger your market share, so you were defined by how much you roasted. There was no further differentiation or reference to your roasting style or ethical considerations. Roasters were for the most part "conventional commercial buyers." They acquired their coffee from the green coffee trade, the importing community that managed the movement of green coffee from producing countries to warehouses in consumer countries.

The technology that enabled this buying model to explode onto the world stage was the simple telex machine and then the fax machine. Offers for multiple containers of coffee on a 6-month spread at a fixed price were telexed/faxed to buying offices around the world. These offices were waiting for the futures market to move in the proper direction so the buyers could do business. Coffee buyers purchased the coffee in packages of multiple optional grades. For example, a roaster would buy three Horseman[4] optional, or five Horseman optional contracts, allowing the seller to choose among three or five different origins and grades for final approval and delivery.

This reflected that type coffee industry in general and its focus on quality and traceability. But at this stage the roaster quality focus was on creating a mainstream consistent product, not the unique aroma profiles specialty coffee

4. A Horseman contract represents a type of contract with a basket of origins. The number stands for the number of origins the seller is free to deliver chose from, e.g., a five Horseman allows to choose from a basket of five different origins.

buyers require today. In addition, traceability simply noted the country of origin, and rarely traced the coffee all the way back to its source.

Conventional commercial buyers are still very similar today in that they focus less on quality, but on creating a uniform mass market product. They are also interested in creating efficiencies in the supply chain through negotiating freight and landing costs, obtaining price discounts for large volume contracts, and increasing blend options to maximize contribution margins. Although these considerations are standard for all roasters, they are more important for this group than for others.

These same roasters tend to be well informed and technically very competent on their use of futures and other instruments used to manage their risk exposure. For them, coffee buying is less of an art, than a business to be run efficiently. They rely almost exclusively on the importing trade for their procurement needs. Some of these importing traders or green coffee traders would, at the same time, consider the exchange both as appropriate source and client. Tendering physical coffee to the exchange in case there is no industry buyer (or the exchange is a better client) or using certified coffee from the exchange to fulfill a contract obligation vis-à-vis a roaster is part of their expertise. Although roasters from time to time travel to countries of origin to get the feel of the deal, typically they simply want deliveries to the port closest to their roasting plants or to their plants themselves.

These buyers' value to the coffee industry is immeasurable, often using E-auction platforms as one of the main means for buying coffee, they are often belittled as cold, calculating, and are thought of as providing lower quality coffee than the other buying models. This is perhaps an unfair assessment as it is not based on a true understanding of the liquidity and depth of the market support they provide.

5.2 The Second and Third Wave

Buyers of second and third wave roasters seek coffee from the exporters or co-operatives located in each specific producing country, some do buy even directly from individual farmers. They are notable buyers on a Free on Board basis. For example, they organize their own freight, and purchase insurance to cover the physical coffee transport. They also import their own shipping containers with the coffee, and normally have a logistics team as well as a buying and quality team. This model has developed based on a clearer understanding of the traceability and transparency around their green coffee purchase.

These companies represent what has been described as second and third wave and they want a clear understanding of who is producing the coffee they buy and how much that person is paid, which is even more true for the third wave which often describes their purchasing as direct trade. Both are quite likely to support one or more third-party coffee certifications, which could also include their own internal certification programs, developed at a significant cost. Although buyers in the Direct Procurement model do also consider costs and efficiencies, their attention to the identification of the supply chain differentiates them from the true conventional buyer.

5.3 The Coffee Investor

As the consumption of coffee globally has evolved from enjoying a satisfying morning beverage to a social consumption experience or even a status symbol, coffee has caught the interest of the private equity investment community.

From retail points of sale to single-serve providers, coffee companies across North America in particular have been absorbed into companies that have no previous exposure to the coffee business. Unlike other consolidation efforts in the early 1990s, these new participants are not traditional coffee industry players. They tend to be privately held and are extremely well capitalized. These are major deviations from past consolidations, which were mostly industry centric.

Investors have purchased coffee companies, both public and private, at a record pace with over $30 billion dollars in acquisitions from 2009 to 2015. From the United States to Canada, one common characteristic has held true; their supplier base has provided the financing of their inventory and by default their working capital.

This Coffee Investor Model consolidates suppliers to a limited few. It is restricted to those merchants who can allow their capital to be tied up for as long as 270 days, or more. This model quickly eliminates direct buying from the producer, small exporter, or independent trader who usually has limited capital, but may have unique product knowledge or abilities.

The coffee culture as we know it today is one that at first slowly evolved with time and then exploded onto the world stage. As you walk into a specialty coffee shop anywhere in the world, you are partaking in a ritual that can trace its Western roots back to the 17th century, but is thoroughly modern and changing all the time. You are also participating in a global marketplace that ties people and nations together in the search for the perfect cup.

6. OUTLOOK

Is the future about hedging or direct trade? Has the consolidation in the industry come to an end? What is the most sustainable approach across the different business models? Most likely the various existing trading methods will continue to coexist and remain present even in the future. There are many developments in the coffee industry that are challenging the way coffee pricing works. Over time, this will hopefully push the industry into creating new ways of pricing coffee and making an impact on coffee growers.

For decades, coffee growers have struggled to increase their share of the total value added in the coffee industry. The reasons for this are multiple but include: lack of economies of scale, lack of capital, and lack of knowledge and vision. Roasters on the other hand have successfully increased their share of the total value added as the coffee industry has matured. They have started to change their business models by investing further down the supply chain to be able to achieve the sustainability they need. Such trend can only be welcomed and numerous examples exist proofing that farmers do benefit a lot from industry which is engaged in its own supply chain.

The relevance of in cup quality in particular for the third wave of coffee is discussed in the earlier Chapter 8 on Cupping and Grading as well as Chapter 19 which will look more into the consumer. Chapter 18 emphasizes on sensory and initiatives such as Cup of Excellence. However, the movement has prompted developments in both consumption and roasting. For instance, consumers are now more interested in the traceability of their coffee and the impact of their coffee purchases on coffee growers and the environment, whereas roasters have openly embraced sustainability. The coffee industry finds itself at a stage where the objectives of the coffee roasters have significantly changed from the ones they had in the first wave. Large volumes and steady quality are now not the objectives. Instead, coffee roasters are producing very high quality coffee that yields better avenues for product differentiation and comparatively smaller volumes. It is not unusual to see roasters offer coffee from one particular geography or even from one particular farm. When the degree of specialization reaches this point one has to question whether the reference price set by the exchange has a high degree of correlation to the actual coffee the roaster is buying, as this mindset has shifted away from a code which has been established to regulate business between anonymous actors. Roasters and farmers are becoming long-term partners, both having established a relationship that is based on the creation of shared value.

The roaster is also concerned about investing in his supply chain, and addressing the issue of sustainability, both for the coffee grower and for his own business. We, therefore, see coffee roasters of very high quality coffee paying high premiums for excellent coffee, so farmers can invest in developing their farms. These coffee roasters also often promote education and investment programs so coffee growers can develop their business.

Farmers themselves also have significant incentives to trade directly, as for some of them reaching the futures exchanges is almost impossible, as they do not have the size and means to do so, and trading directly with the roaster provides a potential way to reduce the uncertainty.

Over the next few years, as the coffee industry continues to develop and themes like sustainability, impact investing, triple bottom line, and other new business models take further hold of the industry, coffee roasters may trade directly with coffee growers and therefore bypass the exchange. As the coffee industry evolves and matures there will be greater incentives to trade directly.

To conclude, coffee culture as we know it today is one that at first slowly evolved with time and then exploded onto the world stage. As you walk into a specialty coffee shop anywhere in the world, you are partaking in a ritual that can trace its Western roots back to the 17th century, but is thoroughly modern and changing all the time. You are also participating in a global marketplace that ties people and nations together in the search for the perfect cup.

REFERENCES

Commodity System Inc., 2016. End of the Day Market Data. Commodities. Available from: http://www.csidata.com/?page_id=10.

Daviron, B., Ponte, S., 2005. The Coffee Paradox. Zed Books CTA, p. 72.

International Trade Center (ITC), 2011. The Coffee Exporters Guide, third ed. Geneva, Switzerland (Chapter 8). http://www.intracen.org/The-Coffee-Exporters-Guide—Third-Edition/.

Kolb, R.W., Overdahl, J.A., 2006. Understanding Futures Markets, sixth ed. Blackwell publishing, UK.

Chapter 10

Decaffeination—Process and Quality

Arne Pietsch
University of Applied Sciences Lübeck, Lübeck, Germany

1. TRENDS IN CONSUMPTION OF DECAFFEINATED COFFEE

1.1 What Is Caffeine?

When asked "what does caffeine look like?" most people tend to answer that they have not actually seen caffeine, but that it is probably a brown substance. This is understandable but this wrong assumption shows the association between caffeine and coffee. A handful of people may answer that they remember something in chemistry laboratory class in high school and that there was something about the term sublimation. Besides these aspects just about everybody knows that caffeine has a stimulating and awakening effect when consumed, well known not only from coffee and different teas but also from caffeinated soft drinks. Summarizing, caffeine is a well-researched chemical substance with interesting properties and at least its name is widely known.

Caffeine was first isolated from coffee beans in 1820 by the chemist F. Runge at the request of the German author Johann Wolfgang von Goethe (Sivetz and Desrosier, 1973). At room conditions, pure caffeine is a white, odorless crystalline powder bitter in taste. It exhibits two different crystal forms in pure state and when crystallized in presence of water very typical whiskers are formed. Fig. 10.1 shows caffeine crystals extracted from coffee. Caffeine crystals can also be found frequently as sediments in coffee processing factories especially around roasting machines. This is due to the previously mentioned effect of sublimation: at elevated temperatures caffeine can change from the solid to the vapor state directly without liquefaction as intermediate step. The name caffeine does not give any information about the chemical nature of the substance. It belongs to the group of methyl xanthines and carries the name 1,3,7-trimethylpurine-2,6-dione. The chemical structure

The Craft and Science of Coffee. http://dx.doi.org/10.1016/B978-0-12-803520-7.00010-4

FIGURE 10.1 Caffeine crystals and chemical structure of caffeine.

(Fig. 10.1) shows the high content of nitrogen in the caffeine molecule. The physiological effects of caffeine have been investigated for a long time and research is ongoing. Simplifying matters, positive and negative health effects have been declared, obviously depending on individual condition and consumed quantities. Important is, that the US Department of Health and Human Services classifies caffeine as a GRAS substance (generally recognized as safe). Recently, the European Food Safety Authority stated, that "habitual caffeine consumption of 400 mg/day does not give rise to safety concerns for non-pregnant adults" (EFSA, 2015). This amount corresponds roughly to five cups of regular drip coffee. Nevertheless, consumed quantities must be observed as the lethal amount in man is estimated as 10 g (Ramalakshmi and Raghavan, 1999). Further information on the effects of caffeine on health refer to Chapter 20 in this book.

Caffeine content in green and roasted beans is roughly the same: mean values are 1.1 wt% for Arabica and 2.2 wt% for Robusta beans. It is often believed that caffeine content is reduced in roasted coffee due to sublimation. However, as weight of the bean decreases the total concentration in the bean remains roughly unchanged. Caffeine content in coffee beverages is dependent on the blend composition (% Robusta), the water to coffee ratio and extraction yields. Typical values (Arabica) are 80–120 mg per cup of drip coffee (150 mL) and 50–100 mg for espressos.

1.2 Consumers and Trends

The economic impact of decaffeinated coffee is generally underestimated. First, the caffeine and its effects is intrinsic to coffee in the same way that alcohol is to beer or wheat is to bread, and thus the removal of such a key component can be felt as an unnatural downgrading of the product. Second, decaf consumers often have some sort of personal limitation, like sleeplessness or a genetic higher sensitivity to caffeine. Although these are good reasons to

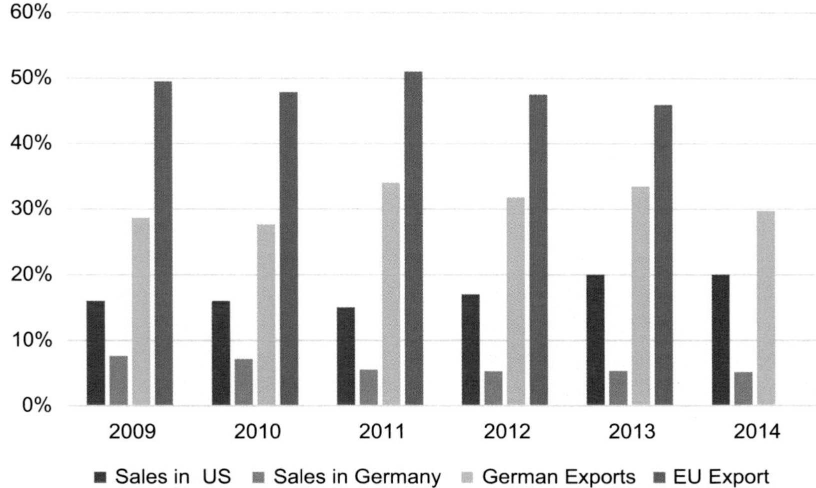

FIGURE 10.2 Trends in decaffeination—fraction of decaffeinated coffee on a mass basis. *Data references: Europe: European Coffee Federation (2014); Germany: Deutscher Kaffeeverband e.V. (2014). Sales in US: NCA (2015) etc.; Export data for green and roasted beans.*

refrain from caffeine it is nothing to impress others with. Thus decaffeination is talked about little. But the understated decaf consumers form a large and reliable group. Fig. 10.2 shows consumer and export trends in recent years.

2. DECAFFEINATION PROCESSES

2.1 Basic Aspects of Decaffeination Processes

The degree of required caffeine reduction for coffee to be labeled "decaffeinated" is not consistent throughout the world. Most European laws restrict coffee products marketed as "decaffeinated" to a maximum of 0.1 wt% anhydrous caffeine based on dry matter in green and roasted coffee and a maximum of 0.3 wt% (basis dry matter) in solid, pasty, or liquid coffee extracts (German By-Law, 2001; EU Directive 1999/4/EC, 1999). Strangely, only *extracts* have been universally regulated in all European (EU) countries; roasted coffee remains subject to national laws. In the United States and Canada the maximum caffeine content in roasted coffee is identical 0.1 wt% (USDA CID coffee, 2004). The less precise information "97.5% initial caffeine removed" seems to be a mathematically derived threshold without formal regulation using average caffeine contents. Other countries can have different regulations, e.g., Hungary, China and Taiwan.

Although during the past century various ideas and solvents have been used for decaffeination, four extraction solvents are in operation today:

TABLE 10.1 Trend of Estimated Annual Decaffeination Capacities in 1000 t/a

Country	2001[a] Capacity	2005[b] Capacity	2015 Capacity	Solvents Used
Europe	362	252	219	DCM, EA, $scCO_2$, $liCO_2$, water
North America	137	132	64	EA, $scCO_2$, water
Central and South America	50	50	50	DCM, EA, water
Asia	9	6	21	DCM, EA, water
World total	**558**	**440**	**354**	

[a]*Ref. Heilmann (2001).*
[b]*Ref. Vitzthum (2005).*

dichloromethane (DCM), water, ethyl acetate (EA), and carbon dioxide in its supercritical ($scCO_2$) and liquid ($liCO_2$) states. Decaffeination of large quantities of coffee beans (e.g., 5000 t/year) requires industrial-sized plants. Table 10.1 lists estimated annual industrial decaffeination plant capacities. A general decline in the last years indicates that there have been excess capacities, and subsequently several decaffeination plants have closed down in the last decade. Just recently JDE announced to shut down its traditional decaffeination plant in Bremen (former Kaffee HAG) beginning of the year 2017. Several of the decaffeination plants also offer other green bean pretreatment processes like steaming (for flavor improvement) or mild treatment, which often use the same equipment. Capacity data on the latter processes are not included here.

These processes have been in operation for decades now, and monographs on coffee science deal with the technologies as well as their history (Sivetz and Foote, 1963; Sivetz and Desrosier, 1973; Katz, 1987; Heilmann, 2001; Vitzhum, 2005). Not much new information on decaffeination processes has been published in the last years and process details are still often in-house know-how and kept confidential.

The ideal decaffeination process removes the caffeine from the bean cells without any other alteration to the bean. Due to the nature of the caffeine molecule and its location inside of the coffee bean cells, it is apparent that such an ideal removal faces severe obstacles. Unwanted side effects can include a loss of aroma or precursor molecules, changes in bean structure and size, mass loss, solvent residues, and changes of the beans appearance. In the beginning plenty of unsuccessful attempts were made to extract caffeine out of the beans with various organic solvents directly. The breakthrough came when Ludwig Roselius, the founder of the company Kaffee HAG in Germany, added an additional preprocess: wetting and swelling of the green beans with water and steam. In 1908, Meier, Roselius, and Wimmer patented the first useful decaffeination process (US Patent 897,763). The bean volume increases up to 100% allowing easier penetration of the solvents and mass transfer in the bean matrix. Later research revealed that the elevated water content liberates the caffeine from a structure with chlorogenic acids. After the extraction process, the initial water content of the beans has to be restored by drying, an unfavorable and costly process step. In summary the process sequence in all industrial processes is bean prewetting with water, caffeine extraction, and subsequent bean drying.

2.2 Extraction With Organic Solvents

The choice of an extraction solvent is subjected to various demands like solubility potential (see Table 10.2), cost, manageability, legal bounds, and availability, among others. Solubility is solvent and temperature dependent. Caffeine extraction from the water-swollen beans requires a solvent which is virtually immiscible with water in order not to lose other water soluble components from the bean. Naturally, the flavor precursor components should remain in the beans as extensively as possible. A variety of organic solvents have been found to be suitable, but only two are commonly used: DCM (methylene chloride; CH_2Cl_2) and ethyl acetate (ethyl ethanoate; $CH_3{-}COO{-}CH_2{-}CH_3$).

With organic solvent decaffeination, beans are first contacted with steam and water to increase their moisture content from roughly 10 to 25 or even 40 wt% (Fig. 10.3). The beans are then decaffeinated by extraction with the organic solvent, either in fixed beds (e.g., percolation column batteries, carousel extractors) or in agitated systems (e.g., rotating drums). The diffusion of caffeine in the beans is rather slow and thereby rate controlling, and thus intense bean agitation, is not imperative. For diffusion coefficients and mass-transfer modeling see e.g., Bichsel et al. (1976), Roethe et al. (1992), Schwartzberg (1997), and Espinoza-Perez et al. (2006). The limited solvation capacity enforces multiple extractions with fresh organic solvent or a vessel sequence operating in countercurrent mode. After completion the residual organic solvent is drained and stripped with steam. Finally, the initial moisture content is reestablished by drying with air (often a sequence of different drying technologies, e.g., fluidized beds, vibrating belts, vacuum dryers). Low air temperatures, oxygen exclusion, and completion by cooling can be helpful to minimize aroma decomposition. The used organic solvent is recycled via distillation (e.g., continuous natural circulation evaporator) producing a solid residue (~60% caffeine, ~40% others, mostly lipids). Caffeine is separated from the residue and refined, with the remains named coffee waxes due to their appearance, although triglycerides are predominant. Various uses of coffee waxes have been proposed and investigated (e.g., mole repellant), but nevertheless most of it is used today as either a binding agent in pelletization or as fuel. Sivetz gave detailed technical information on DCM plants back in 1973,

TABLE 10.2 Solubility of Caffeine in Various Solvents

Solvent	*T* (°C)	Solubility (%wt/vol)		Solvent	*T* (°C)	Solubility (wt%)	
Ethanol	20	1.5	Holscher (2005)	Water	20	1.65	Camenga and Bothe (1982)
Dichloroethane	25	1.8	Rahmalakshmi and Raghavan (1999)		80	27.2	Camenga and Bothe (1982)
Trichloromethane	20	15	Holscher (2005)	Supercritical CO_2, 28 MPa, dry	80	0.3	Johannsen (1995)
Trichloroethane	25	1.5	Rahmalakshmi and Raghavan (1999)	As above, wet CO_2	80	1.38	Kurzhals (1986)
Acetone	20	2	Holscher (2005)	As above, in contact with water containing 1% caffeine	80	0.05	Pietsch (2000)
Ethylacetate	20	2	Holscher (2005)	As above, with coffee solubles	80	0.02	Pietsch (2000)
DCM	20	8	Holscher (2005)				
Coffee oil	20	0.4	Pietsch (unpublished)	Liquid CO_2, maximum	21	0.04	Stahl et al. (1987)

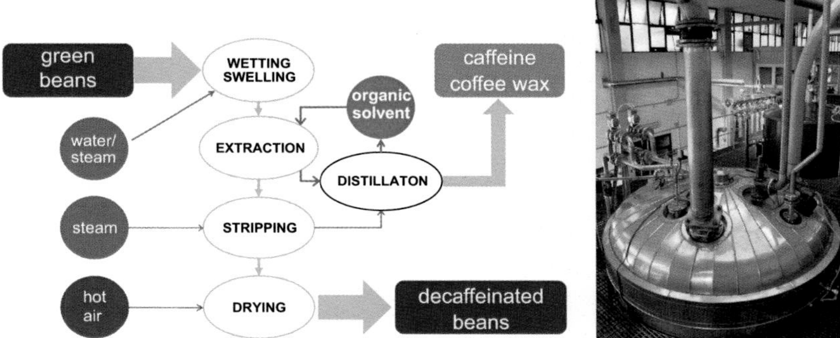

FIGURE 10.3 Simplified process flow diagram for solvent decaffeination (left); decaffeination plant operating with dichloromethane, with courtesy DEMUS, Italy (right).

including necessary steam consumption, cooling power, and other needs. Refined caffeine can be sold as a valuable byproduct to pharmaceutical and beverage industries. It is of natural origin but this fact is seldom used in marketing. Methods and process parameters for caffeine recovery are patented and published in detail (Heilmann, 2001). The two organic solvents currently in use have their individual advantages. DCM is flame resistant and therefore does not require explosion proof installations, but is subject to strict emission and health protection laws while EA is not a chlorinated solvent but flammable. Besides these fundamental technical differences the extraction processes are alike and some of the industrial installations can even handle both organic solvents.

2.3 Extraction With Water

Caffeine solubility in water increases significantly with temperature and therefore extraction is carried out with hot water at atmospheric pressure. Swelling of the beans is also necessary and is attained either by a preliminary wetting/steaming process or simply within the extractor. Choice of extraction systems resembles one of the solvent processes. Plain percolation columns have successfully been in use for decades and typical extraction time is 8 h. Problematic is the limited selectivity with regards to the other coffee components. Aroma precursor components like sugars will be at least partially coextracted when water is used. There are two principal ways to deal with this obstacle: either the water is somehow hindered to extract noncaffeine coffee solubles, or the extracted solubles have to be reincorporated into the green bean.

The latter process is discussed first. The process sequence is as follows: the bed of green beans is extracted with fresh water. The extract stream containing caffeine (abbreviated as Cf in Fig. 10.4) as well as various other coffee solubles passes an absorber bed of activated carbon (AC). A special type of AC

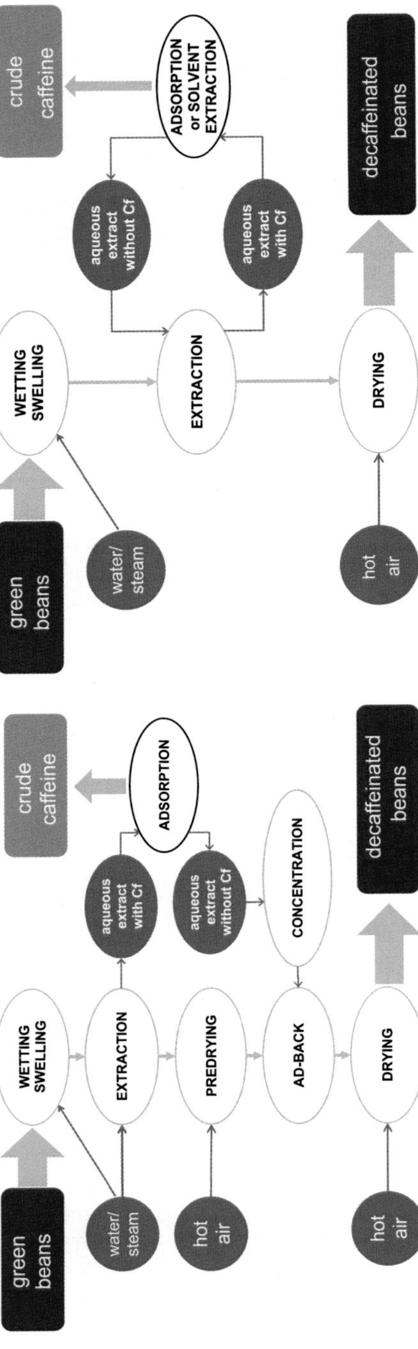

FIGURE 10.4 Simplified process flow diagrams for water decaffeination: fresh water and concentrate ad-back to the beans (left); using saturated water recycling (right).

displays a high selectivity for caffeine so that most other substances stay in the water phase. The now nearly caffeine-free solution is concentrated (e.g., up to 30%) for later addition to the green beans and thereby recycles most of the previously extracted noncaffeine substances. To make the decaffeinated beans absorb the concentrate they need a marked drying step. The processed concentrate waits approximately 1−1.5 h before being added back to the appendant bean charge. After adding back the concentrate, final bean drying completes the process.

The other principal water decaffeination process limits loss of noncaffeine soluble solids by using preloaded water for the extraction: water from a previous extraction process is caffeine free but equilibrium preloaded with all other coffee solubles (Heilmann, 2001). In the preloaded water process the extract flow needs to be stripped from caffeine in order to be reused for the extraction. This is realized either by extraction with an organic solvent like DCM (countercurrent liquid−liquid extraction) or by passing an absorber bed. The first version requires professional solvent handling while the second needs suitable absorbents. Various absorbents have been proposed and patented (ACs, imprinted polymers, molecular sieves, amorphous silica, adsorbent resins). Successful absorbents include ACs either by tailored high selectivity or preloading of the AC with sucrose and/or glucose to block adsorptive sites, which would otherwise absorb coffee solubles of resembling structure and size. A water decaffeination process combining recirculation and readdition was patented and installed recently (WO, 2015/059722).

2.4 Extraction With Pressurized Carbon Dioxide CO_2

Although commonly known as a gas, carbon dioxide exists at elevated pressures either as liquid or a so-called supercritical fluid. In the liquid and even more so in the supercritical state ($T > 31°C$, $p > 7.39$ MPa), CO_2 is able to dissolve some caffeine but less than other solvents. Its advantage is a superior selectivity. Decaffeination with scCO_2 operates at around 25 MPa and 100°C. CO_2 is readily available, physiologically harmless, and nonflammable. This allows for a decaffeination process without any of the other drawbacks, but requires costly installation and maintenance and the use of rather special high-pressure technology. Zosel patented the application of supercritical CO_2 extraction to decaffeinate green coffee beans in 1971. In 1982 the German company HAG started sales of CO_2-decaffeinated coffee and sales increased by 22% (Wittig, 2008). The general process sequence is like in the other processes: beans are swollen with water and then extracted in percolation columns, which are in this case massive high pressure vessels rated to pressures such as 30 MPa (Fig. 10.5). Several regeneration methods for the caffeine loaded CO_2 flow have been proposed, like the use of membranes to separate caffeine from the scCO_2 (Gehrig, 1984). Two methods are used today: either adsorption with AC or stripping in a high-pressure wash-column with

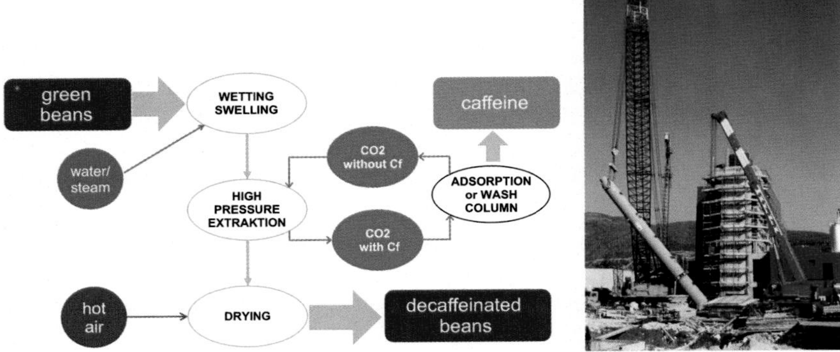

FIGURE 10.5 Flow diagram of CO_2 decaffeination (left). Decaffeination plant with $scCO_2$: mounting of high-pressure wash column for regeneration of caffeine loaded CO_2; with courtesy NATEX Prozesstechnik GesmbH, Austria (right).

water. In the second case, the caffeine-loaded water is said to be recycled via membranes—a nontrivial process (Pietsch et al., 1998). Atlantic Coffee Solution in Houston, USA, operates a unique quasicontinuous $scCO_2$ decaffeination with lock-hopper systems to feed and unload the beans from the high pressure extraction vessel (Mc Hugh and Krukonis, 1994). Lack and Seidlitz (1993) report in detail about realized large scale industrial processes which have been operating since the 1980s. Some of the high pressure vessels from those times have already reached their original maximum number of load cycles and will need either extended approval or replacement investments in the future. Extraction with liquid CO_2 at 6.5–7 MPa and 20–25°C is also possible but needs longer processing times due to the marked lower caffeine solubility at those conditions (Hermsen and Sirtl, 1988). Beneficial is the significantly lower processing temperature minimizing thermal stress.

2.5 Other Process Methods for Decaffeination

It has been attempted to use the sublimation property of caffeine to decaffeinate green beans in pilot scale plants. Camenga and Bothe (1982) showed that this process is not feasible due to the extremely low vapor pressure of caffeine. Earlier reported industrial processes using fatty matter as organic solvents have gone out of operation. Borrel (2012) discusses ideas to change the coffee plant by hybridization, breeding, mutagenization, or genetic engineering (silencing the caffeine biosynthesis pathway). Research has been in progress but commercialization has not taken place yet. Newer publications in this field are from Kumar and Ravishankar et al. (2009), Mazzafera et al. (2009), Benatti et al. (2012), Mohanan et al. (2014), and Summers et al.

(2015). In the context of plant engineering, it is important to consider that caffeine acts as a natural insecticide in the plant.

While patenting activity in the field of coffee decaffeination was intense in the eighties and early nineties, it has nearly died away by now. Some patent publications from the last decade deal with at-home decaffeination (WO, 2006/108292 A1), use of a special fungus for decaffeination (US, 2007/0036880), a process to recover antioxidants from decaffeination processes (WO, 2010/051216), a clay filter to adsorb caffeine (JP, 2013/123662A).

3. PRETREATMENT PROCESSES FOR GREEN COFFEE BESIDES DECAFFEINATION

It is possible to alter the composition and flavor of coffee by steam treatment. Green beans are again swollen with water and then a defined flow of steam passes the fixed or agitated bed of beans in such a way that no condensation occurs (Fig. 10.6). Holscher (2005) explains these technologies and various impacts on flavor in detail. Upgrading of Robusta coffees, general flavor alteration, and mild-treatment are common applications today. A reduction in caffeine content was not found in steam-treated coffees (Stennert and Maier, 1994). Other processes remove the coffee wax or fungal metabolites like Ochratoxin A (US 6,376,001, 2002) which are potentially harmful to human health.

4. IMPACT OF DECAFFEINATION AND PRETREATMENT PROCESSES ON CUP QUALITY

4.1 Aroma

In contrast to tea leaves, which are also decaffeinated to a certain percentage, aroma formation in coffee occurs mainly during the roasting process, which is after the decaffeination process. This is a large advantage as compared to tea decaffeination. But as discussed, caffeine extraction processes may of course influence aroma components. It is well known that caffeine tastes bitter but it has been shown that the bitterness of caffeine is not dominant in coffee flavor and thus its substantial reduction has no major altering effect.

Vitzthum (2005) lists typical defects when the processes are not carried out under optimal conditions:

- DCM and EA decaffeinated coffee can have a "cooked" flavor
- Coffee decaffeinated with supercritical CO_2 can be flat
- Water-decaffeinated coffee can suffer from a loss of soluble solids leading to a thin taste

Cup testing of properly decaffeinated coffees shows little impact of today's processes and only slight differences are noticeable. In case of Robustas or some off-tastes they can even be beneficial. The answer to the

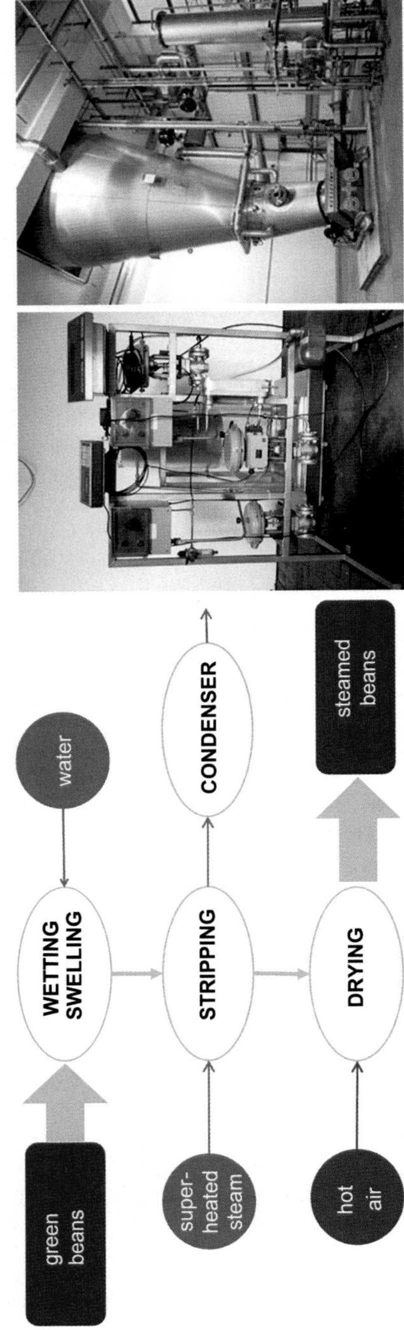

FIGURE 10.6 Steaming of green coffee flow diagram (left); lab test unit (middle); industrial plant, with courtesy SPX Flow Technology, Denmark (right).

common question "which process is the best?" is not evident. Assuming the processes are carried out under optimal conditions there are three main challenges for a valid comparison. Taste differences can be slight and moreover superimposed by roasting. Due to changes in bean structure and composition the roasting process for decaffeinated coffee must be adapted (Eggers and Pietsch, 2001). While roasting profiles generally remain unmodified, roasting times can be shorter, longer, or even unchanged. Suitable target roasting color can be different from non-decaffeinated beans and sometimes mass loss is the better control parameter. The third reason for missing process comparisons with regard to taste is the plain fact that different green coffees are normally treated in different industrial decaffeination processes. It is the same obstacle substantial comparisons of industrial roasting processes face: every company chooses their own coffee provenance, process technologies, and cupping methods, with each striving toward their own brand flavors. Disregarding all these aspects, a free choice of a decaffeination process for industrial sized coffee batches is not realistic anyway due to marked price differences and limited economic suitability of some processes for certain coffees.

Little has been published on the flavor impact of decaffeination processes. Cup testing shows that acidity increases in most processes and that agreeably some mustiness can be reduced. EA processes can lead to an enhanced fruity note. A pronounced hardness (Rio flavor) is generally not removable. SPME-GC-MS analytics (solid-phase microextraction gas chromoatography-mass spectrometry) is applied for headspace determination of volatile coffee flavor (aroma) components by various research groups today, but scarcely for decaffeination. For decaffeinated commercial coffees a principal component analysis showed a reduction of the following largest GC peaks: pyrazine, furfural, N-methylpyrrole, and 5-methyl-2-furancarboxyaldehyde (Ribeiro et al., 2013). For a deeper insight, substances with high odor activity values must be selected.

Fig. 10.7 illustrates results of quantified headspace SPME-GC-MS aroma analytics of two different Brazilian arabicas decaffeinated with DCM. Samples were decaffeinated in industrial scale and lab-roasted at 245–250°C to the same color. The determined aroma quantities of the regular samples were in the range as published (Grosch, 2001; Belitz et al., 2008). The changes by decaffeination are displayed in relation to the initial content (gray circle). None of the measured substances doubled in content or disappeared. Correspondingly, the beverage odor of untreated and decaffeinated was close. DCM decaffeination in both Brazilian coffees led to reduced formation of earthy components (pyrazines) matching cup impression of reduced mustiness and hardness.

Quite similar effects (pyrazine reduction) can be detected for DCM decaffeinated robustas. Smaller bean sizes and longer processing times lead to higher thermal impact thus influencing aroma precursors. In combination with altered acidity this can lead to an increase in roasting products from Maillard

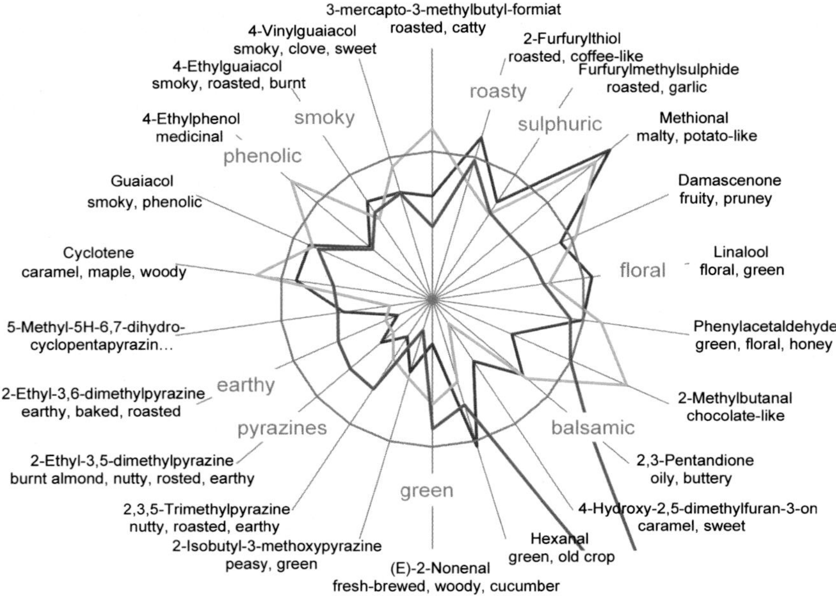

FIGURE 10.7 Change of some aroma substances by dichloromethane (brown, cream) and liquid CO_2 (red) decaffeination (Arabicas from Brazil).

reactions, e.g., an increase in cyclotene. Overall, a moderate reduction of mustiness and robusta taste can be found. Investigation of a liquid CO_2 decaffeinated Arabica sample showed an agreeable more uniform preservation of the aroma profile when compared to DCM samples with two exceptions: IBMP is reduced (matching cup testing) and furaneol increased. The latter could be an indication that aroma precursors (saccharides, amino acids, lipids) are affected in different ways in different decaffeination processes. Chen et al. reported findings that indicate formation of melanoidins in $scCO_2$ decaffeinated beans, unique from those formed in regular roasting (2011). Fig. 10.7 shows first findings and for an analytical investigation of the full flavor

FIGURE 10.8 Appearance of several unroasted coffee bean samples. From left: untreated, $scCO_2$ decaffeinated, $liCO_2$ decaffeinated, dichloromethane decaffeinated and two steamed coffees. All coffees are Arabicas except for the sample to the right, which is a steamed Robusta.

impression other less volatile substances (e.g., acidic components) must be taken into account in the future.

4.2 Solvent Residues in Green and Roasted Beans

Solvent residues in decaffeinated coffees are up to discussion from time to time. Water and carbon dioxide, both physiologically harmless and ubiquitous, are solvents that of course have the edge over organic solvents and can be used as marketing instruments, although not all CO_2-treated products advertise this fact. Processes with the solvent ethyl acetate are occasionally marketed as "natural" processes because EA exists in minute quantities in natural organisms, like ripening fruits. There are no regulations regarding maximum content of EA in decaffeinated coffee. The content of DCM is limited to 2 ppm in roasted coffee and 0.02 ppm in ready-to-eat products like foods containing aroma extracts (EU Directive, 2009/32/EC, 2009). In the United States, 10 ppm is the maximum allowed concentration for decaffeinated roasted coffee and coffee extract (FDA, 2014). In practice only traces around 0.1−1 ppm are typical for commercial European green samples and quantities in the final coffee beverage are estimated to 0.1 ppb, an insignificant amount compared to the allowed daily intake quantities according to maximum admissible aerial workplace concentrations.

4.3 Mass Loss and Water Content

Removal of caffeine naturally leads to a loss in bean mass. Besides this unavoidable effect, further losses can occur by extracting other coffee substances, e.g., the previously mentioned coffee wax. This can lead to reduced solubles yield in regular brewing as well as instant coffee manufacturing. Bean mass loss is of significant economic importance for all coffee processing industries but masked by the superimposed water content of the coffees. Reliable mass loss data is rarely recorded and comparison of the decaffeination processes is unpublished so far. Just by comparing solubility data, extraction with CO_2 should be beneficial due to its selectivity to caffeine. Besides mass loss, the water content is of importance for process control and efficiency. Fast operating moisture balances (using either ultraviolet or infrared radiation) are well-established. Water activity (aw) plays an important role for microbial stability and other water-related effects like sorption, powder stickiness, and plasticizing, and is helpful in coffee processing as well (Iacceri et al., 2015). Water activity is equal to the partial vapor pressure of water directly above a substance divided by the saturation vapor pressure.

4.4 Various Components

Analysis of protein, tannin, ash, and nitrogen content as well as petrol ether extract of regular and decaffeinated coffees (DCM and $scCO_2$) has been published by Udayasankar et al. (1986). Petrolether extract was higher for CO_2 decaffeinated while the other values did not show significant deviations. Lercker et al. (1996) determined the absolute reduction in lipids content by DCM or $scCO_2$ decaffeination (0.5–1.8 wt% in green beans) as well as sterol and fatty acid distribution of the triglycerides. Toci et al. (2006) reported major changes in saccharose (reduced from 9.65% to 3.84% in green Arabica), protein, and trigonnelline content by DCM decaffeination. Acrylamide forms during roasting and its quantity is not altered in decaffeination or steaming processes. The mycotoxin ochratoxin A is more or less removed in all extraction processes (e.g., Nehad et al., 2005).

4.5 Other Effects

Shelf life of treated coffee beans is generally shorter than regular beans due to an effect attributed to removal of surface components and process heat impact. Bean appearance can be altered by decaffeination processes. Color alterations are well known, with treated beans predominantly lacking their fresh green look and instead showing dull and rather yellowish or even brownish colors. This effect is less pronounced in processes with vacuum drying and extra polishing processes can help to improve a dull surface. Additionally, dark spots on decaffeinated coffee beans have been reported, which are especially unwanted for whole-bean products like espresso beans for coffee machines. Bean appearance is generally of lesser concern for in-house roasting. Fig. 10.8 shows examples of unroasted beans after different pretreatment processes. It is important to notice that the appearance of some products comes close to roasted beans, thus complicating roaster control. The freshest and greenest appearance show beans decaffeinated with liquid CO_2.

The mechanical stability of decaffeinated beans can be reduced and therefore higher attention in fluidized bed roasters is required. It also must be considered that the altered fracturing behavior due to steaming or decaffeination processes influences the grinding process. Decaffeinated qualities generally show raised brittleness. Industrial multiroll mills commonly need to be set to adapted gap widths in order to obtain the required particle size distribution when changing to decaffeinated beans. The influence of lab-scale decaffeination (EA and water) on breaking strength of roasted Robusta beans is published by Ramalakshmi et al. (2000). They conclude that breaking strength increases roughly by a factor of 1.1–1.7 depending upon different extraction parameters.

5. FUTURE TRENDS IN GREEN BEAN TREATMENT

Generally all modern decaffeination processes in use have reached a professional and satisfactory level. Of course, various in-house optimizations like energy-saving programs or up-ratings will also take place in the future. The discussion about toxicity of organic solvents arises from time to time but a general turning-away from DCM seems unlikely particularly due to the attained high processing standards. Flavor analysis will improve further but will have limited effects in decaffeination owing to the good cup qualities already achieved today. Improvements in fine-tuning of steam-treated coffees to maintain long running, stable aroma profiles of premium products are to be expected. Enzyme treatments for green beans have been developed but have so far been scarcely used. A general constraint of decaffeinated coffees is the limited choice for the consumer. Needless to say, virtually all industrial manufacturers offer a decaffeinated coffee product, but due to the significantly lower sales volumes most companies understandably decide to market a single decaffeinated quality only. This is of course frequently a medium or standard taste; specialties are seldom found in this segment. Perhaps this will change in the future—a first step in this direction was recently made by the concept of one brand to offer a wider variety of coffees in both regular and decaffeinated forms.

REFERENCES

Belitz, H.-D., Grosch, W., Schieberle, P., 2008. Lehrbuch der Lebensmittelchemie. Springer, Berlin Heidelberg, p. 976.

Benatti, L.B., Silvarolla, M.B., Mazzafera, P., 2012. Characterisation of AC1: a naturally decaffeinated coffee. Bragantia 71 (2), 143–154.

Bichsel, B., Gál, S., Signer, R., 1976. Diffusion phenomena during the decaffeination of coffee beans. Food Technology 11, 637–646.

Borrell, B., 2012. Make it a decaf. Nature 483, 264–265.

Camenga, H.K., Bothe, H., 1982. Coffein, Theophyllin und Theobromin: physikalisch-chemische Untersuchungen für Lebensmitteltechnologie und Pharmazie, 41. Diskussionstagung Forschungskreis der Ernährungsindustrie e.V. Würzburg 4./5. Nov. 1982. Forschungskreis der Ernährungsindustrie e.V, Hannover, Germany.

Chen, Y., Brown, P.H., Hu, K., Black, R.M., Prior, R.L., Ou, B., Chu, Y.-F., 2011. Supercritical CO_2 decaffeination of unroasted coffee beans produces melanoidins with distinct NF-kappa B inhibitory activity. Journal of Food Science 76 (7), H182–H186.

Deutscher Kaffeeverband e.V, 2014. Kaffeemarkt 2014 and Previous Years, Hamburg, Germany.

EFSA, 27 May 2015. Scientific Opinion of the Safety of Caffeine. European Food Safety Authority, Parma, Italy.

Eggers, R., Pietsch, A., 2001. Technology I: roasting. In: Clarke, R.J., Vitzthum, O.G. (Eds.), Coffee – Recent Developments. Blackwell Science, Oxford, pp. 90–107.

Espinoza-Pérez, J.D., Vargas, A., Robles-Olvera, V.J., Rodriguez-Jimenes, G.C., Garcia-Alvarado, M.A., 2006. Mathematical modeling of caffeine kinetic during solid–liquid extraction of coffee beans. Journal of Food Engineering 81, 72–78.

EU Directive 1999/4/EC, 1999. Directive 1999/4/EC of the European Parliament and of the Council of 22 February 1999 Relating to Coffee Extracts and Chicory Extracts.

EU Directive 2009/32/EC, 23 April 2009. Directive on the approximation of the laws of the Member States on extraction solvents used in the production of foodstuffs and food ingredients. European Parliament.

European Coffee Federation, 2014. Coffee Report 2013/14 (and Previous Years). ECF, The Hague, The Netherlands.

FDA, 2014. Code of Federal Regulations Title 21, Chapter I, Subch. B, Part 173, Sec. 173.255.

Gehrig, M., 1984. Verfahren zur Abtrennung von Coffein aus verflüssigtem oder überkritischem Kohlendioxid. German. Patent DE 3443390.

German By-Law, November 2001. Verordnung über Kaffee, Kaffee- und Zichorien-Extrakte vom 15 (BGBl. I S. 3107).

Grosch, W., 2001. Chemistry III. Volatile components. In: Clarke, R.J., Vitzthum, O.G. (Eds.), Coffee – Recent Developments. Blackwell Science, Oxford, pp. 90–107.

Heilmann, W., 2001. Technology II: decaffeination of coffee. In: Clarke, R.J., Vitzthum, O.G. (Eds.), Coffee – Recent Developments. Blackwell Science, Oxford, pp. 90–107.

Hermsen, M., Sirtl, W., 1988. Verfahren zur Rohkaffee-Entcoffeinierung. European Patent 0316694.

Holscher, W., 2005. Rohkaffeebehandlung im Verbraucherland. In: Rothfos, J.B. (Ed.), Kaffee die Zukunft. Behr's Verlag, Hamburg, pp. 88–109.

Iacceri, E., Laghi, L., cevoli, C., Berardinelli, A., Ragni, L., Romani, S., Rocculi, P., 2015. Different analytical approaches for the study of water features in green and roasted coffee beans. Journal of Food Engineering 146, 28–35.

Johannsen, M., 1995. Experimentelle Untersuchungen und Korrelierung des Löseverhaltens von Naturstoffen in überkritischem Kohlendioxid (Ph.D. thesis). TU, Hamburg-Harburg, Germany.

Katz, S.N., 1987. Decaffeinationo of coffee. In: Clarke, R.J., Macrae, R. (Eds.), Coffee Volume 2: Technology. Elsevier Applied Science, London, pp. 59–71.

Kumar, V., Ravishankar, G.A., 2009. Current trends in producing low levels of caffeine in coffee berry and processed coffee powder. Food Reviews International 25 (3), 175–197.

Kurzhals, H.A., 1986. Decaffeination of coffee and tea. In: Hirata, M., Ishikawa, T. (Eds.), The Theory and Practice in Supercritical Fluid Technology. New Technology & Science, Tokyo, Japan.

Lack, E., Seidlitz, H., 1993. Commercial scale decaffeination of coffee and tea using supercritical CO_2. In: King, M.B., Bott, T.R. (Eds.), Extraction of Natural Products Using Near-Critical Solvents. Blackie Academic & Professional.

Lercker, G., Caboni, M.F., Bertacco, G., Turchetto, E., Lucci, A., Bortolomeazzi, R., Frega, N., Bocci, F., 1996. La frazione lipidica del café Nota 1: influenze della torrefazione e della decaffeinizzazione. In: Industrie Alimentari XXXV 10/2009, pp. 1057–1058.

Mazzafera, P., Baumann, T.W., Shimizu, M.M., Silvarolla, M.B., 2009. Decaf and the steeplechase towards Deacffito – the coffee from caffeine-free Arabica plants. Tropical Plant Biology 2, 63–76.

Mc Hugh, M.A., Krukonis, V.J., 1994. Supercritical Fluid Extraction. Principles and Practice. Butterworth-Heinemann, Boston.

Mohanan, S., Satyanarayana, K.V., Sridevi, V., Kalpashree, G., Giridhar, P., Chandrashekar, A., Ravishankar, G., 2014. Evaluating the effect and effectiveness of different constructs with a conserved sequence for silencing of *Coffea canephora* N-methyltransferases. Journal of Plant Biochemistry and Biotechnology 23 (4), 399–409.

NCA, 2015. National Coffee Drinking Trends Report 2014 and Previous. National Coffee Association of USA, Inc.

Nehad, E.A., Farag, M.M., Kawther, M.S., Abdel-Samed, A.K.M., Naguib, K., 2005. Stability of ochratoxin A (OTA) during processing and decaffeination in commercial roasted coffee beans. Food Additives and Contaminants 22 (8), 761–767.

Pietsch, A., Hilgendorff, W., Thom, O., Eggers, R., 1998. Basic investigation of integrating a membrane unit into high-pressure decaffeination processing. Separation and Purification Technology 14, 107–115.

Pietsch, A., 2000. Die Gleichstromversprühung mit überkritischem Kohlendioxid an den Beispielen Hochdruckentcoffeinierung und Carotinoidaufkonzentrierung (Ph.D. thesis). TU, Hamburg-Harburg, Germany, Shaker Verlag, Aachen.

Pietsch, A., unpublished data. GCMS SPME Aroma Analytics of Untreated and Decaffeinated Coffees.

Ramalakshmi, K., Raghavan, B., 1999. Caffeine in coffee: its removal. Why and how? Critical Reviews in Food Science and Nutrition 39 (5), 441–456.

Ramalakshmi, K., Prabhakara Rao, P.G., Nagalakshmi, S., Raghavan, B., 2000. Physico-chemical characteristics of decaffeinated coffee beans obtained using water and ethyl acetate. Journal of Food Science and Technology 27 (3), 282–285.

Ribeiro, J.S., Teófilo, R.F., Salva, T., Augusto, F., Ferreira, M.M.C., 2013. Exploratory and discriminative studies of commercial processed Brazilian coffees with different degrees of roasting and decaffeinated. Brazilian Journal of Food Technology 16 (3), 198–206.

Roethe, A., Rosahl, B., Suckow, M., Roether, K.P., 1992. Physikalisch-chemisch begründete Beschreibung von Hochdruckextraktionsvorgängen. CHEM Technik 55 (7/8), 243–249.

Schwartzberg, H.G., 1997. Mass transfer in a countercurrent, supercritical extraction system for solutes in moist solids. Chemical Engineering Communications 157, 1–22.

Sivetz, M., Foote, H.E., 1963. Coffee Processing Technology. Avi Publishing Company, Westport.

Sivetz, M., Desrosier, N.W., 1973. Coffee Technology. Avi Publishing Company, Westport.

Stahl, E., Quirin, K.W., Gerard, D., 1987. Dense Gases for Extraction and Refining. Springer-Verlag, Berlin.

Stennert, A., Maier, H.G., 1994. Trigonelline in coffee. II. Content of green, roasted and instant coffee. Zeitschrift für Lebensmittel-Untersuchung und -Forschung 199, 198–200.

Summers, R.M., Mohanty, S.K., Gopishetty, S., 2015. Genetic characterization of caffeine degradation by bacteria and its potential applications. Microbial Biotechnology 8 (3), 369–378.

Toci, A., Farah, A., Trugo, L.C., 2006. Efeito do processo de descaffeinação com dichlorometano sobre a composição química dos cafés arábica e robusta antes e após a torração. Quimica Nova 29 (5), 965–971.

Udayasankar, K., Manohar, B., Chokkalingam, A., 1986. A note on supercritical carbon dioxide decaffeination of coffee. Journal of Food Science and Technology 23 (6), 326–328.

USDA CID Coffee, August 17, 2004. US Department of Agriculture. Commercial item description coffee. FSC 8955 A-A-20213B, US Department of Agriculture.

Vitzthum, O.G., 2005. Decaffeination. In: Illy, A., Viani, R. (Eds.), Espresso Coffee – the Science of Quality. Elsevier Academic Press, Amsterdam, pp. 142–148.

Wittig, U., 2008. Mergers&Acquisitions -Voraussetzungen, Ablauf und Folgen von Fusionen und Übernahmen bei Kraft Foods in Deutschland von 1978 bis 1998. LIT Verlag Dr. Wilhelm Hopf, Berlin.

Zosel, K., 1971. Verfahren zur Entcoffeinierung von Rohkaffee. German. Patent 2 005 293.

Chapter 11

The Roast—Creating the Beans' Signature

Stefan Schenker[1], Trish Rothgeb[2]
[1]*Buhler AG, Uzwil, Switzerland;* [2]*Wrecking Ball Coffee Roasters, San Francisco, CA, United States*

1. INTRODUCTION

Roasting is the key unit operation in converting green coffee beans into flavorful roast coffee. It is the heart and soul of any coffee manufacturing operation because it is the roasting process during which flavor is created and physical bean properties are determined. Roasting is generally defined as dry heat treatment. More particular, hot air roasting of coffee beans is a traditional thermal process with the primary objective to produce roast coffee with the desired flavor, but also to generate a dark color and a brittle, porous texture ready for grinding and extraction. During roasting, coffee beans are exposed to hot air. The increasing product temperature induces extensive chemical reactions, dehydration, and profound changes of the microstructure. The process generates the delightful flavor compounds that eventually may be transferred into the liquid phase during careful extraction and finally produce a delightful cup of coffee.

The roasting process of coffee has been studied by numerous authors in academia as well as in research and development departments of leading coffee manufacturing companies for decades (e.g., Sievetz and Desrosier, 1979; Clarke and Macrae, 1987; Illy and Viani, 1995; Schenker, 2000; Eggers and Pietsch, 2001; Geiger, 2004; Yeretzian et al., 2012). Although public knowledge on coffee roasting has increased tremendously, much remains to be discovered and elucidated. The art of a skilled roast master persists to be an essential prerequisite in creating a perfect cup of coffee. As the coffee shop and barista scene currently sees a new sustained trend of artisanal roasting, new freshness concepts, and intriguing in-shop roasting experience for consumers, roasting finds its way to a wider audience and an ever increasing number of followers. This chapter intends to summarize current understanding of the coffee roasting process in brief.

The Craft and Science of Coffee. http://dx.doi.org/10.1016/B978-0-12-803520-7.00011-6

2. PHYSICAL AND CHEMICAL CHANGES OF THE BEAN DURING ROASTING

2.1 Product Temperature

Compared to other roasting processes in food applications (nuts, cocoa, etc.) roasting of coffee requires the highest product temperature for developing the desired product characteristics. In general, the coffee bean temperature is required to exceed 190°C for a certain minimal duration to trigger the typical chemical reactions of roasting. The evolution of product temperature during traditional coffee roasting is characterized by a steady increase up to the final maximum at which the process is then stopped by abrupt optional precooling (water-quenching) and cooling. A typical final product temperature may be in the range of 200–250°C. The typical duration (roasting time) may be from 3 to 20 min.

The term "product temperature" should always be used with due care and attention. Measurement of real surface or core bean temperature during roasting is difficult to achieve. Although bean core temperature measurement has been accomplished in small-scale experiments (Schenker, 2000) it is usually not possible in industrial-scale roasting operations. For practical reasons, most roasting systems use a temperature probe that is located in a preferred spot inside the roasting chamber where it is continuously in contact with the beans but also with hot air. Therefore, this temperature reading always represents a mix temperature of bean surface and hot air. Although this is sufficient and appropriate for process control, it remains highly specific to machine design. This makes it problematic to compare temperature values from one roasting system to another.

2.2 Color Development

The color change during roasting is the most obvious and visible indication of the increasing degree of roast. Coffee beans change color from greenish-gray-blue (color of the green bean) to yellow, orange, brown, dark brown, and finally to almost black. The color development is very much interlinked with flavor development. Therefore, the bean color is the best indicator of the degree of roast and a most important quality criteria. Although baristas often refer to simplifying particular terms such as "city roast," "espresso roast," "French roast" to express various degrees of roast, industrial operators and scientists prefer to measure. For most reliable results, beans are ground and prepared in a standardized way and then measured using a commercial optical color measurement device. In the widely used, scientific L*a*b* color space a lightness value of $L = 26$ would correspond to a medium degree of roast (corresponding to a value of approximately 66 on the Agtron "Gourmet Scale").

2.3 Volume Increase and Structural Changes

The structure of the coffee bean seems to be essential for the creation of the typical roast flavor of coffee. Experiments showed that ground green coffee powder exposed to similar temperature histories as in bean roasting does not produce the desired flavor compounds. The intact bean acts as an essential "minireactor" for the chemical reactions. It controls the reaction environment in a way that the right precursors can react with each other in the right sequence. Temperature, water activity, pressure, as well as mass transfer phenomena are very much related to structure and govern the kinetics of chemical reactions that produce flavor (more details can be found in Chapter 12).

Coffee beans swell during roasting and may increase the volume up to factor 2. The microstructure changes from a dense to a very porous structure (Schenker et al., 1999). By contrast to the popcorn roasting process with the sudden burst-type expansion, coffee beans swell continuously in a steady process (Fig. 11.1). The increasing gas pressure inside the bean is the main driving force for expansion, whereas the thick plant cell walls hold against it. According to glass transition theory, the polysaccharides of the cell walls may be in a "glassy" or "rubbery" state, depending on actual moisture content and temperature. The state change occurs gradually and blurred due to system complexity (various state diagrams for the individual polysaccharides present in the cell wall). However, in principle the beans pass from "glassy" to "rubbery" and finally back to "glassy" state during roasting. The volume increase takes place during "rubbery" state at which the physical resistance of the cell wall material is reduced. Therefore, bean swelling is the result of complex dynamics between gas formation and cell wall resistance. Since the

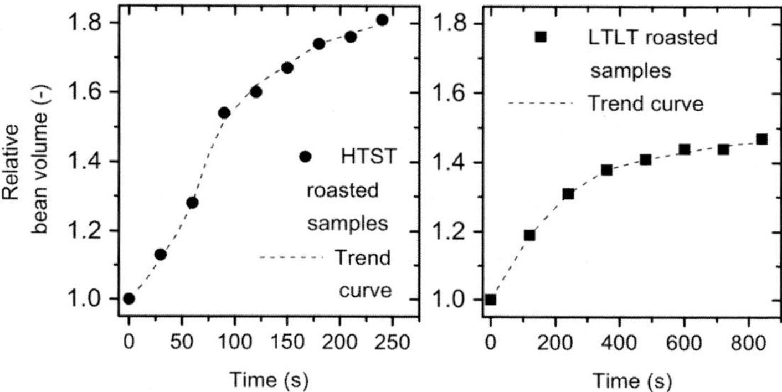

FIGURE 11.1 Development of bean volume increase during fast roast (left) and slow roast (right) conditions. *HTST*, isothermal high temperature; *LTLT*, low temperature roasting temperature.

state diagram is governed by temperature and moisture, the dehydration kinetics during roasting plays a key role. Consequently, the roasting profile is crucial for the extent of bean volume increase.

Volume increase, dehydration, and chemical reactions during roasting result in profound changes in the microstructure of the bean tissue. The green bean is characterized by a very compact and dense structure and the sophisticated intracellular organization of native biological cells (Fig. 11.2). The cell walls of coffee beans are unusually thick as compared to plant material of other species. They are equipped with reinforcement rings that give them the typical nodular appearance in the cross-sectional view. Roasting destroys this native structure and gradually leads to formation of excavated cells. Although the framework of cell walls remains intact, the diminished cytoplasm is pushed toward the wall giving way to a large gas-filled void occupying the center (Fig. 11.3). Some of the remaining denatured cytoplasm stretches along the cell walls. This layer becomes thinner on continuation of roasting, because more and more cell mass are converted into gases and water vapor and the cell sizes increase. In parallel to volume increase, measured porosity increases also gradually during roasting.

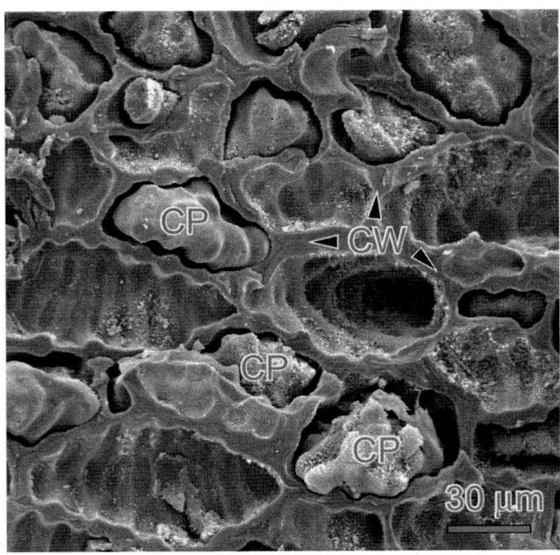

FIGURE 11.2 Scanning electron microscopy micrograph of the tissue of the green coffee bean from a chemically fixed specimen. The cytoplasm (CP) is visible in some cells, whereas it is removed in some others due to fractioning during specimen preparation. Surrounding cell walls (CW) are of remarkable thickness with typical reinforcement rings (Schenker, 2000). *Image: S. Handschin.*

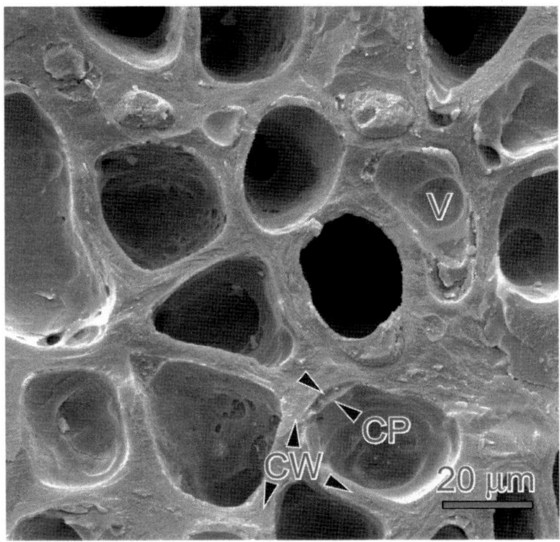

FIGURE 11.3 Cryo-scanning electron microscopy micrograph of the tissue structure of a roasted coffee bean. The biological cells show considerable changes of the cytoplasm induced by roasting (Schenker, 2000). Modified remaining of the cytoplasm (CP) stretch along the cell walls (CW). A large gas-filled space occupies the center of each cell. A number of smaller voids (V) are embedded within these layers of cytoplasm. *Image: B. Frey and S. Handschin.*

2.4 Dehydration

Green coffee beans enter the roasting process with a typical moisture content of about 10–12% [g/100 g, wet base (wb)]. During roasting dehydration takes place. Depending on the roasting conditions the roasted beans may leave the process with a final moisture of about 2.5%. Of course, the final moisture content of roasted beans may also be influenced by water quenching conditions because the beans may partially absorb water that is sprayed onto the bean surface during the precooling step. Beans with higher initial moisture content usually lose more water during the first roasting phase and end up with similar final moisture that is reflected in higher roast loss. Although dehydration during isothermal roasting takes place in a steady and continuous manner (Fig. 11.4), the dehydration kinetics in nonisothermal conditions (multistep process conditions) depend on the roasting profile. In addition to the water present in the green bean there is also a considerable amount of water that is generated as a result of chemical reactions. This water is also vaporized in the course of roasting. The actual total water content and water activity at different stages of the roasting process play a key role for the kinetics of chemical and physical changes in the bean. Apart from temperature, the speed of important flavor-generating chemical reactions depend on the availability of water. Some chemical reactions slow down when the moisture content falls below a certain critical value.

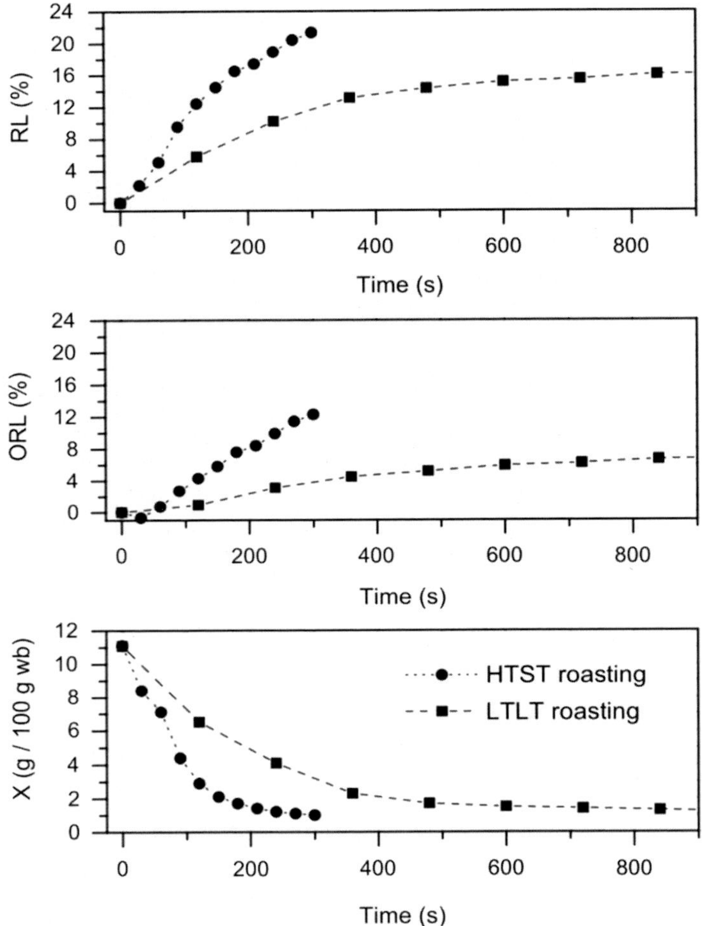

FIGURE 11.4 Development of roast loss (RL), organic roast loss (ORL) and bean water content (X) during isothermal high temperature (HTST) and low temperature (LTLT) roasting conditions.

2.5 Roast Loss

During roasting, water is vaporized and dry matter is partially transformed into volatiles. In general, coffee beans may lose 12–20% weight during roasting, depending on green bean quality, roasting parameters, and final degree of roast. Roast loss (RL, in %) is defined as,

$$RL = \frac{m_{green} - m_{roast}}{m_{green}} \times 100$$

where m_{green} is the weight of green coffee beans (kg); m_{roast} is the weight of roasted coffee beans (kg).

This roast loss consists of several parts, such as water evaporation, transformation of organic matter into gas and volatiles, physical loss of silverskins (tegument), dust, and bean fragments or other light material. The roast loss is always product specific. It increases in a steady and continuous manner during roasting (Fig. 11.4). The highest rate of roast loss is usually found in the early process stages and is mainly caused by dehydration, whereas loss of organic matter is initiated later during the more advanced stages. Dark roasted beans experience a higher roast loss than light roasted beans.

As long as the quality and in particular the moisture of raw material prior to roasting remain constant, the roast loss may serve as an indicator of roast degree. Since green coffee in reality is always subject to natural quality fluctuations (e.g., fluctuating green coffee initial moisture content), the roast loss may also fluctuate from one batch to the next, even when roasted to the identical bean color.

The loss of pure organic dry matter would provide more precise information on roast degree because it takes the varying loss of water into account. This organic roast loss (ORL, in %) is defined as,

$$\mathrm{ORL} = 100 - \left[(100 - \mathrm{RL}) \times \frac{dm_{\mathrm{roast}}}{dm_{\mathrm{green}}}\right]$$

where RL is the roast loss, dm_{green} is the dry matter of green beans (g/100 g, wb) dm_{roast} is the dry matter of roasted beans (g/100 g, wb).

The ORL correlates well with the lightness value from color measurement.

Green beans are still partially coated with the silverskin prior to roasting. These silverskins come off naturally during roasting due to the bean swelling and are carried away with the air. Depending on green bean quality, the loss of silverskins may account for approximately 1% weight loss. Additionally, in any commercially available roasting equipment, beans are also exposed to some mechanical stress. The design of the roasting chamber and the movement of coffee beans need to be optimized to prevent bean breakage. Bean breakage would generate small fragments that can also be lost. Silverskins, dust, small bean fragments, and other light materials are carried away with the air and will be separated in the hot air cyclone of the roasting system.

2.6 Oil Migration to the Bean Surface

Coffee beans may contain up to 18% lipids (coffee oil). Lipids are embedded in the cytoplasm of the native plant cell within separate membrane-protected oil bodies located along the cell walls. Structural changes in the coffee bean tissue during roasting destroy the native biological cell organization, break up the oil bodies, and mobilize the coffee oil. Roasted coffee beans exhibit occasionally a more or less severe "oil sweating." The gas pressure inside the bean pushes the coffee oil through tiny microchannels in the cell wall to the bean surface. During the initial stages of the oil migration, numerous small

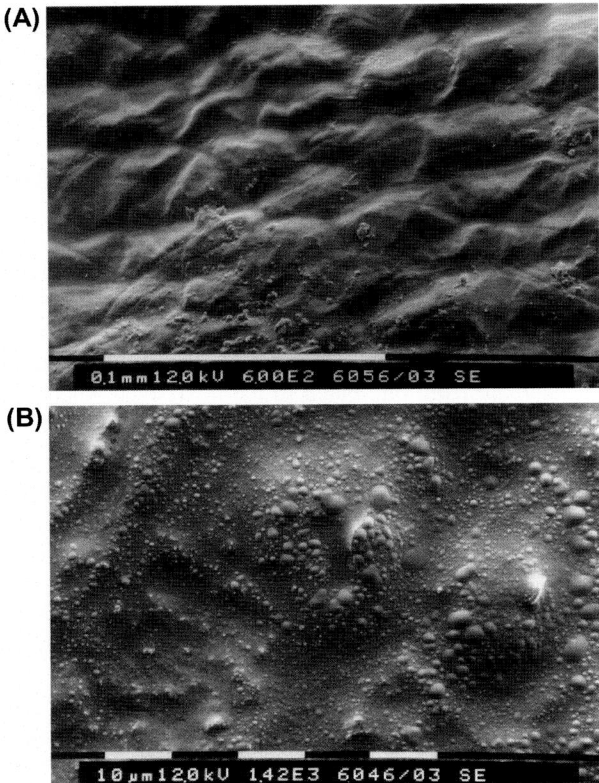

FIGURE 11.5 Cryo-scanning electron microscopy micrographs of the surface of a roasted coffee bean; illustrating the initial stages of oil migration process. (A) Immediately after roasting; smooth epidermal cell surfaces. (B) After one day of storage; numerous very small oil droplets migrated to the outside and cover the surface (Schenker, 2000). *Image: B. Frey and S. Handschin.*

oil droplets appear on the bean surface (Fig. 11.5). Oil droplets may coalesce and become more visible, eventually covering the entire bean with a shiny oil film.

3. CHEMICAL CHANGES DURING ROASTING

3.1 Endothermic and Exothermic Roasting Phase

The increasing bean temperature during roasting induces complex chemical reactions that finally result in severely altered composition of the roasted bean. Some of the most important chemical reactions affecting carbohydrates include Maillard reaction, Strecker degradation, pyrolysis, and caramelization. Roasting also leads to protein denaturation and degradation. Many acids present in the green bean are also degraded. During the initial stages of roasting a considerable energy input is required to drive the evaporation of water and to induce chemical reactions (endothermic phase). At one point during roasting, the energy balance of chemical reactions becomes autocatalytic (exothermic). The beans eventually start to generate heat on their own (Raemy and Lambelet, 1982). Hence, the final stages of the roasting process are characterized by increasing rate of process advancement and gradually approach conditions of a combustion process. Process control becomes crucial at this phase. A few seconds can make the difference between a correctly roasted product with the desired degree of roast and an over-roasted product. Roasting needs to be stopped abruptly at the desired degree of roast with an efficient precooling or cooling step. If roasting continues in an uncontrolled way the beans may catch fire and produce unsafe conditions in a roaster.

3.2 Gas Formation

Roasting generates a considerable amount of gas as a result of pyrolysis and Maillard reaction. The gas formation rate during isothermal roasting is low at the beginning of the process, but accelerates forcefully in the second half of the process. However, it is highly dependent on the roasting conditions. The predominant gas formed upon roasting is carbon dioxide (CO_2). Other important components include CO and N_2. One part of the gas is released to the atmosphere during roasting. Another major part remains entrapped inside the beans and is only released later during storage in a slow desorption process and during subsequent processing steps (e.g., grinding). For this reason, coffee processing lines often include "tempering silos" (bean), degas silos (roast and ground), and degas machinery for gas desorption. These unit operations release gas and avoid overpressure in subsequent steps (e.g., extraction) or in the packed product (e.g., in a hermetically closed single-serve capsule). Packaging materials for beans usually include one-way valves for gas release. This ability to hold back a large amount of gas is a remarkable characteristic of roasted beans. The entrapped gas must cause high pressure inside the bean. Gas measurements and model calculations come to the conclusion that the gas pressure inside the bean upon roasting may exceed values higher than 10 bars (Schenker, 2000). The thick cell walls of coffee are prepared to stand this pressure without breaking, but get gradually stretched and span an increasing pore volume. However, some minor structural break down and cracks occur during the final roasting stages, releasing a tiny quantity of gas in a sudden microburst and manifest in cracking and popping sounds. The gas together with the water vapor constitutes the driving force for bean expansion during roasting.

3.3 Formation of Aroma Compounds

The volatile fraction of roasted coffee is highly complex and consists of more than 1000 compounds. Much scientific work has been devoted to the identification of key aroma impact compounds (Grosch, 2001; Poisson et al., 2014), as described in detail in Chapter 12. The formation kinetics of aroma compounds during roasting is determined by the specific conditions for chemical reactions (e.g., temperature, water activity, pressure) as controlled by the process parameters (e.g., heat transfer over time). Therefore, different time—temperature conditions during roasting lead to specific flavor profiles obtained from the same raw material. Quantitative development of key aroma impact compounds in function of process conditions have been studied using various methodologies (Schenker et al., 2002; Wieland et al., 2011; Zimmermann et al., 2014). Schenker et al. (2002) analyzed the volatile fraction of coffee samples taken at different stages of the roasting process, using six different roasting profiles. The first roasting stage does not produce large aroma quantities, but may be important for the formation of aroma precursors. A majority of aroma compounds showed the highest formation rate at medium stages of the roasting process and medium stages of bean dehydration with the water content ranging from 7% to 2% (wb). The majority of important aroma compounds (e.g., most pyrazines) start to decay at high temperature during advanced stages of the process due to thermal degradation. The concentrations of these volatiles decrease with increasing degree of roast. By contrast, a limited number of aroma compounds continue to be created at high temperature (e.g., guaiacol).

3.4 Evolution of the Acidity/Bitterness Ratio With Increasing Degree of Roast

A good cup of coffee is characterized by a balanced acidity/bitterness ratio. Therefore, the skilled roast-master needs to take care about the often desired and appreciated acidity and keep an eye on the evolution of bitterness compounds during roasting. As a rule of thumb, increasing degree of roast leads to decreasing acidity and increasing bitterness. Therefore, selecting the optimal degree of roast is crucial for a balanced taste profile.

Chlorogenic acids are strongly degraded during roasting. However, their contribution in overall sensory perception is very limited. By contrast, citric and malic acids are highly relevant for sensory perception (Balzer, 2001). These acids are present already in the green bean and are then also gradually reduced during roasting. Acetic acid and formic acid are also strong contributors to total sensory perceived acidity. Their concentration in green coffee is very low. These acids are generated during the initial stages of roasting from a carbohydrate precursor, but then degraded at higher temperatures during the final stages of roasting. The concentrations of quinic acid and some volatile acids are slightly increasing during roasting. Overall, the sensory perceivable total acidity is clearly decreasing during the course of roasting. Light roasted beans unfold more acidity in the cup than dark roasted coffee.

Roasting generates bitter taste in coffee. The identification and formation pathways of bitterness components in roasted coffee have been elucidated only in recent years and are still subject to ongoing scientific research. Although caffeine—which is present in the green bean—has a strong bitter taste, it contributes only some 10–20% to the sensory perceived bitterness in coffee. The main contributors to bitterness are formed by roasting. The class of chlorogenic acid lactones—a break down product of chlorogenic acids—has been identified as one of the main contributors to bitterness in coffee (Hofmann, 2008). A lingering harsh bitterness taste in dark roasted coffee is caused by phenylindanes, a break down product of chlorogenic acid lactones. In general, the perceived bitterness increases with higher degree of roast.

4. INDUSTRIAL COFFEE ROASTING

4.1 Roaster Design Classification Criteria

Although alternative technologies such as infrared, microwave, superheated steam, and others have been developed and tested, hot air roasting technology is still the only widespread technology applied in industrial operations. Hot air roasting machines may be classified regarding various criteria, such as product flow (batch or continuous), mechanical principle, heat transfer, air-to-bean ratio (ABR), air flow (open system and air recirculation system), and automation principles.

4.1.1 Product Flow

Roasting machines employ either continuous product flow or batch roasting concepts. Although continuous roasting systems used to be popular some decades ago, they are nearly extinct today. The advantages of batch principles have led to absolute predominance of industrial batch roasters. Batch roasters provide more process flexibility and are easier to control.

4.1.2 Mechanical Principle

The beans must be kept constantly in motion inside the roasting chamber to assure homogeneous heat transfer from the hot air to the coffee. From rotating drum or bowls to stirring devices, various mechanical principles have been introduced to fulfill this task. By contrast, fluidized-bed roasters use sufficiently high air velocities instead of moving parts to agitate the beans. However, any means of bean movement exposes the beans to some mechanical stress. An optimized design avoids and minimizes bean breakage.

4.1.3 Heat Transfer

In any hot air roasting system heat is always transferred by convection, conduction, and radiation at the same time. Convection transfers heat from the hot air directly to the bean surface. Conduction occurs when heat is transferred from the hot walls of roasting chamber to the beans. The proportions of contribution may vary from one system to another. The contribution of radiation (comparable to the warmth you can feel when you put your hand close to a hot surface without touching it) is usually very limited and negligible. Concerning conduction and convection, the amount of process air used for roasting (ABR) plays a key role. In a fluidized-bed roaster convection is the predominant way of heat transfer, whereas in a drum roaster a substantial amount of heat may be transferred via conduction. A precise calculation or measurement of the conduction/convection ratio is difficult to achieve.

4.1.4 Air-to-Bean Ratio

The same amount of heat can be transferred to the beans using either a low quantity of air at higher temperature or using a larger amount of air at lower temperature. The amount of hot air used in a roasting process in relation to the batch size of coffee beans is defined as ABR. This dimensionless number (kg air per kg green coffee) applies to a specific blend, roasting time, and degree of roast and may vary considerably from one roasting system to another.

4.1.5 Air Flow

Smaller roasting systems usually suck in the process air at one end and emit the off-gas at another end (open system). Since the emitted off-gas is still at high temperature, open systems are not energy efficient. This is why most large-scale operations make use of air recirculation for substantially improved energy efficiency. In recirculation systems a major part of the off-gas stream (e.g., 80%) is led back to the heating unit and then reinjected into the roasting chamber. However, another part of the off-gas stream (typically 20%) must leave the system to avoid accumulation of problematic gas concentrations with the potential for explosion. This off-gas stream may pass a more or less sophisticated cleaning step for off-gas pollution control and compliance with air pollution regulations.

4.1.6 Water Quenching Device

Most medium- to large-scale roasting machines are equipped with a water quenching device. As soon as the beans reach their final temperature the roasting process may optionally be terminated through a sudden precooling step by spraying a predefined amount of cold water onto the beans (water quenching). Water evaporates on the bean surface and cools the beans.

Although this precooling step is optional, it helps to achieve a consistent degree of roast, batch by batch. Moderate quantities of quenching water will not affect the flavor or the physical bean properties. By contrast, excessive amount of quenching water will result in water absorption and therefore increase the final bean moisture. Hence, water quenching may also be used to adjust and control the final bean moisture. High water activity in the roasted bean results in an accelerated staling process and may affect flavor stability and shorten the product shelf-life.

4.1.7 Process Automation Principles

Although small-scale roasting machines are often operated manually, larger systems usually use more sophisticated process control systems. The traditional way of process automation is to set and control an appropriate hot air temperature, either in a single (isothermal) or in multistage process (profile roasting). In a conventional machine, the control system adjusts the burner power to reach and maintain the pre-set hot air temperature. However, the disadvantage is that the actual product temperature evolution may not be consistent from one batch to the next and remains subject to numerous factors that affect roasting, such as, for example, fluctuating initial green bean moisture, varying weather conditions, and cold start behavior of the roaster. By contrast, more advanced process control systems are guided by the actual development of product temperature rather than hot air temperature (real or true profile roasting). The desired product time—temperature master curve is registered in the recipe and is then precisely reproduced in the roaster batch by batch by continuous and meticulous fine tuning of the energy input. This type of process control results in superior quality consistency because the beans experience always the same temperature development. It requires a sophisticated hardware and software design for continuous, rapid, and accurate modulation of energy input into the roasting chamber.

4.2 Industrial Roasting Equipment

Industrial roaster design has been illustrated and described by various authors (e.g., Eggers and Pietsch, 2001). The most recent standards and norms in design of industrial roasting systems are described by the *Verein Deutscher Ingenieure* in VDI guideline 3892 (2015). It documents also the enormous progress that has been achieved by leading equipment manufacturers in terms of pollution control and energy efficiency.

4.2.1 Drum Roasters

The most widespread batch roaster design is the drum roaster. In this traditional design the batch of beans is kept in a horizontal rotating drum. Hot air enters at the drum back-end through a screen, flows through the drum and leaves at the front-end via an expansion chamber. The drum rotation as well as baffles installed in the interior of the drum keep the beans in motion and assure a thorough mixing of beans with the hot air for uniform heat transfer. After completion of roasting and precooling the batch is transferred trough an opening gate or gap at the front-end of the drum and falls into the cooling section. The rotation of the drum and the baffles help for rapid drum discharge. Most often the cooling section comprises a round-bed cooler with a rotating gentle stirring device. Depending on the air handling, the cooling air may flow through the coffee bed either in bottom-up or top-down direction.

Drum roasters usually operate at a relatively low ABR. The maximum applicable amount of air is limited by the maximum exit air velocity at which beans may be carried away with the air. Typical roasting time is in the range of 8–20 min. The convection conduction ratio is largely influenced by the selection of direct or indirect drum heating. In direct drum heating the furnace is located directly underneath the roasting drum. The resulting wall temperature is relatively high and conductive heat transfer of beans in contact with the hot wall becomes substantial. By contrast, in indirect drum heating the drum is insulated and the hot air is not used to heat the bottom of the drum directly. The hot air is guided to the back of the drum for more convective heat transfer inside the drum.

4.2.2 Paddle Roaster (Tangential Roasters)

In this design the roasting chamber is fixed and contains a rotating mixing device with paddles (Fig. 11.6). Hot air enters in the lower part of the roasting chamber, very often tangentially to the half cylindrical-shaped contour of the roasting chamber. It passes then in bottom-up direction across the batch of beans into a broad expansion chamber at the upper part of the roasting chamber before it exits. In the expansion chamber the air velocity is reduced considerably so that no beans are carried away with the exit air, even at high air-to-bean ratio. Optionally, at the end of the roasting process water quenching can be applied as a precooling step. The beans are then discharged at the bottom of the roasting chamber through an opening gate. They fall by gravity into the cooling section. The cooling section may consist of a round-bed cooler with a gently rotating agitator or a rectangular cooling sieve without any mechanical agitation devices. The cooling air usually flows in bottom-up direction across the coffee bed.

Since the beans are kept in motion in the roasting chamber by the rotating paddles relatively independent from the air flow, the roaster design allows to operate within a wide range of ABR. Consequently, the conduction convection ratio is also variable. Depending on the needs, the roasting time may vary in a range from 2 to 20 min.

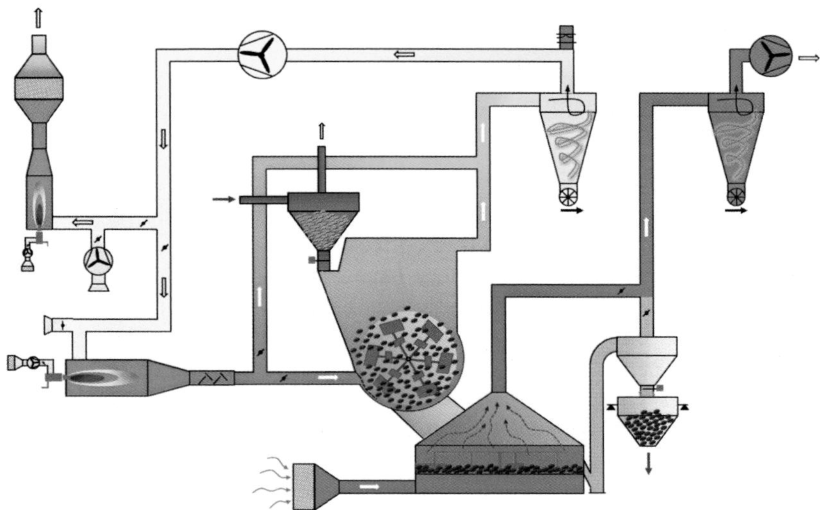

FIGURE 11.6 Schematic drawing of Bühler industrial roasting system InfinityRoast. *Courtesy of Bühler AG, Uzwil, Switzerland.*

4.2.3 Bowl Roaster

A rotating bowl keeps the batch of beans in motion. Centrifugal forces cause the bean movement to the bowl periphery where the beans encounter with stationary guiding baffles that bring them back to the center of the bowl in a spiral-shaped circuit. The hot air is guided top-down in a vertical shaft along the rotation axis and enters the roasting chamber at the bottom of the bowl where it converts into bottom-up direction. After having passed the coffee beans it exits the bowl on top. When the beans have reached their final temperature an optional precooling step may be applied (water quenching). The bowl then moves to a lower position, opening a gap at the bowl edge for bean discharging into the cooling section. The design allows to operate within a certain range of ABR. Typical roasting time may be in the range of 3–12 min.

4.2.4 Fluidized-bed Roasters

There are no moving parts inside the roasting chamber of a fluidized-bed roaster. The beans are kept in motion exclusively by the current of the hot air. A relatively high air velocity is required to generate sufficient buoyancy for fluidization of coffee beans. The air enters at the bottom of the roasting chamber through a perforated plate. Optionally, a specific geometry of the roasting chamber may be used to create a rotation whirl in the air stream (rotating fluidized-bed). Finally the hot air exits on top of the roasting chamber. Convection accounts for the main heat transfer. However, the roasting chamber geometry may also include a zone with inclined walls on which beans slide down and experience a phase of higher share of conduction before they get back to the zone of high air velocity. At the predefined final product temperature the beans are transferred by gravity into the cooling unit.

5. BLENDING

Although roasting of single-origin beans has become popular, most coffee products are still based on a blend of coffees from different origins. The wide variety in flavor potential and physical bean properties of coffees of different origins and species (Arabica and Robusta) leaves endless possibilities for blending to the skilled product developer. However, once the blend composition has been defined, the next and most crucial decision to be taken by the roast master is whether to roast the entire blend in one go (blend-before-roast) or to fractionate the blend and roast individual fractions separately (blend-after-roast, also known as split roast) and finally blend the roasted beans.

5.1 Blend-Before-Roast

Most industrial-scale roasting operations use a blend-before-roast approach. All components of the blend are roasted together in one roasting process. Simplicity and lower cost in operations is the main advantage. However, the different bean varieties may end up with visible difference in degree of roast (inhomogeneous appearance of the roasted beans).

5.2 Blend-After-Roast (Split Roast)

Since the roasting behavior of Arabica and Robusta beans differs considerably, it may often make sense to apply individually optimized roasting conditions to these fractions. Moreover, fractionation can be used to optimize the roasting conditions in a way to push a specific flavor characteristic (e.g., desired acidity) in one fraction and to optimize another flavor note in a different fraction (e.g., strong roast note). This approach may be more common for high quality products. Since it requires more silos and a blending unit after roasting, split roast operation is more demanding and adds complexity to the operation.

6. ROASTING PROFILES

The roasting conditions must be optimized to convert the full green bean potential of a given blend into the desired flavor and physical bean properties. It is the "roasting profile" that gives the beans its signature. For a given blend or roasting fraction the skilled roast master is required to focus on the following main processing parameters of major importance: degree of roast (final product temperature); roasting time; shape of the time–temperature curve; and ABR.

6.1 Degree of Roast

Since most product characteristics are changing continuously during roasting, the achieved degree of roast in the final product is the most important process control criteria. As roasting is continued, more water is evaporated and more organic matter is converted into gas and volatiles. The roast loss increases. Structural changes get more pronounced with increasing degree of roast. The bean density decreases continuously. The darker the roast the more bean volume and porosity will be created. However, bean swelling will level off at one point. The increasing gas formation with increasing degree of roast results in greater gas quantities that are released in the gas desorption process during bean storage. The oil migration process proceeds faster in dark roasted beans due to the stronger driving force. In extremely dark roasted beans the oil may appear on the bean surface already during the last stages of roasting.

The roast flavor becomes more intense with increasing degree of roast. Acidity decreases and bitterness increases. Light roasted coffee brings more acidity to the cup profile. Very simplified, the pleasant and delightful coffee aroma builds up to a certain optimal degree of roast, but then decreases again upon continued roasting. In a similar way the sensory attribute "body" or "mouthfeel" increases to a certain point and then decreases on continued roasting beyond this point. An over-roasted coffee may yield strong intensity of "roasty note" at cup tasting, but very often lacks "body" and produces a sensory perception known as "thin" or "weak mouthfeel." Depending on the origin and characteristics of green beans, different degrees of roast are required to exploit the natural flavor potential to its best.

6.2 Roasting Time

The roasting time plays a key role for the development of flavor and physical bean properties. Long roast and short roast do not result in the same bean properties. Since fast roast is coupled with greater heat transfer rates, the bean temperature increases faster, dehydration and chemical reactions proceed at greater pace. Gas formation rates are higher during fast roasting. Comparing beans of identical degree of roast, fast roasted beans generate larger gas quantities than slow roasted beans. Consequently, the bean expansion also progresses faster (Fig. 11.7). For a given bean color, fast roasted beans exhibit much greater bean volume and porosity and lower density than slow roasted beans. The differences in structure influence also the yield in most types of extraction processes. In general, more soluble matter can be extracted from fast roasted beans. This may be due to greater generation of soluble matter or to better accessibility for the water in high porosity structure or to both. Fast roasted beans finish the roasting process with a slightly higher final moisture content. The water redistribution within the bean takes time and may limit the

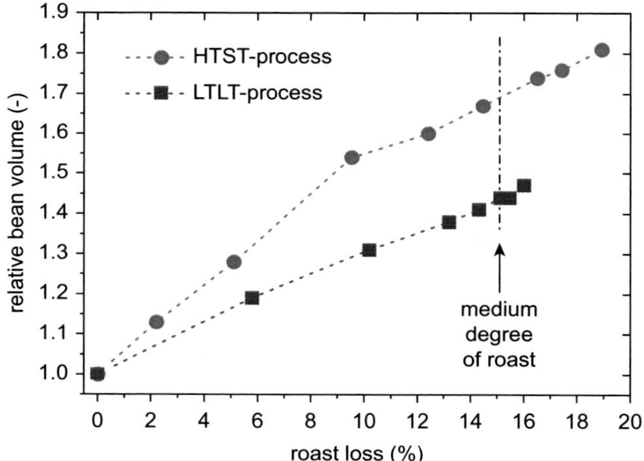

FIGURE 11.7 Characteristic development of bean volume increase as a function of the degree of roast (roast loss) during isothermal fast roast (HTST) and slow roast (LTLT) conditions.

dehydration process in short roast conditions. Fast roasted beans exhibit also a considerably stronger tendency for oil sweating.

Although the overall flavor intensity may be stronger for fast roasted beans compared to slow roasted beans of identical color, it does not mean that the cup profile is necessarily better. The cup profiles are simply different. Consumer preferences alone may decide whether a shorter or longer roasting time is more appropriate for a given raw material. The same aroma impact compounds are formed regardless of the roasting time. However, the quantities of individual compounds or compound groups depend on the roasting time in various ways. Some aroma compounds are preferably generated in fast roasting conditions whereas others are enhanced at slow roast conditions. Consequently, variation of roasting time leads to distinguished profiles of flavor compound concentrations. Fast roasted coffee usually delivers more acidity in the cup profile and often a stronger "roasty" note. Slow roasted coffees often show higher intensity in sensory attributes such as "balanced," "fruity," "nut-like," and "toasty" notes.

One possible problem that may occur in extreme cases of fast roasting is related to the heat transfer within the bean. High rates of heat transfer at the bean surface may result in a substantial temperature gradient within the bean, from the bean surface to the core. In effect, the bean may become over-roasted in the zone close to the surface while still remaining under-roasted in the core. In the cup profile this could lead to "burnt" notes and "greenish" notes at the same time.

6.3 Shape of the Time–Temperature Curve

Traditional roasting profiles apply more or less isothermal heat transfer conditions, sometimes including a stepwise reduction of the heat in the second part of the process. The hot air temperature is set higher for faster roasting time and lower for slow roast conditions. The actual product temperature develops as a function of hot air temperature settings and machine design. In the past, this established, widespread traditional way of profile shaping was due to technical limitations of the roasting equipment. By contrast, modern roasting equipment allows for tailor-made sequences of varied heat transfer over the total roasting time (either multistage process or real profile roasting). In these machines the heat transfer to the beans can be extensively modulated and be controlled in a way to follow a desired product temperature master curve with a preferred time–temperature profile (Fig. 11.8).

Keeping final bean color and roasting time constant, different pathways to reach the final point make a difference to the development of flavor and the physical bean properties. The dehydration kinetics inside the bean depend on the profile of heat transfer and result in distinguished dehydration curves. For example, more or less water can be evaporated during the first roasting stages. Different water activity at different stages of the process influence the chemical reactions and finally also the structural changes. New combinations of bean temperature and water activity may occur during the course of roasting. Prolonged stages at lower temperature may leave more time for the

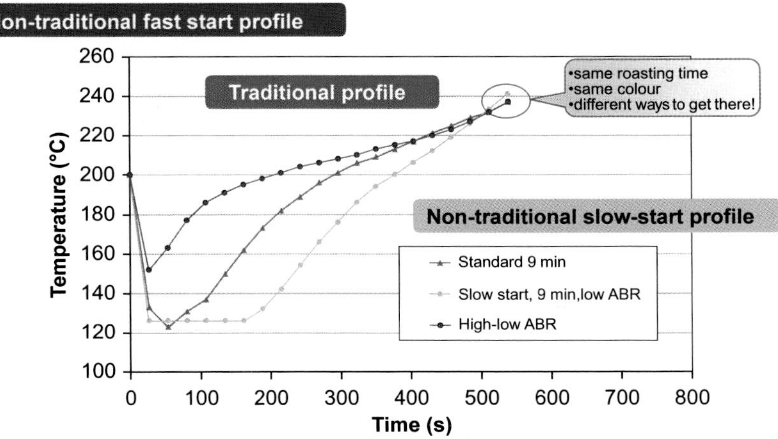

FIGURE 11.8 Examples of nontraditional time temperature curves. Slow-start-fast-final roasting profile (red) and fast-start-slow-final profile (green) compared to a standard profile with a roasting time of 9 min (blue).

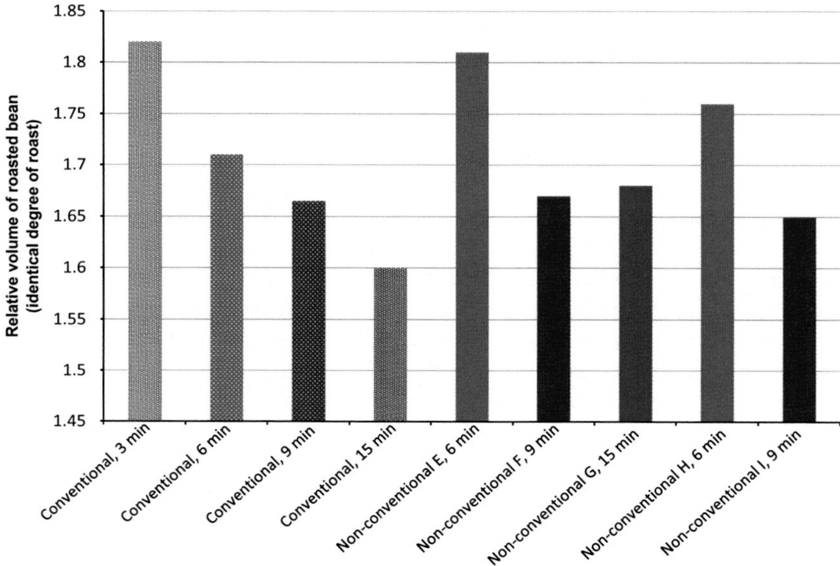

FIGURE 11.9 Final bean volume achieved by various conventional and non-conventional roasting profiles (indicated as E, F, G, H and I) from identical raw material and roasted to identical degree of roast in 3 min (■), 6 min (■), 9 min (■), or 15 min (■) roasting time.

generation of specific flavor precursors and influence subsequent chemical reactions.

For example, a slow-start-fast-final profile keeps the product temperature low at the beginning of the process and increases the heat transfer rate substantially during the final roasting stages. Compared to beans of identical final color and obtained from a traditional roasting process with identical roasting time, the slow-start-fast-final beans usually achieve more bean volume and porosity (Fig. 11.9). In some cases, the structure difference is also reflected in higher extraction yield. The two products also result in statistically significant different flavor profiles.

6.4 Air-to-Bean Ratio

The ABR is widely given by the design of the selected roasting equipment. However, modern roasting equipment very often leaves room for modification of ABR by adjustment of fan speed and flap settings. Following a similar rational as with the nonstandard time–temperature curves, the ABR can make a difference in the development of flavor and structure of the beans. High ABR may lead to greater air velocity at the bean surface and consequently to accelerated mass transfer in evaporation of water from the bean. Bean dehydration progresses faster with increased ABR. Combinations of bean temperature and water activity during the course of roasting are influenced by the ABR. Keeping all other processing parameters constant, variation of ABR can make a limited but statistically significant difference in flavor. Variations in ABR also influence the structure, but not as much as the roast degree and roasting time. In general, high ABR conditions result in slightly decreased final bean volume.

7. ROASTING FROM AN ARTISANAL ROASTERS PERSPECTIVE

Much has been said about the art of roasting coffee, and although it has its artistic elements, it is most certainly a craft driven by the artisan. A roaster would not be able to satisfy even the most devoted customer if they experiment with every roast, changing parameters and profiles on a whim. Delicious coffee and consistency are the ultimate goals, as opposed to constant experimentation or the occasional "happy accident." The artisan is more than simply the machine operator, although they may have begun their career in roasting as one. The individual should possess a core of discipline toward the craft of roasting—not only for the sake of the product but also for the development of their own point of view as an artisan as well.

7.1 The Choice of a Roaster

Roasting machines come in all shapes and sizes and it should be said here that any sized machine could produce a well-crafted product. For artisan roasting, the most popular choices are small-batch, gas-heated drum roasters. These types of roasting machines are prized for possessing two main features, among others, that an artisan needs in their work: (1) the burner controls and (2) the trier, also known as "tryer". The former directly influences the heat transfer to the beans as they roast, and the latter makes the tracking of the process possible.

Any operator manipulates heat controls on a roasting machine, though the artisan roaster is most concerned about the sequence and pattern of the heat applied to achieve a unique and perhaps "signature" cup quality. A gas-powered machine will feature burner sets that apply direct heat to the drum. These can be employed individually or in unison to ramp or reduce heat. Controls will typically be situated near a portal through which the flames can be observed. Infinite adjustment controls to the gas flame are perhaps more popular with artisan roasters, though, as they lend a greater degree of precision to the flame level. Infinite adjustment controls can also be connected to a gas manometer gauge, which allows the artisan to note and then repeat successful gas settings as desired.

The trier, usually situated on the face of the drum, catches small samples of beans during the roasting process. The artisan will take the opportunity to "try," (sample) the roast at a number of key points to assess the development of the beans. Changes in bean color, size, and texture are most notable during the process, and the aroma that the beans emit at each try should be correlated to the stages of roast as well. As the artisan logs more time at the roaster, the messages that the trier delivers will be the most valuable and can be crucial to developing a roasting style or point of view. Temperature probes and gauges may fail or fall out of calibration, or the machine may malfunction in some other way during a roast, but if the artisan can accurately "read" the beans with the trier, the desired roast can still be achieved. At each try, depending upon the size and model of the machine, cool air is introduced to the roasting drum as the trier is pulled from its housing and can affect the momentum of the final stages of the roast. An artisan must try the beans judiciously and thoughtfully with this in mind.

Other notable features of a roasting machine are the airflow and drum speed controls. With both of these, the transfer of heat to beans can be influenced. The artisan may decide to quicken the drum speed to "toss" the beans up and therefore limit the contact to the conductive heat of the drum surface in favor of more overall convection time in the center of the drum. This decision could contribute to a cup with softer acidity, for example, enveloping each bean in a greater opportunity for convective heat, which should homogenize the development throughout each bean. In contrast, too quick of a drum speed could increase the centrifugal force and cause the beans to "hug" the drum sides for a bit longer, and subsequently overly develop the outside of the bean, and at worst, scorch the bean flats.

Airflow controls are used to restrict or increase the velocity of air through the drum, which again, is an adjustment of convective heat transfer to a roast. This allows the artisan to influence convection almost instantaneously, as a flame adjustment will always take more time to change the roasting environment. A roaster can use this feature to develop the outside of the beans, resulting in a desired "roasty" character, or bring the roast in quickly to capture a bright fruity cup. Conversely, an airflow adjustment that stretches out the roast time will typically mellow the acidity in the cup.

Manipulation of the flame and sampling a roast via the trier, and the less often used drum speed and air flow adjustments, are skills the artisan develops over time. As a beginner or apprentice, the artisan must become familiar with the impact each adjustment has on the roast, and by extension, the cup quality. Here, as with all other players in the seed to cup chain, tasting the cup is crucial to the success of the endeavor. A well-developed palate, combined with an understanding of the machine and a unique point of view, will set an apprentice operator on the path to artisan.

7.2 The Feel, Sound, and Smell of Roasting and How This Corresponds to the Physical Changes

Before the tasting can begin, an artisan must rely heavily on their senses of sight and smell as the coffee progresses in the roaster. The changes in the bean that correspond to time and temperature application must be observed and logged. Data logging of temperature from a bean probe and time intervals, as well as the first and temperature of first crack is highly recommended. Logging of data may be done on a laptop computer, or simply with notebook and pencil in hand, whereas a simple stopwatch will measure time from the "charge" (coffee entering the roasting environment), to "drop" (coffee exiting the roasting environment into cooling tray) (Fig. 11.10). The roast plan and intended outcome are commonly referred to as the "profile." A profile can mean both the course of development for the beans inside the roasting machine and the intended flavor characteristics realized in the cup.

Ultimately, logging roast temperature at established time intervals helps set the intended profile for a coffee. For example, a toasted grain aroma from the trier may suggest the coffee's moisture has evaporated the green coffee at perhaps 8 min—and so the roaster may note that temperature from the bean probe and time from the stopwatch into the log, observing the change in color from green to yellow and on to orange. Visually, the color changes from orange through to brown can be observed and logged as the trier emits an aroma of caramel, floral, or fruit notes. The artisan notes these sights and smells to stay on target with the end profile as planned. During the beans' first cracking point, which is marked by a sound similar to crackling or the sound of popcorn popping, the artisan can enjoy the initial aromas of true coffee character and observe the smoothing of the once dense and uneven bean surface texture. The first crack of the roast also corresponds to the beans' switch from endothermic

FIGURE 11.10 Freshly roasted coffee beans in the cooling tray of a small drum roaster.

to exothermic and so the artisan might make an adjustment to flame, and log it to finesse the roasting environment's rate of heat rise. The sense of sight and smell become increasingly more crucial after first crack as this is when the artisan must decide on, and execute, the correct drop time at the desired temperature. A properly developed color and texture can be observed while trying the final stages of the roast, as well as the sweet and heady aromas of sugars being caramelized. If the artisan chooses to continue into second crack, sugar caramelization will cease and roasted notes will be imparted to the coffee. These roasted notes are not inherent in the coffee themselves. Arguably, the profile is constructed of all the data points that came before the drop time. Still, the time logged and the temperature of the beans at the drop will determine the final roast level, and consequently, whether or not the artisan was successful with the intended profile.

7.3 Blend Before Roast Versus Blend After Roast

Although some hotly debate the validity of blending coffees green or after roasted, the final decision must always remain with the individual. Blends are typically used for espresso preparation; single-origin coffees are popular as filtered slow-brews, but tend to be more unbalanced when prepared as a concentrated espresso. Traditionally, blends are created by selecting different coffees as "components," to mitigate overbearing characteristics. Once the raw material has been selected and a cup profile has been crafted, the blending technique must only answer to that individual's vision. The arguments for roasting each component of a blend separately are compelling, but those reasons may make sense only as theory. It could be that the color variations and slightly uneven development resulting from a green blended coffee imparts the flavors that are exactly what the artisan intended. It is once again incumbent upon the artisan to taste the roast and blend trials with an open mind as well as a commitment to the intended cup quality.

7.4 Knowing When the Coffee Is "Just Right"

Perhaps the most crucial choice an artisan makes is not about the machine or its manipulation, but about the coffee beans they choose to roast. In every case, the state of the raw material will dictate the decisions and adjustments necessary to capture the character they hope to achieve. Green bean density, moisture content, screen size, and shape, all factor into the roast plan and all have a great impact on the profile. It may go without saying that the artisan selects coffee through tasting and grading that will suit their purposes as well as delight their senses. An artisan may feel a kinship to the notes from Indonesian coffee, for example, and incorporate it regularly into their menu and blends. Perhaps over many years of trial and error, an espresso blend emerges from their roast log that must always contain a naturally processed

Brazil. These are the building blocks toward an artisan's style. The aforementioned profile should not only compliment the potential the green coffee possesses, but it should also reflect the artisan's style.

The style, or the point of view of the artisan, can be honed, highlighted, and even marketed to spark consumer interest and loyalty. In most cases, the company for whom the artisan works establishes a roast style. In rare cases, a "boutique" coffee roaster may promote the style as unique to the individual artisan to promote the overall coffee brand. The loyal customer may even begin to refer to a company's coffee as the "roast profile" (borrowing the term directly from the artisan) that they prefer over others.

How does an artisan know when the coffee is "just right"? All efforts to understand the roasting machine and its features, the changes that take place in the coffee while roasting, and the choices during selection will lead to this answer. It may take many years to amass this level of understanding of the roasting craft. In the end, if the artisan has succeeded in establishing an individual style as well as a means to execute their desired profile, they have gotten it "just right."

8. OUTLOOK

In industrial roasting, the roasting machine as a tool for flavor generation will become more sophisticated. As roasting equipment becomes more flexible and process control more advanced, we will see increased use of nonconventional, truly innovative roasting profiles in the future. The heat transfer to the beans at various stages of the roasting process will be tailor-made to fit the specific blend and application. Time—temperature conditions during roasting will be optimized to achieve enhanced dehydration kinetics and flavor formation kinetics as well as structural changes. Nonconventional roasting profiles will be applied with a specific objective in mind, such as achieving distinguishing flavor characteristics or creating the best bean pore structure for extraction. Industrial roasting equipment will also become more precise in reproducing an existing optimized roasting profile batch by batch and improve quality consistency. Since consumers link brands to certain quality expectations, consistent in-cup quality will remain a primary objective in industrial roasting. Most likely, the future will also see more advanced and sophisticated split roasting concepts. A new generation of innovative electronic sorting machines will enable the sorting of individual beans according to criteria of chemical bean composition. The blend could for example be split in a bean fraction of high sucrose content and in another fraction of lower sucrose content. Two different roasting profiles would then be applied to deliver the best out of these distinct coffee fractions before they are combined again (blend-after-roast).

Small roasters will continue finding their market niches by roasting specialty coffees, microlots and single-origin coffee. An increasing percentage of consumers also cares about sustainability, which opens a whole new forum for ethically sourced coffee. As societies become increasingly individualistic, more consumers may seek customized coffee products with personalized blends and roasting profiles. In being close to their customers and getting direct consumer insight small artisanal roasters are set to benefit the most from this great opportunity. Intelligent coffee machines (Internet of things) and personalized orders via Website may achieve a new level of real-time consumer insight. Ever-changing consumer trends (darker roast, lighter roast, etc.) will become apparent without delay and small roasters can react immediately to meet customer needs.

Whether an industrial or small-scale operation, it will always need the skills of an experienced roast master for composing the ultimate blend and optimizing the roasting parameters in the first place to get the best out of the bean. This is why roasting remains an art.

REFERENCES

Balzer, H.H., 2001. Acids in coffee. Chemistry. In: Clarke, R.J., Vitzthum, O.G. (Eds.), Coffee. Recent Developments. Blackwell Science Ltd., London (Chapter 1b).

Clarke, R.J., Macrae, R., 1987. Coffee. In: Technology, first ed., vol. 2. Elsevier Applied Science, London.

Eggers, R., Pietsch, A., 2001. Technology I, roasting. In: Clarke, R.J., Vitzthum, O.G. (Eds.), Coffee. Recent Developments. Blackwell Science Ltd., London (Chapter 4).

Geiger, R., 2004. Development of Coffee Bean Structure During Roasting; Investigations on Resistance and Driving Forces (Ph.D. thesis number 15430). Swiss Federal Institute of Technology (ETH), Zurich, Switzerland.

Grosch, W., 2001. Chemistry, volatile compounds. In: Clarke, R.J., Vitzthum, O.G. (Eds.), Coffee. Recent Developments. Blackwell Science Ltd., London (Chapter 3).

Hofmann, T., Frank, O., Blumberg, S., Kunert, C., Zehentbauer, G., 2008. Molecular insights into the chemistry producing harsh bitter taste compounds of strongly roasted coffee. In: Hofmann, T., Meyerhof, W., Schieberle, P. (Eds.), Recent Highlights in Flavor Chemistry and Biology. DFA, ISBN 3-9807686-7-8, pp. 154–159.

Illy, A., Viani, R., 1995. Espresso Coffee; the Science of Quality. Academic Press Limited, London.

Poisson, L., Kerler, J., Davidek, T., Blank, I., 2014. Recent developments in coffee flavour formation using biomimetic in-bean experiments. In: Proceedings of the 25th International Conference on Coffee Science. ASIC, Paris, France.

Raemy, A., Lambelet, P., 1982. A calorimetric study of self-heating in coffee and chicory. Journal of Food Technology 17, 451–460.

Schenker, S., Handschin, S., Frey, B., Perren, R., Escher, F., 1999. Structural properties of coffee beans as influenced by roasting conditions. In: Proceedings of the 18th ASIC Colloquium (Helsinki). ASIC, Paris, France, pp. 127–135.

Schenker, S., 2000. Investigations on the Hot Air Roasting of Coffee Beans (Ph.D. thesis number 13620). Swiss Federal Institute of Technology (ETH), Zurich, Switzerland.

Schenker, S., Handschin, S., Frey, B., Perren, R., Escher, F., 2000. Pore structure of coffee beans affected by roasting conditions. Journal of Food Science 65 (3).

Schenker, S., Heinemann, C., Huber, M., Pompizzi, R., Perren, R., Escher, F., 2002. Impact of roasting conditions on the formation of aroma compounds in coffee beans. Journal of Food Science 67 (1).

Sievetz, M., Desrosier, N.W., 1979. Coffee Technology. AVI Publishing Company Inc., Westport, Connecticut.

VDI-Richtlinie 3892, 2015. Emission Control Roasted-Coffee-Producing-Industry. VDI-richtlinie 3892. Verein Deutscher Ingenieure, Beuth Verlag, Berlin.

Wieland, F., et al., 2011. Online monitoring of coffee roasting by proton transfer reaction time-of-flight mass spectrometry (PTR-ToF-MS): towards a real-time process control for a consistent roast profile. Analytical and Bioanalytical Chemistry 401, 1–13.

Yeretzian, C., Wieland, F., Gloess, A., 2012. Progress on coffee roasting: a process control tool for a consistent roast degree – roast after roast. Newfood 15 (3).

Zimmermann, R., Streibel, T., Hertz-Schünemann, R., Ehlert, S., Schepler, C., Yeretzian, C., Howell, J., 2014. Application of photo-ionization time-of-flight mass spectrometry for the studying of flavor compound formation in coffee roasting of bulk quantities and single beans. In: Proceedings of the 25th International Conference on Coffee Science. ASIC, Paris, France.

Chapter 12

The Chemistry of Roasting—Decoding Flavor Formation

Luigi Poisson[1], Imre Blank[2], Andreas Dunkel[3], Thomas Hofmann[3]
[1]*Nestec Ltd., Nestlé Product and Technology Center Beverages, Orbe, Switzerland;* [2]*Nestec Ltd., Nestlé Research Center, Lausanne, Switzerland;* [3]*Technical University of Munich, Freising, Germany*

1. KEY FACTORS THAT INFLUENCE COFFEE FLAVOR QUALITY

Coffee is highly appreciated for its delightful flavor, which is composed of two distinct sensory modalities, i.e., aroma and taste. Further benefits are the stimulating effect, the enjoyment and indulgence in having a cup of coffee, and sociocultural aspects associated with the consumption of coffee (Caldwell, 2009). The out-of-home consumption runs through steady changing trends, and the connoisseur's community is strongly increasing around the world demanding a high quality cup of coffee. In general, consumers are more aware of quality, with origin and roast degree being a critical part of their coffee selection.

Coffee can be seen both as a craft and science. Researchers and coffee connoisseurs make the analogy to wine making (Mestdagh et al., 2013), where the final product quality is the result of art and knowledge, depending on many factors such as terroir, climate, variety, ripening, fermentation, and aging. Similar to wine, the consumption of coffee is increasingly linked to the consumption occasions and emotions. In general, there is an opportunity for knowledge and technology transfer between coffee and other beverages such as wine and beer.

As discussed in this book (Chapters 1−3), many factors can influence the coffee quality. At the beginning of the coffee value chain stand the agricultural factors, such as the cultivar, climate, and postharvest methods. They determine the green coffee composition, which can modulate the flavor quality (Farah, 2012; Sunarharum et al., 2014; Variyar et al., 2003; Viani and Petracco, 2007). However, roasting (see Chapter 11) is undoubtedly the most important factor in the coffee value chain where important physical and chemical changes lead to

The Craft and Science of Coffee. http://dx.doi.org/10.1016/B978-0-12-803520-7.00012-8

the development of the characteristic roasted coffee attributes (Clarke, 1987). It is only during coffee roasting at temperatures higher than 200°C (Dalla Rosa et al., 1980) at which the green coffee precursors are transformed into roasted coffee constituents giving rise to color, aroma, and taste. However, the coffee's intrinsic quality is predetermined in the green bean by its precursor composition and the roaster only can unlock the full potential by applying the appropriate and optimized roasting conditions. Roasting degree, roasting profile, and the technology are the decisive factors for the final product. The knowledge of flavor precursors as well as the formation mechanism and kinetics of the key flavor compounds is essential in the development of high quality products with desirable sensory attributes. This information should help develop better quality coffees through a targeted selection of raw material and an improved understanding and applications of roasting technology. Therefore, an in-depth molecular science of the roasting chemistry is the key to build up knowledge and to apply it into the daily routine of a roaster's work.

In this chapter, we will first describe the flavor precursors occurring in green coffee and then present relevant aroma and taste compounds obtained by using sophisticated analytical methodologies. In the next section, we will elaborate on the chemical reactions and processing parameters leading to the characteristic coffee flavor. Finally, we will discuss the kinetics of flavor formation and ways to modulate the flavor note.

2. FLAVOR PRECURSORS OCCURRING IN GREEN COFFEE BEANS

The composition of the green beans determines the aroma and taste quality formed during roasting. Therefore, the green coffee constituents were investigated in much detail to draw conclusions on the quality obtained from a specific coffee and to leverage this knowledge for the optimization of coffee processing (Arya and Rao, 2007; Fischer et al., 2001; Nunes and Coimbra, 2001; Redgwell et al., 2002). Roasting can be described as a dry heating food process starting with a drying phase (up to c. 100°C, endothermic), followed by an exothermic phase (c. 170–220°C) resulting in most of the flavor components, and finally a cooling phase. Table 12.1 shows the main green coffee constituents of Arabica and Robusta coffee (Belitz et al., 2009), demonstrating the complexity of green coffee composition. Green coffee is primarily composed of carbohydrates, nitrogen (*N*)-containing compounds (mainly proteins, trigonelline, and caffeine), lipids, organic acids, and water. Almost all green coffee components are potential precursors for flavor and color or involved in their development. Even the water content can play a crucial role for the final coffee quality (Baggenstoss et al., 2008b). From this pool of green coffee constituents, however, the principal flavor precursors are sugars, proteins, free amino acids, trigonelline, and chlorogenic acids (CGA). These precursor classes will be discussed in more detail in this chapter.

TABLE 12.1 Chemical Composition of Raw *Arabica* and *Robusta* Coffee Beans

	Content (% Based on Dry Weight)		
Constituents	**Arabica**	**Robusta**	**Components**
Soluble carbohydrates	9–12.5	6–11.5	
Monosaccharides	0.2–0.5	0.2–0.5	Fructose, glucose, galactose, arabinose (traces)
Oligosaccharides	6–9	3–7	Sucrose (90%), raffinose (0–0.9%), stachyose (0–0.1%)
Polysaccharides	3–4	3–4	Heteropolymers from galactose (55–60%), mannose (10–20%), arabinose (20–35%), glucose (0–2%)
Insoluble carbohydrates	46–53	34–44	
Hemicellulose	5–10	3–4	Heteropolymers from galactose (65–70%), Arabinose (25–30%), mannose (0–10%)
Cellulose, β(1–4) mannan	41–43	32–40	
Acids and phenols			
Organic acids	2–2.9	1.3–2.2	Citric acid, malic acid, quinic acid
Chlorogenic acids	6.7–9.2	7.1–12.1	Feruoylquinic acid, mono- and di-caffeoyl quinic acid
Lignin	1–3	1–3	
Lipids	15–18	8–12	
Coffee oil	15–17.7	8–11.7	Major fatty acids: linoleic acid $C_{18:2}$ and palmitic acid $C_{16:0}$
Wax	0.2–0.3	0.2–0.3	
N-compounds	11–15	11–15	
Free amino acids	0.2–0.8	0.2–0.8	Major amino acids: glutamic acid, aspartic acid, asparagine
Proteins	8.5–12	8.5–12	
Caffeine	0.8–1.4	1.7–4.0	Traces of theobromine and theophylline

Continued

TABLE 12.1 Chemical Composition of Raw *Arabica* and *Robusta* Coffee Beans—cont'd

	Content (% Based on Dry Weight)		
Constituents	Arabica	Robusta	Components
Trigonelline	0.6–1.2	0.3–0.9	
Minerals	3–5.4	3–5.4	

Adapted from Belitz et al. (2009).

Although the overall composition of Arabica and Robusta species is very similar, their relative proportions differ considerably (Table 12.1). Arabica coffees are characterized by higher contents in carbohydrates (i.e., sucrose, oligosaccharides, mannans), lipids, trigonelline, organic acids (malic, citric, quinic), and 3-feruoyl-quinic acid (3-FQA). On the other hand, Robusta coffees contain more caffeine, proteins, arabinogalactans, CGA (except 3-FQA), total phosphate, ash (i.e., Ca-salts), and transition metals (i.e., Fe, Al, Cu). These important differences in composition, as it will be seen later, are decisive for the differences in the roasted coffee qualities and characteristics.

2.1 Carbohydrates

The carbohydrates represent about 40–65% of the dry basis of green coffee, consisting of water-soluble and water-insoluble carbohydrates (Table 12.1). Polymers of arabinose, galactose, glucose, and mannose constitute both the soluble polysaccharides and the insoluble fraction, which form the structure of the cell walls along with proteins and CGA (Bradbury and Halliday, 1990). Cellulose, galactomannan, and arabinogalactan represent around 45% of dry weight of coffee beans (Trugo, 1985) all of them showing complex structures. The soluble disaccharide sucrose accounts for the rest.

The soluble fraction of the green coffee is supposed to be the most important precursor pool in the formation of coffee aroma, taste, and color (De Maria et al., 1996a; Nunes and Coimbra, 2001). The precursors are readily available for manifold reactions, which is demonstrated by their rapid consumption in the early stage of roasting. The water-soluble constituents are divided into two fractions, i.e., high molecular weight (HMW) and low molecular weight (LMW) fraction (De Maria et al., 1994). Water-soluble HMW polysaccharides are primarily represented by galactomannans and arabinogalactans, the latter accounting for 14–17% dry matter (Bradbury and Halliday, 1990; De Maria et al., 1994; Illy and Viani, 1995). The green coffee arabinogalactans are highly branched and covalently linked to proteins to form arabinogalactan proteins (AGPs). Roasting causes structural modifications of the AGPs including depolymerization of main and side chains, thus releasing free arabinose, which acts as an important sugar precursor (Oosterveld et al., 2003; Wei et al., 2012). The main part of released arabinose is involved in melanoidin formation occurring during coffee roasting (Bekedam et al., 2008, 2006; Moreira et al., 2012; Nunes et al., 2012). In addition, the arabinose residues of arabinogalactan side chains might play a role in acid formation, i.e., formic and acetic acid (Ginz et al., 2000). Free galactose, the other constituent of arabinogalactans, could only be detected in significant amounts in green beans and is rapidly degraded (Redgwell et al., 2002).

The water-soluble LMW fraction contains important flavor precursors such as free sugars, trigonelline, and CGA. Mono- and disaccharides are minor constituents, however, they are essential for aroma formation by caramelization and Maillard-type reactions. Sucrose (disaccharide composed of glucose and fructose) is by far the most abundant and important sugar in the green coffee with c. 8% in Arabica and only about half of the amount found in Robusta (3–6%). The more complex aroma and overall flavor of Arabica coffee has been explained by its higher sucrose level (Farah, 2012). In addition, oligosaccharides (stachyose, raffinose) and monosaccharides (fructose, glucose, galactose, arabinose) are found in trace amounts in green coffee. The concentration of glucose and fructose rises in the early roasting stage due to the constant degradation of sucrose. Almost all free sugars are lost upon roasting due to Maillard reaction and caramelization, giving rise to water, carbon dioxide, color, aroma, and taste.

The insoluble part mainly consists of polymeric HMW components. These polysaccharides fraction is located in the rather thick, dense coffee cell wall complex, consisting of three polymers, i.e., mannans, hemicellulose, and cellulose. They are higher in Arabica than in Robusta coffee. Galactomannans are the most abundant polysaccharides in the green coffee bean, representing at least 19% of its mass. They act as storage carbohydrates to form part of the energy reserve of the mature seed, analogous to the role played by starch in cereal endosperms. The structure of the galactomannans consists of a linear backbone of β-1,4-linked mannose molecules with single-unit α-1,6-linked galactosyl side chains at various intervals along the mannan backbone (Fischer et al., 2001; Liepman et al., 2007).

About 12–24% of the polysaccharides are degraded in light roasted coffee, 35–40% upon dark roasting. This can be explained by the degradation of the arabinogalactan side chains to arabinose, whereas cellulose and mannans remain almost intact in the roasted coffee (Bradbury, 2001). Polysaccharides do not seem to specifically contribute to aroma formation during roasting, but impart relevant organoleptic properties of the coffee brew, such as viscosity and mouth feel (Redgwell et al., 2002). In general, the role of the water-soluble fraction in the formation of roasted coffee constituents is much better understood compared to the water-insoluble part.

In summary, monosaccharides and the disaccharide sucrose are very fragile to heat treatment. They are quantitatively degraded under roasting conditions within a few minutes. Depolymerization of polysaccharides and their participation in flavor formation depend on their structure: branched arabinose in arabinogalactans is likely to contribute, skeleton building galactans much less, and supramolecular structures such as cellulose and mannan remain largely unchanged.

2.2 Acids

The acidic fraction in green coffee is composed of volatile aliphatic and nonvolatile aliphatic and phenolic acids present in the raw bean (about 8%). Main nonvolatile acids are CGA, citric, malic, and quinic acid (Maier, 1993). Volatile acids are mainly represented by formic and acetic acids (Viani and Petracco, 2007), which stem from the fermentation process in postharvest treatment, but can also be generated through Maillard-type reactions upon roasting (Davidek et al., 2006).

Coffee is well known for its rich CGA content, one of the highest concentrations of all plants. Robusta green beans contain significantly more total CGA than Arabica coffee. The levels of CGA in green coffee have been reported to vary from 8% to 14.4% dry matter (DM) for Robusta to 3.4–4.8% DM for Arabica (Ky et al., 2001). CGA are a group of phenolic compounds derived from the esterification of hydroxycinnamic acids (caffeic, ferulic, and *p*-coumaric acids) with quinic acid. Caffeoyl-quinic acid (CQA) accounts for about 80% of the total CGA (Farah, 2012). The 5-CQA isomer was found to be the most abundant CGA, which is continuously degraded during roasting, followed by 4- and 3-CQA (Clifford et al., 2003; Farah et al., 2005). Also diesters of hydroxycinnamic acids and monomer quinic acid occur in green coffee (Jaiswal et al., 2012). CGA are important precursors of bitter taste compounds but can also be decomposed to their moieties, quinic acid and hydroxycinnamic acid, which may further degrade to volatile and nonvolatile phenolic compounds (Dorfner et al., 2003; Tressl et al., 1976). Volatile phenols from the class of guaiacols such as 2-methoxyphenol (guaiacol), 4-ethyl-2-methoxyphenol (4-ethyl guaiacol), and 4-vinyl-2-methoxyphenol (4-vinyl guaiacol) provide the typical smoky, woody, and ashy characteristics of dark roasted coffees. Higher amounts of these impact odorants are present in Robusta due to the higher abundance of CGAs.

2.3 Nitrogen (N) Containing Compounds

N-compounds, mainly proteins, represent about 11–15% of dry coffee material. The total protein content is c. 10% for both Arabica and Robusta green coffee. Part of the proteins is linked to the water-soluble polysaccharide arabinogalactan to form the AGPs.

Free amino acids represent only less than 1% of green coffee, however, their importance for the final flavor of roasted coffee is high. They are key reaction partners in the Maillard chemistry as well as in the Strecker degradation to yield many potent odorants. Glutamic acid, aspartic acid, and asparagine are the three main free amino acids. However, it seems the distribution of the single amino acids determines the aromatic profile upon Maillard reaction (Wong et al., 2008). Proteins and peptides may also act as aroma precursors as they can decompose to smaller reactive molecules (De Maria et al., 1996b). Free amino acids are almost completely decomposed upon roasting.

Besides proteins and free amino acids, coffee also contains the alkaloid caffeine, probably the most known and best studied alkaloid in plants. Another nitrogenous component is trigonelline, which upon roasting is partially degraded and converted into nicotinic acid and volatile compounds such as pyridines and pyrroles (Viani and Horman, 1974).

2.4 Lipids

Lipids constitute 15–18% of Arabica and 8–12% of Robusta raw beans (Viani and Petracco, 2007). The lipid fraction is composed of coffee wax coating the bean and triglycerides. Linoleic acid (40–45%) and palmitic acid (25–35%) are the main fatty acids. In addition, diterpenes (cafestol, kahweol) and sterols in free and esterified form are part of the total lipid fraction. During roasting, lipids can form aldehydes through thermal degradation, which can further react with other coffee constituents (Belitz et al., 2009).

3. FLAVOR COMPOUNDS GENERATED UPON ROASTING

Molecular flavor science aims at understanding the impact of flavor compounds, their release and human cognition on a given product or matrix using the knowledge to optimize consumer hedonic responses. Flavor consists of two distinct sensory modalities, i.e., the volatile aroma perceived nasally and the nonvolatile taste perceived in the oral cavity. Both are equally important along with the trigeminal sensation (e.g., cooling, hot, and tingling). We will first focus on analytical considerations with focus on sensorially relevant compounds.

3.1 Flavor Analysis

As coffee flavor is composed of more than 1000 volatile and nonvolatile compounds, a sophisticated analytical approach is required combining sensory and instrumental evaluations to narrow down the number of relevant compounds to focus on. The complexity of coffee flavor requires a high level of analytical expertise to choose the most appropriate methodology for a given task to accomplish, in combination with data treatment and molecular interpretation (Kerler and Poisson, 2011). The detailed knowledge of the impact odorants and taste-active molecules should allow connecting the processing conditions in the coffee value chain to the green bean composition.

Particularly in coffee, the aroma is seen as a key quality parameter, differentiating brewed coffee from a soluble coffee, espresso from an Americano, Robusta from an Arabica, Colombian from a Brazilian coffee. The characterization of coffee aroma is a challenging task as many of the important odorants are just present in trace amounts and/or are reactive and unstable. A major progress in decrypting the coffee flavor has been achieved by the use of sophisticated methods for sensory-directed chemical analysis, i.e., using sensory methods (e.g., sniffing and tasting) in the identification of aroma- and taste-active compounds that really matter for the overall flavor (Blank et al., 1992; Ottinger et al., 2001). The use of gas chromatography—olfactometry and the odor activity calculations, i.e., ratio of concentration to odor threshold (Blank, 2002; Grosch, 2001b), has been essential to identify a rather limited number of aroma-relevant volatiles (odorants). This so-called Sensomics concept has also been successfully implemented in the characterization of the taste-relevant components (Frank et al., 2001). Furthermore, the increasing performance of analytical instruments has been an important breakthrough allowing enhanced separation of components on the chromatographic side and higher sensitivity and selectivity of detection devices, mainly mass spectrometry (MS).

The general sequence of this Sensomics approach is shown in Fig. 12.1. ❶ The isolation of the aroma (volatile fraction) or taste (nonvolatiles fraction) from a coffee product represents the first crucial step. This can be done by various techniques with the aim of obtaining a representative extract of the original aroma (Glöss et al., 2013; Sarrazin et al., 2000) or taste composition. ❷ In the second step the flavor isolates are screened for character impact aroma or taste components by sensory-guided fractionation techniques. Screening of odorants is performed by GC separation of the aroma extract combined with olfactometry where odorants are evaluated by sniffing of the effluent gas (d'Acampora Zellner et al., 2008; Blank, 2002), i.e., the human nose is used as sensitive detector to differentiate potent odorants from the crowd of odorless volatile components. Similarly, various fractionation techniques are used for the characterization of taste components resulting in multiple fractions in which the nonvolatiles are assessed by tasting using the tongue as detector. ❸ The most intense odorants and tastants detected during the screening by Aroma Extract Dilution Analysis or Taste Dilution Assay are then identified by means of MS and nuclear magnetic resonance (NMR) techniques. ❹ The identified putative odorants or taste compounds are quantified and their odor activity values (OAV) or Dose-over-Threshold (DoT) factors are calculated (ratio between concentration and odor/taste threshold in a defined matrix). Any accurate quantification methodology can be used, however,

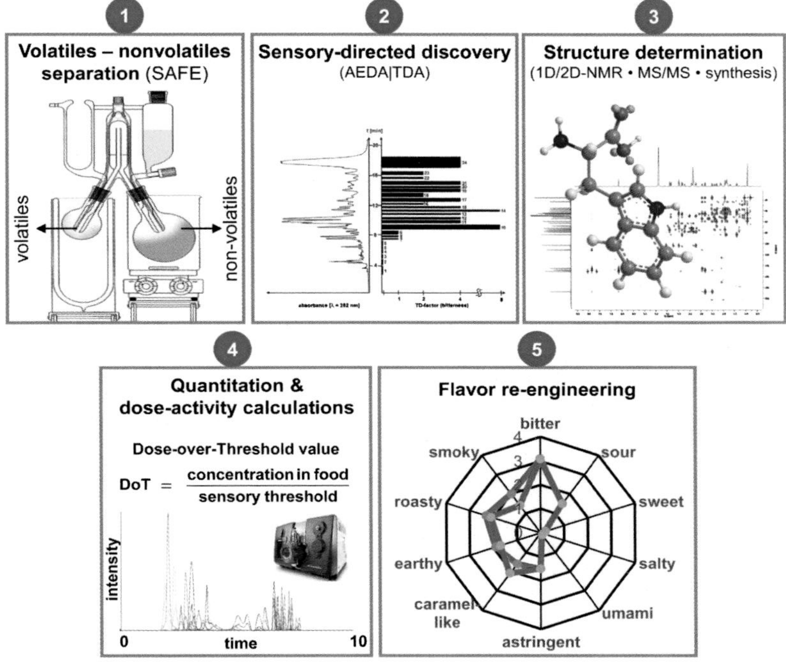

FIGURE 12.1 The five steps of the Sensomics concept.

the method of choice remains the so-called stable isotope dilution assay. This technique involves the use of isotopically labeled molecules (i.e., analytes labeled with stable isotopes ^{13}C or ^{2}H) as internal standards applicable to any of flavor isolation techniques and instrumental assays. The resulting OAV or DoT value is a strong indicator for the relative importance of a flavor compound. ❺ Finally, the OAV (DoT) concept aims at linking the quantitative analytical data to the sensory character of the initial coffee sample (e.g., coffee brew) by mixing the identified key aroma/taste in their natural concentrations and comparing its sensory profile with that of the initial coffee product. The relative importance of an odorant/tastant or a group thereof on the overall coffee flavor model can be evaluated by omitting single compounds or a group of compounds in the test model. Molecules leading to a significant change of the sensory profile can be referred to as impact compounds of coffee.

3.2 Aroma Composition

Most of the aroma research has been performed on roasted coffee or the corresponding beverage. However, the volatile composition of green coffee was studied as well (Cantergiani et al., 2001; Spadone et al., 1990). The aroma of green coffee is described as green, hay- and pea-like (Gretsch et al., 1999), and the taste as sweet, astringent (Viani and Petracco, 2007), which indeed is very different from the product after roasting. The studies on green coffee aroma mainly aimed at identifying off-flavors prior to roasting and predicting the quality of the roasted product (Cantergiani et al., 2001; Spadone et al., 1990). Green coffee beans contain about 300 volatiles in much lower concentrations compared to roasted coffee. Some of the compounds are not affected by roasting and can be found unchanged in the roasted product (e.g., 3-isobutyl-2-methoxypyrazine formed enzymatically in green beans), whereas the content of others decrease during roasting by evaporation (e.g., ethyl-3-methylbutyrate) or degradation. Besides the methoxy pyrazines and esters, further aroma-active compounds are present in green coffee prior to roasting, including linalool, lipid degradation products, and biologically derived alcohols, aldehydes, and organic acids (Cantergiani et al., 2001; Holscher and Steinhart, 1994).

However, the characteristic coffee aroma and taste only arise from a complex network of physical and chemical changes during roasting. Researchers have long been exploring what makes roasted coffee smell so good, and in the meantime the number of volatile compounds identified in roasted coffee exceeds more than 1000 (Nijssen, 1996). The complex mixture and balance of the volatile fraction make up for only about 0.1% of the total roasted coffee weight, with single components ranging from parts per trillion (ppt) levels to higher part per million (ppm) levels, making it to one of the beverages with the richest and most complex flavor content (Hertz-Schünemann et al., 2013).

Meanwhile it is broadly accepted that it is not the number of components that defines the quality, intensity, or characteristic of the overall odor impression of a food stuff. Indeed, for only about 25–30 key odorants a contribution to the overall aroma of coffee has been evidenced (Blank et al., 1992; Grosch, 1998, 2001a; Kerler and Poisson, 2011). Interestingly, a few to about 40 impact odorants were found in most of investigated foods, e.g. roasted beef (Cerny and Grosch, 1992), red wine (Frank et al., 2011), or chocolate (Schnermann and Schieberle, 1997), though hundreds of volatiles were identified. In addition, only a limited number of genuine volatile compounds are spread in most of the food matrices and less than 3% of foodborne volatiles constitute the chemical odorant space (Dunkel et al., 2014).

The aroma-impact compounds belong to the classes of thiols, sulfides, aldehydes, pyrazines, dicarbonyls, phenols, and furanones. Table 12.2 gives an overview of the key odorants in roasted coffees evaluated by different researchers (Blank et al., 1992; Schenker et al., 2002; Semmelroch and Grosch, 1996; Kerler and Poisson, 2011).

3.3 Taste Composition

Roasting generates bitter taste in coffee. The chemicals of CGA present in green coffee beans are strongly degraded during roasting. Intense bitter taste

TABLE 12.2 Description of Key Odorants of Roasted *Arabica* and *Robusta* Coffee and Their Relative Abundance (More Abundant in Arabica Coffee [A], More Abundant in Robusta Coffee [R], Similar in Arabica and Robusta [A/R] as Reported by Different Authors (Blank et al., 1992; Schenker et al., 2002; Semmelroch and Grosch, 1996; Kerler and Poisson, 2011)

Key Aroma Compound	Flavor Quality	Relative Abundance
Methanethiol	Sulfur, garlic	R
Dimethyl sulfide	Sulfur, cabbage	A/R
Dimethyl disulfide	Sulfur, cabbage	A/R
Dimethyl trisulfide	Sulfur, cabbage	A/R
2-Furfurylmercaptane	Sulfury, roasty	R
3-Mercapto-3-methylbutyl formate	Catty, blackcurrant-like	A/R
3-(Methylthio)propionaldehyde (methional)	Potato	A
2-Methylbutanal	Green, solvent, malty	R
3-Methylbutanal	Malty, cocoa	A/R
2,3-Butanedione	Buttery	A/R
2,3-Pentanedione	Buttery	A
2-Ethyl-3,5-dimethylpyrazine	Earthy, roasty	R
2-Ethenyl-3,5-dimethylpyrazine	Earthy, roasty	R
2,3-Diethyl-5-methylpyrazine	Earthy, roasty	R
2-Methoxy-3-isobutylpyrazine	Pea, earthy	A
2-Methoxyphenol	Smoky	R
4-Ethyl-2-methoxyphenol	Spicy, clove-like	R
4-Vinyl-2-methoxyphenol	Spicy, clove-like	R
3-Hydroxy-4,5-dimethyl-2(5*H*)-furanone (sotolon)	Fenugreek, curry	A/R
4-Hydroxy-2,5-dimethyl-3(2*H*)-furanone (furaneol)	Caramel, fruity	A

compounds have been reported in roasted coffee brew based on sensory-guided fractionation, sophisticated analytical techniques, and model roasting experiments with CGAs (Frank et al., 2006). CGA lactones (Fig. 12.2) have been found as intense bitter tastants of coffee. Depending on their chemical structure, the bitter taste threshold concentrations ranged between 9.8 and 180 μmol/L (water). The authors attributed approximately 80% of the bitterness of the decaffeinated coffee beverage to these 10 quinides based on quantification and determination of the DoT factors for the individual bitter compounds. These data also indicate the limited role of caffeine for the overall bitterness perceived in coffee beverages.

Phenylindanes, breakdown products of CQA lactones, have been reported as further bitter tasting compounds, i.e., 1,3-bis(3′,4′-dihydroxyphenyl) butane, *trans*-1,3-bis(3′,4′-dihydroxyphenyl)-1-butene, and eight multiple hydroxylated phenylindanes (Frank et al., 2007) (Fig. 12.2). They have been associated with a lingering harsh bitterness taste in dark roasted coffee. Several bitter compounds identified in coffee brew showed rather low recognition threshold concentrations ranging between 23 and 178 μmol/L (water). In general, the perceived bitterness increases with higher degree of roast. Although caffeine—which is present in the green bean—has a strong bitter taste, it contributes only some 10–20% to the sensory-perceived bitterness in coffee. Its concentration does not change upon roasting. Furthermore, diketopiperazines (DKPs), condensation product of free amino acids, play a role in coffee bitterness. In general, the perceived bitterness increases with higher degree of roast. The taste thresholds (μmol/L water) are in the range of 30–200 (CQAs), 30–150 (phenylindanes), 50–800 (benzene diols), 190–4000 (DKPs), and 750 (caffeine).

Aliphatic acids such as citric and malic acid are highly relevant for the acidic taste (Balzer, 2001). These acids are also present in the green bean and are gradually reduced during the roasting step. Acetic acid and formic acid are strong contributors to total sensory perceived acidity. Their concentration in green coffee is very low, but they are generated by carbohydrate degradation. The concentrations of quinic acid (from CQAs) and volatile acids are slightly increasing during roasting. Overall, the sensory perceivable total acidity is decreasing during the course of roasting. Light roasted beans elicit more acidity in the cup than dark roasted coffee.

4. CHANGES OF PRECURSOR COMPOSITION UPON ROASTING

The roasting process provides the appropriate conditions for the necessary physical and chemical transformations to take place in addition to the change in color from green to brown. In the early drying stage of roasting free water is lost, which is followed by the dry heating stage consisting of multiple chemical processes such as dehydration, hydrolysis, enolization, cyclization, cleavage, fragmentation, recombination of fragments, pyrolysis,

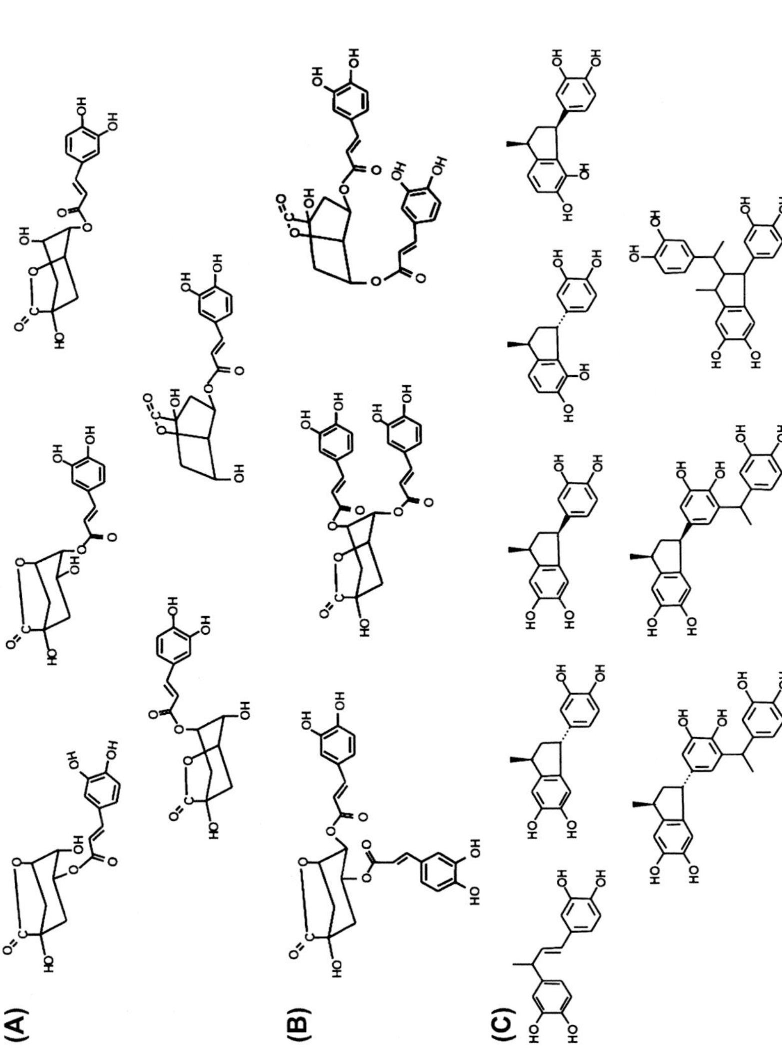

FIGURE 12.2 Chemical structures of selected bitter taste compounds in roasted coffee: (A) monocaffeoyl quinic acid lactones, (B) dicaffeoyl quinic acid lactones, and (C) bis(dihydroxyphenyl) butane and −1 butene, and multiply hydroxylated phenylindanes.

and polymerization reactions. With the rising temperature in the exothermic phase (above 170°C) and the drying of the beans a size expansion is observed under increased pressure accompanied by flavor generation from the precursors.

4.1 Major Chemical Reactions

Many of these changes are associated with the so-called Maillard reaction, also referred to as nonenzymatic browning, leading to the formation of lower molecular weight compounds, such as carbon dioxide and flavor components providing roasted, caramel, earthy, toasted aroma to the roasted beans. The change in the bean color is due to the generation of higher molecular weight–colored melanoidins. Furthermore, Maillard-type reactions may lead to an antioxidative effect and chemoprevention (Somoza, 2005; Summa et al., 2007; Yilmaz and Toledo, 2005), but also to undesirable components to mitigate, e.g., acrylamide and furan (Tamanna and Mahmood, 2015).

The Maillard reaction is a complex cascade of reactions starting with an amino-carbonyl coupling of reducing sugars and amino acids (or peptides and proteins) leading to N-substituted glycosylamines (Schiff base), which are rearranged to the first stable intermediates, i.e., Amadori and Heyns products. These activated sugar conjugates are reactive species and readily decompose leading to smaller fragments, which may react further forming a multitude of volatile and nonvolatile reaction products (Ledl and Schleicher, 1990). Basically, the Maillard reaction is an amino-catalyzed sugar degradation leading to aroma, taste, and color (melanoidins). Sucrose, the most abundant free sugar in green coffee beans, needs first to be decomposed into glucose and fructose by heat treatment to undergo Maillard-type reactions.

Resulting volatiles developed from the Maillard reaction are pyridines, pyrazines, dicarbonyls (e.g., diacetyl), oxazoles, thiazoles, pyrroles and imidazoles, enolones (furaneol, maltol, cyclotene), and many others (Fig. 12.3). The spectrum of the Maillard products and consequently the composition of the roasted coffee can vary depending on the composition of the educts in the green bean. In addition, the flavor formation is influenced by many important parameters such as the type of sugars and amino acids involved, reaction temperature, time, pressure, pH, and moisture content (Ho et al., 1993; Ledl and Schleicher, 1990).

Acetic acid and formic acid strongly contribute to total sensory perceived acidity. They are generated during the initial stages of roasting from carbohydrate precursors in the course of the Maillard reaction and caramelization (Davidek et al., 2006). However, they degrade or evaporate at higher

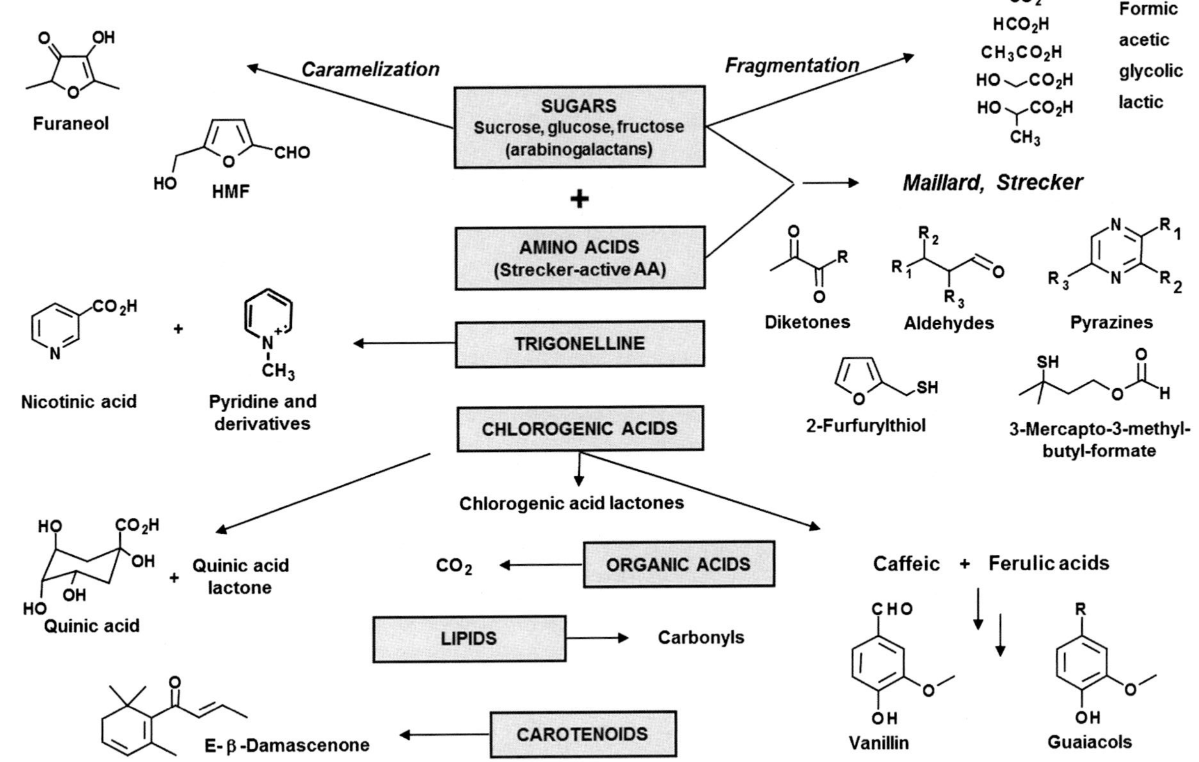

FIGURE 12.3 Schematic presentation of the most important flavor precursors in green coffee and the transformation into key aroma compounds (Yeretzian et al., 2002).

temperatures during the final stages of roasting. Even though the concentrations of short-chain volatile acids are slightly increasing during roasting, the sensory perceivable total acidity is decreasing during the course of roasting, thus indicating their limited role in coffee acidity.

As part of the Maillard reaction network, the Strecker degradation is of outstanding importance for flavor formation contributing to the coffee aroma spectrum with volatile aldehydes having malty (3/2-methylbutanal), potato (methional), and honey-like (phenylacetaldehyde) notes. Basically, it is a deamination and oxidative decarboxylation of the amino acid resulting in the Strecker aldehyde (R−CHO). The Strecker degradation also yields alkyl pyrazines (Amrani-Hemaimi et al., 1995) contributing to the earthy, roasty notes of the coffee aroma. Sulfur-containing amino acids (cysteine, methionine) react to thiols and sulfides. Some of them have rather low odor thresholds, thus contributing at very low concentrations to the coffee aroma, such as 3-mercapto-3-methylbutyl formate, 3-methyl-2-butene-1-thiol, 2-furfurylthiol, and methanethiol (Holscher and Steinhart, 1992; Holscher et al., 1992). Thiols are likely to be oxidized to sulfides. Due to the Maillard reaction, only traces of free amino acids and free sugars can be found after roasting, whereas the crude protein content and most of the polysaccharides change only very slightly upon roasting.

In addition, heat treatment at high temperatures induces the caramelization of sugars giving rise to caramel and seasoning notes. However, aroma formation is favored in the Maillard route due to lower activation energy in the presence of reactive nitrogen species (e.g., amino acids). Both, Maillard reaction and caramelization are the main routes to the formation of brown polymers (i.e., melanoidins).

CGA are strongly degraded upon coffee roasting. The degradation of CGA leads to the hydrolysis products such as quinic acid and the phenolic acids such as ferulic acid, which further degrades forming important phenolic odorants such as guaiacol and 4-vinylguaiacol. After 9 min of roasting, about 90% of total CGA (i.e., 7% green coffee solids) have reacted (Farah et al., 2005). The identification and formation pathways of bitterness components in roasted coffee have recently been elucidated. CGA lactones, breakdown products of CGA (CQAs), have been identified as one of the main contributors to bitterness in coffee. In addition hydroxylated phenylindanes have been reported as intense bitter-tasting compounds (Frank et al., 2007). They are breakdown products of CQA lactones with caffeic acid being a key intermediate in the generation of harsh bitterness reminiscent of the bitter taste of a strongly roasted espresso-type coffee. The structures of these bitter compounds show strong evidence that they are generated by oligomerization of 4-vinylcatechol released from caffeic acid moieties upon roasting.

Lipid oxidation of unsaturated fatty acids produces different highly potent aldehydes such as hexanal, nonenal, other enals, and dienals. However, they do not belong to the key compounds of coffee aroma. The aldehydes can further react through cyclization or with other coffee constituents (Belitz et al., 2009). Hexanal represents a suitable indicator of lipid oxidation in various foods (Sanches-Silva et al., 2004). Moreover, hexanal was held responsible for the coffee staling among other compounds (Spadone and Liardon, 1989).

Most of the polymeric carbohydrates, lipids, caffeine, and inorganic salts survive the roasting process. Alkaloids such as caffeine are relatively stable and only trigonelline is partially degraded to volatile compounds (Viani and Horman, 1974).

4.2 Flavor Generation as Affected by Green Coffee Constituents

Modulation of coffee aroma and taste by roasting remains a highly empirical approach, some roaster would say a handcraft, but for others it is rather an art. Nevertheless, a scientific base on the green coffee constituent functions in the generation of the desired coffee flavor profile might help coffee roasters in developing higher quality products. For this purpose more work is needed on in-depth mapping of precursor changes during the roasting process, as well as the mechanisms by which green bean components are transformed into functional coffee compounds.

The unique structural properties of the green coffee bean (i.e., the thick cell walls without intracellular spaces), the large number of precursor compounds present, together with the enormous physical changes in the bean during roasting (i.e., structural changes, internal pressure build up, increase of bean volume) at high temperatures make coffee roasting a very complex process to study. Researchers applied different strategies to evaluate the role of different precursors or precursor groups on the formation of functional molecules. The roasting chemistry can be traced by model reactions, performed by heating mixtures of precursor molecules to temperatures similar to those during the roasting process to reduce the quantity and complexity of the reaction products formed. Special emphasis has been devoted to studying the generation of Maillard-derived aroma compounds such as thiols, diketones, and pyrazines in model systems under dry heating conditions (Amrani-Hemaimi et al., 1995; Grosch, 1999; Hofmann and Schieberle, 1997, 1998; Tressl et al., 1993; Yaylayan and Keyhani, 1999).

However, most of the research on the formation of Maillard-based flavor compounds is performed in simple sugar and amino acid systems, hardly in sugar—protein or sugar—peptide mixtures. Heating mixtures of a limited number of putative precursors will neither consider other constituents nor possible interactions in green coffee beans. A more comprehensive picture provides the isolation of specific coffee fractions and the subsequent heat treatment under real roasting conditions. This approach has been used to study the role of water-soluble fractions (De Maria et al., 1994, 1996a). The low molecular fraction (main precursors are sucrose, trigonelline, and CGA) showed a much richer profile upon roasting, yielding high amounts of pyrroles, furans, pyridine, cyclic enolones, acetic acid, furfural, phenols, and 2-furfurylalcohol. The HMW fraction depicted less diversity and intensity, however, eliciting high aroma activity by alkyl pyrazines with earthy, roasty, and nutty characters. Roasting of the remaining insoluble part (De Maria et al., 1996b) revealed its function as precursor in the formation of additional alkyl pyrazines through protein-bound amino acids. This approach was also used to elucidate the role of arabinogalactans and cysteine in the formation of the important coffee odorant 2-furfuylthiol (Grosch, 1999).

The tremendous physical processes in the bean during roasting have crucial influence on the equilibriums of flavor formation mechanisms. It was shown that roasting of green coffee powder and green coffee bean fragments was different from roasting of whole coffee beans (Fischer, 2005). Indeed, the coffee bean can be seen as a pressurized reactor. Therefore, the conclusions from model systems have to be taken with care and cannot simply be extrapolated to complex food products. Other authors came to the same conclusion when studying the formation of furan under roasting conditions (Limacher et al., 2008). Thus, the formation of coffee aroma cannot always be explained by model reaction systems, as the extreme reaction conditions during coffee roasting can lead to different reaction pathways.

Hence, to study the importance of precursors for the formation of key aroma compounds during coffee roasting under real conditions, the coffee bean itself can be used as a reaction vessel (Milo et al., 2001). In this so-called biomimetic in-bean experiments, green coffee beans are water extracted, and the resulting bean shells and green coffee extract are freeze-dried (Fig. 12.4). The impact of specific precursors can be estimated by spiking green coffee beans or by selectively reconstituting extracted green coffee beans. Spiking of green beans and reincorporation of precursors in exhausted green beans are carried out by soaking the beans in an aqueous solution containing the precursor compounds. Results obtained from these systems are promising; for

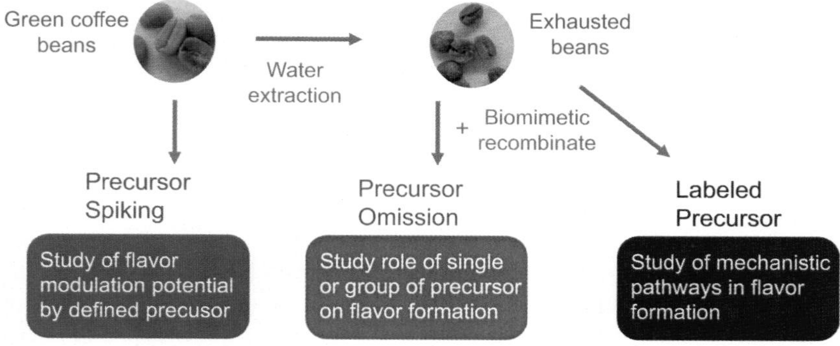

FIGURE 12.4 In-bean concept to study flavor formation upon coffee roasting.

example, it enabled to elucidate the role of the water nonextractable green bean fraction as precursor source for the generation of 2-furfurylthiol since it increased significantly when the exhausted beans were roasted (Milo et al., 2001). In addition, water extraction of whole green beans resulted, after roasting, in a coffee with strongly decreased amounts of key components such as 2- and 3-methylbutanal, α-diketones, and guaiacols. This is most likely the consequence of the removal of free amino acids and CGAs. The formation of other odorants (e.g., alkyl pyrazines, dicarbonyls) was studied in biomimetic in-bean experiments by Poisson et al. (2009), including the incorporation of isotopically labeled precursors.

The in-bean approach was successfully applied to study the mechanism of coffee melanoidin formation (Nunes et al., 2012) and to investigate the presence and nature of thiol binding sites in raw coffee beans (Müller and Hofmann, 2005; Müller et al., 2006). CGA as well as resulting thermal degradation products such as caffeic acid and quinic acid were identified as important precursors for low-molecular-weight thiol-binding sites, leading to a rapid irreversible binding, and thus, reduction of the key odorant 2-furfurylthiol in a coffee beverage.

The combinations of omission, spiking, and mechanistic experiments (Fig. 12.4) under real food matrix conditions are very useful in providing further and more precise insights into Maillard-type reactions and formation mechanisms. The results of different studies clearly indicate that due to the great diversity of precursors and other coreaction agents present in the green bean, competing and even completely different pathways take place for the formation of flavor compounds.

5. KEY ROASTING PARAMETERS INFLUENCING FLAVOR FORMATION AND CUP QUALITY

The flavor of roasted coffee depends on (1) the green bean constituents and (2) the way the roasting operation is conducted. In general, the precursor composition determines which flavor compounds are formed, whereas the physical parameters mainly influence the formation kinetics (Boekel, 2006).

5.1 Flavor Profile and Roast Degree

The quality of the green coffee is the main determinant of the aroma and taste developed during the roasting process, and it is a snapshot at a defined roasting degree. However, there is no concise definition of the roasting degree, although expressions such as "optimum degree of roast" are frequently used in the literature. It seems obvious that the optimum degree of roast is a function of green bean origin, intended brewing method, and personal taste preference. It is widely accepted that characterizing the quality of roasted coffee only by means of weight loss and/or roast color is not sufficient, since these attributes do not make a statement about the individually obtained aroma profiles.

The experienced coffee roaster knows that the aroma evolves from sweet, fruity, floral, bread, and nutty character in light roasts, through more complex aroma profiles in medium roast. Darker roast levels are characterized by cocoa, spicy, phenolic, ashy, pungent, and dark roast flavors. The bitter taste increases during roasting, whereas the acidity decreases during initial stages of roasting (Fig. 12.5). This sensory perception is substantiated by sensory and instrumental analysis of aroma and taste components throughout the roasting course. The flavor composition continuously changes all along the roasting progress. This means, at each time a new aroma/taste profile is delivered. However, perceived aroma goes through an optimum, i.e., over-roasting will lead to unbalanced burnt, harsh off-notes comprising both aroma and overall flavor (body). In general, the sensory perceivable total acidity is decreasing during the course of roasting, whereas bitterness is steadily increasing. Similarly, color development is increasing with roast degree from light brown [color test number (CTN) 150] to almost black (CTN 50).

Taking the main constituents of coffee beans into account, Table 12.3 compares the composition of the green beans to the roasted beans and points out which compounds of the green beans are particularly degraded during roasting. Sucrose and free amino acids are immediately available and highly reactive. Their high reactivity is explained by the presence of a free functional amino group and the rapid thermal hydrolysis of sucrose into reducing sugars. Arabinose branches and CGA are degraded in a later stage. This delayed reactivity can be explained by the additional energy needed for depolymerization or hydrolysis to liberate the reactive functions. Other polymeric carbohydrates (i.e., galactans, mannans, and cellulose) or bound amino acids are less prone to hydrolysis and depolymerization, and thus only contribute at a later stage of roasting to the Maillard reaction (Arya and Rao, 2007).

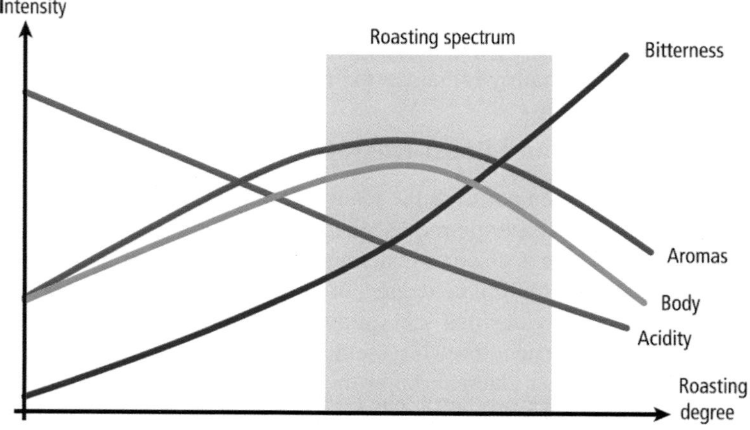

FIGURE 12.5 Schematic presentation of the kinetics of flavor evolution during roasting.

TABLE 12.3 Changes in the Chemical Composition of Green and Roasted Coffee Upon Roasting

	Arabica		Robusta	
Component	**Green (% dm)**	**Roasted (% dm)**	**Green (% dm)**	**Roasted (% dm)**
Caffeine	1.2	1.3	2.2	2.4
Trigonelline (incl. roasted by-products)	1.0	1.0	0.7	0.7
Proteins and Amino Acids				
• Proteins	9.8	7.5	9.5	7.5
• Amino acids	0.5	0.0	0.8	0.0
Sugars				
• Sucrose	8.0	0.0	4.0	0.0
• Reducing sugars	0.1	0.3	0.4	0.3
• Other sugars	1.0	n.a.	2.0	n.a.
• Polysaccharides	49.8	38.0	54.4	42.0
Acids				
• Aliphatic	1.1	1.6	1.2	1.6
• Quinic	0.4	0.8	0.4	1.0
• Chlorogenic	6.5	2.5	10.0	3.8
Lipids	16.2	17.0	10.0	11.0
Caramelization/condensation products (by diff.)		25.4		25.9
Volatile aroma	Traces	0.1	Traces	0.1
Minerals	4.2	4.5	4.4	4.7
Water	8–12	0–5	8–12	0–5

Adapted from Illy and Viani (1995).

Excessive roasting (CTN <50) generally leads to decreasing of many volatile substances such as diketones, furfurals, or 4-vinylguaiacol. These compounds are formed in the early stage of roasting, mainly parallel to free sugar and amino acids consumption, with a maximum at a medium roasting degree. For example, the fruity, blackcurrant-smelling sulfur odorant 3-mercapto-3-methylbutyl formate can completely disappear in a dark roasted coffee. Alkyl pyrazines (e.g., 2-ethyl-3,5-dimethylpyrazine) responsible for earthy, roasty, nutty character of coffee are as well generated in an early stage, but levels remain almost constant throughout the final stage. In contrast, 2-furfurylthiol, the trigonelline degradation products pyridine and *N*-methylpyrrole, as well as dimethyl trisulfide were shown to continuously increase with roasting time (Baggenstoss et al., 2008a). All these molecular changes have their impact on the sensorial characteristics. A masking effect by odorants formed in the later stages of roasting, covering the sweet and earthy notes, was described by Gretsch et al. (1999), and fruity, floral notes appearing at beginning of roasting are replaced by roasty and burnt notes (Schenker, 2000).

5.2 Time—Temperature Profile

As discussed before, a precise control of roasting time and temperature is required to reach specific flavor profiles. The generation of flavor compounds depends on the duration and the final bean temperature reached (Huschke, 2007). Both parameters are constituted by the slope of the roasting curve, i.e., the time—temperature (t—T) profile. Based on the same raw material and same roaster, roasting coffee to the same roasting degree leads to different flavor profiles depending on the time—temperature roasting conditions (Schenker et al., 2002). Short-time roasting at high temperature has been shown to result in considerable differences in the physical properties and kinetics of aroma formation compared to long time roasting at lower temperatures (Baggenstoss et al., 2008a; Glöss et al., 2014). Fast roasting yields more soluble solids while causing less degradation of CGA, and lower loss of volatiles (Nagaraju et al., 1997). This is accentuated in higher quantities of roasty, buttery diketones, and furfurals, whereas much lower concentrations of phenols are formed, leading to less burnt, smoky flavor. On the other side, fast roasted coffee is presumed to be more affected by lipid oxidation due to higher oil migration from the inner bean to the surface (Schenker et al., 2000).

The impact of different roaster technologies on the aromatic composition is less well understood. A similar temperature profile applied on two roasters (laboratory scale fluidizing-bed roaster and traditional drum roaster) resulted in similar physical properties and aroma formation in the assessed coffees (Baggenstoss et al., 2008a).

5.3 Precursor Composition *Robusta* Versus *Arabica*

The evaluation of aroma formation in Arabica and Robusta coffees upon roasting is quite similar, but concentration differences between the two coffee varieties are crucial for the final organoleptic characteristic (Holscher and Steinhart, 1992). Green Arabica coffee contains more oligosaccharides, lipids, trigonelline, and organic acids. On the other hand, Robusta is significantly richer in caffeine and CGA and also exhibit a larger amount of free amino acids than Arabica (see also Table 12.3).

Slightly lower amounts of the precursor isoleucine and leucine in Arabica green beans lead to smaller final amounts of the corresponding Strecker aldehydes (i.e., 2-methylbutanal and 3-methylbutanal) in roasted Arabica beans compared to Robusta coffee. Accordingly, higher amounts of amino acids in Robusta green beans result in higher final amounts of earthy, roasty, nutty smelling pyrazines. Together with the high level of phenols resulting from the CGA degradation, Robusta coffee exhibits their typical smoky, earthy, roasty, and phenolic aroma profile. Tryptophan, an amino acid strongly present in Robusta green beans, result in higher final amounts of undesirable, animalic smelling 3-methylindol (skatole), which is almost not present in Arabica coffee. The higher levels of diketones, furfurals, and cyclic enolones (i.e., furaneol) in roasted Arabica coffees are a consequence of the higher abundance of sucrose in the initial green beans.

Interestingly, variability in aroma formation kinetics is not only found between Arabica and Robusta formation, but also within a coffee species as recently reported for different Arabicas from Colombia, Guatemala, and Ethiopia (Glöss et al., 2014). One way of explaining the varying behavior of coffee of different origins under the same roasting conditions might be the individual changes in physical structures.

5.4 The Effect of Moisture

The initial moisture content represents another important parameter, which has an influence particularly on light roasted coffee, whereas in dark roasted coffee most differences in aroma are leveled off (Baggenstoss et al., 2008b). The effect of moisture in coffee roasting has been discussed previously in detail in this book (see Chapter 11).

Steam treatment of green coffee beans represents a method to improve flavor quality of Robusta coffees (Becker et al., 1989; Darboven, 1995). They get sweeter, more acidic, and less bitter than untreated coffee, lowering significantly the Robusta character (Theurillat et al., 2006). Water treatment with saturated steam at elevated pressure provokes some changes in the precursor composition, mainly on sugars and amino acids, influencing the color formation and diminishing formation of undesired substances (e.g., catechol, pyrogallol, and hydroquinone, pyrazines, volatile phenols) (Becker et al., 1989; Darboven, 1995). These changes can be explained by a mobilization of precursor resulting in partial extraction and removal of water solubles such as free amino acids (Steinhart and Luger, 1995), or CGAs (Milo et al., 2001), or partial hydrolysis of the sucrose to fructose and glucose (Steinhart and Luger, 1995). Decreased amounts of feruloylquinic acid result in lower generation of 4-vinylguaiacol; extraction of free amino acids lowers the amount of alkyl pyrazines in roasted coffee. Steam-treated coffee reaches target color faster, and hence, roasting time for the same degree of roast is shorter (Theurillat et al., 2006). Actually, monsooned coffee (see Chapter 3) passes through similar mechanisms of precursor changes. The partial hydrolysis of CGAs and loss of LMW components in the humidity result in an increased spicy character of the product (Variyar et al., 2003).

The air-to-bean ratio as a further roasting parameter was discussed as well (Schenker, 2000). It seems that a higher air-to-bean ratio leads to a rather less complex and flat flavor, which was linked to an increased aroma stripping at higher air flows.

The major compositional changes and chemical processes are summarized below that affect the development of flavor compounds in coffee upon roasting:

- loss of water ⇨ drying of the bean, low moisture reaction system
- release of carbon dioxide ⇨ expansion of the bean
- migration of lipids to the bean surface ⇨ retaining aroma components generated
- loss of sugars (including sucrose) ⇨ flavor and color formation (Maillard chemistry and caramelization)
- decrease of free amino acids ⇨ flavor and color formation (Maillard and Strecker chemistry)
- partial decomposition of polysaccharides (e.g., arabinogalactan) ⇨ release of arabinose which in turn reacts leading to flavor formation (e.g., Maillard reaction)
- partial decomposition of proteins ⇨ release of amino acids which in turn reacts leading to flavor formation (e.g., Maillard reaction)
- loss of CGA ⇨ formation of bitter taste and color
- decrease of trigonelline ⇨ formation of *N*-containing products (aroma, taste, color)
- formation of melanoidins ⇨ color formation (polymerization of polysaccharides, proteins, and polyphenols)
- partial lipid degradation ⇨ aroma active aldehydes
- interaction between intermediate decomposition products

A simplified overview of those reactions and the corresponding flavor compounds formed are listed in Table 12.4. It shows the variety of flavor qualities that can be generated upon coffee roasting providing the coffee with its particular aroma character. Interestingly, there is only one aroma compound eliciting a note resembling that of the product, i.e., 2-furfurylthiol (see also Table 12.2), however, only in a certain concentration range (Blank, 2015). At elevated concentrations it turns to rubbery perceived as an off-note.

6. KINETICS OF FLAVOR FORMATION

Monitoring the roasting process in view of a repeatable and sustainable quality supply and a better understanding of the roasting kinetics has been a target for a long time by scientists as well as coffee roasters. The generation of flavor

TABLE 12.4 Key Reactions of Coffee Roasting that Impact Flavor Quality

Reactions	Precursors Involved	Compounds Formed (Flavor Quality)
Maillard reaction	• Reducing sugars • *N*-compounds	• Diketones (buttery) • Pyrazines (earthy, roasty, nutty) • Thiazoles (roasty, popcorn-like) • Enolones (caramel-like, savory) • Thiols (sulfury, coffee-like) • Aliphatic acids (acidic)
Strecker degradation	• Amino acids • Diketones deriving from Maillard reaction	• Strecker aldehydes (malty, green, honey-like)
Caramelization	• Free sugars (sucrose after inversion)	• Enolones (caramel-like, savory)
Degradation of chlorogenic acids	• Chlorogenic acids	• Phenols (smoky, ashy, woody, phenolic, medicinal) • Lactones (bitter) • Indanes (bitter, harsh)
Lipid oxidation	• Unsaturated fatty acids	• Aldehydes (fatty, soapy, green)

upon coffee roasting is a highly dynamic process, in particular in the exothermic phase in which many reactions take place with increasing molecular complexity. One way of getting a more precise insight into the formation processes are kinetic studies describing the flavor composition as a function of time.

The evaluation of specific chemical compounds has been proposed to monitor the roasting degree, apart from physical attributes. As examples the ratios of free amino acids (Nehring and Maier, 1992), alkylpyrazines (Hashim and Chaveron, 1995), or CGA (Illy and Viani, 1995) were used to monitor the course of roasting. Research on the formation of volatile organic compounds (VOCs) in coffee roast gases has traditionally relied on chromatographic techniques, most often GC—MS (Baggenstoss et al., 2008a,b, 2007; Schenker et al., 2002), but also high-performance liquid chromatography—MS (HPLC—MS) (Clifford et al., 2006). A continuous monitoring of the roasting course, however, is difficult with off-line GC—MS or LC—MS techniques, where sample preparation, analyte isolation, and analysis are typically decoupled.

As previously mentioned, coffee roasting results in bitterness represented by caffeine and many other compounds belonging to the chemical classes of

CQA lactones, phenylindanes, and DKP. Sensory evaluations indicate a distinct change in the bitterness note from mild bitter at lower roasting degree to harsh bitter at very high roasting degree (Fig. 12.6). This is due to changes in the composition of bitter tastants during the roasting process. As shown in Fig. 12.6, harsh-bitter tasting compounds (phenylindanes and DKP) are continuously formed whereas CQA lactones having a mild bitter note go through a maximum being degraded at around CTN 100 (CTN stands for Color Test Neuhaus determined with the help of Colortest II based on infrared reflectance expressed relative to a standard sample, Neuhaus Neotec, Reinbek, Germany). Actually, CQA lactones with mild bitter notes are transformed into harsh-bitter phenylindanes. Therefore, it is crucial to control the late stage of the roasting process.

6.1 Real-time Monitoring of Aroma Formation During Roasting

Online mass spectrometric techniques were developed to follow the fast dynamics of flavor development during the roasting process. In particular, online chemical ionization mass spectrometric techniques (e.g., proton-transfer-reaction mass spectrometry, PTR–MS) (Yeretzian et al., 2002, 2003) and online photo ionization mass spectrometric approaches (i.e., photon ionization, resonance-enhanced multiphoton ionization time-of-flight mass spectrometry; Dorfner et al., 2003, 2004; Schramm et al., 2009) have been applied for real-time monitoring of coffee

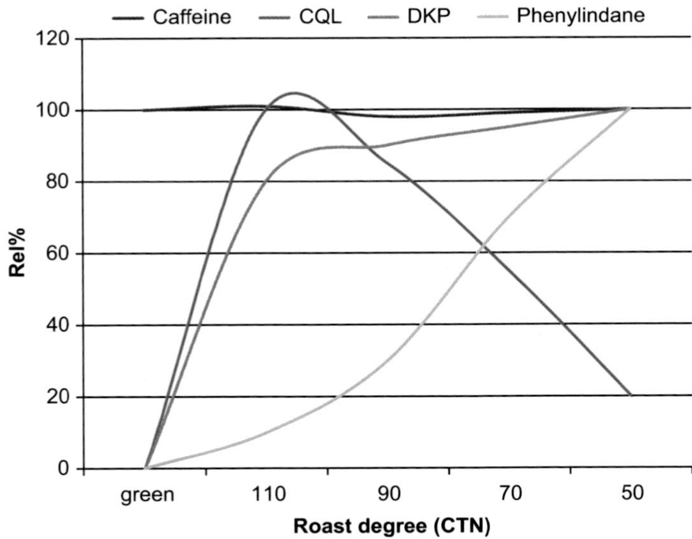

FIGURE 12.6 Formation kinetics of tastants as function of roasting degree expressed as color test neuhaus (CTN). *CQL*, chlorogenic acid lactones; *DKP*, diketopiperazines.

roasting to predict the roast degree of coffee by online analysis of the roast gas. Even the aroma formation in a single bean was studied (Hertz-Schünemann et al., 2013), which allowed the assessment of chemical reaction pathways, for instance, the degradation of CGA. The authors used the photon-ionization techniques to study permeability of the cell walls with regards to different chemical compounds, and suggested the methodology for the rapid determination of the relative Arabica content in coffee blends. PTR—MS techniques were also applied to follow the formation of VOCs in different origins (Colombia, Guatemala, Ethiopia, Indonesia), and different time—temperature roasting profiles (Glöss et al., 2014), interestingly observing varied release dynamics of the online monitored VOCs depending on the origin or the different precursor compositions and physical aspects in beans from different origins, meaning a timely shifted start of VOCs formation.

Wieland et al. measured online the concentrations of volatiles released by the beans into the headspace during roasting (Wieland et al., 2012). The released gas obtained from the roasting chamber was fed into a highly sensitive PTR time-of-flight (ToF) mass spectrometer for real-time measurement. The time-intensity patterns of compound traces confirmed two typical formation behaviors. The first pattern showed strong formation rates in the second half of roasting time, an intensity peak at medium degree of roast, followed by a fast decrease toward the end of the process when roasting to dark degree. The second pattern showed stepwise continuous increase of intensity during the second part of total roasting time. A principle component analysis discriminated the degree of roast along the roasting process and predicted successfully the bean color.

Fischer et al. (2014) used a similar methodology to compare the formation of organic flavor compounds in Arabica and Robusta beans, respectively. Although the basic average single photon ionisation-TOFMS spectra obtained from roast gas of both species looked similar, a more detailed data analysis permitted to discriminate species, degree of roast, and roasting conditions.

Alternatively, the application of chemosensor array monitoring suitable marker substances (2-furfurylalcohol and hydroxy-2-propanone) was proposed (Hofmann et al., 2002), as well as coupling e-nose technique to artificial neural networks (ANNs) to evaluate the roasting degree (Romani et al., 2012). The latter may represent an effective possibility to roasting process automation and to set up a more reproducible procedure for final coffee bean quality characterization. A different approach was applied by Wei (Wei et al., 2012), using ^{1}H and ^{13}C NMR-based comprehensive analysis to monitor the roast degree. Based on this composition-based holistic method they suggested different appropriate chemical markers to control and characterize the coffee-roasting process.

6.2 Sensomics Heat Maps Illustrating Flavor Formation Kinetics

Changes in the composition of aroma compounds can be depicted in a Sensomics heat map that indicates the relative changes in concentration after normalization of the data. Fig. 12.7 illustrates the normalized development and

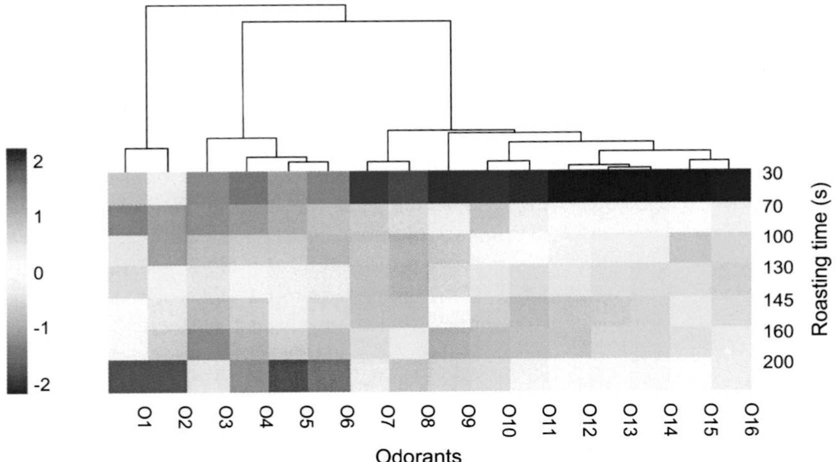

FIGURE 12.7 Heat map of aroma formation in the early stage of coffee roasting (260°C, isotherm). Dimethyl sulfide (DMS, O1), hexanal (O2), 3-mercapto-3-methylbutyl formate (MMBF, O3), *N*-methylpyridine (O4), 2-furfurylthiol (FFT, O5), pyridine (O6), 2,3-butanedione (O7), 2,3-pentanedione (O8), dimethyl trisulfide (DMTS, O9), methyl mercaptane (O10), 3-methylbutanal (O11), 2-ethyl-3,5-dimethylpyrazine (O12), 2-methylbutanal (O13), 2,3,5-trimethylpyrazine (O14), methylpropanal (O15), 4-vinylguaiacol (O16).

degradation of various coffee aroma compounds, roasted isothermally at 260°C for 4 min in a drum roaster. It is shown that the grassy smelling hexanal (O2) is already present in the very early stage of the roasting, to some extent even in the green beans, and then degraded slowly under more severe roasting conditions. The roasty smelling 2-furfurylthiol (O5) is generated after about 200 s of roasting, similar to pyridine (O6) and *N*-methylpyrrole (O4), whereas the blackcurrant−like−smelling 3-mercapto-3-methylbutyl formate (O3) is already being degraded. The buttery smelling diketones O7 and O8 are already formed at mild conditions (120 s) and then slightly degraded again. The malty smelling Strecker aldehydes O11, O13, and O15 are formed at 2−3 min at 260°C isotherm roasting, similar to the earthy smelling pyrazines O12 and O14. The smoky smelling 4-vinylguaiacol (O16) is formed after about 70 s and, after running through a maximum at 130 s slightly degraded again.

Similarly, changes in the composition of bitter-tasting compounds are heat mapped in Fig. 12.8, showing the relative changes in concentration after data normalization. The bitter taste precursors caffeic acid (T32), 5-*O*-caffeoyl quinic acid (T33), 3-*O*-caffeoyl quinic acid (T34), and 4-*O*-caffeoyl quinic acid (T35) are rapidly degraded within the first 120−180 s of roasting and transformed into the bitter-tasting compounds such as the lactones T44 and T49-T52, and esters T39, T42, T43, and T45-T47. With increasing roasting time (120−360 s), further CGA lactones (T12, T17, T30), nicotinic acid, and

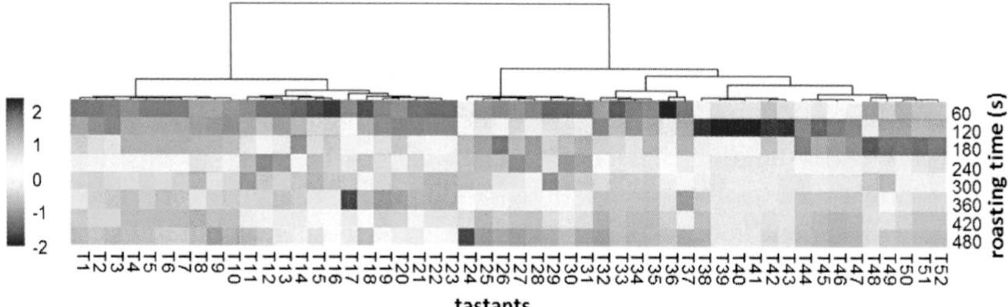

FIGURE 12.8 Heat map of taste formation up to 8 min roasting (260°C, isotherm). Ferulic acid methyl ester (T1), catechol (T2), 1,2-dihydroxy-4-(furan-2-ylmethyl)-3 methylbenzene (T3), *cis*-4,5-dihydroxy-1-methyl-3-(3′,4′-dihydroxyphenyl) indane (T4), *trans*-4,5-dihydroxy-1-methyl-3-(3′,4′-dihydroxyphenyl) indane (T5), *trans*-5,6-dihydroxy-1-methyl-3-(3′,4′-dihydroxyphenyl) indane (T6), *cis*-5,6-dihydroxy-1-methyl-3-(3′,4′-dihydroxyphenyl) indane (T7), nicotinic acid 5-fomylfuran-2-yl ester (T8), 1,2-dihydroxy-4-(furan-2-ylmethyl) benzene (T9), 1,2-dihydroxy-4-(furan-2-ylmethyl)-5-methylbenzol (T10), pyrogallol (T11), 5-*O*-caffeoyl-muco-γ-quinid (T12), ferulic acid furan-2-yl ester (T13), endo-6-(3,4-dihydroxphenyl)-5-(hydroxymethyl-8-oxabicyclo [3.2.1]oct-3-en-2-one (T14), caffeic acid methyl ester (T15), 3-methylcatechol (T16), 3,5-*O*-dicaffeoyl-epi-δ-quinide (T17), nicotinic acid methyl ester (T18), nicotinic acid furan-2-yl ester (T19), *N*-methylnicotinic acid furan-2-yl ester (T20), nicotinic acid furan-2-yl thioester (T21), 4-methylcatechol (T22), 1,2,3-trihydroxy-4-(furan-2-ylmethyl)benzene (T23), ferulic acid (T24), caffeic acid methyl ester (T25), chlorogenic acid furan-2-yl ester (T26), *trans*-1,3-bis-(3′,4′-dihydroxyphenyl)-1-butene (T27), caffeic acid furan-2-yl thioester (T28), furfuryl alcohol (T29), 5-*O*-caffeoyl-epi-δ-quinide (T30), caffeic acid furan-2-yl ester (T31), caffeic acid (T32), 5-*O*-caffeoyl quinic acid (T33), 3-*O*-caffeoyl quinic acid (T34), 4-*O*-caffeoyl quinic acid (T35), 1,3-*bis*-3,4-dihydrophenyl-butane (T36), ferulic acid ethyl ester (T37), trigonelline methyl ester (T38), chlorogenic acid 5-formylfuran-2-yl ester (T39), 5-Hydroxymethylfurfural (T40), *N*-methylnicotinic acid 5-formylfuran-2-yl ester (T41), ferulic acid 5-formylfuran-2-yl ester (T42), caffeic acid 5-formylfuran-2-yl ester (T43), 3,4-O-dicaffeoyl-muco-γ-quinid (T44), 3-*O*-caffeoyl quinic acid methyl ester (T45), 5-*O*-caffeoyl quinic acid methyl ester (T46), 4-*O*-caffeoyl quinic acid methyl ester (T47), exo-6-(3,4-Dihydroxphenyl)-5-(hydroxymethyl-8-oxabicyclo[3.2.1]oct-3-en-2-one (T48), 4,5-O-dicaffeoyl-muco-γ-quinide (T49), 4-*O*-caffeoyl-muco-γ-quinide (T50), 3-*O*-caffeoyl-γ-quinide (T51), 4-*O*-caffeoyl-γ-quinide (T52).

N-methyl-nicotinic acid esters (T18-T21) are formed, followed by their decline after 360 s. Under more severe roasting conditions, particularly aromatic condensation products generated from di- and trihydroxybenzenes are formed as the most important bitter compounds such as T4, T5, T6, T7, T9, and T10.

7. OUTLOOK

The chemistry of coffee flavor formation during roasting has long been a mystery. This is mainly due to the multitude of possible reactions taking place at high temperatures when coffee biopolymers are degraded to smaller, usually more reactive entities. Today thanks to the development of more sophisticated analytical methods allowing for higher sensitivity, many key aroma and taste components have been identified and their sensory relevance evaluated. However, still much needs to be further discovered. Although the myriad of aroma compounds in roasted coffee has been elucidated and their impact on the overall aroma quite well described, the taste dimension still requires more molecular understanding. The coffee taste cannot be fully reconstituted by using known taste-active compounds, indicating that there are additional tastants to be identified, in particular those contributing to bitterness, harshness, and acidity. In addition, still a lot of experimentation is needed to better understand the reactions leading to flavor formation, i.e., from precursors to aroma and taste.

With focus on aroma and taste formation, it seems desirable to determine the amounts of precursor molecules of aroma compounds (such as amino acids, sugars, CGA, etc.) before and after roasting to be able to provide objective statements about the roasting degree, which are transferable from one roasting profile to another and do not depend on raw material fluctuations of the green beans.

The complexity of the green coffee composition as well as the chemical and physical transformations of the bean during roasting are difficult to reproduce in model systems. The approach of in-bean experiments using green beans as minireactors is a more realistic reaction environment.

The combinations of omission, spiking, and mechanistic experiments under real food matrix conditions are very useful in providing further and more precise insights into Maillard-type reactions and formation mechanisms. The results of different studies clearly indicate that due to the great diversity of precursors and other coreaction agents present in the green bean, competing and even completely different pathways may take place in the formation of flavor compounds.

More research is needed on flavor formation kinetics under controlled conditions, potentially using real-time analytical approaches combined with characterization of the changes in flavor precursors to relate green coffee composition to flavor profile as affected by roasting parameters.

Science is helping to understand the mystery around coffee roasting. Description of chemical changes occurring in the coffee bean at a molecular level is of increasing value for adapting the roasting conditions, which today is largely based on empirical knowledge of the barista. We are convinced that better understanding of science is of equal importance to the barista and relevant to "the craft of coffee".

REFERENCES

Amrani-Hemaimi, M., Cerny, C., Fay, L.B., 1995. Mechanisms of formation of alkylpyrazines in the Maillard reaction. Journal of Agricultural and Food Chemistry 43 (11), 2818–2822.

Arya, M., Rao, L.J.M., 2007. An impression of coffee carbohydrates. Critical Reviews in Food Science and Nutrition 47 (1), 51–67.

Baggenstoss, J., Poisson, L., Kaegi, R., Perren, R., Escher, F., 2008a. Coffee roasting and aroma formation: application of different time-temperature conditions. Journal of Agricultural and Food Chemistry 56 (14), 5836–5846.

Baggenstoss, J., Poisson, L., Kaegi, R., Perren, R., Escher, F., 2008b. Roasting and aroma formation: effect of initial moisture content and steam treatment. Journal of Agricultural and Food Chemistry 56 (14), 5847–5851.

Baggenstoss, J., Poisson, L., Luethi, R., Perren, R., Escher, F., 2007. Influence of water quench cooling on degassing and aroma stability of roasted coffee. Journal of Agricultural and Food Chemistry 55 (16), 6685–6691.

Balzer, H.H., 2001. Chemistry I: non-volatile compounds. In: Coffee. Blackwell Science Ltd, pp. 18–32.

Becker, R., Schlabs, B., Weisemann, C., 1989. In: Process for Improving the Quality of Robusta Coffee. Jacobs Suchard AG, Patent US 5019413A.

Bekedam, E.K., Schols, H.A., Van Boekel, M.A.J.S., Smit, G., 2008. Incorporation of chlorogenic acids in coffee brew melanoidins. Journal of Agricultural and Food Chemistry 56 (6), 2055–2063.

Bekedam, E.K., Van Boekel, M.A.J.S., Smit, G., Schols, H.A., 2006. Isolation and characterization of high molecular weight coffee melanoidins. In: Proceedings of the 21st International Scientific Colloquium on Coffee, Montpellier, France, ASIC, Paris, France, pp. 156–160.

Belitz, H.D., Grosch, W., Schieberle, P., 2009. Food Chemistry, fourth ed. Springer-Verlag Berlin Heidelberg.

Blank, I., 2002. Gas chromatography-olfactometry in food aroma analysis. In: Marsili, R. (Ed.), Flavor, Fragrance, and Odor Analysis. CRC Press, New York, pp. 297–331.

Blank, I., 2015. Understanding flavour as a major driver of product quality. In: Taylor, A.J., Mottram, D.S. (Eds.), Flavour Science, Proceedings 14th. Weurman Flavour Research Symposium, Cambridge, UK, pp. 383–390.

Blank, I., Sen, A., Grosch, W., 1992. Potent odorants of the roasted powder and brew of Arabica coffee. Zeitschrift für Lebensmittel-Untersuchung und Forschung 195 (3), 239–245.

van Boekel, M.A.J.S., 2006. Formation of flavour compounds in the Maillard reaction. Biotechnology Advances 230–233.

Bradbury, A.G.W., 2001. Chemistry I: non-volatile compounds. In: Coffee. Blackwell Science Ltd, pp. 1–17.

Bradbury, A.G.W., Halliday, D.J., 1990. Chemical structures of green coffee bean polysaccharides. Journal of Agricultural and Food Chemistry 38 (2), 389–392.

Caldwell, M.L., 2009. Food and Everyday Life in the Postsocialist World. Indiana University Press.

Cantergiani, E., Brevard, H., Krebs, Y., Feria-Morales, A., Amado, R., Yeretzian, C., 2001. Characterisation of the aroma of green Mexican coffee and identification of mouldy/earthy defect. European Food Research and Technology 212 (6), 648–657.

Cerny, C., Grosch, W., 1992. Evaluations of potent odorants in roasted beef by aroma extract dilution analysis. Zeitschrift für Lebensmitteluntersuchung und Forschung A 194, 323–325.

Clarke, R.J., 1987. Roasting and grinding. In: Clarke, R. (Ed.), Coffee, Volume 2: Technology. Elsevier Science Publishers LTD, London, UK, pp. 73–107.

Clifford, M.N., Johnston, K.L., Knight, S., Kuhnert, N., 2003. Hierarchical scheme for LC-MSn identification of chlorogenic acids. Journal of Agricultural and Food Chemistry 51 (10), 2900–2911.

Clifford, M.N., Marks, S., Knight, S., Kuhnert, N., 2006. Characterization by LC-MSn of four new classes of p-coumaric acid-containing diacyl chlorogenic acids in green coffee beans. Journal of Agricultural and Food Chemistry 54 (12), 4095–4101.

d'Acampora Zellner, B., Dugo, P., Dugo, G., Mondello, L., 2008. Gas chromatography-olfactometry in food flavour analysis. Journal of Chromatography A 1186 (1–2), 123–143.

Dalla Rosa, M., Lerici, C.R., Piva, M., Fini, P., 1980. Processi di trasformazione del caffè: aspetti chimici, fisici e tecnologici. Nota 5: Evoluzione di alcuni caratteri fisici del caffè nel corso dei trattamenti termici condotti a temperatura costante. Industrie delle Bevande 9, 466–472.

Darboven, A., 1995. In: Process for Improving the Quality of Green Coffee by Treatment with Steam and Water. Darboven GmbH & Co, Patent EP 0755631 A1.

Davidek, T., Devaud, S., Robert, F., Blank, I., 2006. Sugar fragmentation in the Maillard reaction cascade: isotope labeling studies on the formation of acetic acid by a hydrolytic beta-dicarbonyl cleavage mechanism. Journal of Agricultural and Food Chemistry 54 (18), 6667–6676.

De Maria, C.A.B., Trugo, L.C., Moreira, R.F.A., Werneck, C.C., 1994. Composition of green coffee fractions and their contribution to the volatile profile formed during roasting. Food Chemistry 50 (2), 141–145.

De Maria, C.A.B., Trugo, L.C., Neto, F.R.A., Moreira, R.F.A., Alviano, C.S., 1996a. Composition of green coffee water-soluble fractions and identification of volatiles formed during roasting. Food Chemistry 55 (3), 203–207.

De Maria, C.A.B., Trugo, L.C., Neto, F.R.A., Moreira, R.F.A., Alviano, C.S., 1996b. The GC/MS identification of volatiles formed during the roasting of high molecular mass coffee aroma precursors. Journal of the Brazilian Chemical Society 7 (4), 267–270.

Dorfner, R., Ferge, T., Kettrup, A., Zimmermann, R., Yeretzian, C., 2003. Real-time monitoring of 4-vinylguaiacol, guaiacol, and phenol during coffee roasting by resonant laser ionization time-of-flight mass spectrometry. Journal of Agricultural and Food Chemistry 51 (19), 5768–5773.

Dorfner, R., Ferge, T., Yeretzian, C., Kettrup, A., Zimmermann, R., 2004. Laser mass spectrometry as on-line sensor for industrial process analysis: process control of coffee roasting. Analytical Chemistry 76 (5), 1386–1402.

Dunkel, A., Steinhaus, M., Kotthoff, M., Nowak, B., Krautwurst, D., Schieberle, P., Hofmann, T., 2014. Nature's chemical signatures in human olfaction: a foodborne perspective for future biotechnology. Angewandte Chemie International Edition in English 53 (28), 7124–7143.

Farah, A., 2012. Coffee constituents. In: Coffee. Wiley-Blackwell, pp. 21–58.

Farah, A., de Paulis, T., Trugo, L.C., Martin, P.R., 2005. Effect of roasting on the formation of chlorogenic acid lactones in coffee. Journal of Agricultural and Food Chemistry 53 (5), 1505–1513.

Fischer, C., 2005. Kaffee - Änderung physikalisch-chemischer Parameter beim Rösten, Quenchen und Mahlen. Unpublished Dissertation, TU Braunschweig, Germany.

Fischer, M., Reimann, S., Trovato, V., Redgwell, R.J., 2001. Polysaccharides of green Arabica and Robusta coffee beans. Carbohydrate Research 330 (1), 93–101.

Fischer, M., Wohlfahrt, S., Varga, J., Saraji-Bozorgzad, M., Matuschek, G., Denner, T., Zimmermann, R., 2014. Evolved gas analysis by single photon ionization-mass spectrometry. Journal of Thermal Analysis and Calorimetry 116 (3), 1461–1469.

Frank, O., Blumberg, S., Kunert, C., Zehentbauer, G., Hofmann, T., 2007. Structure determination and sensory analysis of bitter-tasting 4-vinylcatechol oligomers and their identification in roasted coffee by means of LC-MS/MS. Journal of Agricultural and Food Chemistry 55 (5), 1945–1954.

Frank, O., Ottinger, H., Hofmann, T., 2001. Characterization of an intense bitter-tasting 1H,4H-quinolizinium-7-olate by application of the taste dilution analysis, a novel bioassay for the screening and identification of taste-active compounds in foods. Journal of Agricultural and Food Chemistry 49 (1), 231–238.

Frank, S., Wollmann, N., Schieberle, P., Hofmann, T., 2011. Reconstitution of the flavor signature of Dornfelder red wine on the basis of the natural concentrations of its key aroma and taste compounds. Journal of Agricultural and Food Chemistry 59, 8866−8874.

Frank, O., Zehentbauer, G., Hofmann, T., 2006. Screening and identification of bitter compounds in roasted coffee brew by taste dilution analysis. In: Wender, L.P.B., Mikael Agerlin, P. (Eds.), Developments in Food Science, 43. Elsevier, pp. 165−168.

Ginz, M., Balzer, H.H., Bradbury, W.A.G., Maier, G.H., 2000. Formation of aliphatic acids by carbohydrate degradation during roasting of coffee. European Food Research and Technology 211 (6), 404−410.

Glöss, A.N., Schoenbaechler, B., Klopprogge, B., D'Ambrosio, L., Chatelain, K., Bongartz, A., Strittmatter, A., Rast, M., Yeretzian, C., 2013. Comparison of nine common coffee extraction methods: instrumental and sensory analysis. European Food Research and Technology 236 (4), 607−627.

Glöss, A.N., Vietri, A., Wieland, F., Smrke, S., Schönbächler, B., López, J.A.S., Petrozzi, S., Bongers, S., Koziorowski, T., Yeretzian, C., 2014. Evidence of different flavour formation dynamics by roasting coffee from different origins: on-line analysis with PTR-ToF-MS. International Journal of Mass Spectrometry 324−337, 365−366.

Gretsch, C., Sarrazin, C., Liardon, R., 1999. Evolution of coffee aroma characteristics during roasting. In: Proceedings of the 18th International Scientific Colloquium on Coffee, Helsinki, Finland, ASIC, Paris, France, pp. 27−34.

Grosch, W., 1998. Flavour of coffee. A review. Nahrung-Food 42 (6), 344−350.

Grosch, W., 1999. Key odorants of roasted coffee: Evaluation, release, formation. In: Proceedings of the 19th International Scientific Colloquium on Coffee, Helsinki, Finland, ASIC, Paris, France, pp. 17−26.

Grosch, W., 2001a. Chemistry III: volatile compounds. In: Clarke, R.J., Vitzthum, O.G. (Eds.), Coffee − Recent Developments. Blackwell Science Ltd, London, UK, pp. 67−89.

Grosch, W., 2001b. Evaluation of the key odorants of foods by dilution experiments, aroma models and omission. Chemical Senses 26 (5), 533−545.

Hashim, L., Chaveron, H., 1995. Use of methylpyrazine ratios to monitor the coffee roasting. Food Research International 28 (6), 619−623.

Hertz-Schünemann, R., Dorfner, R., Yeretzian, C., Streibel, T., Zimmermann, R., 2013. On-line process monitoring of coffee roasting by resonant laser ionisation time-of-flight mass spectrometry: bridging the gap from industrial batch roasting to flavour formation inside an individual coffee bean. Journal of Mass Spectrometry 48 (12), 1253−1265.

Ho, C.T., Hwang, H.I., Yu, T.H., Zhang, J., 1993. An overview of the Maillard reactions related to aroma generation in coffee. In: Proceedings of the 15th International Scientific Colloquium on Coffee, Montpellier, France, ASIC, Paris, France, pp. 519−527.

Hofmann, T., Bock, J., Heinert, L., Kohl, C.D., Schieberle, P., 2002. Development of selective chemosensors for the on-line monitoring of coffee roasting by SOMMSA and COTA technology. Heteroatomic Aroma Compounds 826, 336−352.

Hofmann, T., Schieberle, P., 1997. Identification of potent aroma compounds in thermally treated mixtures of glucose/cysteine and rhamnose/cysteine using aroma extract dilution techniques. Journal of Agricultural and Food Chemistry 45 (3), 898−906.

Hofmann, T., Schieberle, P., 1998. Identification of key aroma compounds generated from cysteine and carbohydrates under roasting conditions. Zeitschrift für Lebensmitteluntersuchung und -Forschung A 207 (3), 229−236.

Holscher, W., Steinhart, H., 1992a. Formation pathways for primary roasted coffee odorants. Abstracts of Papers of the American Chemical Society 204, 114−AGFD.

Holscher, W., Steinhart, H., 1992b. Investigation of roasted coffee freshness with an improved headspace technique. Zeitschrift für Lebensmitteluntersuchung und -Forschung 195 (1), 33–38.

Holscher, W., Steinhart, H., 1994. Formation pathways for primary roasted coffee aroma compounds. In: Thermally Generated Flavors, pp. 206–217.

Holscher, W., Vitzthum, O.G., Steinhart, H., 1992. Prenyl alcohol source for odorants in roasted coffee. Journal of Agricultural and Food Chemistry 40 (4), 655–658.

Huschke, R., 2007. Industrial Coffee Refinement. Moderne Industrie, Freising, Germany.

Illy, A., Viani, R., 1995. Espresso Coffee: The Chemistry of Quality. Academic Press Limited, London.

Jaiswal, R., Matei, M.F., Golon, A., Witt, M., Kuhnert, N., 2012. Understanding the fate of chlorogenic acids in coffee roasting using mass spectrometry based targeted and non-targeted analytical strategies. Food & Function 3 (9), 976–984.

Kerler, J., Poisson, L., 2011. Understanding coffee aroma for product development. New Food Magazine 14 (6), 39–43.

Ky, C.L., Louarn, J., Dussert, S., Guyot, B., Hamon, S., Noirot, M., 2001. Caffeine, trigonelline, chlorogenic acids and sucrose diversity in wild *Coffea arabica* L. and *C. canephora* P. accessions. Food Chemistry 75 (2), 223–230.

Ledl, F., Schleicher, E., 1990. The Maillard reaction in food and in the human body – new results in chemistry, biochemistry and medicine. Angewandte Chemie 102 (6), 597–626.

Liepman, A.H., Nairn, C.J., Willats, W.G., Sorensen, I., Roberts, A.W., Keegstra, K., 2007. Functional genomic analysis supports conservation of function among cellulose synthase-like a gene family members and suggests diverse roles of mannans in plants. Plant Physiology 143 (4), 1881–1893.

Limacher, A., Kerler, J., Davidek, T., Schmalzried, F., Blank, I., 2008. Formation of furan and methylfuran by Maillard-type reactions in model systems and food. Journal of Agricultural and Food Chemistry 56 (10), 3639–3647.

Maier, H.G., 1993. Status of research in the field of non-volatile coffee components. In: Proceedings of the 15th International Scientific Colloquium on Coffee, Montpellier, France, ASIC, Paris, France, pp. 567–576.

Mestdagh, F., Thomas, E., Poisson, P., Kerler, J., Blank, I., 2013. Learning from other industries – insights from coffee on advanced sensory-analytical correlations. In: Proceedings of the 15th Australian Wine Industry Technical Conference, Sydney, pp. 102–106.

Milo, C., Badoud, R., Fumeaux, R., Bobillot, S., Fleury, Y., Huynh-Ba, T., 2001. Coffee flavour precursors: contribution of water non-extractable green bean components to roasted coffee flavour. In: Proceedings of the 19th International Scientific Colloquium on Coffee, Trieste, Italy, ASIC, Paris, France, pp. 87–96.

Moreira, A.S., Nunes, F.M., Domingues, M.R., Coimbra, M.A., 2012. Coffee melanoidins: structures, mechanisms of formation and potential health impacts. Food & Function 3 (9), 903–915.

Müller, C., Hofmann, T., 2005. Screening of raw coffee for thiol binding site precursors using "in bean" model roasting experiments. Journal of Agricultural and Food Chemistry 53 (7), 2623–2629.

Müller, C., Lang, R., Hofmann, T., 2006. Quantitative precursor studies on di- and trihydroxybenzene formation during coffee roasting using "in bean" model experiments and stable isotope dilution analysis. Journal of Agricultural and Food Chemistry 54 (26), 10086–10091.

Nehring, U.P., Maier, H.G., 1992. Indirect determination of the degree of roast in coffee. Zeitschrift für Lebensmittel-Untersuchung und Forschung 195 (1), 39–42.

Nagaraju, V.D., Murthy, C.T., Ramalakshmi, K., Rao, P.N.S., 1997. Studies on roasting of coffee beans in a spouted bed. Journal of Food Engineering 31 (2), 263–270.

Nijssen, L.M., 1996. Volatile Compounds in Food: Qualitative and Quantitative Data. TNO Nutrition and Food Research Institute.

Nunes, F.M., Coimbra, M.A., 2001. Chemical characterization of the high molecular weight material extracted with hot water from green and roasted arabica coffee. Journal of Agricultural and Food Chemistry 49 (4), 1773–1782.

Nunes, F.M., Cruz, A.C., Coimbra, M.A., 2012. Insight into the mechanism of coffee melanoidin formation using modified "in bean" models. Journal of Agricultural and Food Chemistry 60 (35), 8710–8719.

Oosterveld, A., Harmsen, J.S., Voragen, A.G.J., Schols, H.A., 2003. Extraction and characterization of polysaccharides from green and roasted *Coffea arabica* beans. Carbohydrate Polymers 52 (3), 285–296.

Ottinger, H., Bareth, A., Hofmann, T., 2001. Characterization of natural "cooling" compounds formed from glucose and l-proline in dark malt by application of taste dilution analysis. Journal of Agricultural and Food Chemistry 49 (3), 1336–1344.

Poisson, L., Schmalzried, F., Davidek, T., Blank, I., Kerler, J., 2009. Study on the role of precursors in coffee flavor formation using in-bean experiments. Journal of Agricultural and Food Chemistry 57 (21), 9923–9931.

Redgwell, R.J., Curti, D., Fischer, M., Nicolas, P., Fay, L.B., 2002. Coffee bean arabinogalactans: acidic polymers covalently linked to protein. Carbohydrate Research 337 (3), 239–253.

Romani, S., Cevoli, C., Fabbri, A., Alessandrini, L., Dalla Rosa, M., 2012. Evaluation of coffee roasting degree by using electronic nose and artificial neural network for off-line quality control. Journal of Food Science 77 (9), C960–C965.

Sanches-Silva, A., Rodríguez-Bernaldo de Quirós, A., López-Hernández, J., Paseiro-Losada, P., 2004. Determination of hexanal as indicator of the lipidic oxidation state in potato crisps using gas chromatography and high-performance liquid chromatography. Journal of Chromatography A 1046 (1–2), 75–81.

Sarrazin, C., Le Quere, J.L., Gretsch, C., Liardon, R., 2000. Representativeness of coffee aroma extracts: a comparison of different extraction methods. Food Chemistry 70 (1), 99–106.

Schenker, S., 2000. Investigations on the Hot Air Roasting of Coffee Beans. Unpublished Dissertation No. 13620, Eidgenoessische Technische Hochschule Zuerich (ETH), Switzerland.

Schenker, S., Handschin, S., Frey, B., Perren, R., Escher, F., 2000. Pore structure of coffee beans affected by roasting conditions. Journal of Food Science 65 (3), 452–457.

Schenker, S., Heinemann, C., Huber, M., Pompizzi, R., Perren, R., Escher, F., 2002. Impact of roasting conditions on the formation of aroma compounds in coffee beans. Journal of Food Science 67 (1), 60–66.

Schnermann, P., Schieberle, P., 1997. Evaluation of key odorants in milk chocolate and cocoa mass by aroma extract dilution analyses. Journal of Agricultural and Food Chemistry 45, 867–872.

Schramm, E., Kürten, A., Hölzer, J., Mitschke, S., Mühlberger, F., Sklorz, M., Wieser, J., Ulrich, A., Pütz, M., Schulte-Ladbeck, R., Schultze, R., Curtius, J., Borrmann, S., Zimmermann, R., 2009. Trace detection of organic compounds in complex sample matrixes by single photon ionization ion trap mass spectrometry: real-time detection of security-relevant compounds and online analysis of the coffee-roasting process. Analytical Chemistry 81 (11), 4456–4467.

Semmelroch, P., Grosch, W., 1996. Studies on character impact odorants of coffee brews. Journal of Agricultural and Food Chemistry 44 (2), 537–543.

Somoza, V., 2005. Five years of research on health risks and benefits of Maillard reaction products: an update. Molecular Nutrition & Food Research 49 (7), 663–672.

Spadone, J.C., Liardon, R., 1989. Analytical study of the evolution of coffee aroma compounds during storage. In: Proceedings of the 13th International Scientific Colloquium on Coffee, Paipa, Colombia, ASIC, Paris, France, pp. 145–157.

Spadone, J.C., Takeoka, G., Liardon, R., 1990. Analytical investigation of Rio off-flavor in green coffee. Journal of Agricultural and Food Chemistry 38 (1), 226–233.

Steinhart, H., Luger, A., 1995. Amino acid pattern of steam treated coffee. In: Colloq. Sci. Int. Cafe, [C. R.], 16th. Seizieme Colloque Scientifique International sur le Cafe, vol. 1, pp. 278–285.

Summa, C.A., De la Calle, B., Brohee, M., Stadler, R.H., Anklam, E., 2007. Impact of the roasting degree of coffee on the in vitro radical scavenging capacity and content of acrylamide. LWT – Food Science and Technology 40 (10), 1849–1854.

Sunarharum, W.B., Williams, D.J., Smyth, H.E., 2014. Complexity of coffee flavor: a compositional and sensory perspective. Food Research International 62, 315–325.

Tamanna, N., Mahmood, N., 2015. Food processing and Maillard reaction products: effect on human health and nutrition. International Journal of Food Science 2015, 1–6.

Theurillat, V., Leloup, V., Liardon, R., Heijmans, R., Bussmann, P., 2006. Impact of roasting conditions on acrylamide formation in coffee. In: Proceedings of the 21st International Scientific Colloquium on Coffee, Montpellier, France, ASIC, Paris, France, pp. 590–595.

Tressl, R., Helak, B., Kersten, E., Nittka, C., 1993. Formation of Flavor Compounds by Maillard Reaction. In: Hopp, R., Mori, K. (Eds.), Recent Developments in Flavor and Fragrance Chemistry; Proceedings of the 3rd Haarmann & Reimann Symposium, Kyoto 1992. Verlag Chemie, Weinheim, New York Basel Cambridge, pp. 167–181.

Tressl, R., Kossa, T., Renner, R., Koppler, H., 1976. Gas chromatographic-mass spectrometric investigations on the formation of phenolic and aromatic hydrocarbons in food (author's transl). Zeitschrift für Lebensmittel-Untersuchung und -Forschung 162 (2), 123–130.

Trugo, L.C., 1985. Carbohydrates. In: Clarke, R.J. (Ed.), Coffee, Chemistry, vol. 1. Springer, Netherlands, pp. 83–114.

Variyar, P.S., Ahmad, R., Bhat, R., Niyas, Z., Sharma, A., 2003. Flavoring components of raw monsooned arabica coffee and their changes during radiation processing. Journal of Agricultural and Food Chemistry 51 (27), 7945–7950.

Viani, R., Horman, I., 1974. Thermal behavior of trigonelline. Journal of Food Science 39 (6), 1216–1217.

Viani, R., Petracco, M., 2007. Coffee. In: Ullmann's Encyclopedia of Industrial Chemistry. Wiley-VCH Verlag GmbH & Co. (KGaA), pp. 1–32.

Wei, F., Furihata, K., Koda, M., Hu, F., Miyakawa, T., Tanokura, M., 2012. Roasting process of coffee beans as studied by nuclear magnetic resonance: time course of changes in composition. Journal of Agricultural and Food Chemistry 60 (4), 1005–1012.

Wieland, F., Glöss, A.N., Keller, M., Wetzel, A., Schenker, S., Yeretzian, C., 2012. Online monitoring of coffee roasting by proton transfer reaction time-of-flight mass spectrometry (PTR-ToF-MS): towards a real-time process control for a consistent roast profile. Analytical and Bioanalytical Chemistry 402 (8), 2531–2543.

Wong, K.H., Abdul Aziz, S., Mohamed, S., 2008. Sensory aroma from Maillard reaction of individual and combinations of amino acids with glucose in acidic conditions. Journal of Food Science and Technology 43 (9), 1512–1519.

Yaylayan, V.A., Keyhani, A., 1999. Origin of 2,3-pentanedione and 2,3-butanedione in D-glucose/L-alanine maillard model systems. Journal of Agricultural and Food Chemistry 47 (8), 3280–3284.

Yeretzian, C., Jordan, A., Badoud, R., Lindinger, W., 2002. From the green bean to the cup of coffee: investigating coffee roasting by on-line monitoring of volatiles. European Food Research and Technology 214 (2), 92−104.

Yeretzian, C., Jordan, A., Lindinger, W., 2003. Analysing the headspace of coffee by proton-transfer-reaction mass-spectrometry. International Journal of Mass Spectrometry 223 (1−3), 115−139.

Yilmaz, Y., Toledo, R., 2005. Antioxidant activity of water-soluble Maillard reaction products. Food Chemistry 93 (2), 273−278.

Chapter 13

The Grind—Particles and Particularities

Martin von Blittersdorff[1], **Christian Klatt**[2]
[1]*CAFEA GmbH, Hamburg, Germany;* [2]*Mahlkönig GmbH, Hamburg, Germany*

1. INTRODUCTION

Getting the best out of the bean requires grinding. This means transforming the roast coffee bean through application of mechanical forces into roast and ground coffee, which consists of small particles. They exhibit a big surface, open pores, and a small distance for transfer of the soluble substances that are now accessible for extraction. In view of an optimized extraction, more and more attention has been paid to grinding, leading to new developments of industrial and point-of-sale grinders, which are very important tools for coffee manufacturers and baristas, respectively.

The latter does regularly perform the extraction right after grinding. Such regular monitoring of the flow time and in-cup result is fairly suitable to control the grinding process. The extraction must be consistent throughout the day, it must be possible to run it at the required flow rate, and it must last but not least result in the desired quality of the cup in terms of taste and aroma. Yield and dry matter concentration are as important, but they can only be monitored analytically. At least in industrial manufacturing of roast and ground (R&G) coffee, grinding is not directly followed by extraction or preparation of the cup. Here it is necessary to evaluate the grinding process by analysis of the R&G coffee itself. It is therefore important to understand some characteristics and particularities of roast coffee beans and the resulting R&G coffee and to be able to measure them reliably.

2. CHARACTERIZATION OF ROAST AND GROUND COFFEE PARTICLES

2.1 Particle Size Distribution

A single particle of R&G coffee can be approximately described by a characteristic dimension x, which may for example correspond to the mesh size of

The Craft and Science of Coffee. http://dx.doi.org/10.1016/B978-0-12-803520-7.00013-X

a sieve that the particle would just pass. But grinding natural products like coffee beans always induces lots of particles of different sizes and shapes. The spread over the range of all sizes gives the so-called particle size distribution (*PSD*), which can be described by two to four characteristic values with sufficient accuracy. One refers most commonly to the sizes x_{10}, x_{50}, and x_{90}. These are the characteristic dimensions—or mesh sizes—below which 10%, 50%, and 90% of all particles would be found, respectively. x_{50} is then called the *median* particle size.

For extraction processes, the surface of the particle, together with its volume that contains the soluble substances, is even more important. Therefore, one passes from an elongation to the so-called equivalent diameters. The particle is then compared to a sphere that would either have the same surface or the same volume as the particle itself. Averaged over all particles, this will lead to the surface mean diameter (*SMD*) or the volume mean diameter (*VMD*), respectively. The former is highly impacted by the fine particles, whereas the latter will lie relatively close to the *median*. The ratio of surface and volume gives the *specific surface* S_V. The smaller the particle, the bigger is the specific surface, and the better is the extractability.

However, very fine particles put the consistency of the extraction at risk, when they block the pathways for the water. They may also impede the separation of the spent grounds from the extract by clarification or filtration, when they clog the filter pores. Such particles are referred to as *fines*. They are smaller than the particle that can be considered big enough to not cause problems in extraction or filtration. This is a matter of definition from case to case. However, often the particles below 100 μm are referred to as *fines*. Very coarse particles, on the other side, are sometimes called *boulders*.

With one of the equivalent diameters and the percentage of *fines*, a *PSD* is described adequately. In industry, experienced operators can perfectly control the grinding process by monitoring two or three of all of the above parameters. It may even be controlled by a feedback control loop in combination with online analytical instruments or by so-called artificial neural networks (Mesin et al., 2012). The barista, apart from his fingertips, has only the indirect response in the form of flow time and taste, which are combined results from *VMD*, *PSD*, *fines*, *boulders*, and the extraction technique. To optimize the *PSD*, he will have to draw the right conclusions empirically, as will be discussed later.

Each grinding process aims at an adequate average particle size for the respective coffee preparation method. Nowadays, the finest grind of roast coffee is added as a roast coffee component to soluble coffee. In the cup it shall neither form sediment nor be perceptible on the consumer's tongue. Therefore, its average particle size is around 50 μm. Typical particle sizes and *fines* that one can expect from grinding of such coffee through industrial soluble coffee can be read from Fig. 13.1.

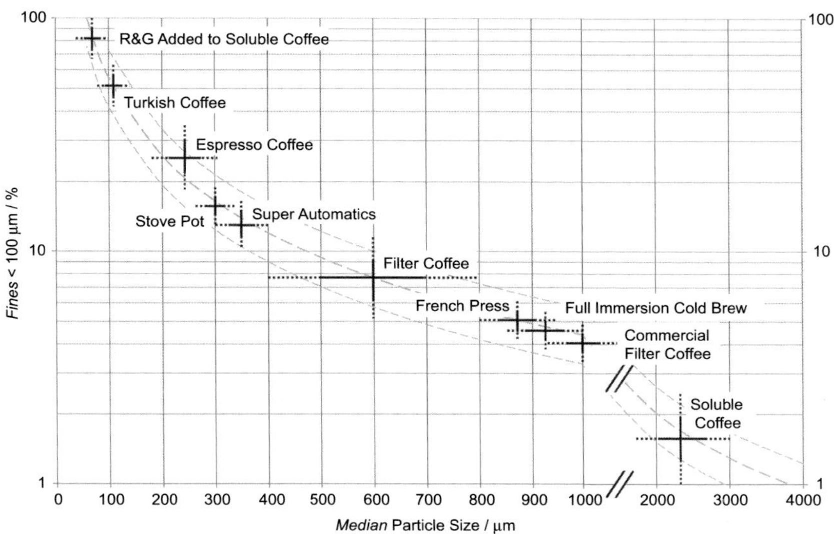

FIGURE 13.1 Typical average particle sizes and fines content for different brewing methods.

2.2 Bimodal Distribution and the Need for Normalization

Mostly, the distribution around the average particle size shall be as narrow as possible, meaning that all particles shall have a similar size. Espresso may be the only exception. Here, a certain percentage of *fines* are deemed necessary to achieve a pressure build-up, body, and a delicate crema, despite the short extraction time. Usually, a narrow, normal distribution has a peak value, referred to as *mode*. This is the most commonly occurring particle size of the *PSD*. For R&G coffee though, the *mode* is not used as a characteristic size for a simple reason. It is not uniquely defined: The *PSD* of R&G coffee is mostly *bimodal*. This means it has two *modes*, or two peak values. One of them is close to the *VMD*. The other, however, heads a peak in the size range of the *fines*.

The presence of this peak and accumulation of *fines* cannot be attributed only to the grinding technique, but also to the reticular structure of the feed material (Petracco and Marega, 1991). No matter how roast coffee beans are tackled. When they crack, very fine particles below 100 µm flake off. This happens again and again with further breakage, so that finally there is an accumulation of *fines*. It can grow to some 10–50% by weight, depending on the average particle size. Even if the average particle size itself is in the range of 100 µm, the two peaks of the *PSD* may still stay apart and reach a similar size when disk grinding. With other techniques, they may merge and make the bimodality disappear, especially when grinding in presence of liquid nitrogen.

In most applications, the *fines* content is as unavoidable as it is undesirable. Hence, there is a need to *normalize* the *PSD*. This means shifting the particles into a narrow, normal distribution around the average particle size by means of post-processing. Such techniques shall be described in Section 3.

2.3 Methods of Analysis

When it comes to determining the average particle size, in the first instance there are the sensitive fingertips of the experienced barista or operator. Empirically, they will be able to “measure” to an accuracy of about 50 μm. Where this is not sufficient, there are three common methods of particle size analysis: sieving, optical methods with laser diffraction, or image analysis. The interpretation of the results of sieving and laser diffraction is mostly based on the assumption of a spherical shape of the particles. Therefore, one obtains the diameters of spheres with equivalent volume, and last the *VMD*. Imaging, on the other hand, results in a realistic view of the particles so that different characteristic dimensions can be derived.

2.3.1 Sieving

For sieving, a stack of test sieves with increasing mesh size is composed according to industrial norms and in function of the expected *PSD*. A defined quantity of R&G coffee is fed to the coarsest sieve on top and driven by gravity and vibration through the sieves. If necessary, an air jet or a sieving aid may help to avoid agglomeration of fine particles. According to its dimensions, every particle will stay on one of the sieves or eventually on the bottom pan. By weighing the remains on every sieve, one can calculate a distribution density q_3:

$$q_3(\overline{x}_i) = \frac{\Delta m_i / m_t}{x_{i+1} - x_i} \tag{13.1}$$

with the mass fraction Δm_i of the total mass m_t, that remains on top of the sieve with the mesh size x_i, and below the next higher one x_{i+1}. The mesh x corresponds to the diameter of the equivalent sphere. The integration of the distribution density results in the *PSD*, from which the characteristic dimensions x_{10}, x_{50}, and x_{90} can be approximately determined as depicted in Fig. 13.2.

2.3.2 Optical Analysis of Particle Size and Shape

For the optical methods, there are special devices to disperse the R&G coffee in an air flow or a solvent for transportation. In case of laser diffraction, it flows across a laser beam, which is then scattered. The diffraction pattern is detected and processed applying numerical algorithms. They calculate what diameters spheres would need to have to lead to the same diffraction pattern. Any of the above characteristic dimensions can then be computed, assuming

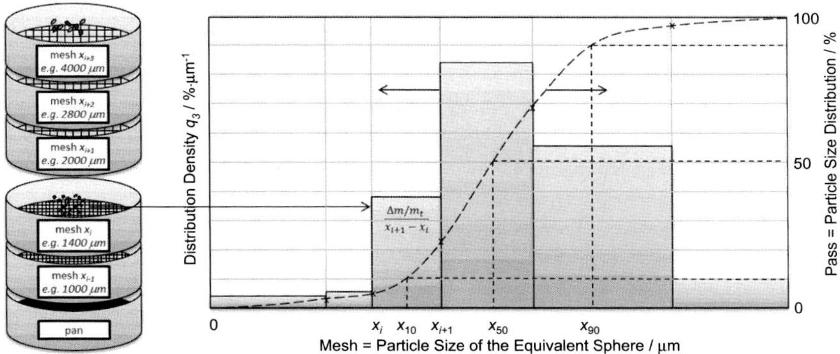

FIGURE 13.2 Sieving analysis: schematic drawing of a tower of test sieves and determination of the particle size distribution of coarse roast and ground coffee.

that they are the dimensions of spheres with equivalent volume. The smallness of such laser applications nowadays allows their online integration into R&G coffee transport pipes directly after grinding.

In imaging analysis, the dispersion of coffee particles is transported in front of a photographic lens. It takes pictures of the particles and produces projections of their outlines as depicted in Fig. 13.3B. Using software, they can be analyzed and any of the characteristic dimensions can be derived. Furthermore, the method allows to check visually and to compute in how far the outlines of the particles resemble a circle. This leads to yet another characteristic number, the *circularity* of the R&G coffee particles. The extremes are small particles with a spherical shape on one side, and bigger, irregularly shaped, or flaky particles on the other side. Section 4.2 will discuss the critical impact that the shape of the particles has, together with the particle size, on downstream processing of R&G coffee.

3. GRINDING TECHNOLOGIES

Grinding technologies differ according to the application of the grinder: On one side there are industrial grinders that are designed to work 24/7 to steadily supply industrial extraction or filling machines for R&G coffee. On the other side there are grinders for small scale, commercial use at the point of sale, and private use. They are designed for grinding on demand. Generally, they shall both be suitable for a whole range of different roast coffee types, and a whole range of average particle sizes of the R&G coffee. Furthermore, they shall meet the following requirements:

- safe and energy-efficient operation,
- temporally and locally consistent intake of the feed material,
- constant and repeatable achievement of a narrow *PSD* with minimum *fines*,

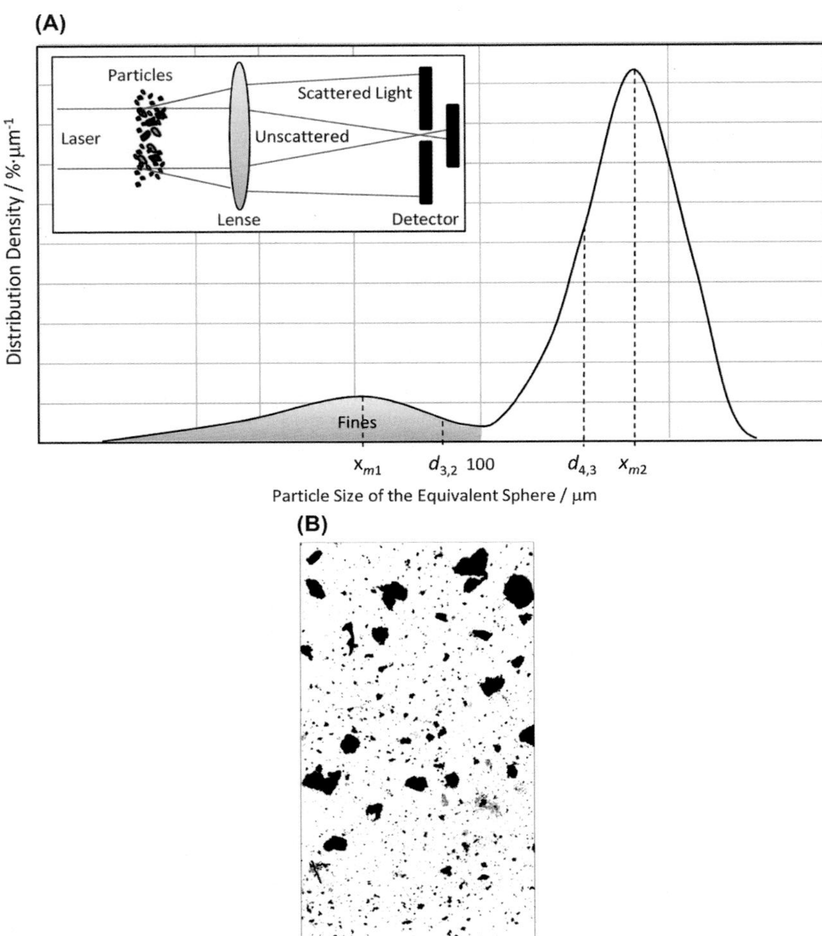

FIGURE 13.3 Optical analyses of particle size and shape. (A) Schematic drawing of laser diffraction analysis and particle size distribution density. (B) Typical shapes of roast and ground coffee particles from imaging analysis.

- avoidance of segregation and accumulation or fouling of *fines*, e.g., through reduction of electrostatic charging,
- gas tightness: keep the gases from the roast coffee inside and avoid intake of oxygen,
- precise and repeatable setting of the parallelism, position, and speed of the grinding tools,
- consistent temperature,
- easy access for cleaning and maintenance,
- minimum wear and quick exchange of grinding tools and wearing parts.

To guarantee physical integrity of the grinder, minimum wear, and a high quality of the R&G coffee, the operator has to secure the purity of the feed material. The greatest adversaries of grinders are stones and metallic foreign bodies. To remove them from the roast coffee, a destoner ought to be part of a roasting plant. For the removal of metallic foreign bodies, either a metal separator or a permanent magnet should integrate the feeding equipment of a grinder.

3.1 Industrial Grinding Technology

The application of grinding as a unit operation reaches far back in history. Grain milling is the forerunner of almost any comminution process, like grinding of coffee beans. Still today, most of the common methods of grinding of roast coffee beans can be compared with grain milling. The roller grinder can still be considered state-of-the-art for industrial coffee grinding, like for example Neuhaus Neotec's revised roller grinder in Fig. 13.4. Due to the particularities of roast coffee, it has evolved toward a very precise machine with very close manufacturing tolerances and special features which distinguish it from grain mills nowadays.

Basically, a roller grinder is a grinder with a number of grinding stages between 1 and 6, each with two rolls facing each other as depicted in Fig. 13.5A. The rolls may have a diameter between 120 and 200 mm, and a length of 200–900 mm or more, to grind from 200 kg/h of fine coffee to several tons per hour of coarse coffee. In the latter case, however, the precision and narrowness of the *PSD* may suffer. The rolls are made from hard casting iron materials with a hardened surface. There are radial or axial grooves milled

FIGURE 13.4 Neuhaus Neotec multi-sectional roller grinder with single drives and compactor for all purposes of industrial coffee grinding. *Courtesy of Neuhaus Neotec.*

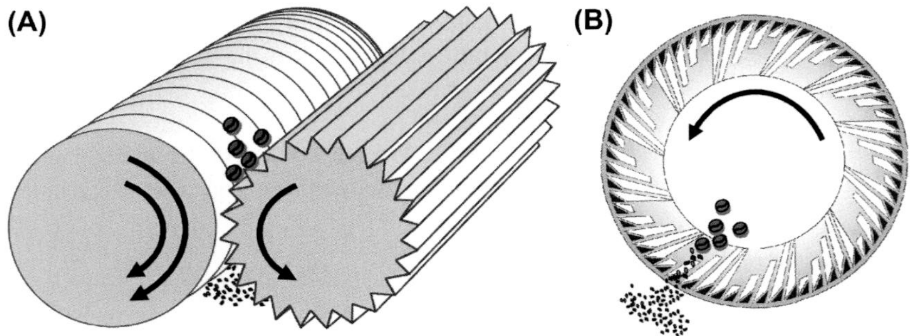

FIGURE 13.5 Schematic drawings of grinding tools. (A) A pair of rolls in a roller grinder. Here: fast roll with radial grooves, slow roll with asymmetrical axial grooves. (B) A grinding disk in a burr grinder.

into this surface unless the rolls are deliberately smooth. The axial grooves are asymmetrical with a sharp and a dull edge. Then, two axially corrugated rolls can face each other sharp to sharp, sharp to dull, dull to sharp, or dull to dull. The rolls may turn with a speed difference generated through a gear box. With the different grooves, their orientation, and the speed one can widely influence the stress on the material. It can be more cutting, or pressing and shearing, which has an impact on the particle shape and *PSD*. In industrial roll grinders, cutting tends to produce more *circularity* and a narrow *PSD*, whereas pressure and shear lead to rather irregular shapes and a wider *PSD* (Clarke and Macrae, 1985). However, the optimum configuration of an industrial roller grinder can best be determined empirically.

For production of a narrow *PSD*, some principal measures can be helpful (Ohresser et al., 2008; Rothfos, 1984):

- Increased number of grinding stages,
- Radially corrugated or smooth rolls,
- Less grooves,
- A certain, but not too high speed difference,
- Aligned grinding gaps, not too wide on top stages, not too narrow on bottom stages,
- A controlled, constant, and not too high temperature,
- A consistent, not too high feeding rate,
- A limited flow of gases inside the grinder.

These measures also reduce the wear of the grinding tools and increase the service time of the grinder, although they may impact the throughput. Still the best way to control the wear of the grinding rolls is monitoring of the *fines*. For a constant average particle size, the *fines* will steadily increase during the service time. Upcoming problems in extraction or filtration may have their origin in the wear of grinding rolls followed by a widening of the *PSD*. This effect can be counteracted by readjusting the grinding gaps. Eventually, after some 4000 h of service time, rolls with fine grooves will have to be resharpened or replaced.

In smaller industrial applications, disk grinders can be applied. However, due to the single pass they are rather limited in capacity, and there are fewer levers to impact the *PSD* around the target particle size. Disk grinders could be used up to 400 kg/h with a need for resharpening after some 250 h.

3.2 Recent Developments of the Roller Grinder, and Alternatives

An important recent development of the roller grinder is the full differential speed. Every single grinder roll is equipped with its own drive. This is supposed to help tailoring the *PSD* and reducing the *fines*, as well as reducing the wear of grinding rolls which can alternately serve as fast or slow roll.

Second, the roller grinder could be transformed into a bimodal grinder. On the lowest grinding stage, the rolls facing each other can be steplessly shifted. Consequently, between 0% and 100% of the roast coffee will pass that stage without being further comminuted. This leads to an intentional amplification of the bimodality of R&G coffee (Kirschner and Heinsz, 2009).

Last but not the least, roller grinders are more and more optimized for the use in the range of finest grinding around 100 μm by further reducing the gap between the grinding rolls. They are equipped with mechanical stops to prevent the rolls touching each other. Therefore, this grinding technique also has its lower limit of particle fineness. The particle size of Turkish coffee can be reached consistently with a roller grinder, provided it has been specially designed, manufactured within close tolerances, and it is operated by experienced staff. For average particle sizes below 100 μm though, other grinding techniques are to be preferred.

For finest grinding, so-called pulverization of roast coffee, it can be fed into a roller grinder, or preferentially into a centrifugal grinder together with liquid nitrogen. Centrifugal grinders basically consist of a rotating disk. It either propels the coffee bean in a single pass against an outer impact ring (impact grinder), or it is equipped with tools, jaws or pins (centrifugal, burr, or pin grinder respectively). These work the coffee particles against an opposite disk, a stator, or a perforated impact ring until the particles are fine enough to leave the grinding chamber through the perforation. Since some particles will stay in the grinding chamber longer than required, these grinders tend to produce more heat and a wider PSD than the roller grinder.

The addition of liquid nitrogen leads to a higher brittleness of the roast coffee bean. It furthermore counteracts the temperature-dependent volatility of aroma compounds and their sensitivity to oxidation. Therefore, there have repeatedly been efforts to add liquid nitrogen, even in coarser grinding during industrial production of R&G specialty (Mathias et al., 2015) or freeze dried coffees (Strobel, 1976). As it is evaporating at normal conditions, liquid nitrogen is suitable for the dry processing of R&G coffee. Even water and edible oil can also serve as grinding additives to help trapping the aroma compounds that are released during grinding (Baggenstoss et al., 2010). Obviously, it is then necessary to process the resulting slurry directly in a wet process like industrial extraction. For wet grinding, with its particular requirements regarding the tightness of shaft seals, etc., special grinders like ball grinders are used.

3.3 Postprocessing: Normalizing, Compacting, and Classification

A normalizer is an integral part of an industrial grinder for Turkish through filter fine R&G coffee. Just below the outlet of the grinder, in a cylindrical tube, the coffee is further processed by powerful mixing, kneading, or beating with paddles or hammers, respectively. On one hand, this aims at further comminuting the big particles and, most importantly, cutting of the chaff or silverskin from inside the fold of the coffee beans that had been released during grinding. During mixing, the silverskins are colored brownish and become rather unremarkable in the R&G coffee. On the other hand, the *fines* shall be reduced through forming agglomerates or through their adhering to bigger particles when hitting each other.

Another benefit of such mechanical postprocessing lies in the controllable densification of the R&G coffee. Since R&G coffee will have to fit into given volumes of silos, percolators, or packaging units like capsules, it is important that it has the right specific weight. For extraction, the R&G coffee bed must exhibit an optimum density and porosity. Both can be adjusted with sufficient accuracy in a normalizer. A similar effect is achieved in artisan espresso making by application of the tamper. The density of R&G coffee is often measured with tapped density analyzers and can be adjusted between 300 and 600 kg/m^3.

There may be applications like soluble coffee production, where normalizing is not adequate or not sufficient so that downstream processes run consistently. There, the finest particles and chaff may have to be altogether removed from the R&G coffee. This is done by air separation, which means transfer of *fines* and chaff from the coffee into a cross-current gas flow. Attempts are being made to integrate this operation into the grinder itself. There are simple technical solutions for detached air separators as well. Their use ought to be careful and restricted though, to avoid the contact with oxygen and to limit the loss of good product.

3.4 Point-of-Sale Grinding, or Grinding on Demand

Coffee grinders for commercial use at a point of sale, or for private use are smaller scale grinders. Most commonly they are operated on demand, i.e., at the moment that the coffee is needed for brewing or extraction. Doing so, the freshness of the coffee can best be preserved, considering that there is no protective atmosphere by adding nitrogen before, during, or after grinding at a point of sale and even less at home. The grinding technology differs depending on the purpose of use, grind speed, hourly capacity, and need of grind quality:

- Flat disk grinder: Produced with diameters between 50 and 180 mm, the disks are positioned horizontally or vertically in the grinder and deliver a throughput of 1–80 g/s. Available burr materials include food-grade tool steel, cast steel, tungsten carbide, and ceramics. In addition, the steel burrs can be hardened and surface treated to extend their lifetime and improve the grinding characteristics. Flat disk grinders are used in all areas of coffee grinding: At home for just a few shots per day, in coffee shops, retail outlets, and small to medium scale roasteries for packaging machines with capacities of up to 400 kg/h.
- Conical burr grinder: Conical grinders usually come with 40–71 mm burrs made of food-grade tool steel or ceramics. They can produce from less than 1 g/s up to 5 g/s of ground coffee. Small conical burrs (<60 mm) are mainly used in home coffee grinders, larger conical burr sets are traditionally used in high volume espresso bars.
- Blade grinder: This is a basic technology for grinding coffee beans that is available for home use and can be compared to the traditional blender. Made with steel blades, these grinders chip the beans apart. A defined grind size can only be achieved when the grinder is equipped with a sieve with a specific mesh size.
- Stone grinder: Stone grinders are traditionally used in flour production and for grinding Turkish coffee. Made of ceramic stones, the disks rub the beans with pressure to a fine powder.

In addition to the mere technical grinding process inside, there are other equally important features characterizing point-of-sale coffee grinders. Primarily, they are grouped according to the dosing system:

- Predosed: A defined mass of coffee beans is taken from a prepacked bag, preweighed cupping bowl, or an automated dosing system. The complete mass of coffee is ground.
- Time dosing: A timer in the grinder activates the motor for a set time for dosing of the R&G coffee. Roughly the required amount of coffee is ground.
- Automatic weight dosing: A load cell weighs the R&G coffee and stops the grinder through an automation loop exactly when the required mass of R&G coffee is reached.

The grinder is further equipped with a dispense system according to the specific needs. This can be the portafilter of an espresso machine, then the grinder comes with a defined portafilter holder system, or a user-defined cup (e.g., cupping bowl, grounds bin) for which the grinder provides a tray, or a bag. In the latter case, the grinder provides a bag holder with or without vibration of the bag, which ensures flow and some compaction of the R&G coffee to prevent overflowing of the bag.

3.5 Recent Developments Toward Quality Consistency

Today's espresso machines are very consistent tools in the preparation of recipes with defined coffee-to-beverage ratios. They provide consistent water temperature, pressure, flow rate, and latest technology using scales to determine the beverage weight. Since then, the focus on consistency has shifted toward the grinders. And the role of the barista to adjust the grinder regarding the brewing method has gained importance. For specialty coffee, the following factors of grinding are considered as most important and attract the research focus of manufacturers and associated labs:

- PSD: The different types of grinders create different *PSDs*. On the part of machine manufacturers, research is focusing on flat burr grinders creating a defined *PSD* by a certain design of the teeth in the burrs, the surrounding burr casing and the spout. Last but not the least the grinder must provide the means to the barista for precise and repeatable adjustment of the distance and the speed of the burrs.
- Grinding temperature: Higher temperature of the R&G coffee results in faster flow times. It was formerly assumed that the reason lies in a coarser grind or higher *VMD* due to the thermal expansion of the grind mechanism. However, recent studies have shown that state-of-the-art grinders stay dimensionally constant. It is rather the behavior of the coffee bean that changes. At higher temperatures and/or humidity, the bean seems to be more plastically deformable which leads to coarser R&G. Finally, it is the barista, who needs to react to changes of ambient conditions, so that grinding remains consistent.
- Weight dosing: State-of-the-art espresso machines push the extraction to a predefined beverage weight, which corresponds to a specific R&G coffee weight in the portafilter. Therefore, grinder development strives for weight dosing rather than running the grinder's motor for a set time. This increases dosing accuracy and helps locking the coffee-to-beverage ratio in the preparation process.

In addition, grinders are further optimized in view of their ergonomics for daily use, reduced noise emission, and electrical components that will allow for precise and repeatable operation.

4. THE IMPORTANCE OF GRINDING FOR COFFEE PREPARATION

The impact that grinding has on coffee processing deserves further discussion. There are highly influential parameters both from the roast coffee beans and from the grinding technology that define the grinding process and the characteristics of the R&G coffee. Mostly, one is aiming at a narrow *PSD*. It has been discussed before what can be done technology-wise to achieve this. Now, the view shall be broadened to the roast coffee and the downstream processes.

4.1 Characterization of the Feed Material and Its Impact on Grinding

With respect to grinding, the roast degree, i.e., roast color or roast loss, is the most decisive characteristic of the roast coffee. With increasing roast degree, especially when obtained in a short roasting time, the porosity and brittleness of the coffee beans increase. They can be ground without difficulty almost right after roasting. When doing so, a relatively wide *PSD* and a high *fines* content are unavoidable, though. For lighter coffees with higher residual moisture, the grinding result is better with respect to the distribution width, but it is achieved with a less favorable energy utilization (Von Blittersdorff, 2010). Coffee beans that have been roasted slowly also possess a more homogeneous matrix, pore sizes, and a higher density. Consequently, grinding them requires more energy, but with a more homogeneous grinding result.

As indicated, the moisture content, and moisture distribution in the roast coffee beans are decisive for homogeneous grinding. The moisture is composed of the residual moisture from the green coffee, which after roasting is around 1–2%. In addition, part of the water that might have been used for quenching is absorbed by the coffee beans. But the quenching water does not instantaneously distribute homogeneously in the coffee bean. It will first be wetting the surface (Fischer, 2005). This is why it is important to respect a resting time between roasting and grinding, within which the water can redistribute evenly inside the coffee bean by diffusion. Coffee oil may also have pushed to the surface, especially after dark and fast roasting with high disruption of the bean matrix. The oil also needs time to move back inside. Studies have shown that the breakage behavior of water-quenched coffee beans becomes consistent after a minimum resting time of 6–12 h. Coffee beans with more than 6% moisture are rather elastic and difficult to grind anyways, even after a long resting time (Baggenstoss et al., 2008).

For grinding consistency, it is required that the coffee beans arrive clean and undamaged at the grinder, without breaking during transport. As mentioned before, particles of different sizes are created at any crack of the beans. Therefore, even initial cracking should be performed in a controlled way in the grinder.

4.2 Impact of Grinding on Downstream Processing

Through production of a defined particle size, *PSD*, and particle shape, grinding has an impact on mass transfer and rheology that play a role in downstream processing. How easy is it to transport, store, drain, and dose the R&G coffee? Can it be easily percolated? How fast will it release the gases that have been formed during roasting? How fast and to what extent do the soluble substances diffuse to the surface of the particles and finally end up in the cup?

Obviously, the first two and last two questions lead to adverse directions regarding the targeted particle size and shape. The bigger the particles and their *circularity*, the better the R&G coffee flows, and the better it can be percolated in extraction processes. Smaller and irregular particles, on the other hand, will release their soluble substances faster and lead to a more intense, more concentrated coffee in the cup. Fortunately or unfortunately, bigger particles are irregular and smaller particles tend to be circular. However, if the particles are too small, i.e., *fines*, there are two risks involved: First, they can clog the pores between bigger particles or in filters, dramatically increasing the percolation time or even blocking percolators or filters. Or they pass the pores and end up in the cup. Second, they tend to be overextracted which shall be further discussed in the next section.

Provided the *fines* creation is under control, it may be of interest to further reduce the average particle size. The following properties will above all be affected:

- The inner friction angle, the repose angle, and the cohesive forces increase (Horta de Oliveira et al., 2014). In general, R&G coffee is rather cohesive than free flowing. Finely ground coffee is less flowable and more difficult to dose, it tends to agglomerate or bridge at the outlet of containers and dosers. A higher moisture content will increase this effect.
- The density increases and the porosity decreases. More solid matter fits in a given volume. But then, percolation leads to a higher pressure drop until blocking of the filter in the extraction or separation process. In case of a wide or bimodal *PSD*, anyways the porosity will be lower and the pressure loss higher than in the case of uniform particles (Stieß, 2013).
- The specific surface increases and diffusion paths are shorter. The mass transfer of diffusion controlled processes like further degassing and extraction of soluble substances will be accelerated. However, the sensitivity to oxidation increases too. All these will further accelerate with increased temperature and increased moisture (Baggenstoss, 2008).

Both high moisture and high temperature have a positive impact on diffusion coefficients. They can accelerate degassing, but also the loss or deterioration of volatile aroma compounds. During grinding, every single particle of the R&G coffee is shortly in direct contact with metallic surfaces of the rolls that can easily be heated or cooled. In case the temperature is deemed important, it shall be adjusted in a grinder. Later, in silos for example, it is hardly possible to change the temperature of the R&G coffee, since its heat conductivity of about 0.11 W/m^2 K is even lower than that of wood (Von

Blittersdorff, 2010). Therefore, the grinder may be cooled to preserve and retain the sensitive aromas, or heated contrariwise, to avoid condensation and accelerate degassing.

Degassing may be necessary since freshly roasted coffee beans retain up to 20 g/kg of gas produced in roast reactions. That is mainly carbon dioxide and some 10–15% of carbon monoxide. Half of it is liberated instantaneously when opening the bigger pores during grinding (Shimoni and Labuza, 2000). Hence, when grinding industrially, e.g., 1 ton of roast coffee per hour, it will release up to 1.3 m^3/h of carbon monoxide. For safety reasons it is recommended to ventilate the environment of a grinder or, in case of industrial grinding, to exhaust the gas from a grinder and from R&G coffee storage containers in a controlled way. Further degassing of the coffee particles will continue by diffusion during some 48 h with a decreasing degassing rate. Before industrial filling or extraction, R&G coffee must simply be given a resting time in an inert atmosphere after grinding. The smaller the particles and the higher the residual moisture and the temperature, the less time is required. The coffee can be further processed as soon as the gas content is uncritical for the following process or the chosen packaging format. A barista, for instance, processing small quantities, will not have to let the coffee rest. Whereas complete degassing is necessary when the R&G coffee is filled to capacity of sealed containers from which no gas could escape.

4.3 Adapted Grinding for Different Brewing Methods and Optimum In-cup Quality

Different brewing methods can be distinguished and characterized by the following:

- contact time between water and coffee, e.g., 30 s for espresso versus 3 min for filter,
- water pressure on the coffee bed, e.g., 9 bar for espresso versus gravity flow for filter,
- water temperature, e.g., 92°C for filter versus 12°C for cold brew.

Apart from time, pressure, and temperature, all different brewing methods require a specific particle size of the R&G coffee as shown in Fig. 13.1. Together with pressure and temperature, it ensures a certain flow of water, i.e., a certain contact time and the desired extraction of coffee into the brew. The longer the contact time of water and coffee will be, the coarser the coffee should be ground. Very fine particles may be overextracted in fact, which leads to higher bitterness of the coffee in the cup (Lingle, 1995). *Fines* may be present, though, when the intended extraction time is very short as for espresso coffee. Then, overextraction is limited through the short contact time. A barista will notice that reproducing the same flow time and the same in-cup result with different grinders requires different settings, different average particle sizes, and different *PSDs*. The following Fig. 13.6. shows different PSDs from different burr grinders. With the same coffee-to-beverage ratio of 1:2, they all lead to the same flow time in espresso extraction.

Curve (a) results from a flat disk grinder *Mahlkönig EK43*. It has large disks that grind to a narrow *PSD* with a small median particle size. The grinder has pushed the specialty coffee industry's standards toward higher extraction without bitterness, which is highly appreciated by the baristas and the consumer. The yield of extraction is high, however, without off-tastes from under- or overextraction, even though the R&G coffee has a relatively high content of fines. Curve (d) results from a conical grinder that grinds to a wider *PSD*. To realize the same extraction time, the median particle size is almost double the size of (a). In this case, the overextraction of the *fines* and the underextraction of the *boulders* result in both more bitter and more acid notes at a lower overall yield, the same flow time notwithstanding. There is a range of different extractions within the same brew and a medley of taste within the cup that sums up to the final in-cup quality. By sieving the R&G coffee prior to brewing, these notes could be avoided. However, sieving means losses in industrial production, and it is impractical in a busy café.

Thus, besides approaches in sourcing and roasting of green coffee, also grinding fosters the movement toward more effective extraction. Consequently, the coffee-to-beverage ratio can be reduced. Nowadays, a barista may even use refractometers to measure the yield in the bar and readjust the grinder accordingly.

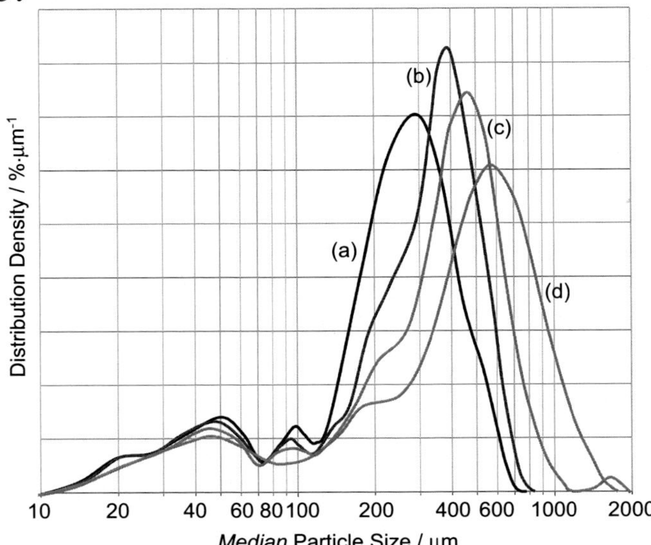

FIGURE 13.6 Different particle size distributions from burr grinders: (a) 98 mm disk (Mahlkönig EK43), (b) 65 mm disk, (c) 75 mm disk, and (d) 71 mm conical burr.

5. OUTLOOK

High quality grinders are required in industry as well as in today's specialty cafés. They will be further developed toward highly precise tools with appropriate interfaces that will allow the operator and barista to tailor the PSD through precise dosing of the coffee and accurate and repeatable control of the grinding tools. The grinders will then deliver a *PSD* of the R&G coffee that leads to the desired extraction. With regard to specialty coffee in particular, more scientific studies are required to substantiate and prove the empirically perceived impact of grinding on extraction and in-cup quality. After all, it needs operators and baristas, who are eager to develop their skilled craftsmanship and who are experienced and hands-on enough so they are in perfect control of their grinders despite possible changes in feed material properties and ambient conditions throughout the day. Then, extraction will result in a sweet, clean, and intense taste and aroma in the cup, and grinding has helped to get the best out of the bean.

REFERENCES

Baggenstoss, J., 2008. Coffee Roasting and Quenching Technology – Formation and Stability of Aroma Compounds (Ph.D. thesis). ETH Zürich.

Baggenstoss, J., Perren, R., Escher, F., 2008. Water content of roasted coffee: impact on grinding behaviour, extraction, and aroma retention. European Food Research and Technology 227, 1357–1365.

Baggenstoss, J., Thomann, D., Perren, R., Escher, F., 2010. Aroma recovery from roasted coffee by wet grinding. Journal of Food Science 75 (9).

Clarke, R.J., Macrae, R. (Eds.), 1985. Coffee Technology, vol. 2.

Fischer, C., 2005. Kaffee. Änderung physikalisch-chemischer Parameter beim Rösten, Quenchen und Mahlen (Ph.D. thesis). TU Carolo Wilhelmina zu Braunschweig.

Horta de Oliveira, G.H., Corrêa, P.C., Santos, F.L., Vasconcelos, W.L., Calil Júnior, C., Machado Baptestini, F., Asdrúbal Vargas-Elías, G., 2014. Caracterização física de café após torrefação e moagem', *Semina: Ciências Agrárias*. Londrina 35 (4), 1813–1828.

Kirschner, J., Heinsz, L.J., 2009. Bimodal Coffee Grinder. US Patent US2009/0145988 A1.

Lingle, T.R., 1995. SCAA Brewing Handbook.

Mathias, P.A., Djamer, A., Villain, O., Guenat, C., Sarrazin-Horisberger, C., Mestdagh, F., Eichler, P., Kessler, U., Closset, E., 2015. Process of Preparing Ground Coffee Ingredient and Capsule Containing Such Ingredient. WO Patent 2015/104172 A1.

Mesin, L., Alberto, D., Pasero, E., Cabilli, A., 2012. Control of Coffee Grinding with Artificial Neural Networks. WCCI IEEE Congress on Computational Intelligence, IEEE, Brisbane, Australia.

Ohresser, S., Eichler, P., Koch, P., Raetz, E., 2008. Method for Delivering a Long Coffee Extract from a Capsule in a Reduced Flow Time. European Patent EP 1 882 431 A1.

Petracco, M., Marega, G., 1991. Coffee grinding dynamics: a new approach by computer simulation. Proceedings of the 14th ASIC Colloquium 14, 319–330.

Rothfos, B. (Ed.), 1984. Kaffee – Der Verbrauch. Gordian-Max Rieck GmbH.

Shimoni, E., Labuza, T.P., 2000. Degassing kinetics and sorption equilibrium of carbon dioxide in fresh roasted and ground coffee. Journal of Food Processing and Engineering 23, 419–436.

Stieß, M., 2013. Mechanische Verfahrenstechnik 1. Springer Verlag.

Strobel, R.G.K., 1976. Process for Grinding Roasted Coffee Beans. GB Patent 1 424 264.

Von Blittersdorff, M., 2010. Kaffeeröstung und Quenchkühlung. Zur Wechselwirkung zwischen Wärme- und Stofftransport und veränderlichen Materialeigenschaften (Ph.D. thesis). TU Hamburg-Harburg.

Chapter 14

Protecting the Flavors—Freshness as a Key to Quality

Chahan Yeretzian[1], Imre Blank[2], Yves Wyser[2]
[1]*Zurich University of Applied Sciences, Wädenswil, Switzerland;* [2]*Nestec Ltd., Nestlé Research Center, Lausanne, Switzerland*

1. THE SECRET TO GREAT COFFEE IS THE PEOPLE WHO MAKE IT

Defining the quality of coffee is by no means a simple endeavor and several renowned publications and coffee experts have offered various definitions and discussions on the subject (Illy and Viani, 2005). However, the definition has remained elusive and with the mounting importance of the specialty coffee community, a rational approach to quality is becoming increasingly important. Indeed, the specialty coffee movement can best be described as the uncompromising quest for the highest quality in the cup.

It all began 50 years ago, when Alfred Peet opened a coffee store in Berkeley, April 1, 1966 (www.peets.com). Noticed by only a few quality aficionados, Alfred Peet can be considered the pioneer of the specialty coffee movement. Although the term "specialty coffee" did not exist at the time, a revolution was brewing. His coffee was unlike anything Americans had ever tasted before—small batches and fresh beans. His philosophy was that there should be the shortest distance possible between the roaster and the customer. Freshness was at the heart of his vision and quality concept. Since then freshness has remained a focus for all those who strive to deliver the highest quality.

Alfred Peet inspired and guided the founders of Starbucks. In 1971, Jerry Baldwin, Gordon Bowker, and Zev Siegl founded Starbucks in Seattle, selling fresh-roasted whole beans to local customers. Freshness was a central motivation to the founders of Starbucks, introducing a larger American public to the specialty coffee movement. More than 15 years later, in the late eighties, Nespresso launched portioned coffee and introduced high quality coffee with a personalized touch to European customer in their homes.

The Craft and Science of Coffee. http://dx.doi.org/10.1016/B978-0-12-803520-7.00014-1

The term specialty coffee was first used in 1978 by Erna Knutsen (www.scaa.org/?page=RicArtp1). She described "special geographic microclimates" that "produce beans with unique flavor profiles, which she referred to as Specialty Coffees." Underlying this idea of coffee appellations was the fundamental premise that specialty coffee beans would always be well prepared, freshly roasted, and properly brewed.

In 1982, the Specialty Coffee Association of America (SCAA) was founded and then in 1998 the Specialty Coffee Association of Europe (SCAE) followed. In 1982, Paul Songer published an article in the Specialty Coffee Chronicles, the Newsletter for members of the SCAA entitled "A Question of Freshness." The closing lines of his articles unequivocally positioned freshness at the center of the specialty coffee movement: "Freshness of the coffee that a roaster or retailer sells and serves is a direct reflection of the standards and abilities of that operation. It will determine one's competitiveness in the marketplace and the ability of the consumer to experience a product that is unique and worth seeking out. The bottom line is flavor. For specialty coffee, flavor means freshness."

Besides addressing the development of the freshness concept from the perspective of the coffee specialist, it is also worth to briefly review freshness from the consumer's perspective and how this may have evolved over the last decade. Although very little has been published on this subject, Péneau (2005) explored the concept of freshness in fruits and vegetables; some of the learnings are of interest also for coffee. She concluded that freshness "is best described by a level of closeness to the original product, in terms of distance, time and treatment." Interestingly, negations (absences of negative attributes) were widely used to describe freshness, whereas people less familiar with the origin and processing of products used negations more often to describe freshness than those more familiar with these aspects of the products. Here we will attempt to develop a concept of freshness that is based on positive attributes rather than defining freshness by the absence of negatives. This also follows the evolution of the concept of quality introduced by the specialty coffee movement that increasingly focuses on positive quality attributes of the cup versus the notion of lack of defects in green bean, which is still the dominant concept of quality in the coffee business in general. It is worth mentioning here that *oxidation* or *oxidized flavor* is not an intrinsic defect that appears as an attribute in cup tastings. However, it is related to freshness, as it may be introduced during storage and aging.

We would also like to mention here that the concept of freshness, as it will be developed and discussed in this chapter, refers to the freshness of roasted coffee, in contrast to freshness in green beans. In fact, the timescale and the chemical and physical processes underlying the loss of freshness in green beans are different to those that occur in roasted beans. Once coffee has been harvested, sorted, and graded, it is often stored as green/raw bean over a prolonged period, from several months up to several years. During storage, there is a distinct decrease in quality, which is expressed by a flattening of the cup quality (Selmar et al., 2008; Scheidig et al., 2007). As a consequence, the provenience-specific characteristic features, especially those of top quality coffees, gradually diminish. In contrast to "off-notes," which may occur during the course of storage and are mainly caused by oxidation processes within the lipid fraction (Speer and Kölling-Speer, 2006), the causes of the progressive reduction in cup quality are still unknown. The appearance of storage-related off-notes can largely be controlled by appropriate storage of the green coffee, with the two critical factors that need to be controlled being moisture content and temperature. However, it must be stated that flattening of the cup quality occurs even under optimal storage conditions. Although it is generally stated in the coffee trade that green coffee beans can last for up to 3 years (if properly stored), increasingly, specialty coffee roasters acknowledge that roasting fresher green beans is beneficial to quality.

2. MEASURING FRESHNESS

Since the advent of the specialty coffee movement in the late 1960s, the concept of freshness of roasted coffee has remained at the center of much of the effort of specialty coffee lovers. However, despite the central role freshness plays in the highest quality coffee and in the specialty coffee movement in general, it appears that much of the discussion on freshness revolves around the process of how freshness can be delivered and assured: freshly roasted, ground within a few days, immediately extracted, and consumed. But when it comes to defining freshness as an objective and scientifically measurable attribute of the cup, things become much less clear. The aim of this chapter is, therefore, to provide a quantitative and scientific answer to the question: *How can we measure freshness?*

The first step to measuring freshness is to clarify what is meant by freshness, i.e., to understand and define freshness. In the context of coffee, we define freshness as coffee that exhibits no impairment to its original qualities. The original point of reference referred to here is coffee that has just been roasted. However, this initial state of an absolutely fresh coffee cannot be defined in absolute terms. Indeed, because coffee is an agricultural product, the initial status of a fresh coffee depends on a large number of factors, such as the green coffee variety (genetics), the altitude, climate and soil composition of the plantation, agronomic and harvest practices, postharvest treatment, and green coffee storage (Yeretzian et al., 2002). This results in green coffee beans with varied chemical compositions and physical properties. Collectively, all these factors lead to vastly differing green beans being roasted and hence affect the chemical and physical properties of the freshly roasted coffee.

During roasting, a range of complex chemical and physical processes occur within the coffee bean (also see Chapter 12), leading to the formation of typical coffee aroma compounds and inorganic gases (mainly carbon dioxide; CO_2).

The sensory characteristics of the volatile compounds shown in Fig. 14.1 are described in Table 14.1, indicating the relevance of their aroma to the roast and ground coffee and the corresponding brew, expressed as the flavor dilution (FD) factor. The FD factor was introduced by Grosch et al. and is defined as the ratio of the concentration of the odorant in the initial extract to its concentration in the most dilute extract in which the odor is still detectable by gas chromatography-olfactometry (Blank et al., 1992; Grosch et al., 1993). Only 2-furfurylthiol (6) elicits a coffee-like aroma in a certain concentration range, all other odorants smell differently from coffee.

Once roasting is completed, a multitude of physical and chemical processes immediately start, leading to an evolution of the coffee over time. Indeed, a freshly roasted coffee is a highly elusive product. Among the many changes that occur over time after roasting, two are of particular importance and related to the quality attributes of coffee. One is (1) the evolution of the aroma profile and the other (2) the degassing of the beans. Once roasting is complete, the clock starts ticking on both of these processes. Although the evolution of both processes can be measured by a multitude of methods, with varying levels of sophistication, the aim here is to establish approaches that are accurate, robust, and simple. The goal is to find an appropriate method for monitoring changes in aroma compositions, i.e., the aroma balance defined as the ratio of volatiles that are relevant from a sensory perspective. These changes may include the loss of certain volatiles, but also the formation of others, which may lead to a

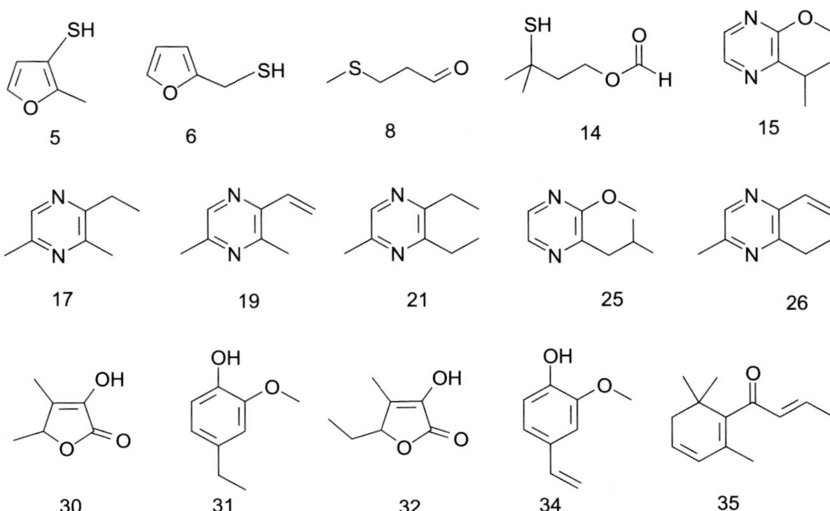

FIGURE 14.1 Selected coffee aroma compounds that represent various chemical classes that contribute to the aroma of roast and ground coffee.

TABLE 14.1 Potent Odorants Found in Arabica Coffee (Blank et al., 1992)

No	Compound	Aroma Quality	Aroma Relevance (FD)[a]	
			Powder	Brew
5	2-Methyl-3-furanthiol	Meaty, boiled	128	<16
6	2-Furfurylthiol (FFT)	Roasty (coffee-like)	256	64
8	Methional	Boiled potato-like	128	512
14	3-Mercpto-3-methyl-butyl formate	Catty, roasty	2048	256
15	3-Isopropyl-2-methoxypyrazine	Earthy, roasty	128	32
17	2-Ethyl-3,5-dimethylpyrazine	Earthy, roasty	2048	1024
19	2-Ethenyl-3,5-methylpyrazine	Roasty, earthy	128	128
21	2,3-Diethyl-5-methylpyrazine	Earthy, roasty	512	128
25	3-Isobutyl-2-methoxypyrazine	Earthy	512	128
26	2-Ethenyl-3-ethyl-5-methylpyrazine	Roasty, earthy	512	32
30	3-Hydroxy-4,5-dimethyl-2(5*H*)-furanone	Seasoning-like	512	2048
31	4-Ethylguaiacol	Spicy	256	512
33	5-Ethyl-3-hydroxy-4-methyl-2(5*H*)-furanone	Seasoning-like	512	1024
34	4-Vinylguaiacol	Spicy	512	512
35	(*E*)-β-Damascenone	Honey-like, fruity	2048	64

[a] The FD factor of 256 for FFT means that the roasty note of FFT in the 256-fold dilution of the original coffee aroma extract was still detected by gas chromatography-olfactometry, i.e., FFT was no longer detectable in the 512-fold diluted extract.

modified aroma composition, which may ultimately be perceived as lack of freshness. Changes in the aroma composition may be due to various phenomena, such as:

1. *Volatilization:* Volatile aroma molecules are lost—the overall coffee aroma fades in and above the cup. This can be limited by protecting the coffee with impermeable packaging.
2. *Intrinsic reactivity:* Aroma molecules are often intrinsically labile reacting with compounds naturally present in coffee—the consequence is again that the aroma in the coffee fades. This cannot really be avoided, even with the

best packaging, but can be slowed down through storage at lower temperature.

3. *Oxidation:* Aroma compounds oxidize—the coffee aroma fades and new volatiles with off-notes are created. This can be prevented by protecting the coffee from oxygen and oxidative processes. The way coffee is being packaged (e.g., the atmosphere inside the pack) and the barrier properties of the packaging material will make a difference here.

In general, these processes are accelerated when beans are ground and when stored at elevated storage temperature.

2.1 Loss of Inorganic Gases

During roasting, coffee beans undergo major chemical transformations, during which a large amount of inorganic gases, mainly CO_2, are generated. Much of these gases remain entrapped within the porous structure of the roasted beans. Approximately 1–2% of the weight of freshly roasted coffee can be attributed to entrapped inorganic gases (excluding water), whereas unroasted, green coffee beans contain no entrapped gases. These gases are mainly released during storage, but the process already starts during the final phase of roasting. Coffee that has been stored for some period of time will have less entrapped gases and consequently a lower rate of gas release. Therefore, one approach for assessing freshness is based on measuring the amount and the kinetics/rate of gas (mainly CO_2 and excluding H_2O) released within a given time window. The approach taken by the research group led by Prof. Yeretzian at the Zurich University of Applied Sciences (ZHAW) group is to measure the weight loss of freshly roasted coffee as a direct means of monitoring the loss of freshness over time, whereas other groups have taken alternative approaches (Wang and Lim, 2014; Wang, 2014). A quantitative discussion of the degassing of freshly roasted whole and ground coffee beans (Arabica and Robusta) that have been roasted to different roast degrees and along different time–temperature roast profiles will be presented in a forthcoming publication.

2.2 Evolution of the Aroma Profile

Probably the most important quality attribute of coffee is its aroma and, therefore, the most appropriate and direct approach for measuring freshness is to examine the aroma and its evolution over time (Sunarharum et al., 2014; Grosch, 2001, 1998; Grosch et al., 1996; Lindinger et al., 2008, 2010; Poisson et al., 2009; Semmelroch et al., 1995; Blank et al., 1992, 1991). However, we must acknowledge that coffee flavor is complex, elusive, and labile (Munro et al., 2003). The most important coffee aroma compounds and nomenclature were shown in Fig. 14.1 and Table 14.1. Once roasting is complete, the aroma has already started to evolve (Gloss et al., 2014). This is due to physico-chemical changes, such as evaporation, as well as chemical reactions, and interactions between aroma compounds and the coffee matrix. For example, a comprehensive mechanistic, chemical study by Müller and Hofmann (2007) explored in detail the degradation of the key coffee odorant 2-furfurylthiol, which contributes to the sulfury–roasty odor quality of a coffee brew and was found to reduce considerably during coffee storage.

As shown in Fig. 14.1, many aroma-active components are reactive species bearing functional groups such as thiols, carbonyls, and enolones. Depending on conditions such as oxygen, moisture, and temperature, they will evolve over time and thus change the perceived aroma of a coffee. Hence, it seems obvious to look for clues of freshness (or loss of freshness) in the evolution of the coffee aroma profile.

As previously outlined, quantitatively assessing changes in aroma compounds during storage will allow the loss of freshness to be determined. The processes involved in such losses of freshness are complex and may occur in two main ways: (1) a loss of highly volatile compounds; (2) through chemical reactions, for example, as a result of oxidation by O_2, or through intrinsic chemical reactions between different coffee components. Many chemical classes (thiols, diones, aldehydes, vinyl derivatives) may react upon storage. This may lead to either a decrease or an increase in headspace concentrations for selected compounds. Hence a loss of freshness can best be described as a progressive imbalance in the aroma profile. Such processes have been extensively discussed in the literature, with the aim of identifying marker compounds for the shelf life of packaged roasted coffee. The first studies on coffee aroma deterioration can be traced back to the 1940s (Shuman and Elder, 1943), followed in the 1950s by work from Merritt et al. (1957) and Buchner and Heiss (1959). Many more groups have examined the shelf life of roasted coffee beans or of roast and ground (R&G) coffee, either from a chemical or a sensory perspective (or both) (Spadone and Liardon, 1990; Nicoli et al., 1993, 2009; Anese et al., 2006; Marin et al., 2008).

Considering the fact that green coffee beans contain more than 10% fat, volatile lipid oxidation products were an early focus in studies on degradation markers in coffee. Such studies have reported a correlation between the process of coffee going stale and the generation of *n*-hexanal after an initiation phase of approximately 7 weeks of storage in air (Spadone and Liardon, 1990). These studies also showed that other products formed by oxidative degradation of unsaturated fatty acids in roasted coffee do not play a significant role in the flavor of roasted coffee. However, the formation of hexanal cannot explain the loss of freshness; at best it may be seen as an early marker of the fading of the coffee aroma which ultimately may lead to a reduction in perceived freshness.

Although several volatile organic compound (VOC) markers have been suggested for monitoring deterioration in freshness of R&G coffee, the major weakness of using absolute concentrations of such marker compounds is the fact that the amount of any single compound depends, among other things, on its initial concentration, which in turn is affected by variables such as blend, roast degree, grinding, extraction, as well as other factors (Kallio et al., 1990; Leino et al., 1992). The use of *ratios* of headspace concentrations of selected VOCs is, therefore, more robust and reflects changes in the balance in the headspace (Spadone and Liardon, 1990; Arackal and Lehmann, 1979; Marin et al., 2008; Kallio et al., 1989). Coffee VOCs that have typically been used in such a ratio are methanethiol, propanal, 2-methylfuran, 2-butanone, 2,3-butanedione, 2-furfurylthiol, dimethyl disulfide, and hexanal.

3. FRESHNESS INDEX

Several VOC have been suggested in the literature as markers for aging or staling. Already in 1992, Holscher and Steinhart reported that methanethiol has a strong impact on aroma freshness and a strong decrease in concentration can be seen just one day after roasting. In 2001, Mayer and Grosch published a study in which they investigated the changes in odorant composition in the headspace of ground coffee based on time after grinding. They reported an approximate 50% reduction in headspace intensities for a range of aldehydes and diketones (methylpropanal, 2-methylbutanal, 3-methylbutanal, 2,3-butanedione, 2,3-pentandione), just 15 min after grinding. Considering the fact that these volatiles are important coffee aroma compounds, it can be inferred that their loss is related to a change in the aroma profile and hence a reduction in freshness. Alternatively, some compounds are hardly present in freshly roasted coffee and only appear when coffee ages (Parliment et al., 1982; Baggenstoss et al., 2008). Their presence may, therefore, serve as a sign of loss of freshness. Compounds such as dimethyl disulfide and dimethyl trisulfide, however, only form during storage due to the oxidation of methanethiol. Therefore, one approach may potentially be to measure the concentration of such compounds and then use these data to estimate the freshness level of the coffee.

3.1 The Concept

A more robust and simple method to assess the freshness of coffee is to monitor ratios of headspace concentrations of selected VOCs (Spadone and Liardon, 1990; Arackal and Lehmann, 1979; Marin et al., 2008; Leino et al., 1992; Kallio et al., 1990). In a previous publication, many of the reported VOC ratios were revisited and the ones that are most robust and suited to assessing the freshness of high quality specialty coffee were selected (Gloss et al., 2014). These were termed "freshness-indices" as they focus on fast changes (in contrast to aging or staling markers). Furthermore, we were aiming to find freshness indices for which the chemistry leading to the observed changes was well understood. Among the many potential ratios of coffee aroma compounds, the freshness index dimethyl disulfide/methanethiol (DMDS/MeSH) was identified and shown to satisfy these criteria and therefore be particularly

TABLE 14.2 Properties of the Two Compounds Used for the Freshness Index Discussed Here

Compound	Odor	Volatility	Reactivity
Dimethyl disulfide (DMDS)	Sulfur, onion, garlic, burnt rubber	Medium	Medium (ox.)
Methanethiol (MeSH)	Sulfur, rotten egg, fish, cabbage, garlic, cheesy	High	Very high (ox.)

suitable (Gloss et al., 2014). Some of the properties of these two compounds are summarized in Table 14.2.

Methanethiol is known to be a highly volatile as well as reactive compound (Grosch, 2001; Steinhart and Holscher, 1991; Sanz et al., 2001), examples of which are oxidation and dimerization to dimethyl disulfide (Belitz et al., 2004; Chin and Lindsay, 1994b). In contrast, dimethyl disulfide has both relatively lower reactivity and volatility. Consequently, the overall evolution of this freshness index is mainly driven by the high reactivity and volatility of methanethiol. A further reaction of dimethyl disulfide to dimethyl trisulfide was not observed in our analyses.

In the following sections we introduce the experimental approach to measuring the freshness in more detail, based on the freshness index DMDS/MeSH, and apply this to three specific applications. (1) The first is the storage of roasted whole beans in packaging made of plastic composite film with a thick aluminum layer, equipped with a CO_2 release valve, and stored at 22°C and 50°C. (2) The second application is analogous to the first, yet with the distinction that the packaging was not equipped with a valve. Since the packaging was fully air-tight, we could introduce variable oxygen contents into the packaging and examine the loss of coffee freshness (i.e., the evolution of the DMDS/MeSH freshness index) as a function of the oxygen content. (3) Finally, the third application deals with the loss of freshness of roasted and ground coffee in single serve capsules.

3.2 The Experiment

The compounds dimethyl disulfide and methanethiol were analyzed by gas chromatography coupled to mass spectrometry (GC–MS). Data analysis and identification of the compounds were performed using MSD Chemstation software (Version G1701 EA E.02.00.493, Agilent Technologies, Switzerland) and an NIST08 spectrum database. Chemical identification was made by comparing the mass-spectra to the database, using the most intensive fragment ion for quantification.

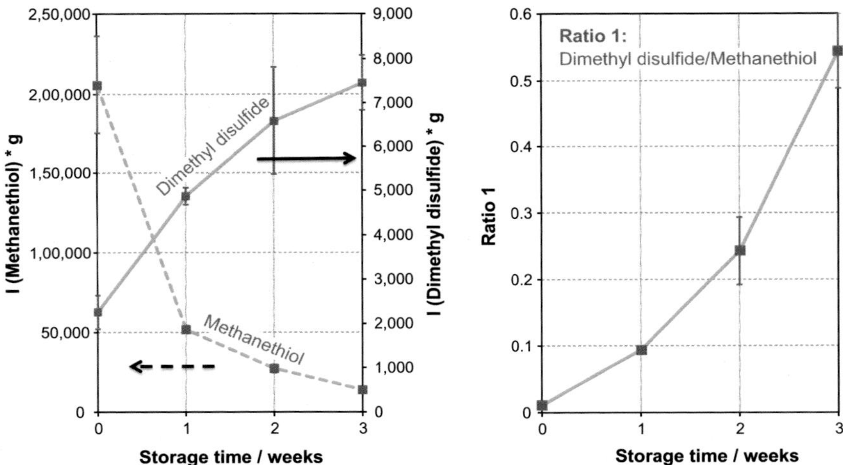

FIGURE 14.2 Demonstration of the freshness index dimethyl disulfide/methanethiol for whole roasted beans stored at room temperature (22°C) in tight packaging with an aluminum barrier (not equipped with a CO_2 degassing valve).

Fig. 14.2 demonstrates the evolution of the GC–MS intensities in single ion mode for dimethyl disulfide and methanethiol (left frame) and the corresponding freshness ratio DMDS/MeSH. One kilogram batches of washed Ethiopian Limu, Grade 2, Arabica were roasted in a Probatino to a roast degree 93 Pt (Colorette), corresponding to a medium roast degree. Sixty-five grams roasted whole beans were packaged immediately after cooling in plastic composite film with a thick aluminum layer. The bag did not contain a degassing valve and was stored at 22°C (room temperature) under inert atmosphere for 3 weeks. At time zero and during each subsequent week, the DMDS/MeSH ratio was determined for five bags. The results are plotted as an average with 95% confidence intervals.

The results show a quick drop in the MeSH signal intensity. After only 1 week the intensity had already dropped to 25% of its initial value, and to 10% of the initial value after just 3 weeks. We can also see an increase in the DMDS signal by more than a factor of three. The right frame shows the corresponding freshness index DMDS/MeSH for the same storage period, where there is a pronounced increase during the 3 weeks of storage.

Before applying this ratio to various storage conditions, it is important to outline the underlying chemical processes that lead to the DMDS and MeSH changes that were observed during storage. Fig. 14.3 summarizes the major steps involved in the formation and degradation of DMDS and MeSH.

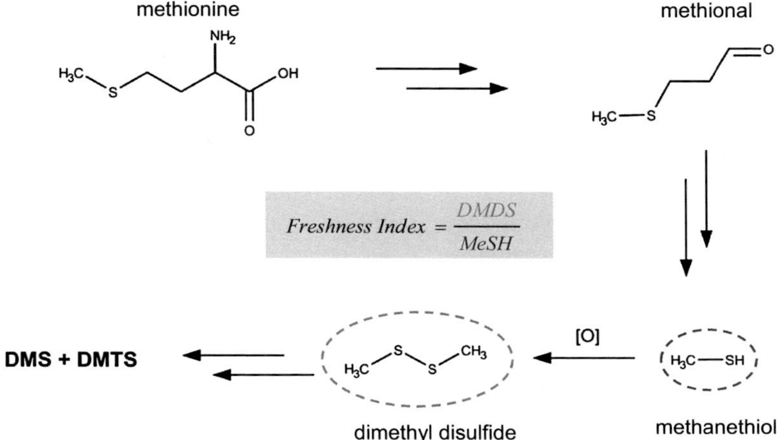

FIGURE 14.3 Reaction scheme underpinning the dimethyl disulfide/methanethiol freshness index.

Methanethiol, also referred to as methyl mercaptan, is a known degradation product of methinonine or its Strecker degradation product methional. It has an unpleasant odor and a low threshold value of approximately 1 ppb (Devos et al., 1990). As a strong nucleophile, it can easily be oxidized to dimethyl disulfide, which has a sulfury odor. Such reactions may take place under mild conditions in the presence of oxygen and transition metals, as shown for 2-furfurylthiol (FFT), one of the character impact compounds of coffee. The dimerization of MeSH is due to oxidative instability, which can be accelerated in the presence of radicals (Blank et al., 2002).

It should be noted, however, that the reaction products of thiol degradation depend on moisture content as well. This means freshness in roast coffee, as discussed throughout this chapter, cannot be simply extrapolated to freshness of a coffee brew. In addition, reactions, initiated by Fenton chemistry, take place in the brew, leading to many more degradation products. The coffee-like smelling compound 2-furfurylthiol decomposes rapidly in the presence of hydroperoxide radicals and transition metals such as of ferrous iron (Blank et al., 2002). In a similar reaction system with methanethiol, methanesulfenic acid (CH_3SOH) has been proposed as an intermediate product during the formation of DMDS and dimethyl trisulfide (DMTS) (Chin and Lindsay, 1994a). However, its existence could not be substantiated, possibly because of the high reactivity of sulfenic acids (Penn et al., 1978), which are known to easily convert to thiosulfinate esters due to their dual electrophilic/nucleophilic characteristics (Block and O'Connor, 1974). Alternatively, thiols may react with phenolic compounds, as shown when FFT is trapped by oxidative coupling to hydroxyhydroquinone in coffee brews (Müller and Hofmann, 2007).

4. APPLICATIONS

4.1 Application 1: Whole Beans in Packaging With Valve

The first application of the freshness index examines the most common method of storage for coffee. Roasted whole beans are stored in packaging composed of a plastic composite film with a thick aluminum layer and equipped with a CO_2 release valve. In this example, a medium roasted washed Arabica was used. The coffee was stored at 22 and 50°C.

Fig. 14.4 shows the evolution of the coffee during storage, seen in the changes in the DMDS/MeSH freshness index. The two left frames represent the evolutions of the two compounds methanethiol and dimethyl disulfide (bottom) and the corresponding DMDS/MeSH ratio at 22°C (top), over a 3-week storage period. The two frames on the right correspond to storage at 50°C, over 4 weeks. Two full replicates of the experiment were conducted, i.e., coffee was roasted twice and the storage experiments conducted separately with the two separate roast batches.

As expected, we can see an increase of the DMDS/MeSH ratio with storage time, irrespective of the temperature. The increase in the freshness index at 50°C is approximately one order of magnitude higher than at 22°C, indicating an accelerated loss of freshness at elevated temperature. The bottom two frames assist in the interpretation of the ratios by also showing the related evolution of the two individual compounds DMDS and MeSH. At 22°C we can see a decrease in the MeSH content, whereas the DMDS content appears to be essentially constant. We have tentatively interpreted this essential constant value for DMDS as a steady-state concentration. Although DMDS is being formed during storage at 22°C, it is further reacting to compounds such as DMTS and dimethyl sulfide (DMS). The result is an apparent constant DMDS content over the 3-week storage period.

At 50°C in contrast we can see a strong decrease in the DMDS during the first week, before its content is stabilized. We have tentatively interpreted this initial increase in DMDS to the higher rate of formation at elevated temperatures. Since DMDS is a relatively stable compound, it does initially accumulate. Once the formation of new DMDS starts to decrease (due to a decrease in its precursor, MeSH), the follow-up reactions of the DMDS to, for example, DMTS and DMS start to establish a steady-state situation, i.e., the formation of DMDS from oxidizing MeSH equals its degradation (further reaction). In comparison to the situation at 22°C, the steady-state content of DMDS is higher at 50°C. After 3 weeks, and once the pools of the MeSH precursors have started to decrease, we observe a decrease in the DMDS content. To elucidate these hypothetical processes, the reaction rate constants and the temperature dependence of the reactions involved need to be experimentally determined, and the underlying processes modeled.

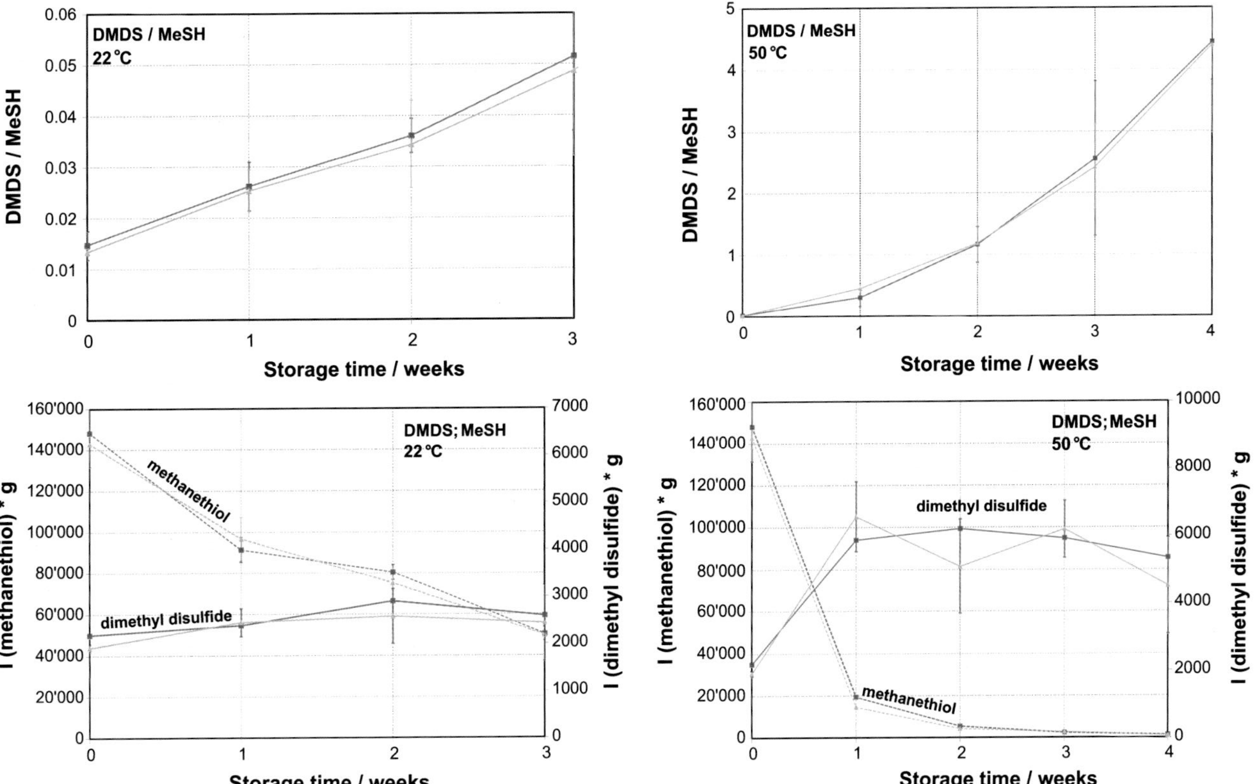

FIGURE 14.4 Loss of freshness of whole beans, stored in bags with a valve. After each week, five samples were analyzed by gas chromatography–mass spectrometry (five packs); each value is given as the mean value with a 95% confidence interval.

4.2 Application 2: Whole Beans in Packaging Without Valve

The second application is analogous to the first, with the exception that the packaging was not equipped with a CO_2 release valve and the coffees were heat sealed with variable oxygen contents in the pack. Storage was limited to only 22°C.

Fig. 14.5 shows the evolution of the DMDS/MeSH freshness index over a period of 3 weeks for three independent experiments, with each charge corresponding to a different roasting batch. Although the three charges show identical DMDS/MeSH freshness indices after 1 week storage, they start to show distinctively different evolutions of the freshness ratios over the following 2 weeks. The higher the oxygen content inside the packaging, the faster the increase during weeks 2 and 3.

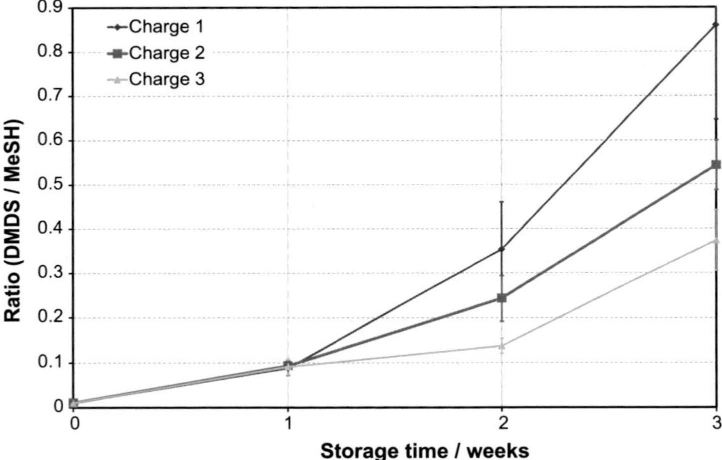

FIGURE 14.5 Loss of freshness of whole beans, stored at different oxygen contents in bags without valves.

4.3 Application 3: Single Serve Capsules

Finally, the third application deals with the evolution of the freshness index for commercial single serve capsule systems. Single serve coffee capsules available on the Swiss market from four different leading commercial brands, labeled C1 to C4 (see Table 14.3), were analyzed (Gloss et al., 2014). The capsules were stored at room temperature for up to 46 weeks. The evolution of the freshness was monitored via the DMDS/MeSH freshness index and is shown in Fig. 14.6.

TABLE 14.3 Description of the Four Different Capsule Systems, Labeled C1 to C4

Capsule	Packaging Materials
C1	• **Body:** PP/EVOH/PP • **Cover:** PP/EVOH/PP; thickness: 0.1 mm • Barrier properties integrated into capsule and cover
C2	• **Body:** PP/EVOH/PP • **Cover:** PP/EVOH/PP; thickness: 0.12 mm • Barrier properties integrated into capsule and cover • Extraction system (perforation points and aluminum foil) and outlet for extract are integrated in cover-material
C3	• **Body:** PP (injection molding without barrier-properties) • **Cover:** Paper with aluminum coating; thickness: 0.03–0.05 mm • **Secondary packaging:** Aluminum; each capsule is individually packed; barrier properties integrated into secondary packaging
C4	• **Body:** 99% aluminum, with thin coating of food-grade shellac • **Cover:** Aluminum foil; thickness 0.03–0.05 mm • Barrier properties integrated into capsule and cover

The second column describes the packaging materials for the cover and the body of the capsule. C3 has in addition a secondary packaging. *EVOH*, ethylene vinyl alcohol; *PP*, polypropylene.

The three main findings are given below:

First, and most importantly, there is an obvious impact of the packaging material. The two capsules C1 and C2, neither of which have any aluminum layer, show the strongest increases in the freshness indices, indicative of a reduction in methanethiol and loss of freshness over time. In contrast, the C4 capsule with a 100% aluminum body and aluminum cover shows hardly any evolution in the freshness index over the 46 weeks of storage. We concluded, therefore, that C4 preserves the freshness of coffee much more efficiently. C3 takes up an intermediate position, with respect to the evolution in the freshness indices. This is in line with the fact that C3 has a PP body (no aluminum) and a cover with only a thin aluminum layer. In addition, it is wrapped into a secondary aluminum packaging that significantly increases the actual headspace and, therefore, the absolute amount of residual oxygen after packaging. Clearly, the absence of aluminum has a strong impact on the reduction in freshness indices for C1 and C2.

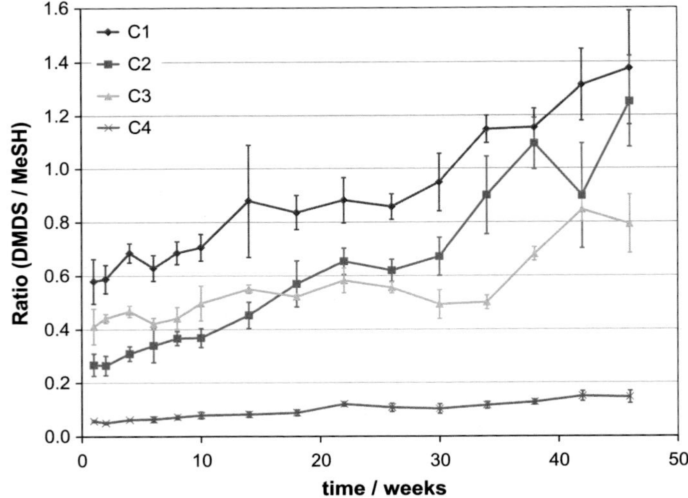

FIGURE 14.6 The evolution of the dimethyl disulfide/methanethiol (DMDS/MeSH) freshness index is plotted as a function of storage time for four different commercial single serve coffee capsules, stored over a period of up to 46 weeks. The error bars correspond to the respective standard deviations of the fivefold measurements.

Second, the starting value of the freshness indices varies between the four different capsule systems. Although C1 already started with a high value, C4 had the lowest freshness index. It is speculated that this is an indication of a certain loss of aroma and freshness from processing prior to packaging into capsules.

Third, the consistency in the capsules appeared to differ greatly. Each capsule was measured in five repetitions and the data are plotted as mean values with a ±68% confidence interval. The standard deviation revealed an unexpected and interesting insight into the consistency of the coffee in the various capsule systems. C4 showed the smallest confidence interval, which is an expression of low capsule-to-capsule variability. In contrast, C1 and C2 show much greater variability between capsules. Particularly for single serve capsules, where a range of capsules are offered by each brand, consistency is an important quality criterion.

Fig. 14.6 gives a comparison of different capsule systems. In such a presentation the axis of the freshness ratio has been chosen such that it fits all four systems. In Fig. 14.7, we show the evolution of the DMDS/MeSH index for one specific aluminum capsule (Nespresso) over 52 weeks. Although the range covered by the freshness index, over the year of storage is approx. 0.02–0.09 (in Fig. 14.6 the Nespresso C4 capsule covers the range 0.05–0.15), it still demonstrates a loss of freshness, albeit over a much smaller range than the other capsule systems.

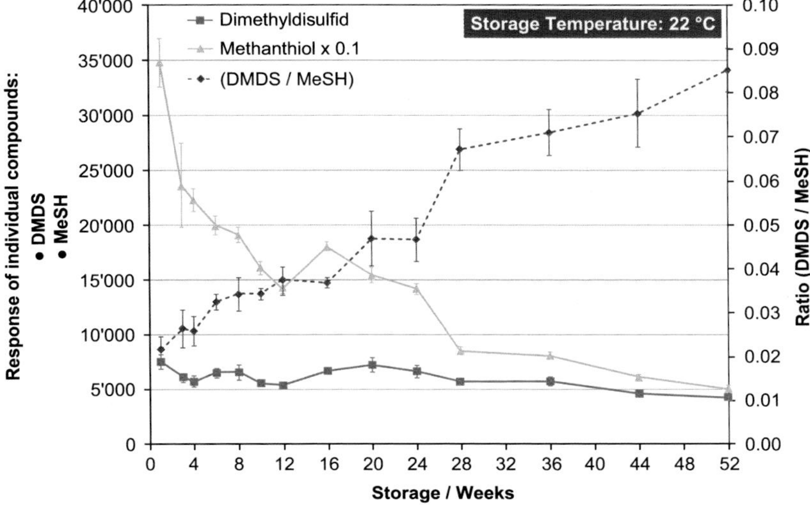

FIGURE 14.7 Evolution of the freshness index for a single serve coffee capsule from Nespresso.

In conclusion, the three examples discussed in Section 4 demonstrate the potential and sensitivity of the freshness index DMDS/MeSH for the monitoring of loss of freshness in whole beans and ground coffee. The dimerization of MeSH is due to oxidative instability and DMDS is a suitable molecular marker for freshness; however, the correlation with the overall coffee aroma needs yet to be established.

Besides the DMDS/MeSH index, there are a series of other indices that have been reported in the literature. However, most often the chemical processes underlying the evolution of these other ratios are not well understood—in contrast to the DMDS/MeSH index. We have chosen to exclusively include the DMDS/MeSH ratio, because it is very sensitive, an excellent early marker of loss of freshness and hence particularly well suited to high-quality coffee applications.

5. ENSURING COFFEE FRESHNESS USING OPTIMAL PACKAGING MATERIALS

As mentioned in the previous sections of this chapter, the freshness of R&G coffee is significantly altered by oxidation. Therefore, limiting the access of oxygen to the product is of key importance for guaranteeing freshness and quality of specialty coffee—an area in which the packaging has a key role to play.

The continuous objective of minimizing the environmental impact on food products means that selecting optimal packaging to meet the protection needs of the product is increasingly important. This process is essential to reduce instances of under- and overpackaging, which can lead to either premature product loss or unnecessary, overengineered packaging.

To optimize the packaging used for a given product, material properties that correspond to the protection requirements of that product must be specified. Therefore, the protection requirements of the product need to be understood and, in the case of R&G coffee, a good understanding of the rate at which it consumes oxygen together with the effect this oxygen consumption has on quality is key. There are several notable studies that have made progress toward establishing methods for determining the oxygen consumption of coffee and the relationship this consumption has with coffee quality.

In his 1997 thesis, Cardelli-Freire used two primary methods to determine the oxygen consumption rate of R&G coffee. In the first, R&G coffee samples were packaged in hermetic containers containing different initial oxygen concentrations. The oxygen concentration within each container was measured during the course of the study, which allowed the oxygen consumption rate to be determined. Using this and the experimental results obtained by Radtke-Granzer and Piringer (1981), a relationship between the oxygen consumption rate and oxygen partial pressure in the headspace was found. This was a first-order relationship for relatively high oxygen concentrations (around 5%), and a half order relationship for lower oxygen levels (around 0.5%). The second method used by Cardelli involved packing coffee samples in permeable pouches. This approach was based on two main assumptions. The first was that the permeation rate of oxygen through the packaging material is directly proportional to the difference in oxygen partial pressure between the headspace and the outside environment. The second assumption was that there is a relationship between the oxygen consumption rate of a product and the oxygen partial pressure in the headspace.

Using this approach, the amount of oxygen permeating into a package decreases as the oxygen concentration in the headspace increases, along with the increase in the oxygen consumption-rate of the product. An equilibrium oxygen concentration will eventually be reached in the package at the point where both mechanisms compensate for each other. The level of this equilibrium oxygen concentration is directly related to the oxygen permeability of the packaging material. The oxygen permeating through the packaging material can therefore be used to calculate the oxygen consumption rate of the packaged product.

Cardelli calculated an equilibrium oxygen partial pressure of approximately 100 mbar at 22°C, when 20 g of roast and ground coffee were packaged in an high-density polyethylene/low-density polyethylene laminate pouch with a surface of 140 cm^2, an oxygen permeability of 380 cc/m^2 day atm and a headspace volume of 34 mL. This enabled an oxygen consumption rate of 3.7×10^{-7} g_{O_2}/g_{coffee} day mbar to be defined for ground coffee.

Determining the oxygen consumption rate of a product provides one indication of how a product evolves over time, however, this value has to be put into perspective with the loss in quality of the product to be able to determine the protection requirements that guarantee its freshness over its complete shelf life.

Cardelli and Labuza (2001) evaluated the sensory profile of roasted and ground coffee stored under different conditions. They showed that R&G coffee was found to be of unacceptable quality after a total O_2 consumption of between 150 and 300 μg/g, depending on the water activity of the product.

More recently, Wyser et al. (to be published) have determined oxygen consumption rates of R&G coffee stored under different conditions to enable the optimal packaging material requirements to be specified for a portioned coffee system. In this piece of work Wyser et al. used optical oxygen sensors (Presens GmbH) to continuously monitor the oxygen partial pressure in hermetic glass containers containing R&G coffee. Several samples were monitored continuously after having been prepared with different initial headspace oxygen levels. A first-order dependence assumption and experimental data obtained at high oxygen partial pressure were then used to define a model to predict headspace partial pressure as a function of time.

Fig. 14.8 shows how experimental data corresponds with that of the model data at the same oxygen level and demonstrates that the first-order dependence

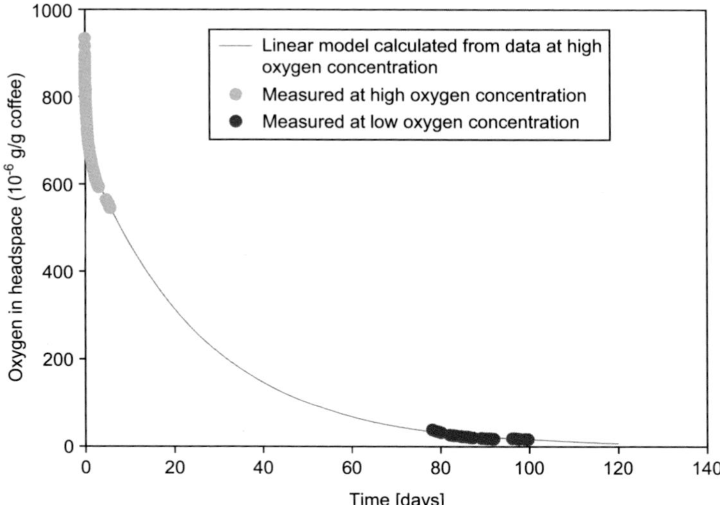

FIGURE 14.8 Headspace oxygen concentration as a function of time in a glass container for roast and ground coffee stored in the dark at 23°C. Experimental results at high (*cream dots*) and low (*brown dots*) initial oxygen concentrations are shown to correspond and thus validate a model (*continuous line*) based on a first-order dependence assumption.

assumption is valid. Based on these data, an oxygen consumption rate of $2.13 \times 10^{-7}\ g_{O_2}/g_{coffee} \cdot day \cdot mbar$ was determined for the coffee tested. Although the value differs from that found by Cardelli, both are of the same order.

As a next step, Wyser et al. used the relationship between oxygen partial pressure, oxygen permeation, and the oxygen consumption rate to calculate the total oxygen consumed by R&G coffee as a function of packaging permeability and initial oxygen concentration. Fig. 14.9 shows this relationship when applied to a typical portioned coffee package containing 5.3 g of coffee with a headspace of 10.3 mL after being stored for 1 year. Having established this relationship, the maximum allowable oxygen uptake for coffee can be used to specify packaging barrier properties required to achieve the desired shelf life of the product, thus ensuring quality and maintaining freshness, while avoiding overpackaging.

It should be noted that the results obtained by Cardelli were for mainstream coffee. It is expected that the critical oxygen uptake value for specialty coffee, where freshness is the key attribute, would be significantly lower, meaning the requirements in terms of permeability and initial oxygen level are even more demanding.

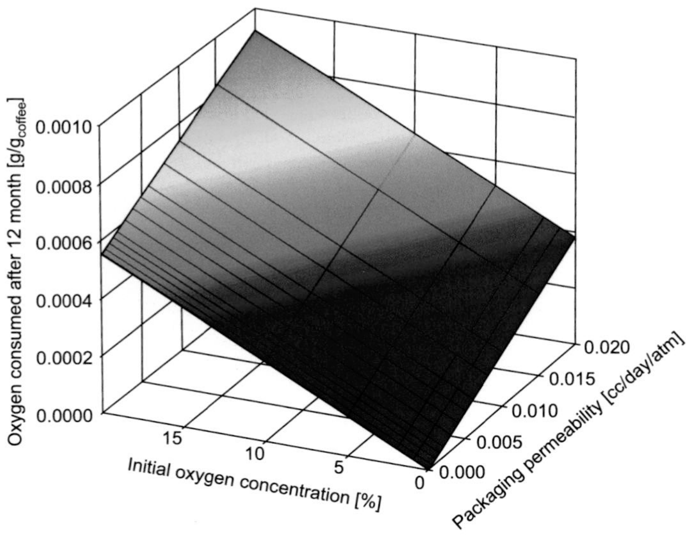

FIGURE 14.9 Predicted oxygen consumed of roast and ground coffee as a function of initial oxygen concentration and package permeability after 1 year of storage.

6. OUTLOOK

Although the concept of freshness is central to the high quality coffee business in general and the specialty coffee movement in particular, a rational description and method for quantitative measurement have long remained elusive. The main motivation for this chapter was to put freshness on a more rational and quantitative basis and to make it more tangible.

Two approaches for measuring freshness of roasted coffee were outlined:

The first refers to changes in the profile of volatile organic aroma compounds of roasted coffee over time and is expressed as "freshness indices"; the ratio of two specific aroma compounds. One particular freshness index was discussed (DMDS/MeSH). Methanethiol is a well-known compound that is found in freshly roasted coffee that decreases within days of roasting. The rate of methanethiol decrease is strongly dependent on the damage caused by contact with oxygen and storage temperature. We demonstrated that packaging, whose role is to protect the product, can have a major impact on the evolution of the freshness index. However, there is no "best" packaging that meets all needs equally. Packaging should be adapted to the expected time between roasting and consumption. For coffee that will be consumed within 1 or 2 weeks, packaging with high barrier properties, which in many cases would have a higher environmental impact, might not be required. However, coffee with extensive distribution channels and a global logistics chain between roasting, the supermarket and the consumer's home requires packaging materials with much higher barrier properties that would contribute to minimizing coffee wastage by guaranteeing the quality of the coffee over its entire shelf life. It should be noted that it is well known that, in most cases, the R&G coffee has a higher overall environmental impact than its packaging.

The second approach is based on weight loss during storage, which is linked to the degassing of roasted coffee (mainly loss of CO_2). Up to 2% of the weight of freshly roasted coffee is made up of entrapped gases, which are released over time. Measuring the rate of weight loss indicates how much gas was released during processing (e.g., grinding, heating) and how much was released after it was freshly roasted.

Preserving the freshness of roasted coffee remains a central dogma of high-quality espresso coffee. The current status of research indicates that the following points should be taken into consideration to preserve freshness during the desired shelf life of coffee:

1. Loss of freshness starts the minute roasting is complete. Hence, exposure to oxygen and humidity during the time between roasting and packaging must be avoided as much as possible.
2. The barrier properties of the packaging, together with the level of residual oxygen after packaging represent the second most important factor to be

considered and adaptations need to be made to achieve the desired shelf life.

3. Finally, the storage temperature further affects the evolution of the freshness index and the degassing kinetics.

Understanding is the basis for creativity. Once we know how to assess and measure freshness we can also start to think beyond freshness. We wish to facilitate and promote a more rational and fact-based discussion on the concept of freshness. To further advance this concept, studies where the analytical data are correlated to the sensory aspects in the cup are needed. At the same time, this will open up the possibility to better assess new and creative concepts of coffee preparation for one of the most aromatic products and to discover new "sweet spots" and novel sensory experiences. What about cold brew, nitro coffee, liquid coffee—can these be considered part of the specialty coffee movement? Questions like, (1) how long coffee should degas prior to grinding and extraction and (2) whether coffee should be extracted immediately after grinding or left to rest for some time, can be discussed in terms of facts and numbers, as well as sensory aspects. Indeed, a better understanding of freshness will ultimately provide better assessment procedures in the world of high quality coffee and open emerging and new avenues to explore methods of coffee preparation and extraction methods.

ACKNOWLEDGMENT

We would like to acknowledge the SCAE for financial support. We also thank Alexia Glöss, Marco Wellinger, Samo Smrke, and Barbara Edelmann for the fruitful discussions that they contributed to.

REFERENCES

Anese, M., Manzocco, L., Nicoli, M.C., 2006. Modeling the secondary shelf life of ground roasted coffee. Journal of Agricultural and Food Chemistry 54 (15), 5571–5576.

Arackal, T., Lehmann, G., 1979. Messung des Quotienten 2-Methylfuran/2-Butanon von ungemahlenem Röstkaffee während der Lagerung unter Luftausschluss. Chemie, Mikrobiologie, Technologie der Lebensmittel 6, 43–47.

Baggenstoss, J., Poisson, L., Kaegi, R., Perren, R., Escher, F., 2008. Coffee roasting and aroma formation: application of different time-temperature conditions. Journal of Agricultural and Food Chemistry 56 (14), 5836–5846.

Belitz, H.D., Grosch, W., Schieberle, P., 2004. In: Food Chemistry, revised ed., vol. 3. V. Springer, Berlin, Heidelberg, New York.

Blank, I., Pascual, E.C., Devaud, S., Fay, L.B., Stadler, R.H., Yeretzian, C., Goodman, B.A., 2002. Degradation of the coffee flavor compound furfuryl mercaptan in model Fenton-type reaction systems. Journal of Agricultural and Food Chemistry 50 (8), 2356–2364.

Blank, I., Sen, A., Grosch, W., 1991. Aroma impact compounds of arabica and robusta coffees. Qualitative and quantitative investigations. In: ASIC-14eme Colloque Scientifique International sur le Café, 1992, Paris, vol. 14, pp. 117–129.

Blank, I., Sen, A., Grosch, W., 1992. Potent odorants of the roasted powder and brew of arabica coffee. Zeitschrift für Lebensmittel-Untersuchung und Forschung 195, 239–245.

Block, E., O'Connor, J., 1974. Chemistry of alkyl thiosulfinate esters. VII. Mechanistic studies and synthetic applications. Journal of the American Chemical Society 96 (12), 3929–3944.

Buchner, N., Heiss, R., 1959. Die Gaslagerung von Bohnenkaffee. Verpackungs-Rundschau 10, 73–80.

Cardelli-Freire, C., 1997. Kinetics of the Shelf Life of Roasted and Ground Coffee as a Function of Oxygen Concentration, Water Activity and Temperature. University of Minnesota.

Cardelli, C., Labuza, T.P., 2001. Application of Weibull Hazard analysis to the determination of the shelf life of roasted and ground Coffee. LWT – Food Science and Technology 34, 273–278.

Chin, H.-W., Lindsay, R.C., 1994a. Ascorbate and transition-metal mediation of methanethiol oxidation to dimethyl disulfide and dimethyl trisulfide. Food Chemistry 49 (4), 387–392.

Chin, H.W., Lindsay, R.C., 1994b. Mechanisms of formation of volatile sulfur-compounds following the action of cysteine sulfoxide lyases. Journal of Agricultural and Food Chemistry 42 (7), 1529–1536.

Devos, M., Patte, F., Rouault, J., Lafort, P., Van Gemert, L.J., 1990. Standardized Human Olfactory Thresholds. IRL Press, Oxford, p. 101.

Gloss, A.N., Schönbächler, B., Rast, M., Deuber, L., Yeretzian, C., 2014. Freshness indices of roasted coffee: monitoring the loss of freshness for single serve capsules and roasted whole beans in different packaging. Chimia 68 (3), 179–182.

Grosch, W., 1998. Flavour of coffee. A review. Nahrung 42 (6), 344–350.

Grosch, W., 2001. Chemistry III: volatile compounds. In: Clarke, R.J., Vitzthum, O.G. (Eds.), Coffee: Recent Developments. Blackwell Science, London, pp. 68–89.

Grosch, W., Czerny, M., Wagner, R., Mayer, F., 1996. Studies on the aroma of roasted coffee. In: Paper Read at Weurman Symposium, 1996, at Reading, UK.

Grosch, W., Semmelroch, P., Masanetz, C., 1993. Quantification of potent odorants in coffee. In: Paper read at ASIC-15eme Colloque Scientifique International sur le Café, 1993, at Montpellier.

Holscher, W., Steinhart, H., 1992. Investigation of roasted coffee freshness with an improved headspace technique. Zeitschrift fuer Lebensmittel -Untersuchung und -Forschung 195, 33–38.

Illy, A., Viani, R., 2005. Espresso Coffee: The Science of Quality, vol. 2. Elsevier, Amsterdam.

Kallio, H., Leino, M., Koullias, K., Kallio, S., Kaitaranta, J., 1990. Headspace of roasted ground coffee as an indicator of storage time. Food Chemistry 36 (2), 135–148.

Kallio, H., Leino, M., Salorine, L., 1989. Analysis of the headspace of foodstuffs near room temperature. Journal of High Resolution Chromatography & Chromatography Communications 174–177.

Leino, M., Kaitaranta, J., Kallio, H., 1992. Comparison of changes in headspace volatiles of some coffee blends during storage. Food Chemistry 43, 35.

Lindinger, C., de Vos, R.C.H., Lambot, C., Pollien, P., Rytz, A., Voirol-Baliguet, E., Fumeaux, R., Robert, F., Yeretzian, C., Blank, I., 2010. Coffee chemometrics as a new concept: untargeted metabolic profiling of coffee. In: Blank, I., Wüst, M., Yeretzian, C. (Eds.), Expression of Multidisciplinary Flavour Science – 12th Weurman Symposium. Wintherthur: Zurich University of Applied Science, pp. 581–584.

Lindinger, C., Labbe, D., Pollien, P., Rytz, A., Juillerat, M.A., Yeretzian, C., Blank, I., 2008. When machine tastes coffee: instrumental approach to predict the sensory profile of espresso coffee. Analytical Chemistry 80 (5), 1574–1581.

Marin, K., Pozrl, T., Zlatic, E., Plestenjak, A., 2008. A new aroma index to determine the aroma quality of roasted and ground coffee during storage. Food Technology and Biotechnology 46 (4), 442–447.

Mayer, F., Grosch, W., 2001. Aroma simulation on the basis of the odourant composition of roasted coffee headspace. Flavour and Fragrance Journal 16 (3), 180–190.

Merritt, M.C., Cawley, B.A., Lockhart, E.E., Proctor, B.E., Tucker, C.L., 1957. Storage properties of vacuum-packed coffee. Food Technology 11, 586–588.

Muller, C., Hofmann, T., 2007. Quantitative studies on the formation of phenol/2-furfurylthiol conjugates in coffee beverages toward the understanding of the molecular mechanisms of coffee aroma staling. Journal of Agricultural and Food Chemistry 55 (10), 4095–4102.

Munro, L.J., Curioni, A., Andreoni, W., Yeretzian, C., Watzke, H., 2003. The elusiveness of coffee aroma: new insights from a non-empirical approach. Journal of Agricultural and Food Chemistry 51 (10), 3092–3096.

Nicoli, M.C., Calligaris, S., Manzocco, L., 2009. Shelf-life testing of coffee and related products: uncertainties, pitfalls, and perspectives. Food Engineering Reviews 1 (2), 159–168.

Nicoli, M.C., Innocente, N., Pittia, P., Lerici, C.R., 1993. Staling of roasted coffee—volatile release and oxidation reactions during storage. In: 15th International Scientific Colloquium on Coffee, vols. 1 and 2(15), pp. 557–566.

Parliment, T.H., Kolor, M.G., Rizzo, D.J., 1982. Volatile components of Limburger cheese. Journal of Agricultural and Food Chemistry 30 (6), 1006–1008.

Péneau, S., 2005. Freshness of Fruits and Vegetables: Concept and Perception, Institute of Food Science and Nutrition. Swiss Federal Institute of Technology Zurich, Zurich.

Penn, R.E., Block, E., Revelle, L.K., 1978. Flash vacuum pyrolysis studies. 5. Methanesulfenic acid. Journal of the American Chemical Society 100 (11), 3622–3623.

Poisson, L., Schmalzried, F., Davidek, T., Blank, I., Kerler, J., 2009. Study on the role of precursors in coffee flavor formation using in-bean experiments. Journal of Agricultural and Food Chemistry 57 (21), 9923–9931.

Radtke-Granzer, R., Piringer, O.G., 1981. On the issue of the quality assessment of roasted coffee via trace analysis of volatile flavour components. Deutsche Lebensmittel Rundschau 6, 203–210.

Sanz, C., Pascual, L., Zapelena, M.J., Cid, M.C., 2001. A New "Aroma Index" to Determine the Aroma Quality of a Blend of Roasted Coffee Beans, 2001, at Trieste, Italy.

Scheidig, C., Czerny, M., Schieberle, P., 2007. Changes in key odorants of raw coffee beans during storage under defined conditions. Journal of Agricultural and Food Chemistry 55 (14), 5768–5775.

Selmar, D., Bytof, G., Knopp, S.-E., 2008. The storage of green coffee (*Coffea arabica*): decrease of viability and changes of potential aroma precursors. Annals of Botany 101 (1), 31–38.

Semmelroch, P., Laskawy, G., Blank, I., Grosch, W., 1995. Determination of potent odourants in roasted coffee by stable isotope dilution assays. Flavour and Fragrance Journal 10, 1–7.

Shuman, A.C., Elder, L.W., 1943. Staling vs. Rancidity in roasted coffee. Industrial & Engineering Chemistry 35 (7), 778–781.

Spadone, J.C., Liardon, R., 1990. Analytical study of the evolution of coffee aroma compounds during storage. In: Paper read at ASIC-13eme Colloque Scientifique International sur le Caf, 1990, at Paris, vol. 13.

Speer, K., Kölling-Speer, I., 2006. The lipid fraction of the coffee bean. Brazilian Journal of Plant Physiology 18.

Steinhart, H., Holscher, W., 1991. Storage-related changes of low-boiling volatiles in whole coffee beans. In: Paper read at ASIC-14eme Colloque Scientifique International sur le Café, 1992, at Paris.

Sunarharum, W.B., Williams, D.J., Smyth, H.E., 2014. Complexity of coffee flavor: a compositional and sensory perspective. Food Research International 62 (0), 315–325.

Wang, X., 2014. Understanding the Formation of CO_2 and its Degassing Behaviours in Coffee, Food Science. The University of Guelph, Ontario, Canada.

Wang, X., Lim, L.-T., 2014. Effect of roasting conditions on carbon dioxide degassing behavior in coffee. Food Research International 61 (0), 144–151.

Yeretzian, C., Jordan, A., Badoud, R., Lindinger, W., 2002. From the green bean to the cup of coffee: investigating coffee roasting by on-line monitoring of volatiles. European Food Research and Technology 214 (2), 92–104.

Chapter 15

The Brew—Extracting for Excellence

Frédéric Mestdagh[1], Arne Glabasnia[2], Peter Giuliano[3]
[1]*Nestec Ltd., Nestlé Product and Technology Center Beverages, Orbe, Switzerland;* [2]*Nestec Ltd., Nestlé Product and Technology Center Beverages, Lausanne, Switzerland;* [3]*Specialty Coffee Association of America, Santa Ana, CA, United States*

1. INTRODUCTION

The previous chapters have described how to obtain a roasted coffee of high quality through adequate pre- and postharvest treatments, roasting, and grinding. Roasted coffee beans must undergo a final transformation to unlock the flavor developed inside the bean during roasting and grinding. A good cup of coffee is characterized by a subtle equilibrium of aroma, taste, and mouthfeel. The brewing step allows extracting the odorants and taste molecules from the roast and ground coffee into the consumer's cup. Various methods of preparation exist, based on origin, culture, and ultimately consumer preference. For each of these, several parameters play a role in delivering a beverage with a balanced flavor. This chapter aims to combine barista experience with molecular understanding of what happens during extraction to broaden the insight into how the in-cup flavor can be modulated through extraction.

2. A GOOD CUP STARTS WITH PROPERLY ROASTED, TEMPERED, AND STORED BEANS

The green coffee variety, origin, processing, and the desired in-cup flavor profile will determine the roasting conditions and final roast color. To prepare the coffee beverage, the beans should be well protected from oxygen till the moment of brewing, but not too fresh. Indeed, roasted coffee beans need at least 12 h, if not a couple of days, to stabilize after roasting (tempering phase). The flavor inside the beans continues to develop as the beans release gas (CO_2 and volatiles), the roasting chemistry stabilizes, and remaining (quenching) water equally distributes. Overall, the character of the coffee further "matures," being less harsh and developing a more balanced flavor profile.

The Craft and Science of Coffee. http://dx.doi.org/10.1016/B978-0-12-803520-7.00015-3

3. HOW TO DEFINE IN-CUP QUALITY?

The main parameter to assess the quality of a cup of coffee remains the sensory experience. However, sensory perception, and even more consumer preference, remains to a certain extent subjective. In order to have more objective means to characterize a coffee brew, some key parameters can be withdrawn from simple physicochemical measurements.

The strength or concentration of a coffee brew is a first indicator of the efficiency of extraction. It can be measured by drying a volume of coffee beverage and weighing the remaining solids. The so obtained total solids or dry matter substance in relation to the volume indicates the strength of the beverage (solubles concentration). This value needs to be put into the context of the used coffee weight for a given cup volume. It is evident that a strong coffee can be obtained by passing a small amount of water through a high amount of coffee and vice versa. In both cases, the obtained beverages may not correspond to the appreciated profile of a good cup of coffee.

Therefore, the yield is another quality criterion that is even more an indicator of brew quality as it reflects the right ratio of solids and brew volume compared to the initial amount of coffee used. The yield is defined as the mass percentage of roast and ground coffee (solids) dissolved in the brew. A barista seeks for the perfect, the most balanced extraction of the roasted coffee to obtain the optimum flavor profile. Yields between 18% and 22% have been generally considered as a good range for quite a long time, when it comes to brew quality. Brews below 18% are thought of as underextracted or underdeveloped, as sensorially being perceived as too sweet and acidic. Brews above this range are thought to be overextracted, leading to bitter and astringent notes judged as unpleasant (Lingle, 1996). However, optimal extraction yields are sometimes reported to be outside this range for several extraction methods (Lopez-Galilea et al., 2007; Navarini et al., 2009; Parenti et al., 2014), mostly with a shift toward higher yields. This is probably due to a change in consumer preference moving from drip filter profile as the standard coffee 50 years ago toward more and more espresso-type coffees as the standard of high coffee quality for many consumers even outside of Italy. Coffee quality, roast color, and extraction techniques may have an effect on optimal yield range, and good research remains to be done on this question. Therefore, though currently established yield ranges are widely accepted by the industry, they are not considered definitive and delicious coffee may indeed be possible outside these ranges.

Next to strength and yield, the balance between the compounds being extracted in-cup also heavily impacts the quality of the beverage. The barista can play with this balance by adjusting the extraction conditions. To start with, the flavor composition of the coffee beverage is very different compared to the roasted bean. When smelling freshly roast and ground coffee powder, a very intense aroma profile is perceived originating from odorants entrapped inside the roasted coffee beans and being released upon grinding. Fig. 15.1 shows how the perception of coffeeness and roastiness increases between intact roasted beans and roast and ground. At the extraction stage, however, the flavor balance and intensity significantly alter. In Fig. 15.1, not only the intensity differs between above-cup and fresh roast and ground coffee, but also the shape of the spider graphs, representing the change in aroma balance upon extraction (Bhumiratana et al., 2011). The water, coming into contact with the coffee bed, selectively extracts molecules from the coffee particles. Due to their different physicochemical properties, not every molecule is extracted the same way. Each extraction parameter allows to play with the extractability and thus the final flavor balance in-cup, as discussed further.

The reason why both intensity and balance are linked to the quality of a cup of coffee can be found in the human perception physiology. Flavor compounds, either aroma or taste molecules, interact with the receptors in the nose or on the tongue. The response of a compound at a given concentration present in the cup depends on several factors. The perception threshold is an important factor for flavor molecules. It describes the concentration at which the compound is recognized for its aroma or taste sensation. Compounds that are present in the beverage in concentrations above these thresholds are

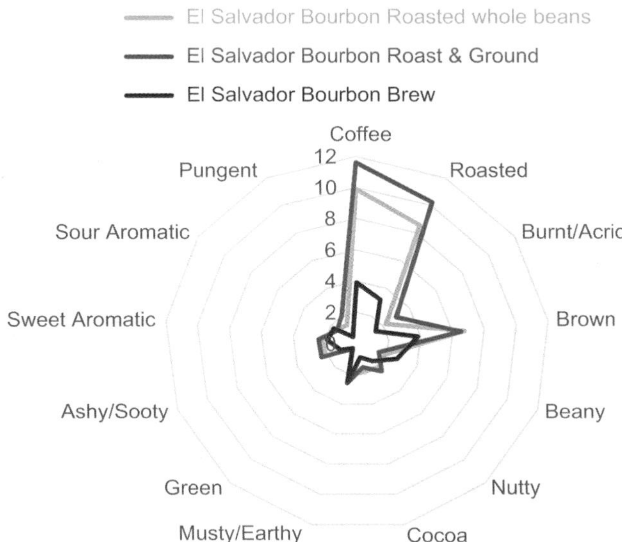

FIGURE 15.1 Change of aromatic sensory attributes between roasted beans, roast and ground coffee and brew. *Reproduced from Bhumiratana et al. (2011).*

considered as key molecules for the coffee flavor. Even though roasted coffee contains more than 1000 volatile odorants, only about 35 are considered as key odorants, responsible for the coffee flavor (Table 15.1) (Blank et al., 1992; Semmelroch and Grosch, 1995). For taste, the situation is very similar. A coffee beverage is of high complexity in view of its nonvolatile composition. Again, only some of these molecules are known to be taste active imparting a certain sensory attribute. Additionally, only a few among those have been shown to really contribute to the taste of a coffee beverage due to their abundancy above the perception thresholds (Table 15.2). A reconstitution of the coffee taste by individual compounds as successfully done for the coffee aroma could not be established yet, most probably due to additive and synergistic effect between the various molecules and interactions that are not fully understood yet.

As mentioned before, a specific balance of all these key flavor molecules will give rise to a characteristic sensory profile in-cup. Lindinger et al. (2008) showed how this analytical flavor balance can indeed be correlated to different sensory directions, such as acidity, bitterness, or roastiness.

Moreover, human perception is not linear, whereas most of the analytical methods are linear. When recording dose—response curves for flavor compounds usually a sigmoid relationship is observed between concentration and sensory response (Fig. 15.2). At concentrations below the threshold, changes in concentration have obviously no impact on the sensory response. Around the range of the detection threshold, sensory response slightly increases first and then turns into a linear concentration—response relationship with a steep slope (impact zone). At concentrations well above the threshold, flavor receptors enter into saturation and sensory response is no longer increasing with ascending concentrations (saturation zone). Although sigmoid relationship is found for all flavor compounds, the threshold concentration and slope can differ significantly between different flavor molecules. The nonlinear relationship is the main reason why the balance of compounds needs to be considered at least as much as the concentration of each molecule on its own.

To better illustrate the complexity of sensory response, we need to imagine a coffee beverage at a given concentration and have a look at three different flavor compounds of a different sensory attributes (Fig. 15.2A). Compound 1 is just above its threshold, compound 2 is in the linear range, and compound 3 is already at the saturation level of the receptor.

If we are changing the extraction conditions toward higher extraction yield (e.g., finer grind size and higher temperature), all three compounds will end up at higher levels in the cup (Fig. 15.2B). Now compound 1 enters the linear range of the response curve with a high impact on the response for this flavor modality. Compound 2 is still increasing linearly and so does the response. In contrast, compound 3, although now present at even higher concentration, is no longer able to induce a higher sensory response due to saturation of the receptor. The sensory profile thus will shift toward the note of compound 1 while that of

TABLE 15.1 A Selection of Key Odorants in Coffee Powder and Brew, With Their Corresponding Quality Descriptors and Sensory Significance (Blank et al., 1992; Semmelroch and Grosch, 1995)

		Sensory Significance (Indicative Values[a])	
Compound	**Aroma Quality**	**Powder**	**Brew**
Acetaldehyde	Fruity, pungent	2	4
Methylpropanal	Fruity, malty	1	2
2-/3-Methylbutanal	Malty	1	2
Phenylacetaldehyde	Honey-like	3	2
(E)-2-nonenal	Fatty	3	1
2,3-Butanedione (diacetyl)	Buttery	1	2
2,3-Pentanedione	Buttery	2	2
Methanethiol	Sulfury, cabbage, gassy	1	1
3-Methyl-2-buten-1-thiol	Amine-like	2	1
2-Methyl-3-furanthiol	Meaty, boiled	4	1
2-Furfurylthiol	Roasty (coffee-like)	5	3
Dimethyltrisulfide	Cabbage-like	1	1
3-Mercapto-3-methylbutylformate	Catty, roasty	8	5
3-Mercapto-3-methyl-1-butanol	Meaty (broth)	2	3
Methional	Boiled potato-like	4	6
2-/3-Methylbutanoic acid	Sweaty	3	3
Trimethylthiazole	Roasty, earthy	1	1
5-Ethyl-2,4-dimethylthiazole	Earthy, roasty	2	1
Trimethylpyrazine	Roasty, earthy	3	2
3-Isopropyl-2-methoxypyrazine	Earthy, roasty	4	2
2-Ethyl-3,5-dimethylpyrazine	Earthy, roasty	8	7
2,3-Diethyl-5-methylpyrazine	Earthy, roasty	6	4
3-Isobutyl-2-methoxypyrazine	Earthy	6	4
2-Hydroxy-3,4-dimethyl-2-cyclopenten-1-one	Caramel-like	3	4

Continued

TABLE 15.1 A Selection of Key Odorants in Coffee Powder and Brew, With Their Corresponding Quality Descriptors and Sensory Significance (Blank et al., 1992; Semmelroch and Grosch, 1995)—cont'd

		Sensory Significance (Indicative Values[a])	
Compound	**Aroma Quality**	**Powder**	**Brew**
4-Hydroxy-2,5-dimethyl-3(2H)-furanone (Furaneol)	Caramel-like	1	5
3-Hydroxy-4,5-dimethyl-2(5H)-furanone (Sotolon)	Seasoning-like	6	8
5-Ethyl-3-hydroxy-4-methyl-2(5H)-furanone	Seasoning-like	6	7
Linalool	Flowery	2	1
(E)-β-damascenone	Honey-like, fruity	8	3
Guaiacol	Phenolic, burnt	2	1
4-Ethylguaiacol	Spicy	5	6
4-Vinylguaiacol	Spicy	6	6
Vanillin	Vanilla-like	2	6

[a]*Indicative values on a scale of 1–8 (1, lowest; 8, highest sensory impact).*

compound 3 is reduced, although the balance of the compounds in view of their concentration has not changed. If compound 1 is a bitter compound this will lead to increased bitterness often associated with overextraction.

In another scenario the balance between the compounds might change. For example, a coarser grind will reduce extraction efficiency. The strength of the cup, defined by the total solid content, can be adapted by higher quantity of coffee. Nevertheless, the balance will change. Highly soluble compounds will be fully extracted even at coarser grind, whereas less soluble compounds will be less extracted. The balance between the compounds will change. In our example, compound 1 is less abundant and present below its detection threshold, whereas compound 2 is now at a higher concentration with a strong sensory impact (Fig. 15.2C). Compound 3 remains at similar levels inducing no change in sensory response. This is a typical example for underextracted coffee. Highly soluble compounds like sugars or acids are proportionally more abundant turning the flavor profile toward sweet/acidic notes.

TABLE 15.2 Known Taste Active Compounds in Coffee and Their Taste Quality and Sensory Significance

Compound Class	Example	Taste Quality	Sensory Significance
Alkaloids	Caffeine	Bitter	+
Chlorogenic acid lactones	3-*O*-caffeoyl-γ-quinide	Bitter	+
Phenylindanes	*cis*-4,5 dihydroxy-1-methyl-3-(3′,4′-dihydroxyphenyl)indane	Bitter	+
Trigonelline derivatives	N-methylpyridinium (NMP)	Bitter	Not reported for coffee yet
Diketopiperazines	Cyclo-Leu–Pro	Bitter	Not reported for coffee yet
Benzenediols	Catechol	Bitter	Not reported for coffee yet
Amino acids	Tryptophan	Bitter	Not reported for coffee yet
Organic acids	Citric acid	Acidic	Not reported for coffee yet
Sugars	Arabinose	Sweet	Not reported for coffee yet
Biogenic amines	γ-Amino butyric acid	Sweet	Not reported for coffee yet
N-phenyl-propionylamides (NPPA)	Caffeoyl-tryptophan	Astringent	Not reported for coffee yet

The compounds described here are fictive compounds for illustration only. The real situation is even more complex also including interaction between compounds and changes in sensory quality with changing concentration found for many aroma compounds. It gives, however, an idea about the interplay of compounds and the possible consequences for the sensory profile to show why coffee brewing is still considered as an art requiring a lot of experience and know-how despite existing machines helping the unexperienced consumer to get a high quality cup of coffee as well.

In summary, strength or intensity of the coffee beverage is as important as the balance of compounds. The strength mainly affects the individual perception of each flavor molecule depending on perception threshold and the dose–response relationship. This defines if and at which intensity a compound

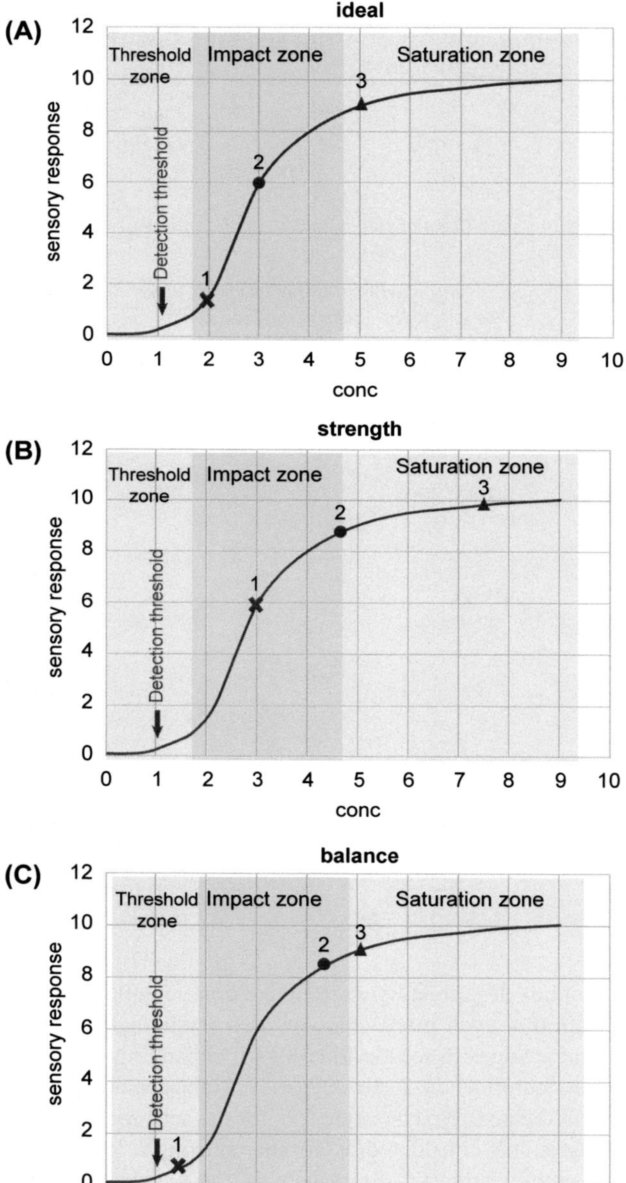

FIGURE 15.2 Dose–response curves of three flavor compounds for ideal (A), increased strength (B), and changed balance (C) profiles.

is perceivable. The balance between all the flavor molecules gives the final overall perception introducing a high complexity of interactions that needs to be mastered to obtain the desired sensory profile. Although developing a balanced cup, the sensory perception of the barista is thus key. To adjust the brewing variables, a good barista might take at each step detailed notes about the resulting flavor results. He may also use a measurement tool, like a refractometer or hydrometer, to quickly assess concentration and yield, which can be used to guide his adjustments quantitatively and serve to make replication of brew conditions in the future easier. However, tasting the coffee remains the key control factor, which makes tasting ability a key skill for the barista.

4. WHICH KEY VARIABLES ARE AVAILABLE TO MODULATE IN-CUP COFFEE FLAVOR?

In practice, a barista may approach a coffee with an understanding of the flavors it may contain: they may understand it to be a coffee with a "classic" flavor profile, like the lemon-jasmine of great Yirgacheffe coffees or the blackberry sweetness of the beautiful coffees of central Kenya. Alternately, they might rely on a coffee roaster's description of the flavors available in the coffee. In any case, the barista seeks to optimize certain flavors sweetness, acidity, aromatics, and body and minimize flavors they find objectionable. Generally, the barista identifies a chosen brewing methodology and creates a brew using familiar parameters as a starting place. They will then taste the coffee critically, asking themselves questions: does the coffee present flavors of over- or underextraction? Would a higher concentration—"stronger" coffee—improve the mouthfeel and balance the acidity better? Are the aromatics well supported by the character of the brew? The answers to these questions may suggest specific adjustments in coffee dose, and also water quantity, water temperature, coffee agitation, grind fineness, and brew time. Table 15.3 gives an overview of some key parameters influencing the final in-

TABLE 15.3 Parameters Influencing Final Cup Quality of Coffee Brews

Water	Coffee	Resulting Variables
Quality	Weight	Water/coffee ratio
Quantity	Particle size and shape	Pressure
Temperature	Particle size distribution	Flow time
	Compaction	Flow rate
	Shape of coffee bed	

cup quality and specificity. The priority of these adjustments varies according to the style and preference of the barista.

4.1 Water Quality

Next to the roasted coffee, water is the second essential ingredient for coffee brewing, as discussed in Chapter 16. Its quality (hardness, acidity, and cation composition) can directly influence the sensorial result of the extract. Crema stability on espresso coffee has also been shown to be reduced by water hardness (Navarini and Rivetti, 2010; Dold et al., 2011). The mechanism can be explained by a change in ion content or by the interaction between the cations and protein/polysaccharide complexes, leading to a destabilization of the foaming mechanism.

4.2 Water Quantity; Extraction Kinetics. What Happens Along Extraction?

The right ratio of water volume to coffee weight is needed to obtain the right yield on the one hand, and the right flavor balance on the other. Stronger coffee (assuming the same extraction yield) can be prepared by increasing the roast and ground/water ratio. This beverage is not more bitter than weaker coffee, but simply has more solids. This is reflected in being darker and in thicker mouthfeel originating from solids and oils. Mouthfeel, however, also depends on suspended solids (very small grinds, so-called "fines"), particularly in French press or espresso as opposed to filter brew.

The organoleptics of a coffee beverage change as more water is allowed to pass through the coffee bed. Indeed, the balance between the molecules in-cup changes continuously during extraction. There are two factors that impact this balance. On the one hand, the solubility of the flavor molecules is the driving force to move from the roasted coffee solids to the liquid brew. The solubility between the compounds is very different depending mainly on their polarity. But also sterical aspects (structure of coffee particles as well as chemical structure of the extracted compounds) play a role that hinder or facilitate the transport from the ground coffee to the brew.

Very soluble compounds such as caffeine, N-methylpyridinium (NMP) (Stadler et al., 2002), sugars, or organic acids are extracted very efficiently in the first seconds of the brew preparation and rapidly reach an extraction yield of >90% (Severini et al., 2015). That is why a lower extraction yield (underextracted coffee) is leading to a cup profile dominated by very water-soluble compounds such as sugars and acids resulting in a sweet–acidic profile. In drip coffee, it is estimated that about 90% of the caffeine is extracted within the first brewing minute.

In contrast, less soluble compounds are only extracted after some time (or volume of water). Among these types of compounds several bitter or astringent tastants are found such as phenylindanes (Frank et al., 2007), chlorogenic acid lactones (Frank et al., 2006), diketopiperazines (DKPs) (Ginz and Engelhardt, 2000; Stark and Hofmann, 2007), or conjugated amides (NPPA) (Stark et al., 2006a,b). In particular phenylindanes are extracted almost continuously over time. Like this, the indanes are proportionally among the most extracted compounds toward longer extraction times. This explains why long cups tend to be both stronger and more bitter. Overextraction favors the extraction of less soluble bitter and astringent compounds. As a consequence, the ratio between early extracted acids or sugars and bitter lactones and indanes will change over time, which imbalances the sensory profile. It will move from a sweet acidic one toward a more bitter-harsh, astringent one. Overextraction can be avoided by stopping the filtration after a planned time and then adding hot water to the brew instead of waiting for all the water to pass through the grounds.

In addition, differences in bitter taste quality have been reported for the known bitter compounds also. Although pleasant coffee-like bitterness attributes are conferred to chlorogenic acid lactones, harsh and lingering bitterness are attributed to the phenylindanes (Blumberg et al., 2010). Thus, not only the bitter intensity is affected but also the quality which will impact consumer acceptance (Fig. 15.3).

As for the nonvolatile tastants, similar phenomena are observed for the (volatile) odorants. Indeed, at the very beginning of espresso extraction, an intense peak of aroma can be perceived above-cup and around the coffee brewing machine (Sánchez-López et al., 2014). Fig. 15.4 shows the extraction kinetics of three coffee odorants from single serve coffee capsules. This extraction behavior could again be linked to the polarity of the extracted molecules as well as chemical interactions with melanoidins (Mestdagh et al., 2014; Sánchez-López et al., 2014, 2016). Highly polar (and also highly volatile) compounds, which easily solubilize in water, are readily extracted at the very beginning of the extraction process. Lower polar odorants (also having a lower volatility) need more time to be extracted from the coffee bed. That is why the share of highly polar compounds gradually decreases during extraction in favor of the less polar compounds. Lee et al. (2011) showed how guaiacol, 4-ethylguaiacol, and 4-vinylguaiacol increased during extraction, which was highly correlated with increasing off-flavors linked to overextraction.

4.3 Water Temperature

Water temperature is one key parameter for the extraction of roasted coffee to obtain a well-balanced brew. The recommended brewing temperature range of (hot) coffee is 91−94°C, which is a compromise to give a good extraction yield and an equilibrated sensory profile (Andueza et al., 2003). This desired temperature range is slightly below the boiling point of water (100 degrees at standard pressure). If temperature is too low, some key compounds will not be extracted efficiently and one will not obtain the desired flavor profile. High

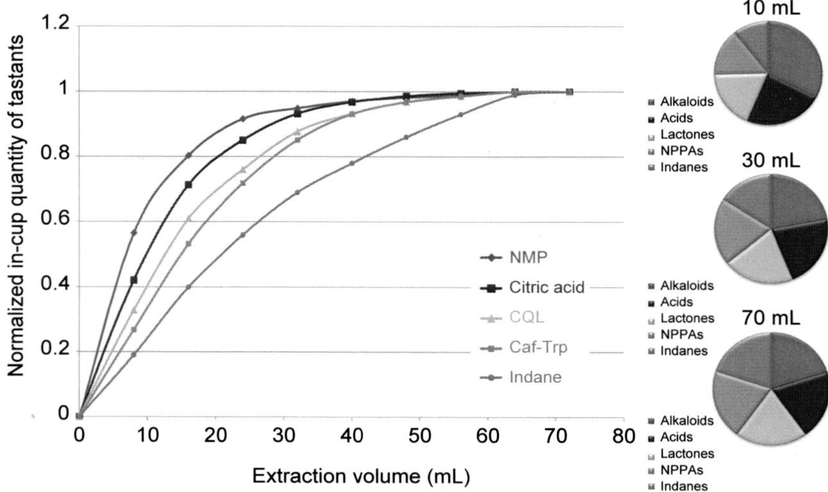

FIGURE 15.3 Extraction kinetics of different bitter taste compounds in espresso (machine preparation). *NMP*, *N*-methylpyridinium; *CQL*, caffeoyl quinic acid lactone; *Caf-Trp*, caffeoyl-tryptophan; *NPPAs*, *N*-Phenylpropenoyl-L-amino acids.

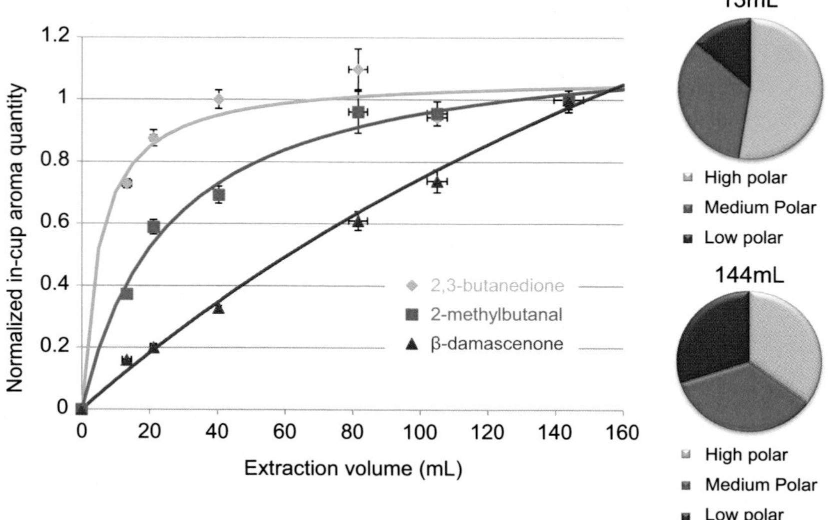

FIGURE 15.4 Extraction kinetics of three coffee odorants, showing the changing aroma balance in-cup.

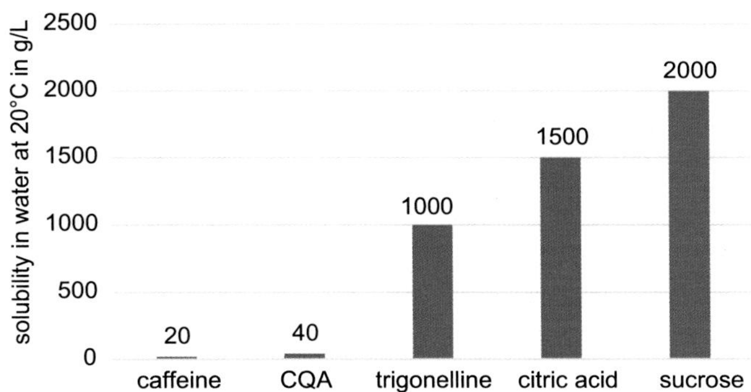

FIGURE 15.5 Solubility of major coffee compounds in water at room temperature. *CQA*, caffeoyl quinic acid. *Merck Index, twelth ed., 1996.*

temperatures will favor the extraction of less polar compounds. As described previously in the chapter, bitter, astringent, phenolic compounds are among those who will profit the most from very high temperatures affecting the flavor balance leading to typical overextracted profiles.

As mentioned before, the extraction is determined by the solubility of the extracted molecules (Fig. 15.5). This water solubility is temperature dependent and generally increases with temperature. However, the relationship between solubility and temperature is not linear, as shown in Fig. 15.6 on two key coffee compounds. The caffeine solubility significantly increases by fourfold between 80 and 100°C. Thus, in that range small temperature changes have a significant effect on the solubility of this compound. In contrast, citric acid has rather a linear increase in solubility with temperature. Consequently, the ratio of solubility between citric acid and caffeine is changing with temperature. Thus temperature changes will affect solubility in a different way and the balance between these two compounds is likely to change with temperature. This also explains why coffee brewed at low water temperatures (e.g., cold brew) often lacks some strength due to lower solubility of taste active compounds leading to less overall solids. This can, however, partially be compensated by a much longer extraction time.

Another effect of the water temperature is more a physical one. At higher temperature the kinetic energy of the water molecules is higher. They, therefore, have a higher mobility increasing the possibility of leaching out compounds from the coffee bed due to higher physical forces. Also the viscosity is temperature dependent and lower at elevated temperatures. Lower viscosity means the water can more easily penetrate the coffee bed (between coffee particles) and access intercellular spaces (even inside coffee particles) to solubilize coffee compounds. This effect explains why, e.g., the amount of lipids in the brew increases with increasing water temperature that is important to impart a certain body and mouthfeel.

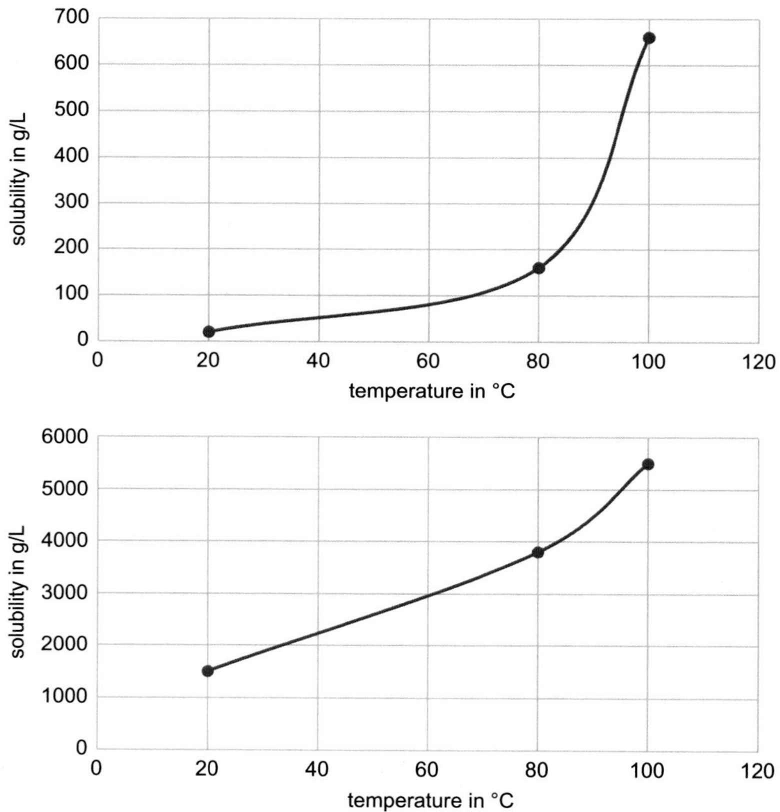

FIGURE 15.6 Solubility of caffeine (top) and citric acid (bottom) in water in relationship to water temperature. *Merck Index, twelth ed., 1996.*

Overall, compounds that are less polar need the higher temperatures to be solubilized from the coffee particles. Most bitter compounds are found among the less polar compounds. This is why overextraction due to too high temperatures is leading to bitter and astringent notes. The polar compounds, which mostly bring sweetness and acidity, are very soluble at room temperature and will not be impacted by higher temperatures in the same way. If coffee is extracted at too low temperatures the acidity and sweetness will remain quite similar whereas some of the aroma and bitterness will be lacking. As a consequence, a typical sweet–acidic profile is obtained for underextracted coffees.

For volatile compounds the situation is a bit more complex. On the one hand, the extraction of the odorants from the coffee particles into the liquid phase will be similar as for the nonvolatiles (Sanchez-Lopez et al., 2016). Yet, once in the liquid phase, the solubility of gases is also temperature dependent but in an opposite way. In line with Henry's Law elevated temperatures force gas molecules into the gaseous phase. Odorants will thus also be released in the air during brewing, and this will be stimulated by higher water temperatures. The aroma release during (and shortly after) extraction of course contributes to the overall aroma perception and experience by the consumer, but decreases the in-cup concentrations of these high volatiles. The solubility in the liquid is thus better at lower temperatures in contrast to the nonvolatiles. Contrarily the above-cup aroma is higher with increasing temperature.

After serving, the coffee begins to cool, and perceived flavors continue to change. Some of this change may be due to continuing chemical development in the coffee beverage. Perception, however, is also influenced by temperature, and tasters will perceive changes in coffee as the coffee cools: often coffee drinkers report an increase in fruity and acid flavors, and a decrease in perceived aromatics as temperatures decrease. Additionally, tasters habituate to the flavors in the coffee, making other flavors seem to appear. Tasting a cooling coffee, therefore, is a dynamic, multidimensional experience.

4.4 Extraction Pressure

One of the driving forces of coffee extraction is pressure, except for immersion brew where solubles merely diffuse from the coffee particles floating freely into the solution. Extraction pressure is not an independent variable, but the result of the equilibrium between the applied force on top of the coffee bed (through the water) and the resistance of the coffee bed against water percolation. Each extraction method has its own driving force and its typical coffee bed properties. As a result, each technique is characterized by its own range of extraction pressures (Table 15.4). To prepare an espresso, the external force is delivered by the pump (allowing to build up higher pressures), whereas for moka pot preparation, somewhat lower pressures are generated by a steam/vapor pressurized chamber. More recently, also centrifugal forces are used in a portioned coffee system to generate extraction pressure over a coffee bed (Nespresso VertuoLine). In a filter brewer, almost no pressure is built up as water flows through the coffee bed by force of gravity.

Next to the applied forces on top of the coffee bed, the coffee bed properties are even more important as these determine the water permeability and the way the water flows through the coffee bed (Corrochano et al., 2015). A too low bed permeability can give rise to (too) high extraction pressures, low overall flow rates, and too long extraction times possibly leading to over-extraction (Corrochano et al., 2015).

Grinding is one of the main parameters influencing the coffee bed permeability, so it is not surprising that it is a key factor to adjust for a barista. Depending on the type (and quality) of the grinder and on the coffee blend and roasting degree, various particle size distributions will be obtained. So not only

TABLE 15.4 General Parameters of Different Coffee Brew Preparation Methods

	Preparation Method					
Parameter	**Boiling Coffee (Turkish)**	**Pour Over (Drip Filter)**	**French Press/Immersion Brew**	**Espresso**	**Moka Pot**	**Cold Brew (Pour Over or Immersion)**
Water-to-coffee ratio (mL/g)	~20	~13–20	~15–20	3.5–6	~9–15	~4–15
Particle size	Fine	Medium	Coarse	Medium	Medium	Coarse
Compaction	No	No	No	Yes	No	No
Pressure (bar)	0	0.1	0	8–19	1–2	0–0.1
Brewing time		3–10 min	2–5 min	25–30 s	3–5 min	2–24 h

the average grind size, but also the particle shape and size distribution will determine how the water flows through the bed, as discussed further in Chapter 13.

Each preparation method requires its own optimal grind size. Brews prepared with too coarse coffee grounds may show a lower aromatic profile compared to a finely ground coffee bed (Severini et al., 2015). Small particles increase the surface exposed to water, permitting a more efficient extraction process (Andueza et al., 2003). Andueza et al. (2002) showed that a pressure increase resulted in a higher viscosity of the espresso coffee, related to an increase in body intensity. Too fine grounds may, however, hinder an efficient water distribution and percolation through the coffee bed, possibly leading to overextraction. On the other hand, the interstitial space between the particles is smaller, increasing the pressure over the bed and decreasing the overall flow speed of the water.

The coffee bed permeability also changes during extraction. The initial water invasion induces wetting of the coffee bed followed by the solubilization of more soluble and low molecular weight compounds, as well as volatile aroma compounds. Simultaneously, coffee bed particles swell, due to the presence of water-insoluble polysaccharides in the roasted coffee, and coffee particles geometrically rearrange due to the water flow and/or pressure. This so-called bed consolidation provokes a progressive decrease in the coffee bed porosity, a decreased overall flow and increased pressure drop over the bed (Navarini et al., 2009). In general, the finer the distribution, the more consolidation is experienced under flow. It is thus up to the barista to grind fine enough to allow efficient diffusion of solubles from coffee particles but not too fine to prevent clogging or too long flow times. An excessive amount of coffee on the other hand would not allow sufficient expansion during wetting, causing overcompaction, disturbing percolation, and eventually resulting in deposit of solids in the espresso cup (Andueza et al., 2007).

The most important aspect of the pressure is related to crema formation. The pressure forces part of the CO_2 present in the ground coffee into the water phase from which it is then slowly released taking some solids with it to form a dense and stable crema on top of the beverage. All brewing methods lacking the pressure are not able to form any crema on top of the beverage. Thus, pressure is definitely crucial for crema formation with standard brewing methods.

Also aroma compounds cannot evaporate from the coffee bed when pressure is applied and have shown to end up in-cup to higher extent as compared to unpressurized or lower pressure extraction methods (Sanchez-Lopez et al., 2016). In contrast, nonvolatiles are less sensitive to pressure as such. For these compounds the importance of pressure is mainly linked to a homogeneous extraction and stable flow rate. Depending on bed permeability the pressure is crucial to have the right flow time in order to extract the nonvolatiles in a balanced way. Again, under- and overextraction might occur if the flow rates become too short or too long. Here pressure is one element of optimizing the flow rate toward the desired profile.

Another effect of the pressure applied to the coffee bed is the extraction of lipids. As the coffee bed acts like a sponge keeping the coffee oil inside the cells the pressure forces the oil droplets to the surface where they are taken with the water to end up in the final cup. The role of extracted oil to the body and mouthfeel is not fully clear. Espressos often are not only higher in oil content compared to nonpressurized methods but also higher in strength due to smaller cup volume. Both effects might act on the perceived body/mouthfeel. This becomes evident, in particular, when comparing to a press-filter coffee. Although considerably high in oil due to the pressing, it is not generally perceived as higher in body/mouthfeel. Most probably, because these coffees are consumed as long cups with low total content, respectively lower strength. In contrast, filter paper is very efficiently absorbing oil and the corresponding brews are low in fat content. Consequently, they are evaluated rather low in body/mouthfeel. However, the role of oil on the one hand must be put into context with the cup size effect of filter brews that can further reduce body and mouthfeel for drip filter preparations.

As a barista, it is thus crucial to have an insight into the interplay of all the variables—coffee dose, particle size and shape, particle size distribution, water quantity, temperature and quality, pressure, and flow time. All of these variables have an effect on the extraction, and many will influence each other: for example, increasing the fineness of grind will often increase the efficiency of extraction *and* extend the flow time, having a multiplying effect on total extraction. Small adjustments and frequent tasting are therefore best practices, as well as objective measurement when possible. Environmental changes, such as the temperature and humidity in a coffee shop, can likewise have an effect on extraction parameters, and a barista working in a shop may have to adjust certain variables over the course of a serving day.

Often, the barista will choose specific coffees that are well suited to specific brewing methodologies: certain coffees for filter brewing, others for espresso, etc. Sometimes, however, the skilled barista may be able to optimize a given coffee for multiple extraction methods simply by adjusting variables.

5. OVERVIEW OF EXTRACTION METHODS AND PARAMETERS

Extraction methods are generally characterized by the extraction tools, but can also be grouped by various key parameters influencing the final in-cup flavor profile. Table 15.4 shows different brewing methods, with their corresponding key parameters.

It is difficult to link a specific preparation method with a sensorial direction, for a given coffee. Of course, the higher pressures used for espresso will give a characteristic crema, unique to this preparation method. An espresso coffee is more intense, has more body, but is this linked to the higher extraction pressure, the metal filter, or the low water-to-coffee ratio, or some combination? Pour over coffee tends to be less concentrated, but the flavor profile is maybe more balanced or delicate, and less prone to overextraction. Again, this might be linked to low extraction pressures, longer extraction times, and/or higher water-to-coffee ratios. Turkish coffee or moka pot preparation can have a harsh character because of the high extraction temperatures. However, a skilled barista is able to make a balanced Turkish or moka coffee by moderating temperatures and grind particle size. The barista has, for each method, a range of parameters to play with. Each parameter changes the in-cup profile, but can also influence and interact with other extraction parameters. In a recent comparison study (Gloess et al., 2013), the moka pot preparation showed the highest extraction efficiency compared to Espresso, pour over, or French press preparation methods. This is probably linked to the higher extraction temperatures inherent to this method. This high extraction efficiency, linked to a specific in-cup profile, might however, dramatically change when applying higher or lower water-to-coffee ratios. Pressurized coffee extraction generally allows to extract the components quickly from the coffee bed. In low pressure environments, extraction occurs more slowly, but due to the longer extraction time and higher water-to-coffee ratio, extraction efficiencies (yield) can sometimes be greater than for espresso preparation (Gloess et al., 2013). The barista thus has a very wide range of variables to control, and this is what makes the craft of the barista so technically challenging.

5.1 Boiled Coffee/Turkish Coffee

Turkish coffee is prepared by boiling the coffee in water. The fine ground powder is put in a pot (e.g., cesve), water is added, and heated up to boiling. The extraction is mainly based on the diffusion of solubles at high temperatures. At these high temperatures, also less water soluble compounds are extracted that give the typical intense, bitter, and dark-chocolate flavors.

Once at boiling temperature the heating is usually stopped, but the coffee grounds remain in contact with the hot water and extraction continues. The boiling may happen multiple times, depending on the practice of the brewer. Grounds need to settle before serving. This results in a rather strong coffee with some sediment in the cup.

5.2 Pour Over Brew/Coffee Percolation

For the pour over brew (or drip coffee), the coffee is ground at coarser particle sizes to allow the water to pass through by gravity alone. The ground coffee is put into a holder containing a filtering device. Various filter sizes, shapes, and materials are commercially available, allowing the barista to control the shape of the filter bed and degree of filtration. Controlling particle size by adjusting the grinder allows for changes in percolation speed and contact time between water and coffee. Percolation is performed by applying hot water, sometimes measured very precisely to an optimal temperature for a specific coffee. Water is applied manually or by an automatic drip filter machine, some of which can be programmed to deliver specific amounts of water at specific temperatures over time. The water passes through the coffee bed mainly by force of gravity. The pressure therefore is very limited depending on the water column that builds up above the coffee bed. The water drips into a serving vessel and the grounds are retained by the filter. This method is sometimes referred a drip filter preparation.

There is much discussion about shape of filter and depth of coffee bed, which determines the quality of extraction. A very deep, narrow filter basket will extract the coffee differently than a wide, shallow one. The reason for this is the contact time between the water and the coffee. Many filter coffee systems use cone-shaped filters, which create a conical coffee bed. These variations will have the effect of optimizing the extraction of certain flavors, having an effect on the final brew.

5.3 French Press/Immersion Brew

In the French press coffee grounds having rather coarse particle sizes and hot water are mixed and allowed to remain in contact for a certain amount of time (2–5 min), depending on the intensity of extraction the barista prefers: long extractions might enhance bitterness and intensity, shorter extractions emphasize acidity and sweetness. A plunger containing a filtering device is then used for separation of the grounds and the liquid beverage. The coffee beverage can now be poured into a cup. Because of the presence of "fines" in the ground coffee and relatively inefficient metal-mesh filtration, higher sediment levels are generally obtained by this preparation method compared to the drip filter method. The pressing of the coffee also squeezes oils out from the coffee bed, increasing the oil content of the final brew (Zhang et al., 2012). An alternative immersion brew is called Aero Press, which starts as an immersion brew. At the end pressure is added by a piston to extract more from the coffee bed, and the final brew is filtered through paper giving it elements of both immersion and drip filter methods.

5.4 Espresso

The original intention of the espresso machine was to prepare a coffee beverage on demand, and the first machines operated using steam instead of high pressure. Quality has, however, since significantly improved as steam-powered machines tended to extract the coffee too hot and thus gave a burnt flavor to the cup. Today's definition of the traditional Italian espresso is a beverage prepared on request from roast and ground coffee beans by means of hot

(88 ± 2°C) water pressure (9 ± 1 bar) applied for a short time (25 ± 5 s) to a compact roast and ground coffee cake (7 ± 0.5 g) by a percolation machine, to obtain a small cup (25–40 mL) of a concentrated foamy cup (Petracco, 2001; Istituto Nazionale Espresso Italiano). There are, however, parallel trends to increase the coffee water ratio to 20 g of coffee for 40 mL of water (Rao, 2013). Good quality espressos are also obtained applying variations on these traditional values, e.g., using even higher water pressures. The character of an espresso is connected to this pressure and a low water-to-coffee ratio.

The extraction pressure has a major impact on the obtained flavor profile. As discussed before, the actual pressure built up on top of the coffee bed and pressure drop through the bed depends on the degree of tampering of the bed, the homogeneity, and shape of the bed—to avoid channeling—and the forces applied through the water on top of the bed. For espresso coffee preparation, the type of pump and its characteristics are crucial. These characteristics are given by the pump operating specifications, defined by the manufacturers' capacity curve. An example is given in Fig. 15.7. This curve links the operating water flow rates of a pump with delivered pressures. A given pump can deliver the highest pressures at the lowest flow rates. When the water flow rate increases the pump delivers lower pressures and vice versa. The coffee bed properties (permeability) will decide on the actual operating pressure and the corresponding flow rate across the bed, brewing time (i.e., time required to produce a given beverage volume), water residence time, and ultimately in-cup profile and quality. A too low bed permeability can give rise to (too) high extraction pressures, low overall flow rates, and too long extraction times possibly leading to overextraction (Corrochano et al., 2015).

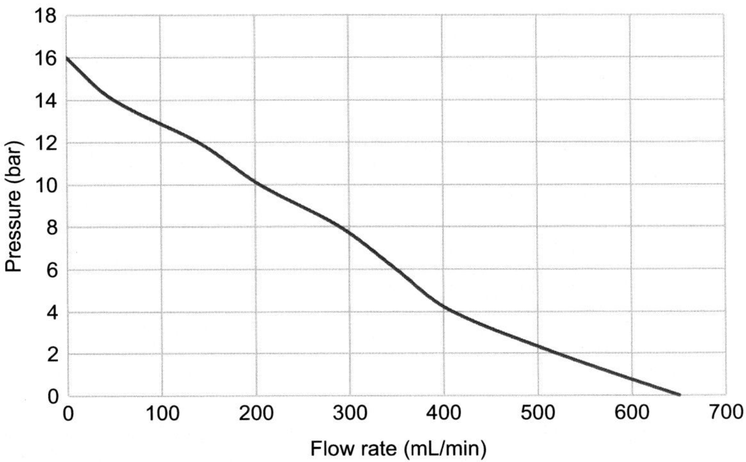

FIGURE 15.7 Example of a capacity curve of an espresso pump.

Recently, sophisticated espresso machines have been developed that allow the barista to control multiple extraction parameters: water flow, variable pressure over the course of the extraction, etc. This trend is allowing baristas to precisely deliver specific flavor attributes in the cup, by focusing on specific and reproducible extraction profiles.

Grind size is highly important for the result in the cup. For example, it was shown for espressos prepared with coarse coffee grounds showed a lower aromatic profile compared to a fine ground coffee bed (Severini et al., 2015). Too fine grounds may, however, hinder an efficient water distribution and percolation through the coffee bed. As a consequence of the different extraction behavior of different odorants and tastants over time, the in-cup and above-cup flavor balance changes continuously during extraction. To obtain an equilibrated flavor profile in-cup, the initial balance between flavor molecules in the coffee powder (as determined by the coffee origin and roasting conditions) needs to be in perfect alignment with the amount of water passed through the bed (see also Chapter 13).

Espresso coffees single-serve systems using pods or capsules have recently grown in market-share, thanks to their shelf life and high quality beverages combined with a high level of convenience for the consumer. Almost all parameters are fixed allowing also very unexperienced users to obtain a very good cup of coffee. Pressure and coffee/water ratio are comparable to those of a barista type espresso preparation resulting also in a strong coffee with good crema on top of it.

5.5 Moka Pot/Stove-Top Coffee Maker

Moka pot is the most popular household coffee brewing method in Italy, invented by Alfonso Bialetti in 1933. It is another pressurized tool to prepare coffee, comprising a three-chamber design. The bottom chamber provides the pressured water/steam that passes through the coffee bed sitting in the middle chamber of the system. The final coffee is then collected in the upper section. The pressure is significantly lower compared to Espresso preparation and no crema is obtained by this preparation method. Despite its quite simple manufacture and functioning, it has been shown that the thermodynamic behavior of the moka is complex in comparison to other coffee-brewing methods (Navarini et al., 2009). Several variables affect the extraction process and are not easy to control, leading rapidly to overextraction. To avoid this, it is important to finish the extraction in time, i.e., when the first amounts of water come up into the top kettle and when a "espresso-like-fluid" is obtained, being dark, syrupy, and almost creamy. This regular extraction phase is driven by increasing air-vapor pressure in the bottom chamber. Pressure increases not only due to increasing water flow rate, but the coffee cake permeability decreases with time (swelling of coffee particles). At the final stages of extraction, when little water is left in the bottom chamber, flow rate and pressure further increase, initiating a phase where *vapor*-liquid—solid extraction occurs, generally deteriorating the quality in-cup. This so-called strombolian phase is announced by a well-known rattling sound. The higher pressure and temperature solubilize more efficiently less soluble (undesired) compounds, increasing bitterness and astringency. In addition, organoleptically unpleasant and less volatile odorants are stripped from the coffee bed and trapped into the beverage.

5.6 Cold Brew and Iced Coffee

Cold brew usually refers to the preparation of the beverage using cold water (room temperature or lower). "Pour over," immersion or even French press cold brew techniques exist. The resulting beverage can be drunk either cold or hot, after dilution with cold or hot water. Iced coffee on the other hand is prepared using hot water, generally using the drip-brewing method. The final beverage is cooled down after extraction, e.g., with ice, and served cold over more ice.

If coffee is brewed with cold water, the force of heat is missing to extract several low polar compounds (such as coffee oil). Higher polar compounds do not "suffer" as much from these low temperatures and are still extracted rather well. As mentioned, solubility of flavor compounds is strongly temperature dependent and differs from compound to compound. Many flavors are therefore extracted, but a longer extraction time is generally needed, up till 24 h, to allow a sufficient extraction yield. On the other hand, losses of volatiles during preparation (evaporation) are much less important as compared to hot extraction, leaving these aromatic molecules "trapped" inside the beverage. Cold-brew preparation thus gives in general a beverage which can be surprisingly different from hot-brewed coffee, provided that the necessary time is allowed for extraction. Cold brewed coffee typically emphasizes body, sweetness, and chocolate notes, and can have a syrupy characteristic. Long brewing times might lead to oxidized flavors in the cup, which can be avoided by controlling exposure to air during the brewing process.

6. OUTLOOK

Every day, new methods of coffee extraction are developed. Most are variations on the basic categories of coffee brewing detailed above, but subtle variations in flavor and technique can make a big difference, and lead to trends in coffee extraction techniques. As an example, AeroPress, a simple combination of immersion and filter techniques, has become very popular within the past decade. In higher-tech brewing devices, innovations in technology are leading to exciting new techniques: the centrifugal-force brewing of the VertuoLine brewer, for example, allows for computer control of pressure due to rotational speed. These new devices also allowed (even less experienced) consumers to prepare barista-like beverages at home. Espresso preparation for baristas has evolved from using equipment based on steam extraction to an advanced extraction technique having precise variable pressure control and weight-sensitive platforms below the coffee spouts (allowing exact measurement of the extracted liquid). This evolution gives baristas ever more control over coffee extractions and also more space to play around. At the moment, these tools have ushered in a period of exploration, while baristas learn to manage multiple extraction parameters to extract specific flavor profiles of their coffee. Gone are the days when mastery of a single extraction technique is sufficient; today's barista moves between brewing devices and techniques fluidly, seeking to match specific coffees and roasting styles to ideal brewing techniques. In the future, this will allow for more intense and idiosyncratic expressions of coffee flavor, which will require increased communication on expectations and preferences between consumers, baristas, and roasters.

Another aspect for the future, beside technical developments, are evolving consumers. Preferences are in constant evolution. Filter brew is historically the most consumed coffee. Depending on culture and preference, many other brewing methods are gaining momentum, both at home and in bars. Today coffee aficionados will not only select the beans to bring variation during the day, but preparation methods may equally play a role to diversity and to create new coffee experiences. Furthermore, many consumers drink their coffees with milk. The milk and milk foam will, however, influence the sensory perception. Do the preparation methods need to be adapted for this? Is a coffee that is well balanced when consumed black still a good coffee when milk is added? There is probably still the need for further optimizing brewing of the coffee to unravel the full potential of the coffee also when consumed as white and to lead toward new sensory experiences.

REFERENCES

Andueza, S., Maeztu, L., Dean, B., Paz de Peña, M., Bello, J., Cid, C., 2002. Influence of water pressure on the final quality of Arabica espresso coffee. Application of multivariate analysis. Journal of Agriculture and Food Chemistry 50, 7426–7431.

Andueza, S., Maeztu, L., Pascual, L., Ibáñez, C., Paz de Peña, M., Cid, C., 2003. Influence of extraction temperature on the final quality of espresso coffee. Journal of the Science of Food and Agriculture 83, 240–248.

Andueza, S., Vila, M.A., de Peña, M., Cid, C., 2007. Influence of coffee/water ratio on the final quality of espresso coffee. Journal of the Science of Food and Agriculture 87, 586–592.

Bhumiratana, N., Adhikari, K., Chambers, E., 2011. Evolution of sensory aroma attributes from coffee beans to brewed coffee. LWT – Food Science and Technology 44, 2185–2192.

Blank, I., Sen, A., Grosch, W., 1992. Potent odorants of the roasted powder and brew of Arabica coffee. Zeitschrift für Lebensmittel-Untersuchung und Forschung 195, 139–145.

Blumberg, S., Frank, O., Hofmann, T., 2010. Quantitative studies on the influence of the bean roasting parameters and hot water percolation on the concentrations of bitter compounds in coffee brew. Journal of Agriculture Food and Chemistry 58, 3720–3728.

Corrochano, B.R., Melrose, J.R., Bentley, A.C., Fryer, P.J., Bakalis, S., 2015. A new methodology to estimate the steady-state permeability of roast and ground coffee in packed beds. Journal of Food Engineering 150, 106–116.

Dold, S., Lindinger, C., Kolodziejczyk, E., Pollien, P., Ali, S., Germain, J.C., Garcia Perin, S., Pineau, N., Folmer, B., Engel, K.-H., Barron, D., Hartmann, C., 2011. Journal of Agriculture and Food Chemistry 59, 11196–11203.

Frank, O., Zehentbauer, G., Hofmann, T., 2006. Bioresponse-guided decomposition of roast coffee beverage and identification of key bitter taste compounds. European Food Research and Technology 222, 492–508.

Frank, O., Hofmann, T., 2007. Structure determination and sensory analysis of bitter-tasting 4-vinylcatechol oligomers and their identification in roasted coffee by means of LC-MS/MS. Journal of Agriculture and Food Chemistry 55, 1945–1954.

Ginz, M., Engelhardt, U., 2000. Identification of proline-based diketopiperazines in roasted coffee. Journal of Agriculture and Food Chemistry 48, 3528–3532.

Gloess, A., Schönbächler, B., Klopprogge, B., D'Ambrosio, L., Chatelain, K., Bongartz, A., Strittmatter, A., Rast, M., Yeretzian, C., 2013. Comparison of nine common coffee extraction methods: instrumental and sensory analysis. European Food Research and Technology 236, 607–627.

Istituto Nazionale Espresso Italiano; http://www.espressoitaliano.org/en/.

Lee, J.-S., Kim, M.-S., Shin, H.-J., Park, K.-H., 2011. Analysis of off-flavor compounds from over-extracted coffee. Korean Journal of Food Science and Technology 43 (3), 348–360.

Lopez-Galilea, I., Paz De Peña, M., Cid, C., 2007. Correlation of selected constituents with the total antioxidant capacity of coffee beverages: influence of the brewing procedure. Journal of Agriculture and Food Chemistry 55, 6110–6117.

Lindinger, C., Labbe, D., Pollien, P., Rytz, A., Juillerat, M.A., Yeretzian, C., Blank, I., 2008. When machine tastes coffee: instrumental approach to predict the sensory profile of espresso coffee. Analytical Chemistry 80, 1574–1581.

Lingle, T.R., 1996. Coffee brewing control chart. In: Lingle, T.R. (Ed.), The Coffee Brewing Handbook. A Systematic Guide to Coffee Preparation. Specialty Coffee Association of America, Long Beach.

Mestdagh, F., Davidek, T., Chaumonteuil, M., Folmer, B., Blank, I., 2014. The kinetics of coffee aroma extraction. Food Research International 63, 271–274.

Navarini, L., Nobile, E., Pinto, F., Scheri, A., Suggi-liverani, F., 2009. Experimental investigation of steam pressure coffee extraction in a stove-top coffee maker. Applied Thermal Engineering 29 (5–6), 998–1004.

Navarini, L., Rivetti, D., 2010. Water quality for Espresso coffee. Food Chemistry 122 (2), 424–428.

Parenti, A., Guerrini, L., Masella, p., Spinelli, S., Calamai, L., Spugnoli, P., 2014. Comparison of espresso coffee brewing techniques. Journal of Food Engineering 121, 112–117.

Petracco, M., 2001. Beverage preparation: brewing trends for the new millennium. In: Clarke, R., Vitzthum, O. (Eds.), Coffee: Recent Developments. Blackwell Science, Oxford.

Rao, S., 2013. Espresso Extraction: Measurement and Mastery. Self-publication, USA.

Sanchez-Lopez, J.A., Wellinger, M., Gloess, A.N., Zimmermann, R., Yeretzian, C., 2016. Extraction kinetics of coffee aroma compounds using a semi-automatic machine: on-line analysis by PTR-ToF-MS. International Journal of Mass Spectrometry. http://dx.doi.org/10.1016/j.ijms.2016.02.015.

Sanchez-Lopez, J.A., Zimmermann, R., Yeretzian, C., 2014. Insight into the time-resolved extraction of aroma compounds during espresso coffee preparation: online monitoring by PTR-ToF-MS. Analytical Chemistry 86 (23), 11696–11704.

Severini, C., Ricci, I., Marone, M., Derossi, A., De Pilli, T., 2015. Changes in the aromatic profile of espresso coffee as a function of the grinding grade and extraction time: a study by the electronic nose system. Journal of Agriculture and Food Chemistry 63, 2321–2327.

Semmelroch, P., Grosch, W., 1995. Analysis of roasted coffee powders and brews by gas chromatography-olfactometry of headspace samples. LWT—Food Science and Technology 28, 310–313.

Stadler, R., Varga, N., Hau, J., Arce Vera, F., Welti, D., 2002. Alkylpyridiniums. 1. Formation in model systems via thermal degradation of trigonelline. Journal of Agriculture and Food Chemistry 50, 1192–1199.

Stark, T., Justus, H., Hofmann, T., 2006a. Quantitative analysis of N-phenylpropeonyl-L-amino acids in roasted coffees and cocoa powder by means of a stable isotope dilution assay. Journal of Agriculture and Food Chemistry 54 (8), 2859–2867.

Stark, T., Hofmann, T., 2007. Structures, sensory activity and dose/response functions of 2,5-diketopiperazines in roasted cocoa nibs (*Theobroma cacao*). Journal of Agriculture and Food Chemistry 53, 5419–5428.

Stark, T., Bareuter, S., Hofmann, T., 2006b. Molecular definition of the taste of roasted cocoa nibs (*Theobroma cacao*) by means of quantitative studies and sensory experiments. Journal of Agriculture and Food Chemistry 54, 5530–5539.

Zhang, C., Linforth, R., Fisk, I., 2012. Cafestol extraction yield from different coffee brew mechanisms. Food Research International 49, 27–31.

Chapter 16

Water for Extraction—Composition, Recommendations, and Treatment

Marco Wellinger, Samo Smrke, Chahan Yeretzian
Zurich University of Applied Sciences, Wädenswil, Switzerland

1. INTRODUCTION

Water has received renewed attention over the last few decades in connection with many points along the coffee value chain. Areas of attention not only include farming and processing methods (Chapters 3 and 4), but also how extraction may affect the sensory properties of the coffee in the cup. Every barista knows that the choice of water either can highlight the specificities of a coffee, or may make it flat and dull. Historically, analysis of water quality has largely focused on safety and technical aspects. This is also reflected by the fact that the most common classification for water is based on hardness, a measure of the maximum amount of scale formation in coffee machine boilers or kettles.

The goal of this chapter is to equip the reader with a basic understanding of the influence water has on coffee and how to change the composition of water to alter its total hardness and alkalinity. Section 2 will start by briefly introducing the basics of the science of water, which apart from in certain particular forms, such as distilled water or after treatment by reverse osmosis, contains dissolved compounds, such as minerals, gases, and organic molecules. Afterward, in Section 3, we will explain the most important aspects of water composition, with a special focus on total hardness and alkalinity—both concepts are central to understanding the interaction between water and coffee. Common technical considerations with regard to scale formation and corrosion, which are a direct consequence of dissolved minerals in water, are also explained. Section 4 provides an overview of the impact of water composition on technical and sensory aspects of coffee extraction. In this section we will also discuss current recommendations for "ideal" water composition and address flavor defects that

The Craft and Science of Coffee. http://dx.doi.org/10.1016/B978-0-12-803520-7.00016-5

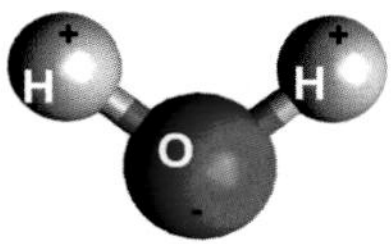

FIGURE 16.1 Structure of a water molecule and its charge distribution.

are caused by unsuitable water compositions. Since water, as we encounter it in its natural form, often does not correspond to what we consider the "ideal" composition for coffee preparation, total hardness and alkalinity first need to be adjusted, before the water is used for coffee extraction. Section 5, therefore, will review the effect of water treatment methods on total hardness and alkalinity by introducing a systematic and novel approach that we believe is particularly useful and suited to the specialty coffee community. We will conclude this chapter by presenting practical examples on the impact of different water compositions in a cupping experiment and espresso extraction.

2. PHYSICOCHEMICAL CHARACTERISTICS OF WATER

Water is a molecule that is composed of one oxygen and two hydrogen atoms (H_2O) that are differently charged, leading to an uneven distribution of the electrical charge (see Fig. 16.1), known as polarity.

Due to its polar nature, water is a very good solvent for polar compounds, as is the case during coffee extraction. In contrast, it hardly dissolves any nonpolar compounds such as oils and fats, which make up more than 10% of the weight of roasted coffee. However, under high pressure and/or high temperature extraction conditions, such as espresso and moka pot, or alternatively over long contact times, such as French press, the solubility of nonpolar compounds is significantly increased (Gloess et al., 2013). The polarity of water is also responsible for a phenomenon called self-dissociation of water, in which a proton is transferred from one water molecule to another. This naturally leads us to the pH, which is calculated as the negative logarithm of the proton concentration,[1] where the higher the concentration of protons the lower the pH. For example, pH 6 corresponds to a concentration of protons in water of 10^{-6} mol/L. For every increase in the pH value of one unit (+1) the concentration of protons decreases by a factor of 10 and vice versa. The pH of pure water is 7, although this is only strictly true at 25°C.

1. The pH is a scale from 0 to 14 that defines how acidic a solution is. More acidic solutions have lower pH. More alkaline solutions have higher pH. For example, coffee has a pH somewhere around 5.5 and is hence slightly acidic. To find out what the pH of a solution is, the concentration of protons (H^+) in the solution must be measured. The formula for calculating pH: $pH = -\log c(H^+) \cdot c(H^+)$ is the concentration of protons in the solution as mol per liter (mol/L). For a definition of mol, please consult a chemistry textbook. In essence mol/L tells us how much of the respective compound is present in 1 L of the solution. Once you know the concentration of H^+ in the solution, the negative value of the logarithm of the concentration of H^+ provides the pH value.

2.1 Carbon Dioxide and Carbonic Acid

Acids are key to coffee quality; specific acids can convey a fresh and lively sensation to the cup, a characteristic of some renowned specialty coffee origins. Chemically, an acid is a substance capable of donating a proton. An acid of particular importance to water and coffee is carbonic acid (H_2CO_3). It plays a central role in understanding various properties of water in relation to coffee. This section lays the foundation for understanding how the different carbonate species are related to each other; the applied aspects of which will be explored later in Sections 3.4, 4, and 6.2.

Water naturally dissolves carbon dioxide (CO_2) from the air and roasted coffee provides another important source of CO_2. During roasting CO_2 is generated (up to 2% of the weight of freshly roasted coffee is entrapped CO_2), which is subsequently released during extraction. Carbon dioxide dissolved in water forms carbonic acid (H_2CO_3), a weak acid that dissociates to hydrogen carbonate $\left(HCO_3^-\right)$ plus one proton (H^+). The proton released into the water makes the water more acidic and the pH is reduced. At natural concentrations of CO_2 in the air, water in equilibrium with the air is slightly acidic with a pH of 5.7.

The hydrogen carbonate $\left(HCO_3^-\right)$ still has a second proton that it could release; however, this only happens at pH values above 8.3, resulting in a doubly charged carbonate ion $\left(CO_3^{2-}\right)$. Eq. (16.1) summarizes the chain of chemical reactions that extends from carbon dioxide dissolved in water (left) through to a carbonate ion that has released two protons (right):

$$CO_2 + H_2O \leftrightarrows H_2CO_3 \leftrightarrows HCO_3^- + H^+ \leftrightarrows CO_3^{2-} + 2\,H^+ \qquad (16.1)$$

3. WATER COMPOSITION

In practice, tap water from a lake or re-infiltrated from a river is on average significantly lower in mineral content than groundwater that has a typical residence time in the order of months to years. Furthermore, areas with bedrock made of carbonates are higher in mineral content than areas of silicate rock. For instance, in Switzerland the mineral content of the groundwater in the Central Plateau (fine sediments rich in carbonates) is about five times higher than in the central and southern Alps (crystalline rock dissolves more slowly and contains little carbonates).

3.1 Alkalinity

Now that we have introduced some basic properties of water, we can address two central concepts linking water to coffee that will guide our discussion—alkalinity and total hardness. Alkalinity represents the capacity of water to buffer acids. The amount of acid that has to be added to a water sample, or even to a coffee brew, until a specific pH is reached is a measure of its alkalinity. If, for example, an acid is added to a coffee brew, its pH will decrease. The more acid that has to be added to reach a specific pH, the higher the alkalinity. Alkalinity should not be confused with a solution that has an alkaline pH, which simply means that its pH is higher than 7 (at 25°C). Alkalinity hence reflects the capacity of water or coffee to resist a change in pH, when acids are added, irrespective of its actual pH value. For freshwater, alkalinity can be calculated by the following equation:

$$\begin{aligned}\text{Alkalinity} = \ & {HCO_3}^{-} + 2 \times {CO_3}^{2-} + OH^{-} \\ & -H^{+} \ (\text{expressed as concentrations, mol/L})\end{aligned} \tag{16.2}$$

For water with a pH below 8.3, alkalinity can be accurately approximated from the ${HCO_3}^{-}$ content [Standard Methods for the Examination of Water and Wastewater (SMWW), 2012]. This is because all other components (${CO_3}^{2-}$, OH^{-}, and H^{+}) are present in concentrations several orders of magnitude lower than that of ${HCO_3}^{-}$. Above pH 8.3, a substantial contribution from the carbonate ion $\left({CO_3}^{2-}\right)$ must also be considered, though this only occurs in regions with very hard water (more than 370 ppm $CaCO_3$ of alkalinity). Although this is rare, it occurs in some regions that are characterized by dominant carbonate bed rock. For example in the region of the canton of Zurich, at least 10% of all tap water has an alkalinity above 370 ppm $CaCO_3$ of alkalinity.

Alkalinity is also very important in relation to its effect on sensory properties; it is the opposite of acidity in chemistry as well as in coffee sensory terminology (see Section 4.2). Additionally, and most importantly, alkalinity can be measured by commercially available test kits. Test kits measure the amount of acid that has to be added to reach a specific pH (most commonly pH 4.5), by counting the number of standard drops that need to be added until the color of the water changes. It should be noted that these test kits are often incorrectly marketed as carbonate hardness test kits—see next section for the distinction.

3.2 Hardness of Water

In this chapter the term "hardness" will be used to denote "total hardness," which is defined in accordance with worldwide industrial norms (SMWW; DIN 38409-6, 1986; ASTM D-1126, 2002; EPA Method, 130.2, 1982) as the sum of amounts or equivalent concentrations of calcium and magnesium (see Section 3.3 for calculation). In contrast, "carbonate hardness" is defined by the maximum amount of scale that can form for a given water composition, and is determined by the common minimum of total hardness and alkalinity, whichever is lower. During scale formation both total hardness and alkalinity are reduced in equal amounts. Most natural water sources have a total hardness that is higher than their alkalinity, as depicted in Fig. 16.2A where the carbonate hardness is equal to the alkalinity. However, water treated with a so-

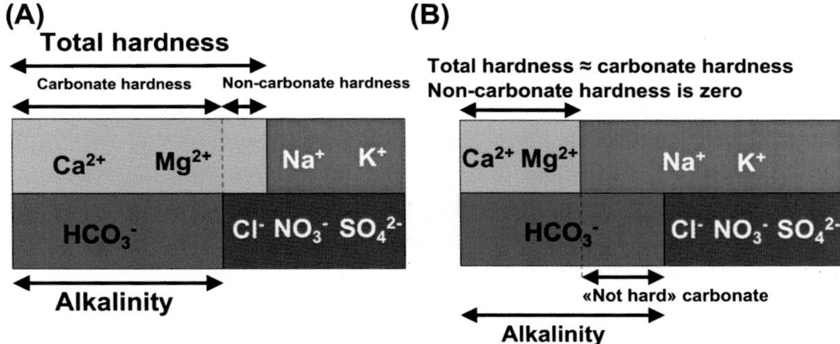

FIGURE 16.2 Terms used with respect to hardness and alkalinity for two sample water compositions—size is proportional to equivalent concentrations: (A) total hardness > alkalinity; (B) total hardness < alkalinity.

called softener that removes calcium or magnesium and replaces them with sodium or potassium (see Fig. 16.2B) has a reduced total hardness, without affecting the alkalinity. In this case, total hardness is lower than alkalinity and carbonate hardness is equal to total hardness.

3.3 Units of Hardness

When calculating the hardness of water, it is crucial to distinguish between amount concentrations, such as mol/L (where 1 mol can be understood to be the "chemical dozen" equaling 6.4×10^{23} atoms, ions, or molecules), and mass concentration (e.g., mg/L on bottled waters). Although amount concentration refers to the number of entities in a given volume of water, mass concentration refers to the weight of compounds in the same water volume. Mass concentration does not reflect the true proportions of different substances and is, therefore, unsuitable for comparing the amount of different compounds in a water sample. This can best be illustrated by a simple example: When considering the compounds that contribute to the total hardness of water, 1 mmol/L of Ca^{2+} is equal to 40 mg/L while 1 mmol/L of Mg^{2+} is equal to 24.3 mg/L. For the alkalinity of water, 1 mmol/L of HCO_3^- is equal to 61 mg/L. In all three cases the amount concentrations and hence the number of compounds are equal (namely 1 mmol/L), whereas the mass concentrations are obviously different, as the compounds all have different weights.

From a chemist's perspective, amount concentrations such as mol/L should be used since the values are proportional to the actual number of atoms, ions, or molecules in a specific volume of water.

It is also important to note that, throughout industry, hardness values are referred to in equivalent units, such as ppm $CaCO_3$ (USA) or German (°d) degrees. Compared to amount concentration (such as mol/L) equivalent units, the molar concentrations of Ca^{2+}, Mg^{2+}, and HCO_3^- scale effectively as they react together. Therefore, the equivalent concentrations of calcium and hydrogen carbonate always decrease by the same amount when scale forms or during specific water treatment methods (see Fig. 16.5).

Table 16.1 provides an overview on the conversion factors for the most commonly used units in water analysis; the values have been calculated based on the IUPAC Periodic Table of the Elements (2013) and are also in agreement with Hem (1985) and the DIN norm for water hardness (1986).

Please note that the standard unit used by the Specialty Coffee Association of America (SCAA) "ppm $CaCO_3$" (or alternatively "mg/L $CaCO_3$") is often misleadingly abbreviated as ppm or mg/L for both total hardness and alkalinity (see Section 4.3 on water standards), since "straight" mass concentrations (as labeled on water bottles) does not correspond to the true proportions of calcium, magnesium, and hydrogen carbonate.

3.4 Scale Formation and Corrosion

Water boilers in the home and industrial equipment can become encrusted with scale residues that are mainly composed of calcium carbonate ($CaCO_3$) and at high pH (>10) also of magnesium hydroxide ($Mg(OH)_2$). The latter occurs in steam boilers, especially with sodium-softened water. The speed of this process is determined by the solubility of the calcium carbonate for the given temperature and pH conditions. The primary technical concerns in terms of scale formation when using hard water are the decrease in efficiency of the heating system (when the layer of scale acts as an insulator) and the blockage of valves and flow restrictors (usually called gicleurs in coffee machines). On the other hand, water that is very low in alkalinity and can easily become acidic (since insufficient acid buffer is present), which can cause corrosion of metal parts. To increase longevity and hence reduce maintenance costs for coffee machines, a number of countries have issued recommendations to minimize the costs of scale formation (high hardness and alkalinity values) and corrosion (generally low mineral content and low alkalinity in particular). In Switzerland, there is only a recommendation with regard to hardness that specifies the optimal hardness as 12–15°fH (SVGW, 2008). In other countries, optimal ranges for alkalinity are provided that are typically between 5 and 6°fH (Navarini and Rivetti, 2010). As an additional measure, coffee machine companies recommend regular descaling using acid dissolved in water (preferably one with little taste), and also offer commercial solutions.

4. IMPACT OF WATER COMPOSITION ON EXTRACTION

Water that has no minerals other than calcium or magnesium carbonate ($CaCO_3$, $MgCO_3$) contains equal amounts of hardness and alkalinity (in hardness equivalent units). A total hardness that is much higher than alkalinity indicates that there is a significant amount of sulfate present in the water.

TABLE 16.1 Conversion Factors for Units of Hardness and Alkalinity—Rounded to Four Significant Digits—Given in Bold Are the Most Commonly Used Conversion Factors

		ppm $CaCO_3$	°dH	°fH	gpg	°e	$Ca^{2+} + Mg^{2+}$ (mmol/L)	HCO_3^- (mmol/L)	Ca^{2+} (mg/L)	Mg^{2+} (mg/L)	HCO_3^- (mg/L)
ppm $CaCO_3$ (=mg $CaCO_3$/L)	1 ppm $CaCO_3$ =	1	0.05603	0.1	0.05842	0.07022	0.009991	0.01998	0.4004	0.2428	1.219
German degrees (°dH)	1°dH =	**17.85**	1	1.785	1.0423	1.253	0.1783	0.3567	7.147	4.334	21.76
French degrees (°fH)	1°fH =	10.00	0.5603	1	0.5842	0.7022	0.09991	0.1998	4.004	2.428	12.19
Grains per US gallon (gpg)	1 gpg =	**17.12**	0.9591	1.712	1	1.202	0.1710	0.3421	6.855	4.157	20.87
English degree (°e)	1°e =	14.24	0.7979	1.424	0.8320	1	0.1423	0.2846	5.703	3.458	17.36
$Ca^{2+} + Mg^{2+}$ (mmol/L)	1 mmol/L =	100.1	5.608	10.01	5.847	7.028	1	–	**40.08**	**24.30**	–
HCO_3^- (mmol/	1 mmol/L =	50.04	2.804	5.004	2.923	3.514	–	1	–	–	**61.02**
Ca^{2+} (mg/L)	1 mg/L =	**2.497**	0.1399	0.2497	0.1459	0.1753	0.02495	–	1	–	–
Mg^{2+} (mg/L)	1 mg/L =	**4.118**	0.2307	0.4118	0.2406	0.2891	0.04114	–	–	1	–
HCO_3^- (mg/L)	1 mg/L =	**0.8202**	0.04595	0.08202	0.04791	0.05759	–	0.01639	–	–	1

Water and its minerals are arguably the most important ingredients in the quality of a coffee beverage after the roasted and ground coffee itself (Navarini and Rivetti, 2010). Due to its potential for scaling, water is commonly treated to reduce its hardness to avoid the precipitation of minerals. A further factor that is equally important as water composition in terms of its impact on extraction is the mass ratio of water to roasted and ground coffee, since this can differ by as much as a factor of 10 for different extraction methods. An extreme example is an espresso, which can be prepared with a dry coffee mass to beverage mass ratio of 1:2 (16 g of coffee to make to two cups, each containing a beverage of 16 g or less). On the other hand, a typical ratio for drip coffee is 1:15 (i.e., 60 g of coffee per 1000 g of water resulting in approximately 880 g of filter coffee). Therefore, the extent of the acid buffering effect of the alkalinity (of the water used for brewing) is much lower in an espresso than in a filter coffee—see Section 4.1. The extraction pressure must also be considered when assessing the extraction of coffee with water, since the solubility of carbon dioxide increases drastically with pressure (Sanchéz et al., 2016). There is mounting evidence (Hendon et al., 2014) that calcium and magnesium are major contributors to the efficiency of coffee extraction. Moreover, it has been suggested that magnesium (per hardness equivalent or mole) is more efficient in extracting constituents from coffee (Hendon et al., 2014). Although the study of the impact of water composition on filter coffee dates back to 1950, espresso has only been under investigation in the last 20 years. In the studies presented below three different situations are described:

1. At high alkalinity (>100 ppm), more of the acids extracted from coffee are neutralized by hydrogen carbonate, thereby producing carbonic acid that can degas as carbon dioxide, depending on the pressure and temperature. The generation of carbon dioxide during extraction creates additional resistance and thereby prolongs the extraction time under otherwise equal circumstances (Fond, 1995; Rivetti et al., 2001). See Section 6.2 for an applied example of espresso extraction with water treated by an ion exchanger that changes either total hardness or alkalinity.
2. High alkalinity and concurrent low hardness lead to increased pH when water is heated—this effect is even more pronounced in the presence of sodium. It has been suggested to be the cause of increased resistance during extraction due to decreased solubility of carbohydrates in water at high pH (Navarini and Rivetti, 2010).
3. At high hardness and equally high (or lower) alkalinity, no increase in percolation time was observed (Fond, 1995; Rivetti et al., 2001).

4.1 Buffering of Coffee Acidity

Perception of acidity in coffee has been studied extensively (for an overview see Gloess et al., 2013). The most commonly cited phenomenon is the correlation of sensory perception of acidity with titrable acidity of coffee extracts (titration to a pH of 6.6, which corresponds to mouth pH). From a chemical point of view this would also make sense as stated by Clifford (1988): "In effect, the reaction of the acid with the receptor is a titration, and thus very similar to the process used in measuring titratable acidity." Because alkalinity of water reacts with the extracted acids from coffee, it effectively neutralizes part of the acids. Based on data from Gloess et al. (2013) water can amount to as much as either 30% or 11% of the titrable acidity measured in filter or lungo espresso extractions, respectively (Gloess et al., 2013; authors' measurements and calculations). These percentages were measured for water with an alkalinity of 50 ppm $CaCO_3$ (1 mmol/L) that was used to extract a beverage from a medium to dark roasted coffee (Guatemala, Antigua, la Ceiba, 80 Pt on Colorette 3b, ~45 Pt on Agtron M-Basic scale). As a result of Clifford's suggestion, a simple equation can be formulated for perceived acidity, where the perceived acidity of coffee equals the acidity extracted from the coffee diminished by the alkalinity of the water.

4.2 Ideal Water Composition for Coffee Extraction

The most common standard in the coffee industry with respect to sensory properties is the "SCAA Standard: Water for Brewing Specialty Coffee" (2009). It defines an optimum extraction at a total hardness of 68 ppm $CaCO_3$ (with an acceptable range of 17–85 ppm $CaCO_3$), an alkalinity of 40 ppm $CaCO_3$, and pH 7 (acceptable range of pH 6.5–7.5). This area is indicated with a red line in Fig. 16.3, on which the optimum is marked with a circle (pH is not considered in the graph). In addition, it provides a target value of 10 mg/L for sodium, although sodium only becomes perceptible at concentrations above 250 mg/L (Pohling, 2015). The statement on "total chlorine," which should be zero only refers to chlorine gas and hypochlorite (OCl^-), which are used as disinfectants and impart an unpleasant flavor. Tasteless ion chloride (Cl^-) on the other hand is not restricted. An optimum total dissolved solids (TDS) value of 150 mg/L is stated; however, the standard method of determining TDS using a conductivity meter has an uncertainty of approximately 30%. This uncertainty is primarily caused by a varying conversion factor for the transformation of the effective measurement of electrical conductivity from μS/cm to TDS in mg/L, which, depending on the mineral composition, can vary between 0.5 and 1.0. Additionally, most so-called TDS-meters do not measure and take into account water temperature, which can lead for example, to a further 20% variation in the measurement for a 10°C change in temperature (e.g., tap temperature to room temperature). Furthermore, it is not clear which anions are suggested to achieve charge neutrality. Fig. 16.3 summarizes the recommended water compositions for coffee extraction according to the SCAA, Colonna-Dashwood and Hendon (2015, given in ppm $CaCO_3$ without transformation), Rao (2008), and Leeb and Rogalla (2006). The figure also contains descriptors illustrating taste

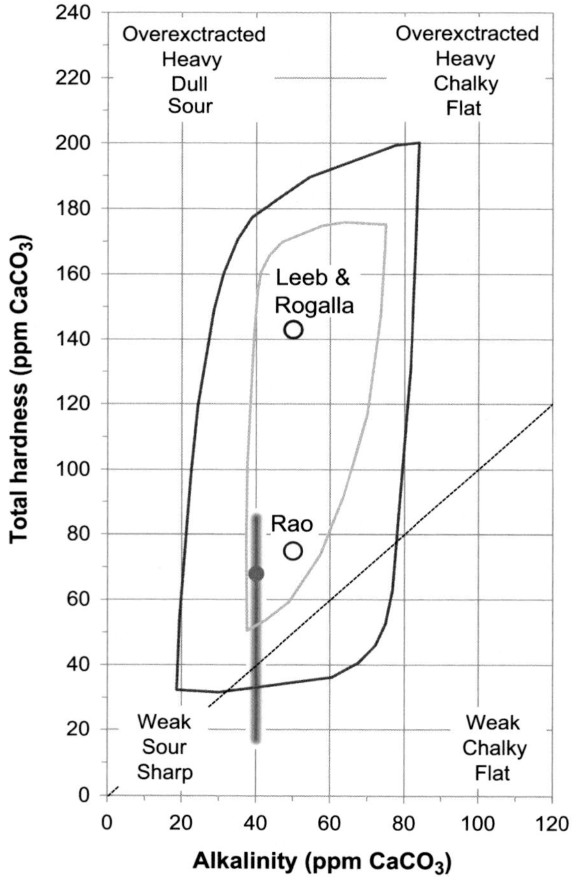

FIGURE 16.3 Recommended water compositions.

imbalances or off-notes that arise from suboptimal values for either total hardness (impact on extraction efficiency) or alkalinity (impact on degree of buffering of the coffee acids) based on reports from Rao (2013) and Colonna-Dashwood and Hendon (2015). As explained in Section 3.4, a deviation from the diagonal line indicates the presence of ions other than magnesium, calcium, and hydrogen carbonate. For water with a total hardness higher than its alkalinity, as in most recommendations, this surplus is associated with the presence of chloride or sulfate.

5. WATER TREATMENT

In this section, we will present a novel and practical way of describing water treatment methods in terms of hardness and alkalinity. In general, water treatment methods can be classified into five common categories:

- Filtration: Removal of particles, microbes, or organic compounds responsible for off-flavors.
- Ion exchanger: Cation exchange [e.g., magnesium and calcium ions are exchanged for hydrogen ions (protons) or sodium ions]—combined cation, and anion exchangers produce deionized, almost pure H_2O, which is then mixed again with tap water to adjust its mineral content.
- Reverse osmosis: Nonselective removal of all dissolved solids by filtration through a semipermeable membrane, i.e., only permeable for water but not for other components present in water.

Other methods that can even further purify water are distillation and precipitation. However, these methods are expensive and/or elaborate and are not commonly used for water for coffee brewing.

With respect to coffee extraction, odor free, clean drinking water can be accurately characterized by its hardness (as a measure of extraction efficiency) and alkalinity (as a measure of its acid buffering capacity). Fig. 16.4 provides an overview of the changes in hardness and alkalinity resultant from the most common treatment methods, using two different initial water compositions **1** and **2**:

- **a**: Softener—Cation exchanger: Ca^{2+} and Mg^{2+} for potassium (K^+) or sodium (Na^+) only affects hardness and is, therefore, vertically oriented in Fig. 16.4. The alkalinity is not affected.
- **b**: Decarbonizer–Cation exchanger: Ca^{2+} and Mg^{2+} for H^+—oriented diagonally with a slope of 1. The net effect is that the change in alkalinity equals the change in hardness. See Section 6.2 for a discussion of the effect of this treatment on espresso extraction.
- **b***: Combination of decarbonizer (**b**-type) with a small fraction of softener (**a**-type).
- **c**: Demineralizer: Reverse osmosis (RO) or deionizer by ion exchange—removing ions irrespective of the initial composition; RO can also be used to increase the mineral content by mixing concentrated water with untreated water before the membrane—oriented toward the point of origin (0/0) or away from it.
- **d**: Dealkalizer: Anion exchanger: HCO_3^- for Cl^- (not yet commercially available for coffee applications) or addition of a strong acid (e.g., HCl). This will lead to a reduction of the alkalinity, without affecting its total hardness.
- e : not shown: Cation exchange of Ca^{2+} for Mg^{2+}—does not change either hardness or alkalinity, so both values remain constant. However, the mineral composition of the water will be altered.

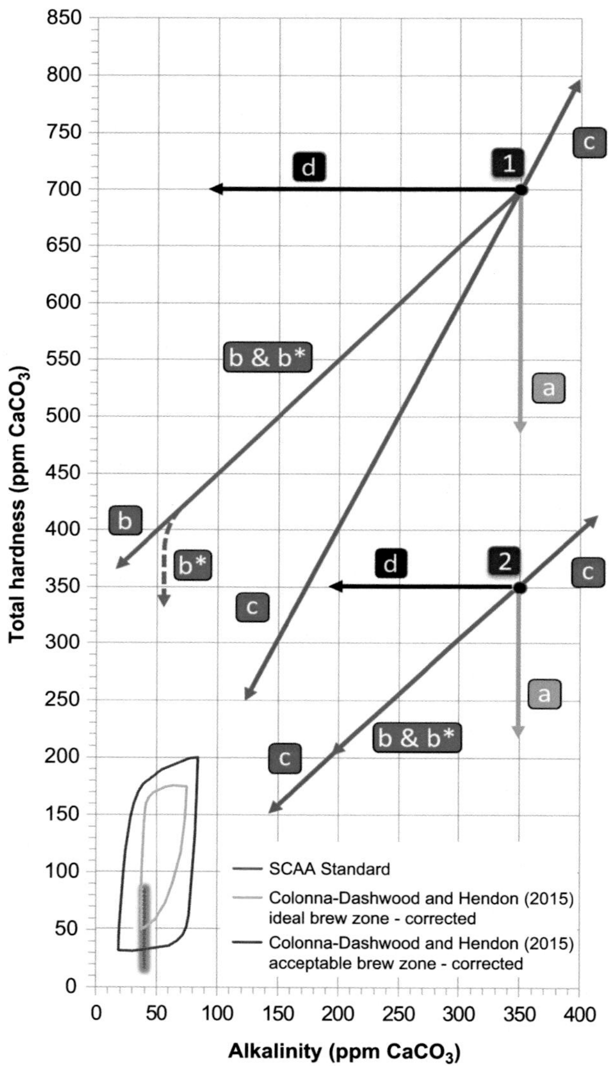

FIGURE 16.4 Changes in hardness and alkalinity for five treatments and two initial compositions (1 and 2). a: Cation exchanger: $Ca^{2+} + Mg^{2+}$ versus K^+ or Na^+. b: Cation exchanger: $Ca^{2+} + Mg^{2+}$ versus H^+ b*: Mixture of mostly b-type and some a-type ion exchanger. c: Demineralizer; d: anion exchanger: HCO_3^- versus Cl^-.

6. APPLIED EXAMPLES

In this section, we explore how the contents of the previous section can be combined to investigate the effect of different water compositions on extraction. This involves the determination of the starting water composition, the choice of a target water composition, and a suitable water treatment method for achieving the desired change in composition.

6.1 Cupping With Different Waters

To investigate the impact of water composition on the sensory properties of coffee extraction, a cupping experiment was conducted (Wellinger and Yeretzian, 2015). This experiment involved three different water compositions prepared by using a decarbonizer-type ion exchanger that reduces total hardness and alkalinity in equal amounts. The water compositions obtained are shown in Fig. 16.5 and the alkalinity is varied by ±50% relative to the SCAA target of 40 ppm $CaCO_3$ alkalinity. For reference, the water compositions as used in the World Barista Championship (WBC) 2013 in Melbourne and 2014 in Rimini are also depicted (data from Colonna-Dashwood and Hendon, 2015).

The experiment used the following two coffees:

1. Colombia, washed process, Caturra and Castillo, screen size 15, La Argelia farm, Tolima region, 1580−1900 m.
2. Brazil, natural process, yellow bourbon, Tupi, Icatu, and Yellow catuai, screen size 16−18, Lagoa Formosa farm, Minas Gerais region, 1000−1200 m.

The coffees were prepared in a blind tasting session according to the SCAA cupping protocol with slight modifications to the sensory attributes: 13.8 g of freshly ground coffee was used with 250 mL water at 93°C. Fragrance, uniformity, and clean cup scores were not evaluated. Sweetness was evaluated on a scale of 1−10. Fig. 16.6 shows the average scores computed from three repetitions and three cuppers: either certified Q grader or WBC judge. When significant differences were observed, the highest score was attributed to the coffee samples brewed using low or medium mineral content water. It was generally observed that the water with the lowest mineral content scored highest for most attributes in both coffees. Across the spectrum, the Colombian coffee scored higher than the Brazilian one. For both coffees, significant differences in the flavor and acidity attributes were observed ($\alpha = 5\%$), with the lowest hardness and alkalinity scoring the highest. Increased hardness and alkalinity resulted in gradually lower scores. Additionally, the Colombian coffee exhibited significant differences in the scores for balance and overall, whereas the scores for aftertaste were significantly different for the Brazilian coffee.

The results show that even relatively small changes in hardness and alkalinity significantly affect the sensory attributes of a coffee. Although

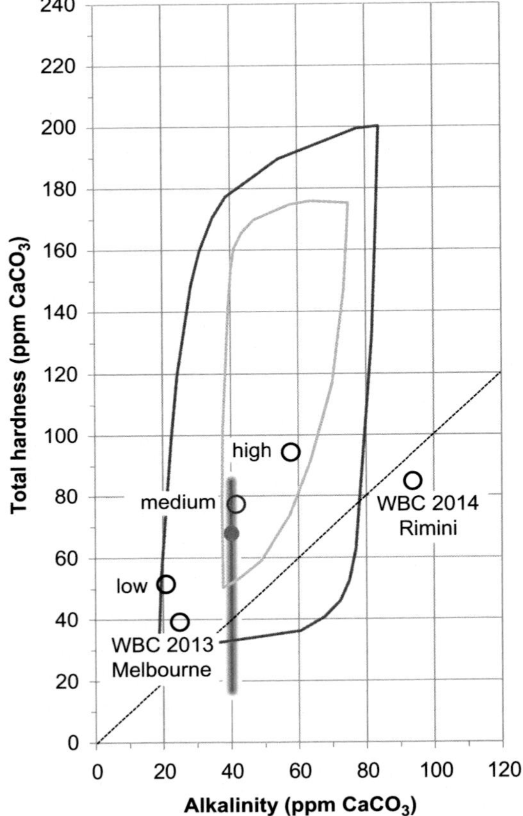

FIGURE 16.5 Water compositions used compared to existing recommendations and WBC competitions water.

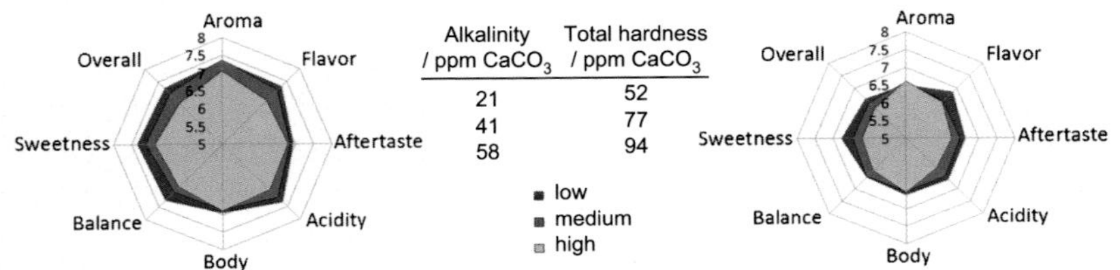

FIGURE 16.6 Cupping scores of two different coffees prepared with water containing three different levels of hardness and alkalinity: low, medium, and high (Wellinger and Yeretzian, 2015).

further research is needed to clarify the mechanisms that cause these attribute changes, we suspect that this is the combination of two effects: (1) the total hardness affecting the extraction efficiency and (2) the alkalinity affecting the perceived acidity. In summary, the results demonstrate that although general tendencies can be formulated for how a specific change in hardness or alkalinity impacts the sensory attributes of a coffee beverage, there are differences between different coffees that are closely linked to their overall flavor profile. The fact that the low mineral content water received the highest score for most attributes could be related to the relatively high extraction yield that is typically achieved by the cupping method (attributed to extended contact time compared to other brewing methods), which also leads to the perception of heavier body compared to drip coffee for instance. Hence, the results shown here might well be quite different if repeated with the same coffees but another extraction method, such as filter.

6.2 Water Decarbonization and Espresso Extraction

A common problem encountered when using a decarbonizer, exchanging Ca^{2+} and Mg^{2+} for H^+, to treat hard water for use in espresso machines is the formation of carbonic acid, which leads to excess dissolved carbon dioxide in the water. The protons that are released, in exchange for Ca^{2+} and Mg^{2+}, neutralize the hydrogen carbonate present in the water forming carbonic acid that can in turn degas as carbon dioxide, depending on the pressure and temperature. In a café where the water from the ion exchange cartridge stays pressurized all the way up to the espresso machine, the carbonic acid cannot escape as carbon dioxide and subsequently can significantly impact the extraction. Carbon dioxide becomes more soluble as pressure increases and less soluble as temperature increases. When the water containing high amounts of carbonic acid enters the extraction chamber (basket) it creates an additional resistance (see Section 4) in an identical way to the carbon dioxide content from the freshly roasted and ground coffee. Therefore, when hard water is treated with a decarbonizer, a similar effect may be observed in espresso extraction as observed for coffee that has been freshly roasted before extraction (typically <3 days postroast). Freshly roasted coffee forms a large amount of crema with untypically large bubbles that collapse much quicker than those formed from the same coffee that has been rested for a longer period. For example, a reduction in the alkalinity by 200 ppm $CaCO_3$ alkalinity will increase dissolved carbon dioxide by 176 mg/L. Therefore, if we consider the extreme example of extracting a double espresso using 15 g of Arabica coffee that has been freshly ground (2 min) just 1 h after roasting using 30 g of decarbonized water, the water would add another 20% to the carbon dioxide already contained in the coffee grounds. The excess carbon dioxide produced by decarbonization could be removed by a process called degassing, which is well established in other industrial applications. Alternatively the water could be treated with a demineralizer that removes all ions without generating additional dissolved carbon dioxide.

7. CONCLUSIONS AND OUTLOOK

Given clean water that is free of particulates and off-flavors, a water sample can be accurately characterized for its use in coffee extraction by the measurement of hardness and alkalinity; both can be measured by titrations, for which simple to use commercial kits exist. Although research and industry recommendations for hardness levels propose a relatively wide ideal range (as much as a factor of 5), the recommended variation for alkalinity is much smaller (factor of 2). As explained in Section 5.3 the conductivity of water, as measured by a conductivity meter (sometimes incorrectly called a TDS-Meter), is not a meaningful parameter on its own, except for obtaining a rough estimate of the total solid content. However, for any given water source (e.g., tap water or bottled water) and a defined water treatment method, measuring the conductivity and correcting for temperature offers the opportunity to detect changes in the output water composition caused by changes in either the tap water composition or the efficiency of the water treatment method. Since conductivity can be measured with inexpensive online sensors, this offers the opportunity to perform continuous real-time monitoring of water quality.

The aim of this chapter has been to provide a scientific and quantitative framework to discuss water, its composition and its impact from the perspective of the coffee industry. This includes a description of the various water treatment methods used within the (specialty) coffee industry and how water quality impacts the extraction process as well as the sensory properties of the brew. In the past, water was often regarded as something exclusively relevant to technical issues, such as maintenance intervals and the like, rather than a relevant parameter for coffee extraction and the quality of the resulting cup. As shown in Section 6.1 the sensory impact of different water compositions is also dependent on the specific type of coffee used. Therefore, the existence of a single optimum value for total hardness and alkalinity is most probably unlikely, because this optimum depends on the coffee (both its green quality and the roasting process applied), the extraction method, and last but not the least the sensory preferences of the targeted consumer group for a specific beverage. As water comes into focus as an important parameter for the sensory properties, this could also impact the way coffee is roasted. Many roasters might have chosen their style of roasting based on the specific water composition they have, since it would take a significant additional effort to test every roast with different waters (Colonna-Dashwood and Hendon, 2015).

To extend our database and experience, areas for future research should include applied studies on the impact of water composition on extraction efficiency and the sensory impact on espresso coffee. Since the ratios of water to coffee for espresso and for filter coffee are very different, much higher total hardness and alkalinity (both over 100 ppm $CaCO_3$) might still result in a high-quality espresso beverage. Nevertheless, the use of such water will lead to scale formation and is, therefore, not recommended on a permanent or long-term basis. Moreover, the impact of water composition on flow speed and therefore contact time must also be considered when designing further experiments to determine the influence of water composition on the resulting cup. As demonstrated with the experiments in Section 6.1, not every coffee will be affected identically by a specific change in water composition. In addition, the impact of, for example, pH or specific substances present in the water on aroma binding or volatility is an unexplored field in coffee. Therefore, further investigations on the effect of different parameters, such as total hardness and alkalinity, on different coffees and their sensory attributes represent a promising and important field of research. The often suggested differences between calcium and magnesium should also be further investigated—it is suggested that the impact of identical concentrations of each on the sensory profile differs. In case there are significant differences, we recommend reassessing the current standard for total hardness as a sum of calcium and magnesium to a new standard where the calcium and magnesium content are measured and considered individually to achieve optimum results in the cup.

ACKNOWLEDGMENTS

The original research presented in this chapter was financed by the Specialty Coffee Association of Europe (SCAE) and the Zurich University of Applied Sciences (ZHAW).

REFERENCES

American Public Health Association, 2012. American Water Works Association, Water Environment Federation; Standard Methods for the Examination of Water and Wastewater, twenty-second ed.

ASTM D 1126, 2002. Standard Test Method for Hardness in Water.

Clifford, M.N., 1988. What factors determine the intensity of coffee's sensory attributes. Tea & Coffee Trade Journal 159, 8–9.

Colonna-Dashwood, M., Hendon, C., 2015. Water for Coffee; Self-Publication. Bath, UK.

International Union of Pure and Applied Chemistry (IUPAC), 2013. Periodic Table of the Elements. iupac.org/reports/periodic_table.

DIN 38409-6, 1986. German Standard Methods for the Examination of Water, Waste Water and Sludge, Summary Indices of Actions and Substances (Group H), Water Hardness (H 6). EPA, United States Environmental Protection Agency. EPA Method 130.2; Hardness, Total (mg/L as $CaCO_3$) (Titrimetric, EDTA); 1999.

Environmental Protection Agency, Method 130.2, 1982. Total Hardness, mg/L as $CaCO_3$, titrimetric by EDTA.

Fond, O., 1995. Effect of water and coffee acidity on extraction: dynamics of coffee bed compaction in espresso type extraction. In: Proceedings of the 16th Colloquium. ASIC, Kyoto.

Gloess, A.N., Schoenbächler, B., Klopprogge, B., et al., 2013. Comparison of nine common coffee extraction methods: instrumental and sensory analysis, European Food Research and Technology, 236, 607–627.

Hendon, C., Colonna-Dashwood, L., Colonna-Dashwood, M., 2014. The role of dissolved cations in coffee extraction. Journal of Agricultural and Food Chemistry 62 (21), 4947–4950.

Hem John, D., 1985. Study and Interpretation of the Chemical Characteristics of Natural Water, third ed. U.S. geological survey water-supply paper 2254.

Leeb, T., Rogalla, I., 2006. Kaffee, Espresso & Barista, fourth ed. TomTom Verlag.

Navarini, L., Rivetti, D., 2010. Water quality for Espresso coffee. Journal of Food Chemistry 122, 424–428.

Pohling, R., 2015. Chemische Reaktionen in der Wasseranalyse.

Rao, S., 2008. The Professional Barista's Handbook. Self-Publication, USA.

Rao, S., 2013. Espresso Extraction: Measurement and Mastery. Self-Publication, USA.

Rivetti, D., Navarini, L., Cappuccio, R., Abatangelo, A., Petracco, M., Suggi-Liverani, F., 2001. Effect of water composition and water treatment on espresso coffee percolation. In: Proceedings of the 19th ASIC Colloquium; Trieste, Italy.

Sanchéz, J.A., Wellinger, M., Gloess, A.N., Zimmermann, R., Yeretzian, C., 2016. Extraction kinetics of coffee aroma compounds using a semi-automatic machine: on-line analysis by PTR-ToF-MS. International Journal of Mass Spectrometry 401, 22–30.

Specialty Coffee Association of America (SCAA), November 21, 2009. SCAA Standard: Water for Brewing Specialty Coffee.

Schweizerischer Verein des Gas- und Wasserfaches (SVGW), 2008. Wasserhärte: Was muss beachtet werden, TWI 13.

Wellinger, M., Yeretzian, C., 2015. Water: why quality matters. Cafe Europa 61, 20–22. SCAE.

Chapter 17

Crema—Formation, Stabilization, and Sensation

Britta Folmer[1], Imre Blank[2], Thomas Hofmann[3]
[1]*Nestlé Nespresso SA, Lausanne, Switzerland;* [2]*Nestec Ltd., Nestlé Research Center, Lausanne, Switzerland;* [3]*Technical University of Munich, Freising, Germany*

1. INTRODUCTION

A large percentage of coffee consumed today is in the form of espresso. When comparing espresso coffee to coffee brewed using other techniques, one of its main characteristics is a dense brown layer of foam bubbles, also called crema, which covers the liquid coffee (Illy and Viani, 2005). Coffee experts use the crema to judge the quality of the extraction. For instance, crema indicates whether all parameters that influence extraction, such as degassing, grinding, tempering, and water pressure, were just right when extracting that one cup. Many consumers prefer the presence of a nice crema layer on their coffee, but it is also part of the consumption ritual. Some will spoon it off, some will stir it in, and some will swirl the cup to mix the crema into the last sip of coffee. At the same time, beautiful visuals of the crema are used to create expectations of an indulgent, smooth, and flavorful cup of espresso.

For most baristas, the formation of the crema is a craft rather than a science. A properly packed coffee bed allows a slow flow of the pressurized water. This pressure, along with the carbon dioxide (CO_2) present in the fresh roast and ground coffee and the carbonates present in the water, is the main driver for the formation of the crema. Baristas know how to adjust grinding and temping to optimize the extraction and obtain the desired color, quantity, and delicacy of the crema.

Although the crema's dark brown color or "tiger skin" and fine bubble size are signs of a good extraction, they are only secondary indications of a good extraction and tasting experience, after the flavor of the coffee. At the same time, crema has been associated with the increased "body" for which Robusta espresso coffees are well known (Navarini et al., 2004a). Studies also suggest

The Craft and Science of Coffee. http://dx.doi.org/10.1016/B978-0-12-803520-7.00017-7

that coffee foam can help distinguish Robusta from Arabic coffee (Maeztu et al., 2001a,b). However, the question is "What is the impact of color, bubble size, and crema quantity on the coffee experience?" Is there one crema that is optimal, or are crema characteristics part of the overall coffee sensory profile? We need to answer these questions before we can reflect on how to extract the perfect cup of espresso.

As a first step, it is important to understand the physico-chemical aspects of crema, and how we can fine-tune crema's properties to optimize its quality. To do this, we need to distinguish crema formation from crema stabilization. In crema formation, energy is required to create an air in liquid dispersion. With espresso coffee, we add this energy by injecting pressurized water onto the coffee bed. A typical freshly prepared crema of an espresso is shown in Fig. 17.1. We also need to stabilize the freshly formed air bubbles, which we can do via the adsorption of interfacially active compounds at the air–liquid interface.

In addition, the conditions we apply when we brew espresso coffee influence surface tension–related phenomena such as foam formation and stabilization. For instance, both the presence of carbonate in the water and the gases in the coffee have been reported as key drivers of crema formation (see Chapter 16).

Unfortunately, systematic chemical and physical studies are currently too scarce to completely understand this complex system. The most recent and comprehensive review has been published by Illy and Navarini (2011), indicating the current gap in molecular understanding in crema formation and stabilization. It is only very recently that the role of crema on the consumer experience has been investigated, with an emphasis on how crema impacts visual aspects, as well as smell and taste.

In this chapter, we will review physico-chemical processes and properties of crema formation and stabilization. We will also explain why we need to

FIGURE 17.1 Photo of an espresso crema extracted using a high pressure machine and an Arabica coffee blend (using the Nespresso extraction system).

consider the characteristics of a good crema in a wider perspective, including the full sensory experience.

2. CREMA FORMATION AND STABILITY

Foam is a complex and challenging phenomenon consisting of four key events (1) bubble formation, (2) bubble rise, (3) drainage, and (4) coalescence and disproportionation (Bamforth, 2004). In essence, foam is a coarse dispersion of gas bubbles in a liquid continuous phase. With espresso coffee, the gas phase consists primarily of the CO_2 generated during coffee roasting, which is partially entrapped within the cell structure. This continuous phase is an oil in water emulsion of microscopic oil droplets (<10 μm) in an aqueous solution of several coffee constituents (e.g., sugars, acids, proteins) as well as small solid coffee cell-wall fragments (2−5 μm) (Illy and Viani, 2005). The bubble shown in Fig. 17.2 is about 100 μm in diameter and appears to be covered by protein rich material.

The espresso crema can be classified as a metastable foam with a specific lifetime (Dickinson, 1992). In most cases, it takes up to 40 min before the crema disappears (Dalla Rosa et al., 1986). As the crema ages, its properties evolve from a liquid fine foam in freshly prepared espresso to a dry polyhedral foam upon aging. Ideally, the crema should represent at least 10% of the

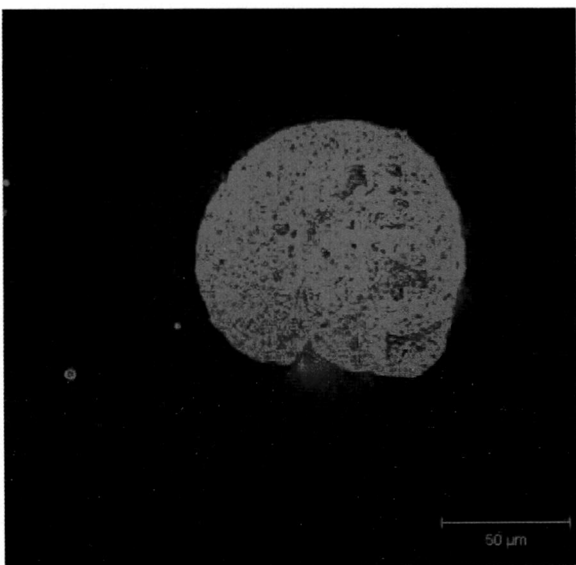

FIGURE 17.2 Confocal microscopy image of an isolated air bubble in crema prepared from a freshly extracted pure Arabica coffee using the Nespresso extraction system. Rhodamine was used for staining of proteins. *Courtesy of E. Kolodziejczyk, Nestlé Research Center, Lausanne.*

volume of an espresso (Illy and Viani, 2005) with a foam density of 0.30–0.50 g/mL (Navarini et al., 2006). The latter authors have shown a linear relationship between CO_2 content and foam weight as well as foam volume (Fig. 17.3).

There have been several attempts to describe how crema is formed in espresso coffee. As water is forced through the coffee under pressure, it emulsifies the coffee oils into the extracted liquid. In addition, roasted coffee out-gases CO_2 for a while (degassing), and the longer the coffee is exposed to ambient pressure, the more CO_2 it will release. This is why packaging materials for espresso coffee are sometimes made such to keep overpressure and CO_2 in the coffee matrix. The remaining CO_2 is then emitted during extraction.

Although CO_2 has been frequently suggested as the gas phase responsible for espresso coffee foaming, coffee experts have not investigated the bubble formation mechanism in detail. The relationship between CO_2 chemistry and foam formation has on the other hand been studied. In fact, research suggests that the bicarbonate–carbonic acid equilibrium plays a role in the dynamics of the transient phase of the espresso extraction (Fond, 1995, Chapter 16).

The transient phase of the extraction is the initial wetting stage of brewing espresso when hot water spreads into the voids in coffee particles, and, inter- and intraparticle gas is simultaneously pushed out of the coffee bed (Petracco and Liverani, 1993). This mass transfer between coffee particles and water occurs simultaneously, and the bicarbonate ions present in the extraction fluid

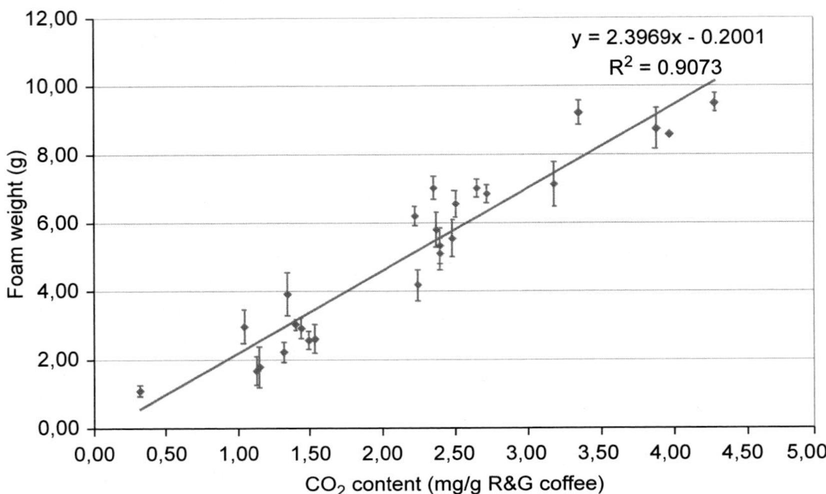

FIGURE 17.3 Espresso coffee foam volume as a function of carbon dioxide content per gram of roast and ground coffee. *From Navarini et al. (2006).*

(water), along with the displacement of its equilibrium according to the pH evolution during brewing (from 7.0 to 7.5 to 5.5−5.0) generate chemical reactions that occur at high temperatures in the coffee bed. The coffee is compacted through the water pressure, swelling of the coffee grounds, and through CO_2 degassing, which in turn generates the foam and emulsion that creates the much appreciated espresso crema.

Other research has suggested that CO_2 supersaturation conditions in coffee are a possible driving force for the formation of espresso coffee foam (Navarini et al., 2006). More specifically, the CO_2 solubilization (present in the coffee bed) in water at a high pressure and temperature may cause supersaturation conditions in the final cup, resulting in nucleation of tiny bubbles.

To explain this further, the CO_2 concentration in water might be below the CO_2 solubility during extraction at high pressure and with a water temperature of 100°C. However, it may also be above solubility at 1 bar and 70°C. These conditions are compatible with the foaming that occurs through bubble formation when there is heterogeneous nucleation and bubble rise following a phase transition (from high pressure to ambient pressure) when the high pressurized water exits the coffee bed and enters the cup. This effect, which is known as effervescence, is also observed with champagne. In this case, the effervescence may apply to the effect observed immediately after espresso preparation. With espresso, the micronic solid particles and submicronic cell-wall fragments that are present in the beverage may act as nucleation sites. In addition, the small volume of an espresso beverage offers a limited length of about 1.5−2 cm for bubble rise, and this leads to characteristically tiny bubbles in espresso foam (Illy and Navarini, 2011).

After the foam is developed, foam destabilization generally occurs through three phenomena (Prins, 1988). First there is a coalescence among the bubbles, which occurs when the film separating the bubbles collapses. Second, there is Ostwald ripening, which takes place when the foam is polydispersed in sizes. In this phase, pressure differences inside different sized bubbles diffuses the smaller bubbles into the larger ones. Third, gravity forces the liquid to separate from the air bubbles. This in turn causes the film to thin, which leads to coalescence and Ostwald ripening.

Although these phenomena are observed in any foam system, the high temperature of crema, and the way it cools from the top down will add another complexity to understanding the physical phenomena in the formation and stability of espresso crema. A previous study had suggested that the relatively high temperature of the beverage may negatively impact foam stability. The premise was that the water evaporated and caused the foam to collapse by reducing the thickness between bubbles (Navarini et al., 2006).

Another study has shown that lipid content can also affect foam stability. In a regular espresso (25 mL), the total lipids range from 45−146 mg for Arabica to 14−119 mg for Robusta (Petracco, 1989; Maetzu et al., 2001). On average, pure Arabica espresso contains a higher content of total lipids than Robusta espresso, and therefore the probability of lipid-induced foam destabilization is higher for Arabica.

Since espresso coffee is well known to contain emulsified lipids, the lipid-induced foam destabilization may also occur through oil spreading at the air-beverage interface (Schokker et al., 2002). This is consistent with the lower surface tension generally observed in pure Arabica espresso as compared to pure Robusta espresso (Petracco, 2001).

Studies suggest solid particles also affect foam stability. Such particles in a pure Arabica dry espresso foam are shown in Fig. 17.4. They are located in the plateau border, suggesting the tendency to be unattached. During drainage, unattached particles predominantly follow the net motion of the liquid, suggesting a stabilizing role of the solid particles within the crema. Hydrophobic particles may destabilize the foam by film bridging, which can cause dewetting of the surface. Although the wetting nature of the solid particles present in espresso coffee has not yet been investigated, the "tiger skin" effect observed in Arabica espresso may suggest that the solid particles have a certain foam stabilizing role (also known as pickering effect). Otherwise the antifoaming effect would be rather rapid (Kralchevsky et al., 2002).

However, the different foam adhesion phenomena observed between the two coffee species may also be due to different transition rates from liquid to dry foam. For instance, the solid-like nature of Robusta espresso foam may come from a higher drainage rate, which seems to be the prerequisite for adhesion. In contrast, the Arabica espresso foam rheology seems to be of the liquid-viscous type for a longer time.

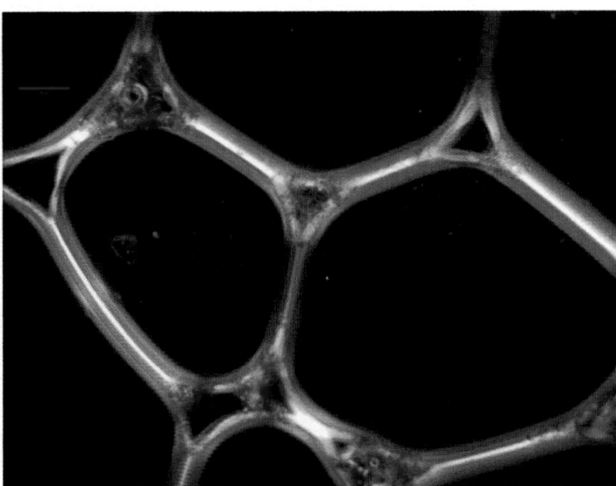

FIGURE 17.4 Optical microscopy image of a dry pure Arabica regular espresso coffee foam with the scale bars representing 50 μm. *From Illy and Navarini (2011).*

3. THE CHEMISTRY OF CREMA STABILIZATION

No detailed studies have been published so far on the chemical compounds responsible for the formation and stabilization of espresso crema. Nunes et al. (1997) found that the foamability increased linearly with the degree of roast. They also found a dependence on the amount of protein in the infusion. Foam stability of espresso coffee was found to be related to the amount of the polysaccharides galactomannan and arabinogalactan. Other dependent variables observed were total solids, pH, lipids, protein, and carbohydrate contents. A strong correlation was found between foam stability and high molecular weight compounds, suggested to consist of complexes between polysaccharide, protein, and phenolic compounds caused by the roasting process (Nunes and Coimbra, 1998).

Recent activity-guided studies on foam-active components in green and roasted coffee has given some new insight into the molecular structures of the complex molecules responsible for crema formation. Although it was found that sucrose fatty acid esters (Fig. 17.5A) are the primary foam-active agents in green coffee, low- and high-molecular weight 4-vinylcatechol oligomers (Fig. 17.5B) are the key foam-active components in the fine crema of roasted coffee (Unpublished data Kornas and Hofmann et al.).

Quantitative analysis of these foam modulators by means of liquid chromatography tandem mass spectrometry (MS/MS) demonstrated the sucrose esters to be natural products in the green coffee and to be gradualy degraded upon roasting with increasing roasting time (Fig. 17.6A). In comparison, the 4-vinylcatechol oligomers were observed to be generated with increasing roasting time (Fig. 17.6B).

As the low- and high-molecular weight 4-vinylcatechol oligomers have been found to be generated upon thermal degradation of chlorogenic acids and caffeic acid, respectively, caffeic acid was thermally treated for 20 min at 220°C and, then, added to espresso coffee beverages in concentrations of 0.007%, 0.035%, and 0.07% prior to ultraturrax-mediated crema generation. The foam volume was significantly increased with increasing levels of roasted caffeic acid, e.g., more than 60% increased foam volume was measured when the coffee beverage was spiked with 0.07% roasted caffeic acid (Fig. 17.7). For the first time, these data demonstrated low- and high-molecular weight 4-vinylcatechol oligomers, generated upon roasting from caffeic acid moieties, as key contributors to the crema of roasted coffee beverages.

To characterize the impact of the foam active components on the foam structure of the crema, the foam was collected from an espresso coffee, that was spiked with sucrose palmitate (0.05%) or roasted caffeic acid (0.007%), and from nonspiked espresso (control) and then analyzed by means of a foam scan instrument equipped with a camera. The foam pictures shown in Fig. 17.8 did not show any differences in the initial phase of roasting, but clear differences were observed in the foam structure after 300 s. The pure espresso

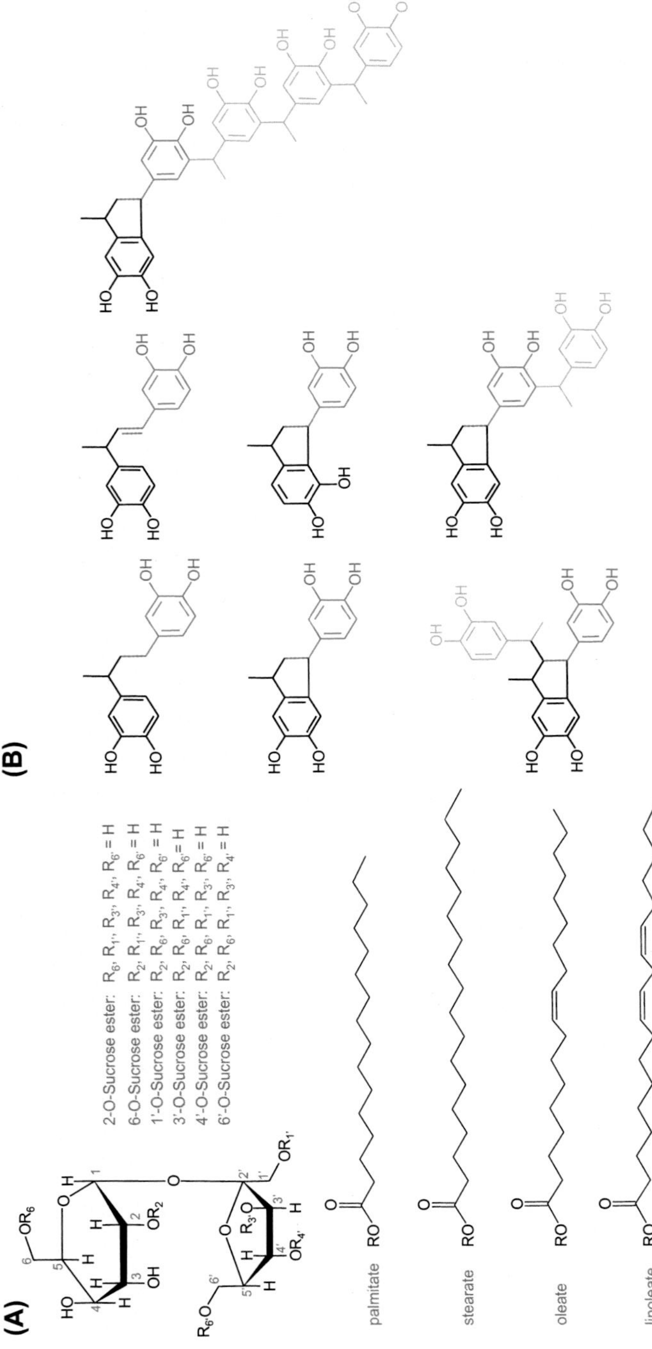

FIGURE 17.5 Chemical structures of (A) sucrose fatty acid esters identified foam from green coffee beans, and (B) 4-vinylcatechol oligomers (monomers highlighted by different color) identified in foam roasted coffee beans (Unpublished data Kornas and Hofmann et al.).

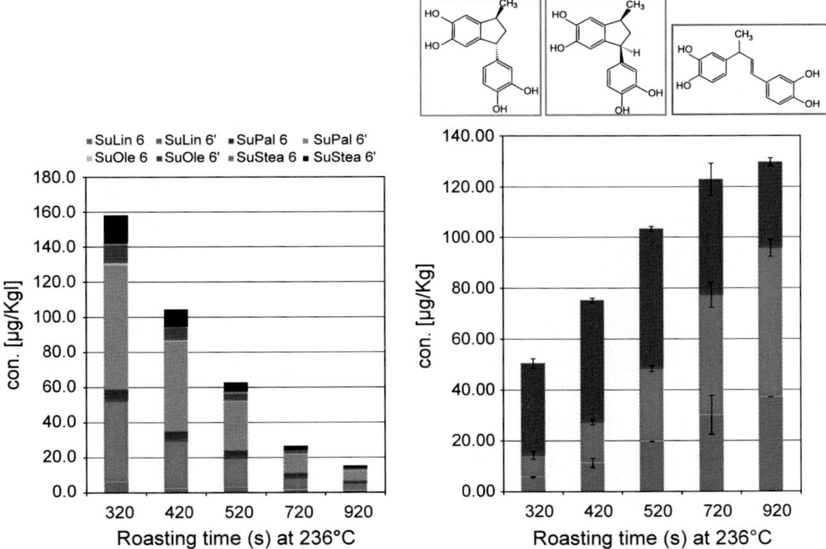

FIGURE 17.6 Influence of roasting time on the degradation of sucrose fatty acid esters (A) and the formation of a selection of 4-vinylcatechol oligomers (B) (Unpublished data Kornas and Hofmann et al.). *Su*, sucrose, *Lin*, linoleic acid; *Pal*, palmitic acid; *Ole*, oleic acid; *Stea*, stearic acid.

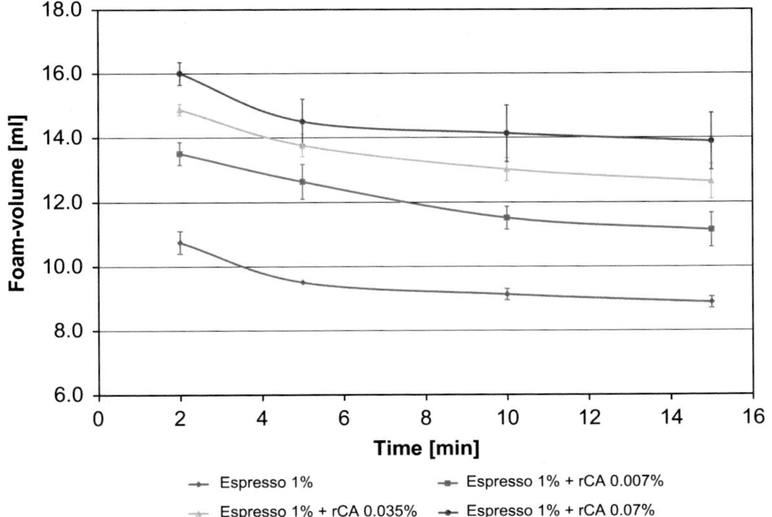

FIGURE 17.7 Influence of roasted caffeic acid on foamability and foam stability of a coffee beverage made from an instant espresso powder (1 g/100 mL) (Unpublished data Kornas and Hofmann et al.).

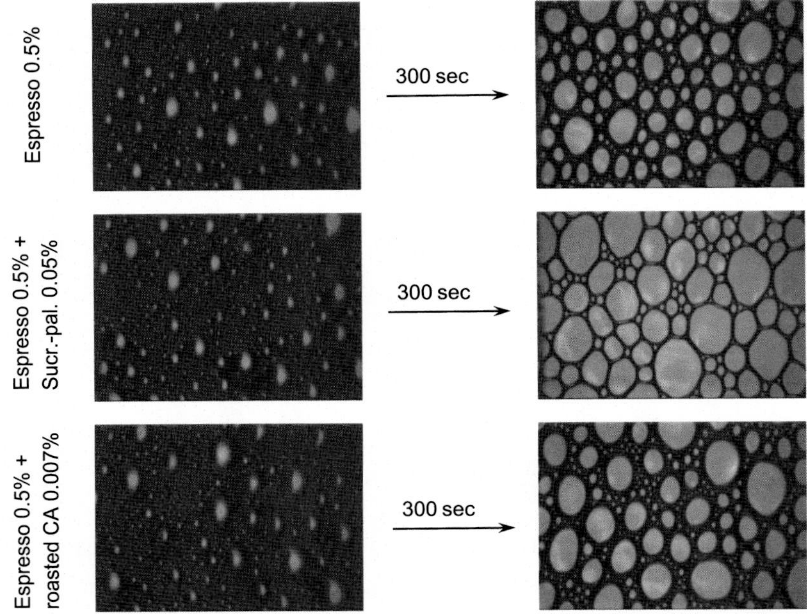

FIGURE 17.8 Foam structure of expresso spiked with sucrose palmitate (0.05%) and roasted caffeic acid (0.007%) measured after 300 s by means of a foam scan (Teclis IT-concept) (Unpublished data Kornas and Hofmann et al.).

coffee as well as the coffee spiked with roasted caffeic acid showed comparable foam structure with smaller bubbles and a higher amount of liquid between the bubbles, whereas the foam generated with coffee spiked with sucrose-palmitate shows a bigger bubble with lower amount of liquid phase between the bubbles. Therefore, the foam of the coffee spiked with roasted caffeic acid was comparable with the original espresso coffee foam in contrast to the foam of coffee spiked with fatty acid sucrose esters, which is a more coarse and dry foam. These data clearly demonstrate the 4-vinylcatechol oligomers, rather than the sucrose esters, to play a key role in determining the fine-structured crema of a premium espresso.

These results indicate that Robusta, being richer in caffeic acid and chlorogenic acid than Arabica coffee, will thus promote the formation of the high- and low-molecular weight 4-vinylcatechol-based surfactants during roasting, which in combination with the role played by the pressure and CO_2 might explain the foam promoting role of Robusta coffee in espresso blends.

4. CREMA AND THE CONSUMER

Many studies exist on how external factors influence the consumer perception or experience of a given product. For example, we know that the information consumers obtain before they consume a product influences their expectations and ultimately how they rate the quality of that product (Siegrist and Cousin, 2009; Lange et al., 2002). A consumer's level of expertise can also influence quality judgements (Sáenz-Navajas et al., 2013).

In addition, with beverages, the material of the cup or the glass that a consumer uses has a major impact on the way they experience and evaluate that beverage (Schifferstein, 2009; Wan et al., 2015). Studies also show that the way the food packaging is designed can influence various aspects of the food experience (Schifferstein et al., 2013). For all these reasons, it is not surprising that the crema has an impact on the overall espresso tasting experience.

4.1 The Impact of Crema on the Visual Perception

In a recent study by Labbe et al. (2016), researchers examined the influence of crema quantity on consumer perceptions. In this study, consumers evaluated products in three conditions, i.e., (1) visually, and researchers assessed the expectations generated by visual cues, (2) in a "blind" condition where all visual cues were suppressed and researchers assessed the in-mouth perception, and (3) by taste and sight where consumers evaluated the coffee in a standard way. Through these various conditions, researchers could separate the expectations consumers had from a product's visual appearance from their overall product perception.

The study showed that the expectations did not influence hedonic and sensory attributes in a similar manner. Although the presence of crema created high expectations on quality (a hedonic indicator), crema quantity did not impact the expected quality (Fig. 17.9). The researchers also found that the perceived quality of the product when they could see and taste it normally was higher than the blind in-mouth condition. This indicates that visual cues positively impacted the in-mouth experience.

If we look at the research in more detail, we can see that smoothness (a sensory indicator) was strongly associated with crema quantity, since coffees with crema were expected (visual condition) to be smoother than coffee without crema (Fig. 17.10). In fact, perceived in-mouth smoothness also increased with crema quantity, which was most likely due to the physical texture properties of crema. Overall smoothness was rated higher when it was seen and tasted than it was when consumers were in a blind in-mouth condition.

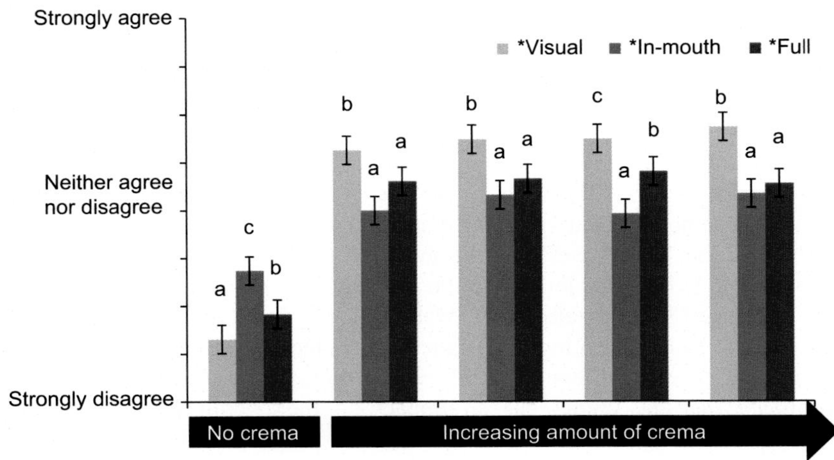

FIGURE 17.9 Average quality score for espressos with increasing crema quantity for the three evaluation conditions.

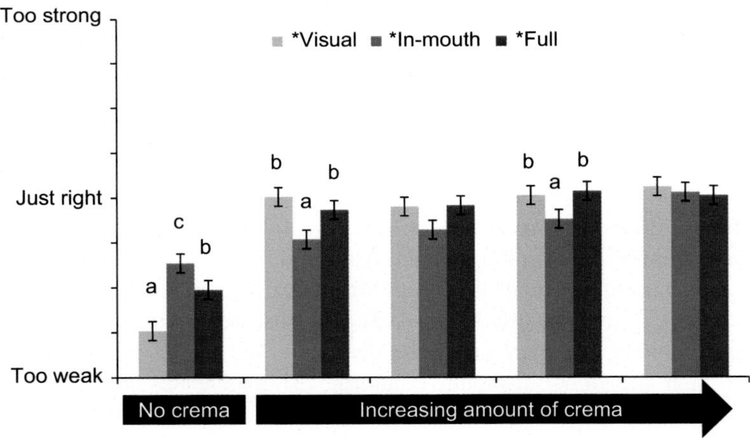

FIGURE 17.10 Average snoothness score for espressos with increasing crema quantity for the three evaluation conditions.

Overall, the study shows that crema is an important component of the espresso coffee experience. In fact, the absence of crema induced low expectations in quality, overall taste, bitterness, and smoothness which in turn reduced perceived quality and sensory attributes.

4.2 The Impact of Crema on the Aroma Release

The term aroma refers to the perception of volatiles through the olfactory system (see also Chapter 18). Volatile organic compounds can reach the olfactory epithelium in the upper part of the nose through two different pathways. Orthonasal stimulation occurs when odor-active compounds enter through the nostrils. This is the pathway that aroma molecules travel when consumers smell the coffee. However, upon tasting the coffee, retronasal stimulation occurs when odor-active compounds enter through the internal nares located inside the mouth. The following section will focus on the mechanism that impacts orthonasal aroma perception.

In one study, researchers used the Nespresso system to investigate the role of crema thickness and stability on aroma release above the cup. They compared coffees with different crema quantities (Barron et al., 2012) and stabilities (Dold et al., 2011) to coffee without crema. Although it had often been suggested that crema acts as a lid that prevents aromas from escaping (Petracco, 2005), the studies instead showed a much more complex mechanism.

Using MS methods, the study traced aroma molecules with different volatilities above the cup as a function of time. Within the first 2.5 min after the start of extraction, the presence of crema generally generated an above the cup volatile concentration that was significantly higher than that found in liquid coffee without crema (Figs. 17.11 and 17.12).

In fact, the rupture of the thin lamella located at the foam surface causes the bubbles to collapse and releases entrapped gas, containing mainly high-volatile aromas. The thinning of the films through evaporation is considered the main cause for this bubble rupture (Dold et al., 2011; Weaire and Hutzler, 1999), but other phenomena described in Section 2 may also play a role. In addition, as the liquid evaporates, the low-volatile aromas also evaporate from the liquid into the air. In contrast, when there is no crema present, all aroma release (high and low volatiles) is driven purely by evaporation from the liquid to the gas phase.

After the initial release of aromas, crema stabilization begins playing a major role in the release pattern. When there is low crema stability, the coffee releases the largest amount of aromas into the headspace. As the same time, crema volume is inversely correlated to the release of low volatiles, most likely because the crema acts as a "lid," preventing aromas from escaping.

The hypothesis that the crema layer acts as a means to liberate aromas above the cup when the crema layer is unstable was confirmed by Parenti et al. (2014) who compared different espresso preparations. This study showed an inverse relationship between crema quantity and stability when the aroma quantity was measured above the cup.

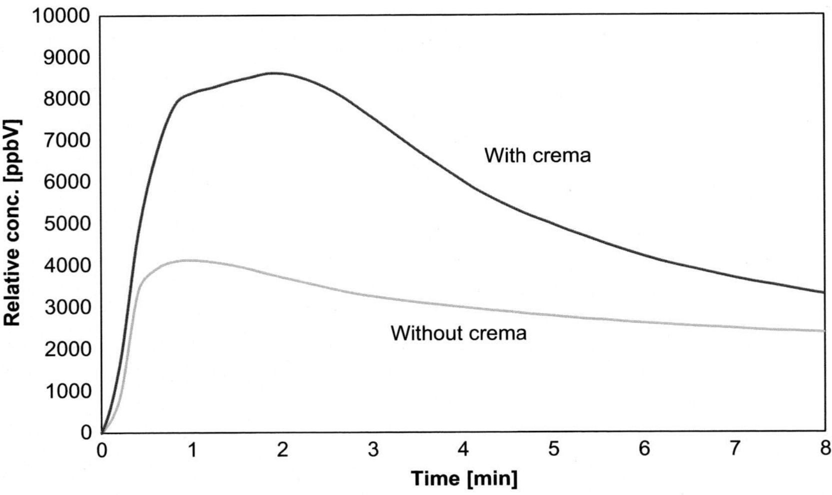

FIGURE 17.11 Volatile release profiles of the sum of selected ion mass traces as measured by proton transfer reaction mass spectrometry ion traces for samples prepared from an Arabica blend long cup coffee with crema and the same coffee without crema (removed by filtration) (t = 0 min equals 1 min after starting coffee extraction) (averages of triplicates).

FIGURE 17.12 Crema helps aromas to be released above the cup.

In conclusion, we need to consider different mechanisms when we look at the impact of crema on aroma release. This should include: (1) the way low crema stability (i.e., high bubble rupture) enhances the release of the high-volatile aromas, which are abundant in the gas phase of the foam bubbles; (2) the way high crema stability retains high volatiles because the crema acts as a barrier that entraps the aromas; and (3) how crema collapse will allow for diffusion and the subsequent release of low-volatile aromas.

For all these reasons, crema, and aroma release is often the focus when developing new espresso brewing systems. For example, the Caffè Firenze is based on pressurized air, which creates espresso coffee with a more persistent foam layer (Masella et al., 2015). This foam layer is considered responsible for a lower volatility of aroma molecules above the cup. This, in turn, confirms previous findings showing that crema stability plays a major role in aroma release (Dold et al., 2011).

With soluble coffee, researchers have also used crema creation to enhance the sensory aspects of the beverage. For instance, one study focused on aroma release above the cup for forming crema in soluble coffees with pressurized internal gas (Yu et al., 2012). However, this study cannot be linked to the mechanisms of aroma release we previously described. This is because the technology to prepare the coffee is based on gas incorporation in the soluble extract and not on the extraction technology.

Another recent technology, Centrifusion™, which Nespresso uses in its VertuoLine system, is based on using centrifugation for extraction instead of using pressure to force the water through the coffee bed. In this case, the system creates crema by combining gas expansion and mechanical formation.

The studies described above are all based on analytical methods used to quantify or reveal the mechanism of the above cup aroma release. To our knowledge, there are currently no sensory or consumer data published that illustrate the effect of crema on the perceived above cup aroma. However, various studies have examined sensory perception in-mouth.

4.3 The Impact of Crema on In-Mouth Perception

Few studies are available on the in-mouth perception of coffee (Charles et al., 2015; Barron et al., 2012; Labbe et al., 2016) and to our knowledge only Barron and Labbe have focused on crema. Researchers often use a method called temporal dominance of sensation (TDS). This methodology enables to monitor temporal evolution of the sensory perception along product tasting (Pineau et al., 2009; Le Révérend et al., 2008; Pineau et al., 2012; Pineau and Schlich, 2015).

The study by Barron et al. (2012) focuses on consumption of a dark roasted intense espresso where the amount of crema has been changed. The TDS was performed on seven sips, allowing consumption of the whole cup. For each sip, the panelists had to choose the dominant attribute of in-mouth perception among a list of 11 attributes: carbony, roasted, cereal, fruity, sweet, bitter, acidic, liquid, thick, gritty, and silky. Fig. 17.13 shows that the roasted dominance increases with increased crema quantity and remains dominant along the consumption of the cup. In contrast, a low crema amount, or no crema, triggered carbony dominance as well as bitterness.

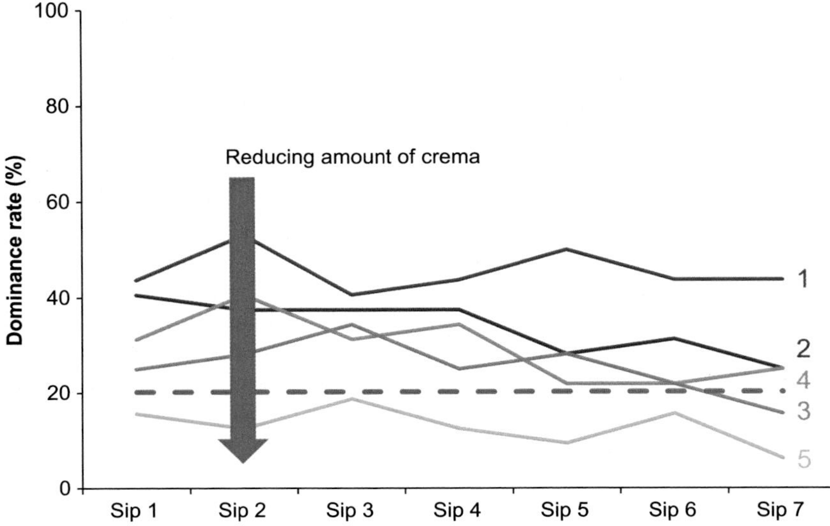

FIGURE 17.13 Comparative evolution of the roasted dominant sensation between espresso coffees with different amounts of crema consumed over seven sips. The *dotted line* indicates the level of significance. The amount of crema reduces from 1 to 5, with 5 having no crema.

The same study confirmed these results using nose-space analysis, a method where aroma release is investigated in vivo using mass spectrometry to analyze exhaled air during the consumption (Barron et al., 2012). The study showed an overall higher intensity for a larger crema amount. Combining the TDS and the nose-space data we can therefore conclude that aromas that are trapped in the crema can also be perceived in sensory perception and measured in the nose-space during consumption.

Obtaining results from nose-space analysis is, however, rather challenging as differences between panelists may be large. In the future, researchers may be able to obtain better results with advancements in analytical methods for nose space analysis. Some studies have reported analytical progress in improving discrimination between samples by coupling time of flight to MS (Romano et al., 2014).

5. OUTLOOK

There is more to crema than just the visual appearance. The statement that a good crema has to be dark brown and of a certain thickness may not be wrong, but may not be exploiting the full potential of what crema can bring to the coffee experience. Although the flavor of espresso coffee is the primary driver for product liking, textural and mouthfeel properties play a key role in the overall appreciation for consumers. Navarini et al. (2004b) already investigated this area with consumers and concluded that mouthfeel attributes can be grouped into components related to viscosity (thick), substances (rich), resistance to tongue (round), after feeling (lingering), coating of oral cavity (mouth-coating), and feeling on soft tissue (smooth). It is suggested that different coffee preparations give different concentrations and a mouthfeel vocabulary should therefore be adopted to describe these different beverages. In addition to visual appearance and texturization of the espresso coffee, crema has also been shown to enhance the aromatic aspects of the coffee, and to create expectations on the consumption of the espresso coffee.

Despite detailed knowledge on coffee composition and foam formation in beer, milk froth, and sparkling beverages, there is still little known about formation and stability of coffee crema. Results published so far are largely descriptive without proving mechanisms or their relative importance in espresso coffee. There is some knowledge on components playing a key role in crema formation and stability. However, it is well known that foams are highly complex systems to study due to its dynamics and multiple phenomena interacting on the foam in parallel. Therefore, most of the research in this field is taking place on model systems, using one or two components which allow a more systematic approach to understanding formation and stabilization. In real food systems, many different molecules are present and complexity increases significantly.

Certainly, espresso is a complex polyphasic liquid composed of (1) emulsion of oil droplets, (2) suspension of solid particles, and (3) effervescence of gas bubbles that evolve into a foam (Illy and Viani, 2005). Therefore, studying espresso crema requires a multidisciplinary approach applying most recent and highly sophisticated analytical methods to better understand this phenomenon at different length and time scales. This calls for more systematic and dedicated research on coffee foam and the parameters influencing its formation and stability. It is only when the molecular compounds, and the mechanisms of formation and destabilization are well understood, that they can become triggers to modify the crema.

Combining the consumer perception of crema with the molecular and physical approach is the prerequisite for developing strategies to improve espresso crema, which may include blending, roasting, grinding, tempering, and extraction parameters including water quality. This novel approach requires considering the espresso coffee in its integrity.

REFERENCES

Bamforth, C.W., 2004. The relative significance of physics and chemistry for beer foam excellence: theory and practice. Journal of the Institute of Brewing 110 (4), 259–266.

Barron, D., Pineau, N., Matthre-Doret, W., Ali, S., Sudre, J., Germain, J.C., Kolodziejczyk, E., Pollien, P., Labbe, D., Jarisch, C., Dugas, V., Hartmann, C., Folmer, B., 2012. Impact of crema on the aroma release and the in-mouth sensory perception of espresso coffee. Food & Function 3, 923–930.

Charles, M., Romano, A., Yener, S., Barnaba, M., Navarini, L., Märk, T.D., Biasoli, F., Gasperi, F., 2015. Understanding flavour perception of espresso coffee by the combination of a dynamic sensory method and in-vivo nosespace analysis. Food Research International 69, 9–20.

Dalla Rosa, M., Nicoli, M.C., Lerici, C.R., 1986. Qualitative characteristics of espresso coffee with reference to percolation process (in Italian). Industrie Alimentari 9, 629–633.

Dickinson, E., 1992. An Introduction to Food Colloids. Oxford University Press, Oxford.

Dold, S., Lindinger, C., Kolodziejczyk, E., Pollien, P., Santo, A., Germain, J.C., Garcia Perin, S., Pineau, N., Folmer, B., Engel, K.H., Barron, D., Hartmann, C., 2011. Influence of foam structure on the release kinetics of volatiles from espresso coffee prior to consumption. Journal of Agricultural and Food Chemistry 59, 11196–11203.

Fond, O., 1995. Effect of water and coffee acidity on extraction dynamics of coffee bed compaction in espresso type extraction. In: Proc. 16th Internat. Sci. Colloq. Coffee (Kyoto). ASIC, Paris, pp. 413–420.

Illy, A., Viani, R., 2005. Espresso Coffee: The Science of Quality, second ed. Elsevier Academic Press, London.

Illy, E., Navarini, L., 2011. Neglected food bubbles: the espresso coffee foam. Food Biophysics 6, 335–348.

Kralchevsky, P.A., Danov, K.D., Denkov, N.D., 2002. Chemical physics of colloid systems and interfaces. In: Birdi, K.S. (Ed.), Handbook of Surface and Colloid Chemistry, second ed. CRC Press, New York, pp. 106–165 (Ref. i–xxiii).

Labbe, D., Sudre, J., Dugas, V., Folmer, B., 2016. Impact of crema on expected and actual espresso coffee experience. Food Research International 82, 53–58.

Lange, C., Martin, C., Chabanet, C., Combris, P., Issanchou, S., 2002. Impact of the information provided to consumers on their willingness to pay for champagne: comparison with hedonic scores. Food Quality and Preference 13, 597–608.

Le Révérend, F.M., Hidrio, C., Fernandes, A., Aubry, V., 2008. Comparison between temporal dominance of sensations and time intensity results. Food Quality and Preference 19, 174–178.

Maeztu, M., Sanz, C., Andueza, S., Paz De Peña, M., Bello, J., Cid, C., 2001a. Characterisation of espresso coffee by statistic headspace GC-MS and sensory flavor profile. Journal of Agricultural and Food Chemistry 49, 5437–5444.

Maetzu, L., Andueza, S., Ibanez, C., Paz de Pena, M., Bello, J., Cid, C., 2001b. Multivariate methods for characterization and classification of espresso coffees from different botanical varieties and types of roast by foam, taste and mouthfeel. Journal of Agricultural and Food Chemistry 49, 4743–4747.

Masella, P., Guerrini, L., Spinelli, S., Calamai, L., Spugnoli, P., Illy, F., Parenti, A., 2015. A new espresso brewing method. Journal of Food Engineering 146, 204–208.

Navarini, L., Barnabà, M., Suggi Liverani, F., 2006. Physicochemical characterization of the espresso coffee foam. In: Proc. 21th Internat. Sci. Colloq. Coffee (Montpellier). ASIC, Paris, pp. 320–325.

Navarini, L., Cappuccio, R., Suggi Liverani, F., 2004a. The body of the espresso coffee: the elusive importance. In: Proc. 20th Internat. Sci. Colloq. Coffee (Bangalore). ASIC, Paris, pp. 193–203.

Navarini, L., Cappuccio, R., Suggi-Liverani, F., Illy, A., 2004b. Espresso coffee beverage: classification of texture terms. Journal of Texture Studies 35, 525–541.

Nunes, F.M., Coimbra, M.A., 1998. Influence of polysaccharide composition in foam stability of espresso coffee. Carbohydrate Polymers 37, 283–285.

Nunes, F.M., Coimbra, M.A., Duarte, A.C., Delgadillo, L., 1997. Foamability, foam stability, and chemical composition of espresso coffee as affected by the degree of roast. Journal of Agricultural and Food Chemistry 45, 3238–3243.

Parenti, A., Guerrini, L., Masella, P., Spinelli, S., Calamai, L., Spugnoli, P., 2014. Comparison of espresso coffee brewing techniques. Journal of Food Engineering 121, 112–117.

Petracco, M., 1989. Physico-chemical and structural characterisation of espresso coffee brew. In: Proc. 13th Internat. Sci. Colloq. Coffee (Paipa). ASIC, Paris, pp. 246–261.

Petracco, M., Liverani, F.S., 1993. Espresso coffee brewing dynamics: development of mathematical and computational models or dynamics of fluid percolation through a bed of particles subject to physico-chemical evolution, and its mathematical modeling. In: Colloque Scientifique International sur le Café, 15. Montpellier, Francia, Juin 6–11.

Petracco, M., 2001. Beverage preparation: brewing trends for the New Millennium. In: Clarke, R.J., Vitzthum, O.G. (Eds.), Coffee Recent Developments. Blackwell Science, London, pp. 140–164.

Petracco, M., 2005. The cup. In: Illy, A., Viani, R. (Eds.), Espresso Coffee, Espresso Coffee: The Science of Quality, vol. 2005. Academic Press, p. 301.

Pineau, N., Schlich, P., 2015. Temporal dominance of sensations (TDS) as a sensory profiling technique. In: Delarue, J., Lawlor, J.B., Rogeaux, M. (Eds.), Woodhead Publishing Series in Food Science, Technology and Nutrition, Rapid Sensory Profiling Techniques. Woodhead Publishing, pp. 269–306.

Pineau, N., Goupil de Bouillé, L., Lenfant, F., Schlich, P., Martin, N., Rytz, A., 2012. The role of temporal dominance of sensations (TDS) in the generation and integration of food sensations and cognition. Food Quality and Preference 26, 159–165.

Pineau, N., Schlich, P., Cordelle, S., Mathonnière, C., Issanchou, S., Imbert, A., Rogeaux, M., Etievant, P., Köster, E., 2009. Temporal dominance of sensations: construction of the TDS curves and comparison with time–intensity. Food Quality and Preference 20, 450–455.

Prins, A., 1988. Principles of foam stability. In: Dickinson, E., Stainsby, G. (Eds.), Advances in Food Emulsions and Foams. Elsevier Applied Science, Essex, p. 91.

Romano, A., Fischer, F., Herbig, J., Campbell-Sills, H., Coulon, J., Lucas, P., Cappellin, L., Biasioli, F., 2014. Wine analysis by FastGC proton-transfer reaction-time-of-flight-mass spectrometry. International Journal of Mass Spectrometry 369, 81–86.

Sáenz-Navajas, M.P., Campo, E., Sutan, A., Ballester, J., Valentin, D., 2013. Perception of wine quality according to extrinsic cues: the case of Burgundy wine consumers. Food Quality and Preference 27, 44–53.

Schifferstein, H.N.J., Fenko, A., Desmet, P.M.A., Labbe, D., Martin, N., 2013. Influence of package design on the dynamics of multisensory and emotional food experience. Food Quality and Preference 27, 18–25.

Schifferstein, R., 2009. The drinking experience: cup or content. Food Quality and Preference 20, 268–276.

Schokker, E.P., Bos, M.A., Kuijpers, A.J., Wijnen, M.E., Walstra, P., 2002. Spreading of oil from protein stabilized emulsions at air/water interfaces. Colloids and Interfaces B: Biointerfaces 26, 315–327.

Siegrist, M., Cousin, M.E., 2009. Expectations influence sensory experience in a wine tasting. Appetite 52, 762–765.

Wan, X., Zhou, X., Woods, A.T., Spence, C., 2015. Influence of the glassware on the perception of alcoholic drinks. Food Quality and Preference 44, 101–110.

Weaire, D., Hutzler, S., 1999. The Physics of Foams. Oxford University Press, Oxford, UK.

Yu, T., Macnaughtan, B., Boyer, M., Linforth, R., Dinsdale, K., Fisk, I.D., 2012. Aroma delivery from spray dried coffee containing pressurized gas. Food Research International 49, 702–709.

Chapter 18

Sensory Evaluation—Profiling and Preferences

Edouard Thomas[1], Sabine Puget[1], Dominique Valentin[2], Paul Songer[3]
[1]Nestlé Nespresso SA, Lausanne, Switzerland; [2]AgroSup Dijon, INRA, Dijon, France; [3]Songer and Associates, Inc., Boulder, CO, United States

1. SENSORY EVALUATION OF COFFEE: INTRODUCTION

When he had his first specialty coffee, years before being head judge for the Cup of Excellence and enjoying among the best coffees in the world, Paul had already some awareness about specialty coffee and he knew it would be special. Even though it was not a "road to Damascus" life-changing moment, his first experience led him to realize this coffee deserved to be experienced fully rather than simply consumed. The sensorial experience revealed to him that specialty coffee was not the one-dimensional beverage with caffeine he was used to but was rather rich in complexity and diversity. When she joined the Nespresso Coffee Team as sensory and consumer scientist, Sabine was introduced to specialty coffee tasting and remembers very well her first perception, different from Edouards 13 years earlier when he joined the team. The fine acidic and delicately citrus AA Kenya prepared in a pour over to impress Sabine turned out to be a light, acidic, and not over the top coffee in her mouth of consumer, far from appreciating the taste of the coffee variety in its terroir! Despite the that fact Sabine and Edouard were tasting the exact same coffee prepared at the same time, or that Paul was in a receptive mind-set, our experiences of specialty coffee (that Sabine definitely appreciates at is right value today) were completely different. This illustrates the fact that sensory perception of foods or beverages is personal and is shaped first by our genetics but also by our culture and past experiences.

Tasting has always been at the heart of coffee growing and production and largely the domain of green coffee cuppers, the expert coffee tasters who learned their trade from senior practitioners as an apprentice. Coffee cuppers first select green coffee and then other expert coffee tasters, the roaster, the brewer, and the barista will all do sensory assessments to judge the taste of coffee within their own disciplines. Expert coffee tasting prevails while applying a correct and structured sensory evaluation still requires improvement because it is too often misunderstood by those experts who taste their

The Craft and Science of Coffee. http://dx.doi.org/10.1016/B978-0-12-803520-7.00018-9

own work in the coffee or because it is seen as too demanding in resources to justify a sensory professional. In reality, the key actors of this coffee sensory assessment all along the value chain are the expert coffee taster, the sensory analyst, and the consumer.

Expert coffee tasters have been trained to analyze their experiences using own procedures based upon empirical models or interpretations of coffee. In implementing this training, they have had multiple experiences and come to know details that may never be perceived or conceived of by the consumer. This consumer, on the other hand, usually expresses their sensory experiences in terms of enjoyment or whether or not their expectations have been met. A major challenge for sensory evaluation in coffee as in food and beverage industry in general is bridging the gap between the expert tasters, valued for their accuracy and consistency, and consumers, who will actually purchase and hopefully enjoy the product but usually lacks extensive knowledge or have an adequate vocabulary to describe the coffee.

The sensory analyst's role is to be the link between green coffee suppliers, roasters, and other production personnel, brewers, barista, and the marketers. They are responsible for specifying a product based upon its intended use using a common language agreed upon among professionals. The accuracy of these specifications is critical because they will be essential to the product's identity throughout the value chain from its manufacture to its marketing to its final consumption. Examining the sensory data in parallel with production parameters such as levels of roast or extraction conditions can reveal the technological source of certain flavor attributes, while examining sensory attributes in parallel with consumer preferences or sales data can reveal what is attractive to the consumer and explain the success of that particular coffee.

Sensory evaluation is defined by Stone and Sidel (2004) as "the scientific discipline used to evoke, measure, analyze, and interpret human reactions to those characteristics of foods and beverages as they are perceived by the senses of sight, smell, taste, touch, and hearing." But why call upon such a discipline?

A coffee business is successful because it provides something that a consumer desires and for which it is willing to pay a certain price. The professional sensory analyst must learn to generate objective judgments about what is tasted and communicate it in a clear and consistent fashion. The consumer purchases a coffee product if it appeals to them in some way, they have found the product to meet their needs in previous purchases, a recommendation has been made, or marketing has succeeded in appealing to them. In objectively reporting perceptions, the professional sensory analyst is not able to ascertain whether or not a consumer will like or continue to buy a coffee product; on the other hand, the consumer rarely has the vocabulary to express what they are looking for and can occasionally be pleasantly surprised by a new unfamiliar experience. The reason to be of any coffee product must echo with robust professional sensory analysis and must be validated in the field by the consumer.

Sensory tests are therefore conducted to bridge the gap between the expert taster and the consumer, playing a key role at each stage of the production and development process. It is especially important in coffee because, as a natural crop product, supplies and the quality of those supplies constantly change, and consumer preferences evolve over time. Thanks to the continuous innovation in the discipline, methods are becoming more flexible, faster, and adaptable to one's business reality but they require more than ever the know-how of the sensory professional.

This chapter will look at the methods used and issues in sensory testing with an emphasis on how they are currently being applied in the coffee industry. In evaluating these methods, bridging the gap between the sensory experiences of experts and those of consumers will be considered in detail. The sensory evaluation of wine and other foods and beverages will often be used as a reference since it has received much attention and extended research.

The initial section investigates how coffee and other foods and beverages are physically perceived in terms of tastes, aromas, and other sensory systems. In the following section, the scientific basis of sensory analysis are examined and how results of tests are interpreted using statistical and other methods. The next section examines methods that were developed in traditional sensory analysis programs and some more recent developments.

With this as a basis, we move to how expert tasters are trained and their roles in the development of coffee products, along with a look at consumers and their motivations for buying and appreciating the coffees. Finally, the chapter concludes with a general outlook of sensory analysis in the coffee industry, how it might develop to better meet the evolving trends and changing consumer preferences, and, most importantly, bridging the gap between expert tasters and consumers.

2. SENSORY EVALUATION: OBJECTIVE MEASUREMENT OF SENSORY PHENOMENA

What is more natural than tasting a cup of coffee? The apparent simplicity of this daily activity hides a great complexity that mainly rests on the complexity of the taste perception. What we call taste in the common language is also called flavor in the scientific language and relies on a larger perception than the simple sense of taste. Flavor is defined as a sensory percept induced by food or beverage tasting. It relies mainly on the functional integration of information transmitted by the chemical senses: olfaction, gustation, oral, and nasal somatosensory inputs (Thomas-Danguin, 2009). Once in the mouth, volatile compounds are retronasally conveyed to the nasal cavity where they are inclined to activate the olfactory receptors located on the top of the olfactory cavity and also by the trigeminal fibers located into the whole nasal mucosa. At the same time, soluble compounds are dissolved into the saliva and some of them can further be detected by the gustatory cells of the taste buds and the trigeminal fibers present in the oral mucosa (Laing and Jinks, 1996; AFNOR, 1992). These three sensory modalities are simultaneously activated and interact to create an integrated unique perception. To better understand the sensory response elicited by chemical stimuli, we will here after describe some basics regarding the anatomy and physiology of our sensory systems.

2.1 Taste or Gustation

Taste perception is mediated by taste receptor cells that are located on the tongue (edge and anterior dorsal part) but also on the soft palate, the pharynx, and the larynx (Breslin and Huang, 2006). These cells predominantly reside within the taste buds that are included into the papillae. These structures contain different types of cells playing a different role in taste perception. Molecular compounds eliciting sweet, bitter, and umami perception are detected by G-protein coupled receptors, whereas ions such as Na^+ eliciting a salty perception or Ca^{2+} eliciting other taste sensations are detected by ion channels (Breslin and Huang, 2006). Beyond the few data available on taste cells, the mechanisms underlying taste coding and perception remain poorly understood and still debated (Chandrashekar et al., 2006; Roper, 2007; Vandenbeuch and Kinnamon, 2009).

2.2 Olfaction

The activation of the olfactory system is responsible of the odor perception. A specificity of olfaction is that the receptors are expressed by neurons rather than by dedicated cells as is the case for other sensory systems such as taste (Buck and Axel, 1991). Each of these olfactory neurons expresses a single type of olfactory receptors (Nef et al., 1991), which can detect several odorant molecules (Duchamp-Viret et al., 1999). In addition, one single odorant molecule can activate several olfactory neurons (Moulton, 1967) and as a consequence several types of receptors (Fig. 18.1). The olfactory coding (i.e., the perceived odor) is based on the combination of olfactory receptors being activated. The population of olfactory neurons expressing a same olfactory receptor converges to the same structure called glomerulus (Monbaerts et al., 1996). Thus, the olfactory response is converted into an odorant map that can be experimentally visualized (Leon and Johnson, 2003).

2.3 Somatosensory Systems

The somatosensory system provides tactile, thermal, proprioceptive (body position, movement), and nociceptive (pain) information. For the face and the mouth, these sensations are conveyed by the trigeminal nerve and are referred to as trigeminal sensations. It is worth noting that the trigeminal

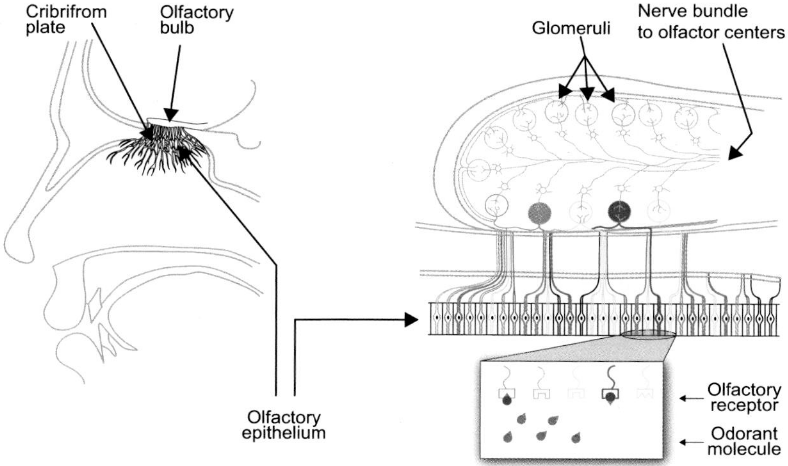

FIGURE 18.1 Schematic illustration of the olfactory system.

nerve also conveys chemical information. Thus, the free nerve endings of the trigeminal nerve detect the chemical compounds dissolved both in saliva and the olfactory mucus. For instance, most of the odorant compounds activate the trigeminal nerve at higher concentrations (Doty, 1975; Commetto-Muniz and Cain, 1994) as compared to the olfactory system. Trigeminal stimulation elicits a few qualitative perceptions such as cool, warm, burning, pungent, tingling, stinging, numbing, or irritation. However, some studies suggest broader trigeminal qualitative discrimination, suggesting that such perception could be more complex (Laska et al., 1997). Several studies evidenced the relationship between perceived intensity and the compound potency to activate the trigeminal nerve suggesting that the trigeminal activation contributes to the perception of odor intensity (Doty, 1975; Cain, 1974, 1976; Murphy, 1987).

2.4 Construction of the Flavor Perception

As exhaustively reviewed by Verhagen and Engelen (2006), interactions occur between senses. As an example, Cain and Murphy (1980) observed a suppression of an odor when mixed with an irritant compound demonstrating that the somatosensory system interacts with olfaction. It has also been shown that olfaction and taste interact. Lately, a large attention was made on odor-induced taste enhancement and especially on sweet enhancement induced by aromas. As a basic study, Frank and Byram (1988) evidenced that the sweetness of whipped cream increased when a strawberry aroma was added. Such cross-modal perceptual interactions were also demonstrated for other tastes. Labbe et al. (2006) demonstrated that the addition of cocoa aroma to chocolate beverage could enhance bitterness but did not affect other tastes. We can easily imagine that similar effect could be observed with coffee. For instance, a fruity aroma in coffee could well increase its perceived acidity. In the same way, we can imagine that roasted notes could contribute to bitterness.

Cross-modal interactions depend on the congruency between taste and smell stimuli. Frank et al. (1991) demonstrated that similarity judgment was a good predictor of the odor-induced taste enhancement. In contrast, incongruent or novel odors suppress taste intensity (Prescott, 1999). For example, Stevenson et al. (1999) showed that a sweet-smelling odor such as caramel enhanced the sweetness of sucrose in solution while it suppressed the sourness of a citric acid solution. The acquisition of taste properties by odors can be learned via combined exposure to the two sensations (Stevenson et al., 1995, 1998).

Mechanisms underlying taste–smell interactions are relatively well understood. It is argued that sensory signals involved in flavor perception are "functionally united when anatomically separated" (Small and Prescott, 2005). In their model, neural substrate for this unitary percept that is flavor is built over time by repeated experience to stimuli.

2.5 Sensation and Perception

Beyond the integration of the olfactory, taste, and somatosensory responses to build the flavor perception, sensory information is more largely integrated with other information such as memory or emotions to shape our perception of a food product. Thus, a clear distinction should be made between a sensation and a perception (Chaudhuri, 2010). The sensation is the actual physical response that results from a stimulus, whereas perception is the conscious experience of one or multiple sensations. As reviewed by Köster (2009) many factors influence our product perception and will shape our food choices such as physiological factors (hunger/satiation), psychological factors (personality traits, memory, past experience, emotions), situation (context), and sociocultural factors (culture, habits, beliefs).

In this regard, coffee as any food product is not easy to comprehend. Trained coffee experts are not necessarily able to estimate whether the product they shape will be accepted by the average consumer. For example, acidity is highly valued by expert tasters to the point where some roasters deliberately accentuate this attribute, and yet the amount of consumers liking this experience has not been studied.

3. THE ROLE OF SENSORY EVALUATION AND THE SCIENCES BEHIND

As highlighted by Meillgard et al. (1999), sensory evaluation started to develop in the 1940s based on the wartime efforts to provide acceptable food to the American forces. This field has for particularity to use human subjects as measuring instrument. It is a scientific method and its objective is to apprehend food products with a view to optimize them and trigger consumer choice. It can be applied at all the stages of the product research and development from the identification of the consumer drivers of liking, to the identification of the sensory characteristics and the variables that impact them, to the quality control of selected sensory attributes.

Tasting, the main task of sensory assessors, is a daily routine for everyone and is hard for some people to be conceived as part of a scientific discipline. The four blocks "evoke, measure, analyze, and interprete" of the Sidel and Stone (2004) are critical to sensory evaluation. The discipline requires to clearly define the objective, present the stimulus (product or its ingredients) in a controlled way, select a methodology relevant to the objective, run the appropriate statistical analyses, and finally interpret the results.

Sensory evaluation leans on several academic disciplines. It is based on the human physiology and psychology to better understand how sensations are perceived, which factors influence this perception and how we elaborate our final judgment. It shares methodologies with psychophysics and statistic to determine the relationship between a stimulus and the resulting sensation in the form of a mathematical function, equation (Meillgard et al., 1999), or representation.

The role of the sensory analyst is therefore not to be the expert taster of the business but to master both the background and the technique of the sensory tests managing a pool of trained tasters with proven performance records. She/he must have listening and speaking skills to dialogue with the stakeholders to perform the most relevant sensory tests. Because she/he tastes a lot off-records to the tests she/he conducts, the sensory analyst eventually becomes a coffee expert taster.

3.1 Conducting a Scientific Sensory Test

To obtain accurate and consistent information, we need to design sensory tests to acquire specific information. We also need to properly conduct both the tests themselves and the process to collect and analyze the resulting data. In experimental design, we define the test goals and determine the best way to get the needed information. To conduct the test, we put conditions in place so the panelists can interact with the samples and produce the needed data. We can then analyze these data using statistics.

The first step in experimental design is to evaluate what we already know and what we want to find out. We can then develop strategies for obtaining the needed information. Most importantly, we need to determine which aspects we can control and which we cannot.

Those inputs we can control are called "independent variables," whereas the information we do not know, but that the experiment will measure is called "dependent variables." In addition, any aspects that we cannot control, but which will affect the experiment are called "extraneous variables." The most accurate tests include a "control" that is not influenced by independent variables and that can help establish a baseline.

To conduct a sensory test and control the test conditions, we need to start by qualifying the panelists based on their known abilities, experience, and training; the most general classifications include the "expert" taster and the "naïve" taster. The panelists must also understand how the test will work, which may require training or a calibration session.

A major challenge in coffee testing is sample preparation and presentation. We must consider all aspects that affect flavor, including the green coffee selection, roasting, grinding, the ratio of water to coffee, water quality and temperature, and the method of brew preparation (Fig. 18.2). Since perceived coffee flavor changes as coffee cools, we must also control the timing of the preparation and analysis. Also to minimize tasting bias, samples must be coded with a three-digit code, presented in a random or balanced order, evaluated in repetitions and individually by panelists.

After the test, we need to scrutinize the raw test data independently of the taster's reputation. We do not learn much from a single measurement because we cannot tell whether the measurement is just a random occurrence. If we take two or three measurements with a similar result, we can reasonably assume that the information is correct. However, we then need to use statistics to quantify how close the measurements are to the actual measurement (accuracy) and how consistent a measurement would be if repeated (consistency).

Statistics measure two aspects of the test: the truth or significance of the findings and the consistency of those findings. They can also measure the possibility of an error. A "significant" finding means that the measurement taken did not happen simply by chance and is reasonably accurate. The

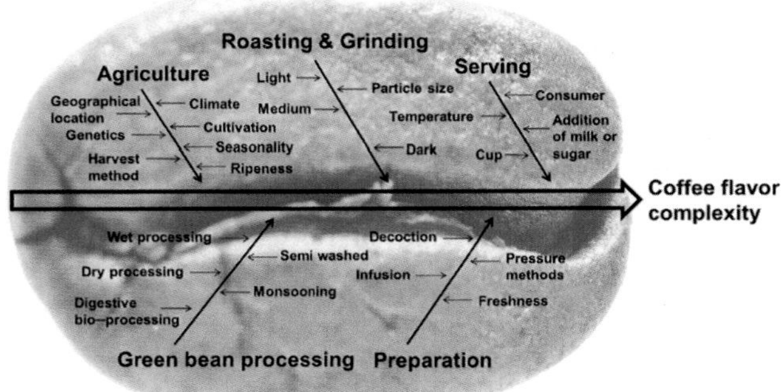

FIGURE 18.2 Factors that influence coffee flavor complexity from farm to cup (Sunarharum et al., 2014).

consistency of the measurement is indicated by the "confidence" level, which quantifies the likelihood that a measurement taken again under the same conditions would be within a certain range.

Another challenge of performing sensory testing is the practical aspect of producing forms, collecting data, and processing that data. In the past 15 years, spreadsheets and computers have made it easier for us to manage and manipulate data. We can use statistics programs for advanced analysis and graphic programs to illustrate results in visual and more elaborate ways.

In the era of social media and instant information, we can also enter data in real time as the test is progressing and responses are submitted. This helps make tests shorter, easier to analyze, and more accurate.

The temptation in analyzing sensory results is to focus solely on inputs and outputs. However, panelists' responses require close evaluation to ensure that they are an accurate rendition of panelists' perception.

In designing a sensory test, after defining the objectives and the strategy for obtaining information, we also need to consider our resources, including number of panelists, sample availability, staff preparation, and the time period in which to conduct the test.

Keeping these requirements in mind, research professionals have designed several sensory tests for specific purposes. Rather than designing a test from the beginning, it is usually more efficient to use one of these standardized tests since they are well known, easy to explain, and come with definite statistical procedures to analyze the data.

3.2 Sensory Tests

The application of formal experimental methods of sensory testing is relatively recent when compared to other sciences such as biology, chemistry, and physics. The methods used have developed in parallel to discoveries in the field of psychology and those studies involving how human beings respond to various stimuli. The "classic" methods are those that have been used for some time; more recent developments will also be discussed.

There are two main approaches for evaluating the sensory experience: qualitative and quantitative (Meillgard et al., 1999). "Qualitative" methods seek to define the sensory experience, often by comparing that experience with a standard or a commonly known experience. Qualitative aspects are also referred to as "attributes, characteristics, descriptive terms, descriptors, or terminology." "Quantitative" methods use various scales that allow the analyst to express the relative level of their experience. The most common of these is intensity (for example, rating the sweetness of a sample on a scale of 1–10) and either numerical or verbal scales (like very much, like slightly, dislike, etc.) that can be easily converted to numbers that are used. Most sensory tests use a combination of qualitative and quantitative evaluations, though one may be emphasized over another.

The choice of a sensory test is based upon the goal of testing. There are many tests designed for specific problems that often present themselves in making a determination about a food or beverage. These are some of the main classifications and the questions they seek to answer:

- Are two or more samples the same or different from one another? In this case, "difference" or "discrimination" tests are used.
- What are the flavor attributes of a coffee and the intensity of those flavor attributes? This is answered using "descriptive" methods comprised of a qualitative description (which flavor attributes are present), and possibly a quantitative rating of the intensity of those attributes.
- Does this coffee meet a standard profile or set of standards? These tests involve the comparison to a known standard or an assumed standard. Grading methods fall into this category, as do quality control difference-from-control/standard tests.
- How well is the coffee liked by those consuming the sample? These are known as "hedonic" tests since they explore the liking of coffee, or how it affects the consumer.

All of these methods can be implemented at all the stages of the coffee value chain to study treatment and effect in agriculture, postharvest, green physics, blend, roast, extraction, packing and shelf life, preparation, serving,...to comprehend better the coffee complex flavor (Fig. 18.2; Sunarharum et al., 2014) including its taste and mouthfeel expression.

A selection of tests only will be described among which the most classic and used as well as the most innovative that have their place in the coffee industry.

3.3 The Classic Methods

3.3.1 Difference or Discrimination Tests—Example of the Triangle Test

There are several versions of difference or discrimination tests. All are based on the comparison of two or more samples, with the analyst requested to make a defined choice. The most familiar is the "triangle test" where one of three submitted samples is different from the other two (Fig. 18.3). Since there is a one-in-three chance of the analyst guessing the correct answer, several iterations of the test must be conducted to determine significance and confidence levels. Other tests of this type are the paired comparison and duo-trio tests. A challenge with this type of testing is palate fatigue; as several sets of samples must be submitted to ensure statistical validity. The assessors may become too tired and desensitized to make an accurate determination.

It was shown in Brazil with regular coffee consumers in a series of triangle tests that they could start distinguishing between clean and unclean coffees when the latter was composed of as low as 2.3–2.5% of a mixture of black,

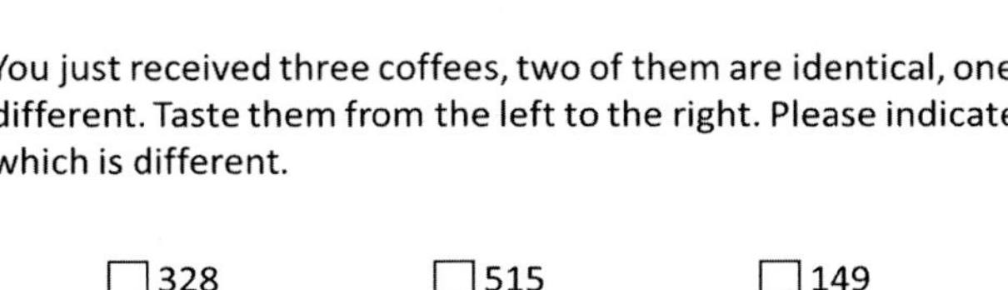

You just received three coffees, two of them are identical, one is different. Taste them from the left to the right. Please indicate which is different.

☐ 328 ☐ 515 ☐ 149

If you don't perceive differences between the samples, you must guess.

FIGURE 18.3 Example of a triangle test questionnaire.

immature, and sour defective beans (Deliza et al., 2006). Another way of using the triangle test could be to evaluate whether two varieties such as Castillo and Caturra are perceived the same or different by a panel of experts or by a panel of consumers.

As obvious as it might appear, discriminative tests are very useful before considering descriptive tests as those exercises become pointless if the panel cannot differentiate the samples.

3.3.2 *Descriptive Testing—The Quantitative Descriptive Analysis*

Variants exist for descriptive quantitative analysis among which the Spectrum™ Descriptive Analysis and the quantitative descriptive analysis (QDA) (Stone et al., 1992).

The QDA is probably the most commonly used descriptive testing method. It begins with the determination of flavor attributes, which can come from a standard vocabulary already developed like the "Sensory Lexicon" by World Coffee Research (2016) and the "flavor wheel" available through the Specialty Coffee Association of America (SCAA) or in a panel brain-storming session of free association. The latter takes several sessions and once a consistent vocabulary is developed those attributes are rated quantitatively in terms of intensity. The goal is to develop a "flavor profile" that lists the flavor attributes in the order of their appearance with the intensity indicated by the distance from a specific point (Figs. 18.4 and 18.5).

For comparison of descriptive results, a principle component analysis is often used. This multivariate method integrates the distance between the sensory attribute across samples on one hand and the distance between samples across sensory attributes on the other hand to create two-dimensional spaces on which the samples and the attributes themselves can be mapped.

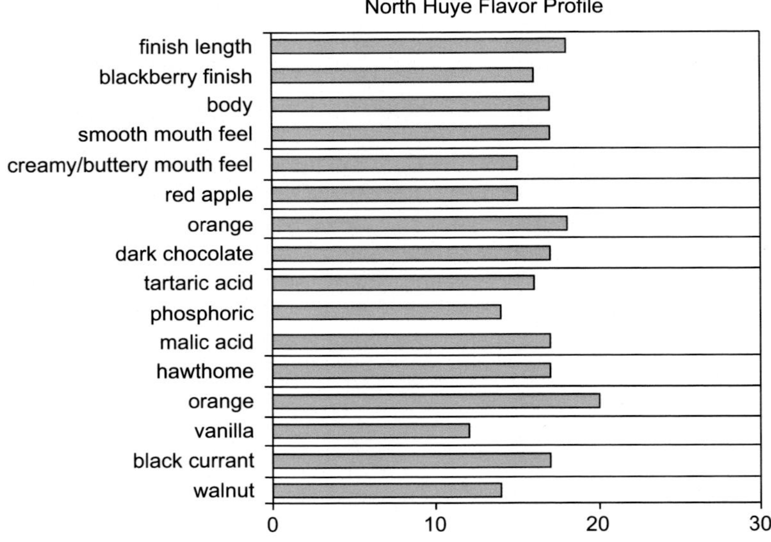

FIGURE 18.4 A bar graph descriptive analysis of a coffee from the North Huye region of Rwanda (Songer, 2008).

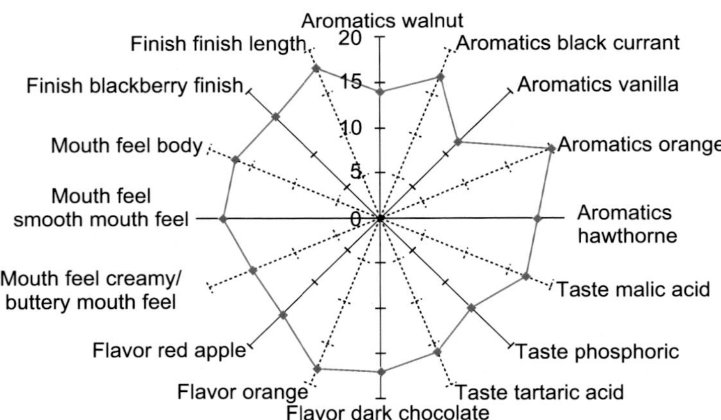

FIGURE 18.5 A radar (also called spider) graph shows the same information as in Fig. 18.5 (Songer, 2008).

An example can be seen in Fig. 18.6 where the QDA is used to compare producing regions in Rwanda and thus show the internal diversity in taste profile within the country.

From this graph, one can conclude that the North Huye and South Huye samples are similar since they are located fairly close together in the lower left

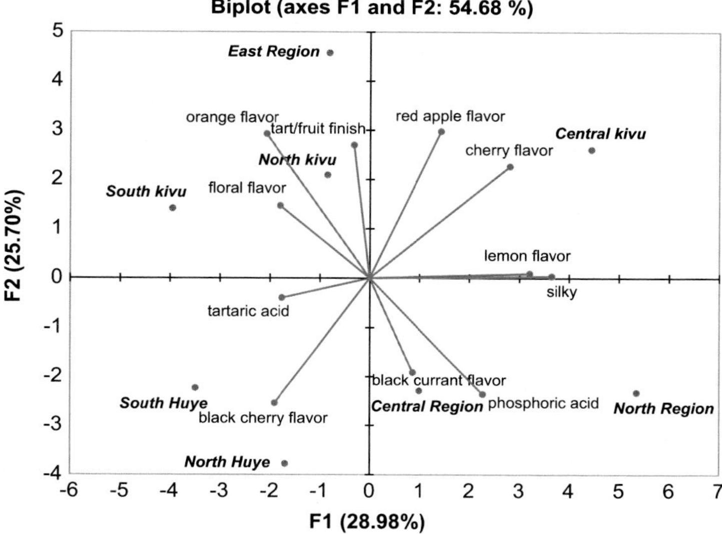

FIGURE 18.6 A principal component analysis can show how samples relate to each other in terms of their descriptive attributes. In this case, 55% of the difference between the samples is explained by the selected flavor attributes.

quadrant, whereas the East Region sample (located in the upper left quadrant) is quite different from both Huye samples since there is a large distance between them. It is also reasonable to assume that when a black cherry flavor is found (lower left quadrant), it is unlikely to find red apple or red cherry flavor attributes in the same sample (upper right quadrant). One must keep in mind that the chart has only accounted for almost 55% (28.98% + 25.70%) of the difference between the samples.

Applications of the QDA are very wide. Wet processed robustas were profiled as more acidic and fruitier than the same robustas when dry processed, which were found more rubbery, burnt, bitter, and woody (Leloup et al., 2004). After determining by triangle tests the threshold of detection of defectuous bean in coffee by Brazilian consumer (Deliza et al., 2006), a QDA describes the clean coffee as sweet with body texture and the coffee blended with defectuous beans as significantly more bitter, astringent, chemical, green, and burnt (Deliza et al., 2008). Focusing on the final mode of preparation at the consumer end, QDA showed the impact of the milk quality added in dash onto the coffee beverage (Steinhart et al., 2006): double homogenized milk with lower fat content and lower fat globules led to an increase of the coffee-related attributes, whereas milk with added casein led to a decrease of those same attributes versus whole milk. Also tracking the shelf life of roasted and vacuum-packed coffee beans, the QDA concludes that changes are possible during storage (Kreuml et al., 2013). The coffee at 9 months was profiled with less intensity on the positive flavor attributes and with higher bitterness and sourness compared to the freshly roasted coffee. At 18 months, this was even more accentuated on top of the apparition of rancid and stale notes typical of oxidation.

Accurate descriptive analysis requires a large amount of time and training to using attribute references. It is usually done when developing a set of standards for a product, for research investigation, precise development, or track regularly a product that really matters. Such a close evaluation of a product is most useful when one has the possibility of adding considerable value to a product, for example, improving a blend by changing a component, isolating a particularly popular flavor attribute preferred by consumers, varying the roast profile, or simply benchmarking your lead product regularly versus others.

3.3.3 Quality Control and Grading

The most basic method of exploring if a sample or set of samples meets a standard is the sample-to-standard test. Similar to the duo-trio test, a "gold standard" representing the ideal sample is provided and the samples of interest compared. Often the most important flavor attributes of the product are listed and the panelist asked to compare a newly produced product to the standard. For example, in a quality control test one may compare recently roasted samples to the "gold standard" to ensure that the proper degree of roast was accomplished and the intended balance of flavor attributes is present.

Grading systems are quality control procedures that are extended to classify a sample based on what is found. As an example, for the "Q" grading procedures, a series of tests are performed on green coffee, including a defect count on a randomly selected sample and a sensory test where a coffee must score an "80" to be qualified as specialty (for details see Chapter 8).

3.3.4 Hedonic Tests

Most affective (also called "hedonic") tests are performed by consumers who may be likely to purchase the product. A common test is that of "liking"; there is a standard 9-point or 7-point scale beginning with "dislike extremely" to "neither like nor dislike" to "like extremely" and consumers are asked to rate each sample according to this scale. There are also comparison tests, such as those where consumers are asked to rank a set of samples in the order of which is preferred from most preferred to least or given a set of two or three samples and asked which is preferred. Some tests use the just about right to focus on a particular quality such as strength or intensity: consumers are asked to determine if the product is too strong, not strong enough, or just about right.

These tests are useful to understand consumer acceptability to new products in a given country like the globalized cappuccino and latte or the emergence of cold coffees. In France, a traditional hot black coffee country, the reaction of consumers to the little known iced-coffee was tested (Petit and Sieffermann, 2007). The two variants tested plain iced coffee and milk iced coffee were rather well accepted. The milk variant was significantly better appreciated and described as sweet, creamy, and milk taste, whereas the plain variants was described as watery, light, and bitter.

3.4 Novel Methods

Several methods have been developed as alternatives to classical descriptive analysis, namely the QDA described before, with the advantages of flexibility, fastness, and aptitude to be performed by experts, trained assessors, or directly by untrained consumers (Varela and Ares, 2012). Those new methods will continue their implementation in coffee where training calibration of tasters is a demanding task, capturing directly consumer's perception becoming an evident must and where many variants with minute sensory differences between products can lead to dramatic consumer change of liking.

3.4.1 Sorting Task and Projective Mapping

The first method developed with this idea in mind was the sorting task. This method originated in psychology (see Coxon, 1999 for a review and historical perspective) and was used to understand how people perceive and categorize the world. It was first applied to a food product by Lawless (1985).

Free sorting tasks are single sessions. All products are presented simultaneously and randomly displayed on a table with a different order per assessor. Assessors are asked first to look at smell and/or taste (depending on the objectives of the study) all the products and then to sort them in mutually exclusive groups based on product-perceived similarities. Assessors can use the criteria they want to perform their sorts and they are free to make as many groups as they want and to put as many products as they want in each group. Once they are done with their groupings, assessors can be asked to provide terms to characterize each group they formed. The first step to analyze the data is to generate a similarity matrix by counting the number of times each pair of stimuli is sorted in the same group.

This similarity matrix is then submitted to multidimensional scaling (MDS). MDS produces a spatial representation of the product similarity in which products are represented by points on a map: two products that are close on the map have been judged as similar by the assessors. The descriptors associated with the different groups of products can then be projected onto the similarity map. This is done by first computing the frequency at which each descriptor was used to describe each product. The descriptors given for a group of products are assigned to each product of the group and descriptors given by several assessors are assumed to have the same meaning. Then the projection of the descriptors on the similarity map is done by computing the correlations between the frequency of descriptors and the product factor scores.

The sorting task was applied on the milk dash, cappuccino, and latte recipes to study the impact of the coffee profile and the quantity of milk used on the final beverage (Puget et al., 2010). For the milk dash, the final beverage overall profile matched that of the coffee used. For cappuccino, four categories of final beverage were identified according to both the coffee intensity level and its aromatic dominance, whereas only two categories were identified for the latte depending only of the aptitude of the coffee intensity to show or not in the high milk quantity.

Developed around the same time as the sorting task, Risvik et al. (1994) proposed an alternative method to describe the similarity between products, the projective mapping, also called napping (Pagès, 2003, 2005). As in the sorting task, all products are presented simultaneously and assessors are asked first to look at, smell, and/or taste all the products and then to position the products on a sheet of paper according to their similarities or dissimilarities. Assessors are instructed that two products should be placed very close to each other if they are perceived as identical and far one from the other if they are perceived as different. There is no further instruction as to how the samples should be separated in this space, and so each assessor chooses his/her own criteria.

After they have positioned the products on the map, assessors can be asked to describe each product by writing a few words directly on the sheet near the products. The X and Y coordinates of each sample are recorded on each assessor map and compiled in a product by assessors table where each assessor contributes to columns representing, respectively, his or her X and Y coordinates. The matrix is then submitted to a multivariate analysis such as generalized procrustes analysis or multiple factor analysis to provide a sensory map of the products. Currently, no publication with coffee is available but they are numerous on wine.

3.4.2 Flash Profile and Check-All-That-Apply Questionnaires

The inconvenience of the rapid methods described earlier is that they are based on similarities and therefore descriptions of the products have a secondary priority. Flash profiling (Delarue and Sieffermann, 2004) is a method that instead focuses on the analytical description of the products. This descriptive method is a combination of free choice profiling (FCP) and ranking methods. It relies on the often noted fact that it is easier and more natural to compare products than to evaluate them on an absolute scale. Flash profile involves two sessions.

In the first session, the whole set of products is presented simultaneously to each assessor who is then asked to observe, smell, and/or taste the products (depending on the objectives of the study) and to generate a set of attributes, which should be sufficiently discriminant to permit ranking these products. The assessors are free to generate as many descriptive attributes as they want. In the second session, the assessors are asked to rank the products in the order from least to most on each of their attributes. Assessors' ranking data are then collected and analyzed using multivariate analysis such as the general procrustes analysis or multiple factor analysis.

The main advantage of the flash profile is the small number of judges (about 10 is enough) required and the very short time to provide a product map. However, the counterpart is the interpretation of the sensory terms because of the large number of terms generated and the lack of definitions and evaluation procedure. To limit this problem, product experts may be favored as assessors. As for napping, no publication with coffee was yet found but there are numerous on wine. For example, Liu et al. (2015) studied the performance of flash profile and napping with and without training for describing small sensory differences in a model wine.

Easy to read and interpret maps can be produced with the check all that apply (CATA) method. This approach originated from the work of Coombs (1964) and was first used in marketing research for studying consumers' perception of different brands. It has been recently introduced in sensory evaluation to understand consumer preference to help optimizing food products. A CATA questionnaire consists of a list of attributes (words or phrases) from which assessors should select all the attributes they consider appropriate to describe a given product during its evaluation.

Products are presented one at a time to the assessors according to a Latin square or a randomized design. Assessors are asked to evaluate each product and to check in the list the attributes that best describe the product for them. Assessors can check as many attributes as they wish. The attributes are not constrained to sensory aspects but could also be related to hedonic and emotional aspects as well as product usage or aptitude to fit a marketing concept.

A frequency matrix is then compiled by counting the number of assessors who used each attribute to describe each product and submitted to a correspondence analysis (CA) (i.e., a factor analysis for qualitative variables). Although it requires a rather important number of participants (usually consumers), the main advantage of CATA is its great simplicity both from the assessor and experimenter points of view. Its main drawback is that it necessitates an a priori list of descriptors. This list can be obtained either from the literature (or from previous studies) or gathered before the test via a focus group.

Table 18.1 gives some examples of CATA terms used for different product categories (Ares and Jaeger, 2015).

3.4.3 Pivot Profile

With the exception of CATA, the methods described previously do not enable data aggregation from different studies as all samples need to be presented at the same time. This might be problematic when the number of samples is too high for one session and when samples should be compared in different spaces

TABLE 18.1 Example of CATA Terms Used for Different Product Categories (Ares and Jaeger, 2015)

Product Category	List of Terms Included in the CATA Question
Milk desserts	Aftertaste, no aftertaste, off-flavor, no off-flavor, vanilla flavor, not much vanilla flavor, sweet, not very sweet, creamy, not very creamy, smooth, rough, heterogeneous, homogeneous, gummy, not gummy, thick, not thick
Orange drinks	Aftertaste, artificial, astringent, bitter, concentrated, diluted, intense color, intense flavor, light color, mandarin flavor, natural, off-flavor, orange flavor, smooth, sour, sweet
Raspberry coulis	Sour, green grape, green stalks, raspberry, strawberry, jammy, fruitiness, plum, sweet, woody, floral, boysenberry
White wine	Sweet, sour, tingly, sharp, dry, fruity, tropical, floral, green, cooked, vegetables, light, intense flavors, typical sauvignon blanc, thick, viscous, fades quickly

CATA, check all that apply.

and times. A way to bypass this problem is to still use a comparative method as it is easier for untrained assessors but instead of comparing all the products together to compare them to a stable reference.

Pivot profile was developed in the field of wine description and build on the idea of free description techniques often used in this field (Thullier et al., 2015). Assessors are provided with pairs of product including the reference (clearly identified as such) and the product to be evaluated. Assessors are asked to observe, smell, and/or taste the reference and the product and to write down each attribute that the product has in smaller or larger amount than the reference product (e.g., less sweet, more astringent).

Data analysis begins by regrouping synonyms and optionally regrouping the terms by categories. Then, negative and positive frequencies are computed for each term and each product and the negative frequency is subtracted from the positive frequency. The resulting score is finally translated so as to obtain positive scores only. The final matrix is submitted to a correspondence analysis to obtain a sensory map of the products. Pivot profile seems to be a very promising method for complex products such as wines and coffees where a tradition of free description is relatively strong among experts. This method might provide a trade-off between experts' practice and sensory evaluation and as such might be well adapted for coffee description.

Fig. 18.7 shows the map resulting from the application of the pivot profile in one session on five Champagne wines varying in their percentage

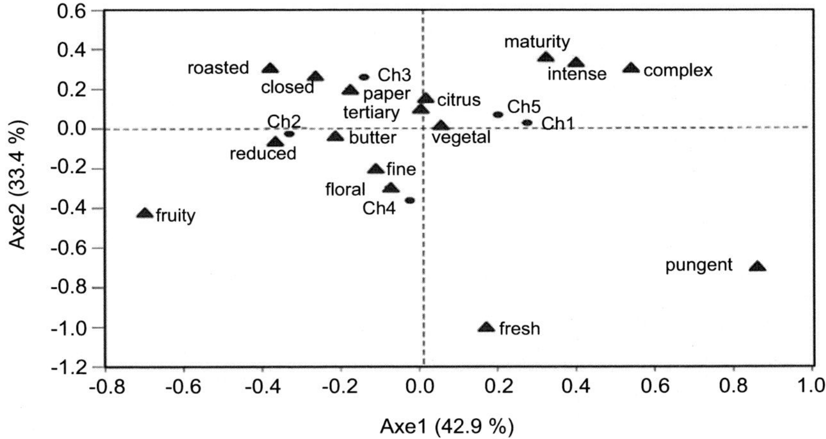

FIGURE 18.7 Projection of Champagnes (Ch1–Ch5) in the CA map (subspace 1–2) (Thuillier et al., 2015).

of wine variety (Pinot Noir, Pinot Meunier, Chardonnay), percentage of reserve wine and year of bottling and evaluated by 10 wine experts (Thuillier et al., 2015).

3.5 Applying Sensory Methods in a Professional Situation

The methods described previously are used in the food and beverage industry to develop new products, concepts, quality control of current products, and product improvement. Many others are available and have their own interest. The temporal dominance of sensation aims at identifying the sensory characteristics and their intensity that prevail as long as the evaluation last in time. The FCP leaves each assessor to select the attributes relevant to him/her before he/she rates their intensity in the products. Ranking methods belong to the classic method; rather an easy task for tasters, they generate robust results to position a product in comparison to known others.

Creating a new coffee starts with the production of several prototypes for sensory evaluation and consumer acceptability. Once it is determined what the consumer likes, the coffee may be produced with confidence of success and quality control procedures are developed. With the implementation of the new sensory methods seen earlier, the consumer understanding gets broader and more precise on whether he/she would like a new coffee profile, why he/she likes a taste, a concept, and how he/she would like it best. Moreover, with those new sensory methods added to the already long time existing, it becomes possible for any type of business size to consider which approach will match its purpose and objective within budget and timing.

4. THE EXPERT COFFEE TASTER

The coffee industry has changed significantly in the past few decades. We used to think of coffee as a one-dimensional beverage that mainly functioned as a caffeine delivery system. Now there is a considerable segment of the market that buys coffee for the sensory experience. In this fast-growing segment, the coffee expert taster plays an important role by making selections to meet consumer expectations.

With mainstream coffee, the coffee just needs to be relatively free of defects and off-flavors. However, in the specialty coffee market, there are many flavor profiles available and the "story" of the coffee and how its sensory profile links back to its origin, terroir, and the creation process is important to marketing the coffee. Consumers who enjoy this coffee are also often interested in the farmer who produced it and whether it is certified.

In this section, we look at the experts and the sensory tests they conduct as they use their expert knowledge to create coffees that match consumer preferences.

4.1 Who Are the Experts; Differences and Similarities Versus Nonexpert Tasters

Expert tasters are professionals whose work is directly linked with the taste of the coffee. Gatchalian (1981) defines these experts as thoroughly knowledgeable about the product to the point that their judgment is used for the decision-making process. The green coffee cupper is the most official taster, although the roasters, brewers, baristas, and sensory tasters also develop tasting skills to help them in their work.

Expert tasters can accurately relate the taste of coffee to the variables under their control. Guatemalan cuppers, for example, rely on their experience to differentiate the taste profiles of the coffee from specific valleys and farms. For Yadessa et al. (2008), experienced cuppers of Ethiopian green coffee can generate overall quality cupping scores correlated with the degree of the coffee cherry maturity and the green coffee physical characteristics. Similarly, many roasters can tell immediately if a coffee is slightly under- or overdeveloped.

Becoming an expert taster takes years of experience and requires building a sensory universe through continuous training and calibration with peers or mentors. Every year, the best coffee experts and the coffee they produce are recognized in the World Barista Competitions and the Cup of Excellence.

4.2 Expert Tasters: Their Role

The expert taster plays a major role in determining coffee quality. However, quality is an abstract concept, which is difficult to define (Sweeney and Soutar, 1995). Experts decide what constitutes quality, and they define and categorize it both to meet customers' preferences and so they can obtain the coffee they want from their suppliers. Their sensory judgment is largely based on many collaborative tasting sessions with their direct partners and stakeholders in the coffee value chain.

Expert tasters are both trendsetters and guardians of tradition. In recent years, coffee producers in many different regions have experimented with different processing methods. For example, natural pulped coffee has gained popularity for the aromatic complexity it delivers, even though in the past this was considered a lower quality product.

Traditionally, Costa Rican coffee is "wet processed" and is characterized by a clean, fruit-like acidity and sweetness. By using other processing methods, such as a natural pulped, honey process, producers can create a heavier bodied but less acidic coffee with more aromatic acids. Previously, the Villa Sarchi was the most typical Costa Rican tree variety, but during the past decade, farmers have planted other varieties from Africa and Asia.

Although Costa Rica has often led coffee innovation, other countries are also conducting their own experiments. Marsh et al. (2010) showed that a panel of 67 international and local specialty coffee cuppers ranked Indonesian natural pulped coffees higher in preference than the fully washed versions.

Donnet and Weatherspoon (2006) claim that cupping juries like those of the Cup of Excellence play a key role in defining the industry's quality standard. However, this raises the issue of "typicity," a controversial subject in wine as well as coffee. When one grows and produces a natural product, initially one uses the most practical and available means to create the product. This may include wine barrels of a certain type of local wood or a variety of plant that is particularly well adapted to the unique conditions of the growing area.

The combination of these aspects often results in a unique flavor that distinguishes the growing region and builds a market as consumers expect that flavor. As technology evolves, producers can increase production, reduce variability, or produce a style of product that is known to appeal to a certain market. These new products may be profitable and appealing; however, they may also risk losing some of the local typicity. This was the case with Bordeaux wines where sophisticated wine-making technology took over the effect of terroir and ultimately the quality ratings of reputable experts such as Parker, Bettane and Desseauve, and Broadent (Gergaud and Ginsburgh, 2008).

At the same time, like anyone expert coffee tasters have preferences and often deeply held convictions about what constitutes an ideal cup of coffee. They may also be business people with a financial interest in certain choices. These different roles can cause a conflict of interest when trying to make an objective sensory judgment.

Most specialty coffee businesses are small organizations where the proprietor fills several roles, including conducting a sensory analysis of samples. Many small coffee companies have become successful because they have identified a particular market niche that they are trying to fill. They make judgments based on their beliefs and business priorities. However, these judgments need to be objective to build accurate sensory tests and responses. If we consider the low level of alignment among renowned wine experts in their ratings of 2009 Bordeaux wines (Cicchetti and Cicchetti, 2014), we can see how greatly expert judgments can differ.

4.3 Expert Tasters: Their Skills

4.3.1 Perceptual Abilities

Sensory experts play an important role in describing and evaluating food and beverages. We rely on experts because we believe they have superior discriminative and descriptive abilities compared with novices. There have been a few studies to try to verify this assumption by comparing wine experts' and consumers' performance in detection and discrimination tasks.

Surprisingly, there was no difference in the detection threshold either for olfactory compounds such as 1-butanol (Bende and Nordin, 1997; Parr et al., 2002) or trigeminal stimuli such as tannin and alcohol (Berg et al., 1955). However, wine experts seem better than novices at discriminating single odors in odor mixtures (Bende and Nordin, 1997) and performing triangle tests (Solomon, 1990). At the same time, the experts' advantage seems to be related to increased exposure to wines, which enables them to learn to focus on the chemical features that distinguish the best wines. As Edmond Roudnizka said, "experts do not have a longer nose, they just know how to use it better."

4.3.2 Descriptive Abilities

We expect sensory experts to be able to describe food or beverages' organoleptic properties using a lexicon acquired during their training. Desor and Beauchamp (1974) and Cain (1979) showed that people can improve their ability to identify odors through practice and feedback, but this does not mean it increases their ability to communicate about them (Lehrer, 1983).

Studies showed that experts are more accurate and more consistent in identifying odors than novices and have a wider variability in their wine vocabulary. However, their ability to communicate is rather poor. An analysis of the vocabulary that experts and novices use indicates that the expert's superiority comes mainly from their ability to use specific words to describe particular aromatic and flavor elements (Solomon, 1990; Lawless, 1984). Although we also see that novice performance improves when they are provided with expert descriptions, their performance still remains below that of experts, so the expert superiority cannot be explained only by the preciseness of their descriptions (Valentin et al., 2003).

4.3.3 Memory Structures

Sensory experts build memory structures that reflect the sensations found within specific foods or beverages. When asked to describe a food or beverage sample, they start by activating characteristics from these types of products in their memory. Because olfactory notes are often ambiguous (a hint of red fruits, a rose-like aroma...), having specific expectations about the product might help in labeling ambiguous notes. For example, Solomon (1997) found that experts who incorrectly identified a pinot gris wine as a chardonnay described this wine with descriptors generally found in chardonnay wines. Likewise, Pangborn et al. (1963) found that experts perceived a white wine that was colored pink like rosé as sweeter.

In coffee, descriptors are also very specific. For example, we might call a green coffee immature or roasted beans scorched or baked. At the same time, the difference between particular descriptors is an endless source of debate, such as defining the difference between an overfermented fruit and a deep fruit. In Cicchetti and Cicchetti (2014) experts' comparison of Bordeaux wine ratings, the authors discussed the big disparity of results on the 2003 Premier Grand Cru Pavie Saint Emilion. One group scored this wine poorly and the other gave it a superior rating. Although the tasters agreed that the wine had a note of fruit, the two groups gave it different quality ratings based on whether or not they personally liked the fruity note.

4.4 Expert Sensory Tests in Current Use

Expert tasters usually develop their own tasting methods. The most commonly used method is the technical tasting where a group of 2–10 tasters simply describes the coffees. This can be coupled with an overall evaluation of the sample based on the expert perception of the sample quality.

Coffee is produced naturally, and therefore the crop can vary from year to year and harvest to harvest. As a result, sampling procedures are important for accurate results. All sampling methods use specific sampling procedures to ensure that the samples under evaluation represent the entire lot. Most often, they use a standard sampling procedure called "cupping." (see Chapter 8).

4.5 Grading Procedures

The New York Board of Trade grades coffee per the "C" rules. Coffee certified as part of a "C" lot consists of 37,500 pounds of coffee that is "free from all unwashed flavors in the cup, of good roasting quality, and of bean size and color in accordance with criteria established by the Exchange" (see Chapter 9). Graders use standard cupping practices and physical tests of the green beans (such as defect counts) to establish whether the lot is free of off-flavors or defects.

The Coffee Quality Institute also has the Q Coffee System "Q," which ensures that the coffee meets a quality standard that would qualify it as "specialty" coffee. Graders also examine the sample physically, and it must receive a score of 80 or higher as scored on the SCAA/Q form during the cupping evaluation.

In both of these procedures, the grade established is pass/fail. The sample either passes the test and is certified as "C" or "Q" or is not certified.

4.6 Scoring Methods

There are several methods to score the coffee, virtually all of which are based on a "100-point" scale. This is similar to the wine ratings systems. These rating systems report the relative quality of the coffee and communicate that quality to potential buyers, including consumers.

Numerical methods are useful in comparing coffees; if one is rated higher than another, we assume it is better, especially if there is a large difference between the scores. The score is often augmented by nonstandardized descriptors.

One scoring method, which George Howell developed, is used in Cup of Excellence competitions. These competitions are designed to choose the highest quality samples among those entered; theoretically, all of them would necessarily pass the Q rating on the Q Coffee System. Panels from the origin country and then cuppers from consuming countries cup the samples several times. They then average and compare the scores to determine rankings and winners. Lots entered in the final phases of the competition are placed in a bonded warehouse and test auditors take samples.

The Cup of Excellence form for a single sample is shown in Fig. 18.8. There are eight categories, each of which is scored on a scale of 0–8. A total of 36 points is added to this score to correspond to the 100-point score. The qualitative categories include "Clean" along with the individual attributes "Sweet," "Acidity," and "Mouthfeel." The final four qualitative categories ("Flavor," "Finish," and "Balance") involve judgments of combined categories or value judgments ("Overall"). As there is rarely a "standard" available, there is sometimes some disagreement over these types of qualities. Currently, a new form for use in the competitions is under development.

The "Q" system of scoring (Fig. 18.9) determines whether a coffee is considered a specialty coffee and combines the scores of qualitative attributes

Cup of Excellence

Name____ Date____ Rnd 1 2 3 Sn 1 2 3 4 5 TLB #____ Country____ Number____

ROAST COLOR	AROMA (DRY CRUST BREAK)	DEFECTS (# x I x 4 = SCORE)	CLEAN CUP	SWEET	ACIDITY	MOUTH FEEL	FLAVOR	AFTER-TASTE	BALANCE	OVERALL	TOTAL (+36)
	3 2 1 / 3 2 1 / 3 2 1	_x_x 4 =	0 4 6 7 8	0 4 6 7 8	0 4 6 7 8 (H M L)	0 4 6 7 8 (H M L)	0 4 6 7 8	0 4 6 7 8	0 4 6 7 8	0 4 6 7 8	

SAMPLE

FIGURE 18.8 Cup of excellence cupping form.

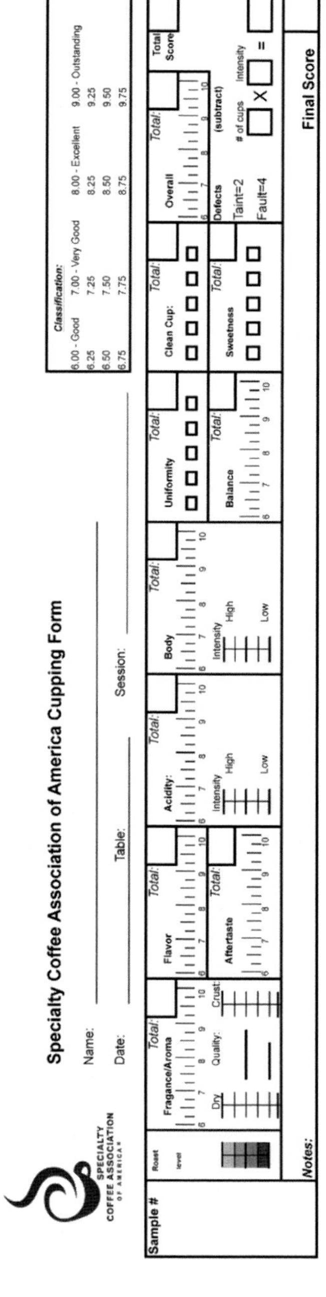

SPECIALTY COFFEE ASSOCIATION OF AMERICA

Specialty Coffee Association of America Cupping Form

Name: ____________

Date: ____________ Table: ____________ Session: ____________

Classification:			
6.00 - Good	7.00 - Very Good	8.00 - Excellent	9.00 - Outstanding
6.25	7.25	8.25	9.25
6.50	7.50	8.50	9.50
6.75	7.75	8.75	9.75

Sample #

Roast level

Fragance/Aroma 6 7 8 9 10 *Total:*
Dry Quality: Crust:

Flavor 6 7 8 9 10 *Total:*

Aftertaste 6 7 8 9 10 *Total:*

Acidity: 6 7 8 9 10 *Total:*
Intensity High Low

Body 6 7 8 9 10 *Total:*
Intensity High Low

Uniformity *Total:*

Balance 6 7 8 9 10 *Total:*

Clean Cup: *Total:*

Sweetness *Total:*

Overall 6 7 8 9 10 *Total:*

Defects (subtract)
Taint=2 # of cups Intensity
Fault=4 ☐ X ☐ =

Total Score

Notes:

Final Score

FIGURE 18.9 The SCAA cupping form used for Q grading.

with some check boxes. A total of five cups are evaluated for "Fragrance/Aroma," "Flavor," "Aftertaste," "Acidity," "Body," and "Balance" using a 10-point category score; "Uniformity," "Clean cup," and "Sweetness" are scored using check boxes (each box represents one cup), and there is a final "Overall" preference category. There is a separate form for Robusta coffees.

4.7 Other Methods for Finished Product to Evaluate the Quality

No standard method is used across the whole coffee industry, probably because finished products have so many different expressions. Fig. 18.10 shows the Nespresso Coffee Score Card where the coffee's core descriptive and qualitative attributes are rated in separate categories to distinguish between the coffee expert judgment of quality and the descriptive evaluation of the product.

5. THE CONSUMERS

5.1 Consumers Are the Opposite of Experts

Consumers approach coffee in a very different manner than experts. For them, coffee is an everyday beverage (as opposed to a luxury or special occasion beverage), and most coffee drinkers enjoy it in the morning. It is often a social beverage, served to groups during meetings and friends "get together for coffee." Some consumers drink coffee mainly for its stimulating effect, whereas others are sensitive to or simply prefer avoiding caffeine and thus choose decaffeinated coffee.

On the other hand, coffee has become an affordable luxury with the emergence of a wide variety of specialty coffee products. In parallel, specialty coffee consumers are becoming more sophisticated in their coffee knowledge. Some will buy a coffee on the basis of its origin, whereas others prefer a blend they know well, or a coffee intended for a particular method of preparation.

However, even the most knowledgeable coffee consumers lack the knowledge of the coffee expert who values the sensory diversity in the final beverage and understands the intrinsic quality of the coffee origins. Instead the consumer's relationship with coffee is shaped by his/her personal cognitive experience, cultural and social environment, genetic heritage, brand knowledge, perception of quality and flavor preferences.

According to Cristovam et al. (2000), a new market for espresso specialty products has emerged, and coffee suppliers do not understand consumer preferences. In addition, they lack the ability to relate to different market segments. They are developing products based on the espresso characteristics solely when sales are dominated by cappuccino and latte drinks. Any coffee expert intending to offer consumers a great coffee experience should always

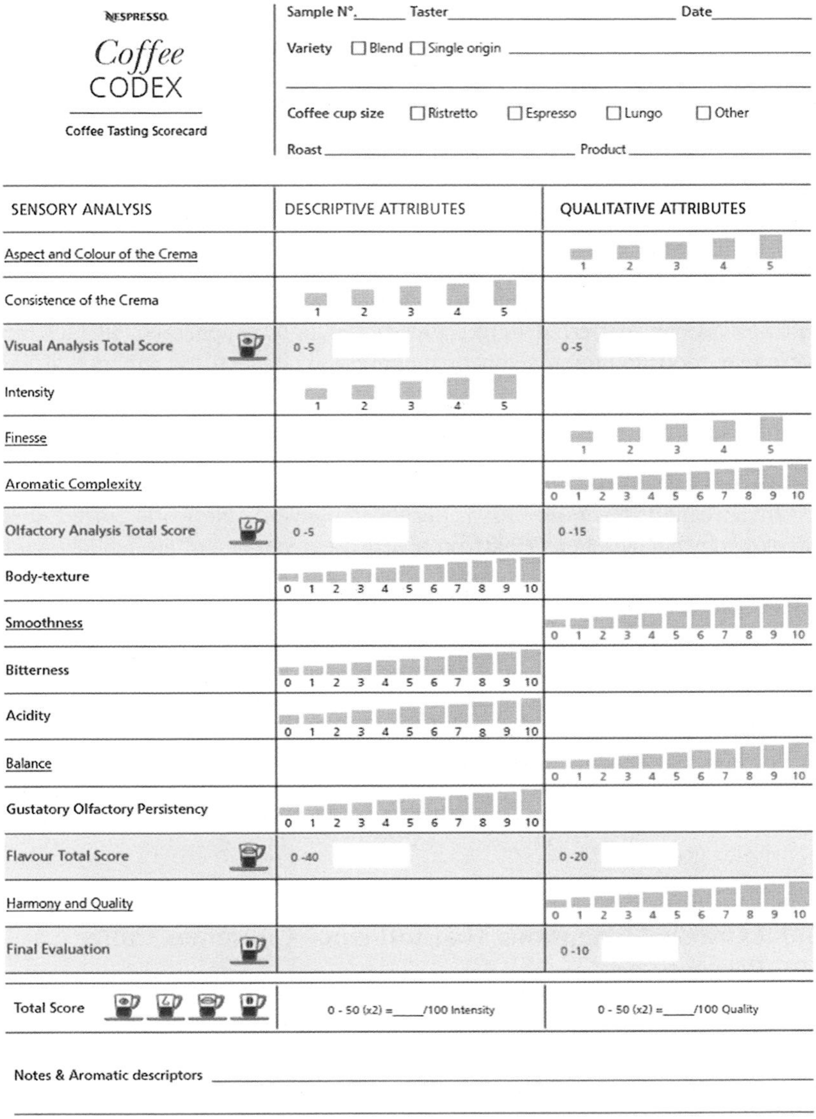

NESPRESSO

Coffee
CODEX

Coffee Tasting Scorecard

Sample N°.______ Taster______________________________ Date___________

Variety ☐ Blend ☐ Single origin ______________________________

Coffee cup size ☐ Ristretto ☐ Espresso ☐ Lungo ☐ Other

Roast______________________________ Product___________________

SENSORY ANALYSIS	DESCRIPTIVE ATTRIBUTES	QUALITATIVE ATTRIBUTES
Aspect and Colour of the Crema		1 2 3 4 5
Consistence of the Crema	1 2 3 4 5	
Visual Analysis Total Score	0-5	0-5
Intensity	1 2 3 4 5	
Finesse		1 2 3 4 5
Aromatic Complexity		0 1 2 3 4 5 6 7 8 9 10
Olfactory Analysis Total Score	0-5	0-15
Body-texture	0 1 2 3 4 5 6 7 8 9 10	
Smoothness		0 1 2 3 4 5 6 7 8 9 10
Bitterness	0 1 2 3 4 5 6 7 8 9 10	
Acidity	0 1 2 3 4 5 6 7 8 9 10	
Balance		0 1 2 3 4 5 6 7 8 9 10
Gustatory Olfactory Persistency	0 1 2 3 4 5 6 7 8 9 10	
Flavour Total Score	0-40	0-20
Harmony and Quality		0 1 2 3 4 5 6 7 8 9 10
Final Evaluation		0-10
Total Score	0 - 50 (x2) = ____/100 Intensity	0 - 50 (x2) = ____/100 Quality

Notes & Aromatic descriptors ______________________________

FIGURE 18.10 The Nespresso Coffee tasting scorecard from the coffee codex (Vaccarini et al., 2015) requires the expert taster to focus equally on rating the intrinsic characteristics of the coffee beverage and on judging its quality.

ask themselves these key questions: What is the level of coffee awareness of their consumers? What are their preferences? What are their motivations for drinking their current coffee(s) and what could motivate them to change or discover other types of coffees?

5.2 Consumer Preferences for Intrinsic Coffee Dimensions

In the fast-moving coffee scene in Brazil, past studies show that consumers prefer their traditional coffee to newer specialty coffee. Even if these preferences gradually evolve over the years, the Brazilian coffee study could possibly be extrapolated to all countries where traditional coffee has a long history.

Preparation modalities vary enormously among consumers. In Spain (Varela et al., 2014), for example, when given the option of instant coffee with varying amounts of milk and sugar, most consumers preferred this sweet, milky coffee beverage. In another study, in Glasgow (Cristovam et al., 2000), most consumers preferred light espresso and light cappuccino. In fact there were two opposite preference patterns observed between consumers that liked light espressos and consumers who liked intense espressos. In this case, adding milk for cappuccino changed the preference as all consumers showed in comparison the black preparation an increased preference for intense espresso in their cappuccino.

In a reanalysis of the 1995 European Sensory Network coffee study, Moskowitz and Krieger (1998) proposed a consumer preference segmentation based on the level of coffee bitterness. Tested across five European countries, consumers were split among those who liked coffee with low bitterness, those who liked intense bitterness, and those who liked intermediate bitterness. A similar study on bitterness in chocolate milk by Harwood et al. (2012) showed that consumers who declared preferring dark chocolate rejected bitterness at a level twice as high than consumers who declared preferring milk chocolate. For the authors, either liking or rejecting bitterness in the chocolate was not based on the subject's sensitivity to the taste, given that the detection threshold for bitterness was the same between the two consumer groups.

5.3 Extrinsic Dimensions That Influence Consumer Coffee Perceptions

5.3.1 Price and Information

_underWe can segment consumers into several categories based on how they view the beverage, how they consume the beverage, and what they expect from the beverage. Some consumers see coffee as a common experience and make their buying decisions mainly on the basis of price. Others view coffee as a lifestyle choice and have specific expectations in terms of quality, social responsibility, or status identification; for these consumers, price is less important.

In Asioli et al.'s (2015) study on the relatively new beverage, iced coffee, Norwegian consumers said they would choose the cheap and low-calorie beverage proposed to them, whether they preferred the sensory profile of the ice coffee made with an espresso or a latte. In a 2002 study with French consumers, Lange et al. (2015) studied the impact of providing both more information and sensory exposure for Fairtrade and regular coffees. Results showed that when consumers received more information about the nature of the coffee, their willingness to pay for ethical products increased compared to regular products. This preference became statistically significant when consumers also had the opportunity to taste the coffee before buying it.

5.3.2 Consumer's Emotions to Coffee

For Bhumiranata et al. (2014), consumers drink coffee primarily for pleasure, and this experience creates deep and distinct emotions, which deserve their own emotion lexicon. According to the authors, consumers have varying preferences for coffee, and they seek different emotional experiences based on the sensory properties of coffee. Some drink coffee to elicit positive lower energy feelings such as feeling comfortable, pleasant, or rewarded. Others drink coffee to experience arousing, positive, high-energy emotions, and feel more active, energetic, or rested. Still others drink coffee to achieve a focused mental state such as feeling more educated or motivated.

For Labbe et al. (2015) the consumers' motivation for drinking coffee helps generate a different emotional state and level of pleasantness. During coffee preparation, consumers with a hedonic motivation were more likely to feel relaxed, satisfied, amused, or energetic. However, consumers with more of a utilitarian motivation were more agitated and excited. After the drinking of the coffee, the emotional state of consumers in a utilitarian motivation rejoined that of the consumers in a hedonic motivation.

5.3.3 Are Consumers and Experts Bound Not to Understand One Another?

For Morales (2002), coffee grading systems and experts have a limited impact on consumer purchasing because there is little input from consumers. There should be a better match between the expert quality claims on coffee packages and consumers' perceptions and willingness to buy a certain coffee. On the other hand, we knew that expert quality opinions influence consumers as illustrated in wines by Vignes and Gergaud (2007). In this study, consumers ranked four champagne wines differently when tasting them blind as compared to tasting them when they had full information on pricing and brands. This showed that there is a disconnect between consumer beliefs and their actual taste preferences, which depend on their oenological levels and personal tastes.

In Brazil, Garcia et al. (2008) study with 54 consumers showed that in blind tests, they preferred their traditional classified coffee over gourmet classified coffee. However, a comfortable majority of the subjects said the brand, the type of packaging, and the Associação Brasileira da Indústria do Café (ABIC) quality seal, as well as the quality category, are priority information to help their purchasing decisions.

How then can we ensure the consumer has adequate information and is willing to pay for the product? The expert should work toward matching the perceived sensory experience with the consumers' expectations and preferences. He should also work to provide better product information so consumers can discover high quality and preferred products.

6. SENSORY AND CONSUMER SCIENCE TO BRIDGE THE GAP BETWEEN COFFEE CONSUMERS AND COFFEE EXPERTS

As seen in Section 4 of this chapter, experts have a better ability than consumers to describe products, although the difference is far from impressive. Experts can also be biased by their own expectations about quality and their personal preferences. As such, experts are divided on the notion of quality and are too often disconnected from the consumer motivations and likings. Sensory and consumer science in the coffee industry have become the mediator between coffee experts and consumers. It is helping to evolve from a situation of "follow me, I am the expert" to "we the experts understand you."

Sensory scientists put consumers at the heart of their discipline. They help translate the kind of coffee experience consumers like to marketers and developers. As the discipline continues to evolve, sensory and consumer scientists tackle questions such as why and how consumers accept or reject a given coffee experience. In essence, sensory science helps enrich and ease communication with consumers to guide them to coffees that will provide pleasurable experiences.

Coffee quality is defined differently by the different people in the coffee value chain (Leroy et al., 2006) and sensory and consumer science has shown that the concept of quality differs both among the experts and between the experts and consumers. When quality is better defined among coffee stakeholders, it becomes easier for researchers and developers to define and measure quality goals. This quality circle uniting experts, consumers, and science helps align methodologies and vocabulary and defines quality results.

6.1 Consumer Segmentation

Today consumers are at the heart of the debate around coffee quality. Preference mapping shows that consumers are segmented in their coffee preferences. For example, whether experts like it or not, dark roasted bitter coffee pleases about one-third of consumers across Europe (Moskowitz and Krieger, 1998). Experts, on the other hand, tend to value highly acidic coffees with exceptional ratings, even though there has not yet been a study to determine the size of the consumer segment that might find this coffee appealing. In addition, shelf life for quality in cup is based on acceptance limits which find their true relevance when involving consumers. For this reason, Manzocco and Lagazio (2008) proposed a prediction model for consumer acceptance and rejection of the shelf life of brewed coffee.

6.2 Common Language

Beyond the experts and how well consumers like a particular coffee, it is important to understand how the description of the coffee can influence the perceived level of quality.

Vocabulary is an important bridge that sensory and consumer scientists have developed to try to help the whole coffee community communicate on the same level. Expert wording, as seen in Section 4, helps consumers better anticipate the experience the coffee will deliver, so expert voices certainly provide a certain level of trust and influence on the consumer perception of product quality. Conversely, the words themselves can also mislead. Herz and von Clef (2001) showed how verbal positive or negative labeling can influence the perceptions of odors in perfume in what she calls "olfactory illusions."

In fact, sensory and consumer science has become the guardian of a common language which all coffee stakeholders can use. Today, cuppers, developers, and marketers use this common language. Extending this language to all stakeholders, from the farmers to the consumers, will improve the coffee experience for everyone. Tools such as "Le Nez du Café, Edition Jean Le Noir" or the Coffee Lexicon by the World Coffee Research also help by defining and providing reference scents for the sensory descriptors related to coffee.

6.3 Alignment Among Experts and With Consumers

Refining the coffee vocabulary to improve alignment among all stakeholders is an ongoing process. Some aspects of this, such as agreeing on the cocoa or jasmine notes or defining acidity on a general basis, are relatively easy. However, it becomes more difficult when trying to draw the line between qualitative and unpleasant acidity, such as the fine acidity of the best green coffee origins and acidity that occurs during coffee staling. Coffee has its pool of general concept terms including quality, sweet perception, intensity/strength, fine acidity, and complexity that needs to be more specifically defined to improve understanding and meaning, not only with consumers, but also among the coffee experts themselves.

The wine industry has led the way with the fairly recently developed concept term "minerality." This term, which references the very low concentration of minerals in wines, makes an improbable direct link between the soil where the vine is grown and the emergence of mineral notes in wines

(Maltman, 2013). In Rodriguez et al. (2015) study, using a free word association exercise with the term minerality, Chablis wine makers produced slightly more words, which were more consensual than those of consumers. They developed a structured representation including the origin (terroir and soil) as well as the sensory properties (chalky, shellfish, freshness) of the wine. In contrast, for burgundy wine consumers, the concept of minerality was limited to the terroir. However, for both groups, minerality was seen as a positive descriptor of quality.

6.4 Analytical Support

Sensory and consumer science is also growing in the coffee industry as a quantitative discipline, completing or correcting the sensitive qualitative experience of the coffee cupper. It connects the abstract analytical data with the more tangible sensory evaluation and tasting descriptors that influence the end cup result. Sensory and consumer science serves, therefore, as part of the process variables related to terroir characteristics, green coffee physics, roasting profiles, and chemical aromatic composition.

Bhumiratana et al. (2011) proposed an experimental design to measure the impact of three green coffee origins at three roasting degrees on the sensory aroma profile from a trained panel. This showed that the roast degree had a higher impact than the coffee origin before concluding that consumer testing is necessary to identify the key attributes that might impact acceptability. A study from Miyai et al. (2010) with ready-to-drink chilled espresso showed that the sensory description from trained tasters correlated with the origin and roast degree as well as with nonvolatile aromatics. In his study, bitter and full body beverages related directly to the quinic acid and pyroglutamic acid found in higher proportion in dark roast. In contrast, light, green perceived beverages related directly to the citric acid, chlorogenic acid, trigollenine, and formic acid found in higher proportion in light roast.

Tolessa et al. (2015) proposed a model that used near-infrared spectra to predict whether Ethiopian green coffee beans would be considered specialty coffee. He based this determination on the high correlation between the near-infrared spectra and the specialty coffee scores generated on 86 samples of dry, wet, and semiwashed Arabica from three districts of the Jimma zone in Ethiopia. The study showed that this instrumental measure can indeed predict and thus replace human sensory evaluation to save time and money. However, we need to remember that these prediction models are only valid within the limits of the sensory space of the samples and conditions selected.

7. OUTLOOK

The expert goal is that consumers appreciate high quality coffee. However, quality is a notion that cannot be left in the hand of experts only. Experts and consumers need one another to produce great coffee experiences and the sensory and consumer scientists are their bridge. Where to draw a clear-cut line between the experts noncompromise for high quality with absence of defects and their customers current usage and linking is worth reconsidering. Applying the pillars and the basis of sensory evaluation and consumer science across the whole coffee value chain will comfort the experts.

In the globalized world, the faces of the cross-cultural coffee product vary enormously from traditional to innovative and from standard to high quality. Marketing and development must rely on a robust updated knowledge of consumer segments for preferences and identification of sensory attributes driving liking. Experimental designs for a disciplined development versus the repeated trial and error iterations approach can certainly maximize acceptance, maintain cost, reduce time to discover product optimization, and set a basis to refer upon in the long term (Moskowitz and Krieger, 1998). The barriers due to the apparent costs and resources of disciplined development should be easily overcome when considering the missed opportunity and total amount of resources spent in the trial and error iteration approach.

Consumers are changing rapidly, therefore, repeated assessments of the behavior and sensory sensitivity become important and not only should we track the consumer unconscious behavior with blind testing for liking but also their conscious behavior toward the information carried by the coffee like its origin, certification, sensory aromatic, and intensity.

The consumer demand for coffee of high quality is on the increase and the production must respond appropriately to keep delivering higher volumes of differentiated coffee of quality. As pointed by Barrère (2007) for champagne wine, the production of quality can be economically viable only if quality is known on the market. More than ever, products of highest quality need to distinguish clearly from the standard offer in markets where quality overall has raised and is cautioned by experts via tasting notes, certifications, or publications while the growing number of new consumers start with limited tasting competencies of discernment. The coffee expert must, therefore, be an educator to raise the level of awareness among all the coffee lovers and particularly among those interested but not yet connoisseurs. He/she also should act as the guardian of quality and stand for its faithful reproduction in the future.

REFERENCES

AFNOR, 1992. Sensory Analysis. Vocabulary. NF ISO 5492. AFNOR, Paris- La Défense.

Ares, G., Jaeger, S.R., 2015. Examination of sensory product characterization bias when check-all-that-apply (CATA) questions are used concurrently with hedonic assessments. Food Quality and Preference 40, 199–208.

Asioli, D., Næs, T., Øvrum, A., Almli, V.L., 2015. Comparison of rating-based and choice-based conjoint analysis models. A case study based on preferences for iced coffee in Norway. Food Quality and Preference 48 (2016), 174–184.

Barrère, C., 2007. Le Champagne, d'un vin aristocratique à un produit de luxe maarchand ? Publié in « Le Champagne, journée internationale d'étude sur le vin de Champagne, approches pluridisciplinaires. Presses Universitaires de Reims, 2007.

Bende, M., Nordin, S., 1997. Perceptual learning in olfaction: professional wine tasters versus controls. Physiology and Behavior 62 (5), 1065–1070.

Berg, H.W., Filipello, F., Hinreiber, E., Webb, A.D., 1955. Evaluation of thresholds and minimum difference concentrations for various constituents of wines. Food Technology 9, 23–26.

Bhumiratana, N., Adhikari, K., Chambers, E., 2014. The development of an emotion lexicon for the coffee drinking experience. Food Research International 61 (2014), 83–92.

Bhumiratana, N., Adhikari, K., Chamberst, E., 2011. Evolution of sensory aroma attributes from coffee beans to brewed coffee. LWT – Food Science and Technology 44 (2011), 2185–2192.

Breslin, P.A., Huang, L., 2006. Human taste: peripheral anatomy, taste transduction, and coding. Advances inOto-rhino-laryngology 63, 152–190.

Buck, L.B., Axel, R., 1991. A novel multigene family may encode odorant receptors: a molecular basis for odor recognition. Cell 65, 175–187.

Cain, W.S., 1974. Contribution of the trigeminal nerve to perceived odor magnitude. Annals of the New York Academy of Sciences 237, 28–34.

Cain, W.S., 1976. Olfaction and the common chemical sense: some psychophysical contrasts. Sensory Processes 1, 57.

Cain, W.S., 1979. To know with the nose: keys to odor identification. Science 203 (4379), 467–470.

Cain, W.S., Murphy, C.L., 1980. Interaction between chemoreceptive modalities of odour and irritation. Nature 284 (20), 255–257.

Chandrashekar, J., Hoon, M.A., Ryba, N.J.P., Zuker, C.S., 2006. The receptors and cells for mammalian taste. Nature 444, 288–294.

Chaudhuri, A., 2010. Fundamentals of Sensory Perception, 2010. Oxford University Press, p. 4.

Cicchetti, D., Cicchetti, A., 2014. Two enological titans rate the 2009 Bordeaux wines. Wine Economics and Policy 3, 28–36.

Commetto-Muñiz, J.E., Cain, W.S., 1994. Sensory reactions of nasal pungency and odor to volatile organic compounds: the alkylbenzenes. American Industrial Hygiene Association Journal 55, 811–817.

Coombs, C.H., 1964. A theory of data.

Coxon, A.P.M., 1999. Sorting data: Collection and analysis, vol. 127. Sage Publications.

Cristovam, C., Russell, C., Paterson, A., Reid, E., 2000. Gender preference in hedonic ratings for espresso and espresso-milk coffees. Food Quality and Preference 11 (2000), 437–444.

Deliza, R., Gonçalves, A.-M.O., Barros, P.R.S., Ribeiro, E.N., Farah, A., 2006. Consumer Detection Limit of Defective Beans in Brazilian Coffee. Asic 2006.

Deliza, R., Martinelli, M., Farah, A., 2008. Sensory Profiling and External Preference Mapping of Coffee Beverages with Different Levels of Defective Beans. Asic 2008.

Delarue, J., Sieffermann, J.M., 2004. Sensory mapping using flash profile. Comparison with a conventional descriptive method for the evaluation of the flavour of fruit dairy products. Food Quality and Preference 15 (4), 383–392.

Desor, J.A., Beauchamp, G.K., 1974. The human capacity to transmit olfactory information. Perception & Psychophysics 16 (3), 551–556.

Donnet, M.L., Weatherspoon, D.D., 2006. Effect of sensory and reputation quality attributes on specialty coffee prices. In: American Agricultural Economics Association Annual Meeting, Long Beach, California, July 23–26, 2006. Copyright 2006.

Doty, R.L., 1975. Intranasal trigeminal detection of chemical vapors by humans. Physiology & Behavior 14 (6), 855–859.

Duchamp-Viret, P., Chaput, M.A., Duchamp, A., 1999. Odor response properties of the rat olfactory receptor neurons. Science 284, 2171–2174.

Feria-Morales, A., 2002. Examining the case of green coffee to illustrate the limitations of grading systems/expert tasters in sensory evaluation for quality control. Food Quality and Preference 13 (2002), 355–367.

Frank, R.A., Byram, J., 1988. Taste-smell interactions are tastant and odorant dependent. Chemical Senses 13, 445–455.

Frank, R.A., Shaffer, G., Smith, D.V., 1991. Taste-odor similarities predict taste enhancement and suppression in taste-odor mixtures. Chemical Senses 16, 523.

Garcia, A.O., Teles, C.R.A., Barbieri, M.K., Pires, S.H.L., 2008. Evaluation of Acceptability of Gourmet, Superior and Traditional Coffees. ASIC 2008.

Gatchalian, M.M., 1981. Sensory Evaluation Methods with Statistical Analysis. College of Home Economics, University of the Philippines, Diliman, Quezon City, Philippines.

Gergaud, O., Ginsburgh, V., 2008. Natural endowments, production technologies and the quality of wines in Bordeaux. Does terroir matter? The Economic Journal 118, F142–F157 (June).

Harwood, M.L., Ziegler, G.R., Hayes, J.E., 2012. Rejection thresholds in chocolate milk: evidence for segmentation. Food Quality and Preference 26 (2012), 128–133.

Herz, R.S., von Clef, J., 2001. The influence of verbal labeling on the perception of odors: evidence for olfactory illusions? Perception 30 (3), 381–391.

Köster, E.P., 2009. Diversity in the determinants of food choice: A psychological perspective. Food Quality and Preference 20 (2), 70–82.

Kreuml, M.T.L., Majchrzak, D., Ploederl, B., Koenig, J., 2013. Changes in sensory quality characteristics of coffee during storage. Food Science & Nutrition 1 (4), 267–272.

Labbe, D., Damevin, L., Vaccher, C., Morgenegg, C., Martin, N., 2006. Modulation of perceived taste by olfaction in familiar and unfamiliar beverages. Food Quality and Preference 17, 582–589.

Labbe, D., Ferrage, A., Rytz, A., Pace, J., Martin, N., 2015. Pleasantness, emotions and perceptions induced by coffee beverage experience depend on the consumption motivation (hedonic or utilitarian). Food Quality and Preference 44 (2015), 56–61.

Laing, D.G., Jinks, A., 1996. Flavour perception mechanisms. Trends in Food Science & Technology 7, 387–389.

Lange, C., Combris, P., Issanchou, S., Schlich, P., 2015. Impact of information and in-home sensory exposure on liking and willingness to pay: the beginning of Fairtrade labeled coffee in France. Food Research International 76 (2015), 317–324.

Laska, M., Distel, H., Hudson, R., 1997. Trigeminal perception of odorant quality in congenitally anosmic subjects. Chemical Senses 22 (4), 447–456. Aug.

Lawless, H.T., 1984. Flavor description of white wine by "expert" and nonexpert wine consumers. Journal of Food Science 49, 120–123.

Lawless, H.T., 1985. Psychological perspectives on winetasting and recognition of volatile flavors. In: Birch, G.G. (Ed.), Alcoholic Beverages. Elsevier Applied Science Publishers, London, pp. 97–113.

Lehrer, A., 1983. Wine and Conversation. Indiannia University Press, Bloomington.

Leloup, V., Gancel, C., Liardon, R., Ryz, A., Pithon, A., 2004. Impact of Wet and Dry Process on Green Coffee Composition and Sensory Characteristics. Asic 2004.

Leon, M., Johnson, B.A., 2003. Olfactory coding in the mammalian olfactory bulb. Brain Research Reviews 42, 23–32.

Leroy, T., Ribeyre, F., Bertrand, B., Charmetant, P., Dufour, M., Montagnon, C., Marraccini, P., Pot, D., 2006. Genetics of coffee quality. Brazilian Journal of Plant Physiology 18, 229–242.

Liu, J., Grønbeck, M.S., Di Monaco, R., Giacalone, D., Bredie, W.L.P., 2015. Performance of flash profile and napping with and without training for describing small sensory differences in a model wine. Food Quality and Preference 48 (2016), 41–49.

Maltman, A., 2013. Minerality in wine: a geological perspective. Journal of Wine Research 24 (3), 169–181.

Manzocco, L., Lagazio, C., 2008. Coffee brew shelf life modelling by integration of acceptability and quality data. Food Quality and Preference 20 (2009), 24–29.

Marsh, A.Y., Mawardi, S., 2010. The Influence of Primary Processing Methods on the Cup Taste of Arabica Coffee from the Indonesian Island of Flores. ASIC 2010.

Meillgard, M.C., Civille, G.V., Carr, B.T., 1999. Sensory Evaluation Techniques 3rd Edition, 1999. CRC Press, Boca Raton, pp. 162–164, p. 45.

Miyai, T., Akiyama, M., Michishita, T., Katakura, T., Ikeda, M., Araki, T., Sagara, Y., 2010. Taste Compounds Affecting Sensory Characteristics of Ready-to-Drink Chilled Espresso. ASIC 2010.

Monbaerts, P., Wang, F., Dulac, C., Chao, S., Nemes, A., Mendelsohn, M., Edmondson, J., Axel, R., 1996. Visualizing an olfactory sensory map. Cell 87, 675–686.

Moskowitz, H., Krieger, B., 1998. International product optimization: a case history. Food Quality and Preference 9 (6), 443–454, 1998.

Moulton, D.G., 1967. Spatio-temporal patterning of response in the olfactory system. In: Hayashi, T. (Ed.), Olfaction and Taste, vol. II. Pergamon Press, Oxford, pp. 109–116.

Murphy, C., 1987. In: Finger, T.E., Silver, W.L. (Eds.), Olfactory Psychophysics. In Neurobiology and Taste and Smell. Wiley, New York, pp. 251–273.

Nef, P., Heineniann, S., Dionne, V.E., 1991. Spatial distribution of OR3, a putative seven transmembrane domain olfactory receptor, reveals an olfactory map. Chemical Senses 16, 562.

Pagès, J., 2003. Recueil direct de distances sensorielles: application à l'évaluation de dix vins blancs du Val de Loire. Sciences des aliments 23, 679–688.

Pagès, J., 2005. Collection and analysis of perceived product inter-distances using multiple factor analysis: application to the study of 10 white wines from the Loire. Food Quality and Preference 16 (7), 642–649.

Pangborn, R.M., Berg, H., Hansen, B., 1963. The influence of color on discrimination of sweetness in dry table wine. American Journal of Psychology 76, 492–495.

Parr, W.V., Heatherbell, D., White, K.G., 2002. Demystifying wine expertise: olfactory threshold, perceptual skill and semantic memory in expert and novice wine judges. Chemical Senses 27, 747–755.

Petit, C., Sieffermann, J.M., 2007. Testing consumer preferences for iced-coffee: does the drinking environment have any influence? Food Quality and Preference 18 (2007), 161–172.

Prescott, J., 1999. Flavour as a psychological construct: implications for perceiving and measuring the sensory qualities of foods. Food Quality and Preference 10, 349–356.

Puget, S., Dugas, V., Pineau, N., Folmer, B., 2012. Milk modifies sensory properties of milk and reveals unexpected notes. Proceedings of the 24th International Conference on Coffee Science ASIC. Costa Rica.

Risvik, E., McEwan, J.A., Colwill, J.S., Rogers, R., Lyon, D.H., 1994. Projective mapping: a tool for sensory analysis and consumer research. Food Quality and Preference 5 (4), 263–269.

Rodrigues, H., Ballester, J., Saenz-Navajas, M.P., Valentin, D., 2015. Structural approach of social representation: application to the concept of wine minerality in experts and consumers. Food Quality and Preference 46 (2015), 166–172.

Roper, S.D., 2007. The role of pannexin 1 hemichannels in ATP release and cell-cell communication in mouse taste buds. Proceedings of the National Academy of Sciences 104, 6436–6441.

Small, D.M., Prescott, J., 2005. Odor/taste integration and the perception of flavor. Experimental Brain Research 166, 345–357.

Steinhart, H., Parat-Wilhelms, M., Hoffmann, W., Borcherding, K., Gniechwitz, D., Denker, M., 2006. Flavour Perception of White Coffee Beverages – Influence of Milk Processing. Asic 2006.

Solomon, G.E.A., 1990. Psychology of novice and expert wine talk. The American Journal of Psychology 495–517.

Solomon, G.E., 1997. Conceptual change and wine expertise. The Journal of the Learning Sciences 6 (1), 41–60.

Songer, P., 2008. Development of Appellation Designations Based upon Flavor Profiles for Rwanda Regional Coffees. Songer and Associates, Inc, p. 25. April 27, 2008.

Stevenson, R.J., Boakes, R.A., Prescott, J., 1998. Changes in odor sweetness resulting from implicit learning of a simultaneous odor-sweetness association: an example of learned synesthesia. Learning and Motivation 29, 113–132.

Stevenson, R.J., Prescott, J., Boakes, R.A., 1995. The acquisition of taste properties by odors. Learning and Motivation 26, 433–455.

Stevenson, R.J., Prescott, J., Boakes, R.A., 1999. Confusing tastes and smell: how odours can influence the perception of sweet and sour tastes. Chemical Senses 24, 627–635.

Stone, H., Sidel, J.L., 2004. Sensory Evaluation Practices, third ed. Academic Press, Redwood City.

Stone, H., Sidel, J.L., Bloomquist, J., 1992. Quantitative descriptive analysis. Descriptive Sensory Analysis in Practice 53–69.

Sunarharum, W.B., Williams, D.J., Smyth, H.E., 2014. Complexity of coffee flavor: a compositional and sensory perspective. Food Research International 62 (2014), 315–325.

Sweeney, J.C., soutar, G.N., 1995. Quality and value: an exploratory study. International Journal of Business Studies 3 (2), 51–66.

Thomas-Danguin, T., 2009. Flavor, Encyclopedia of Neuroscience, pp. 1580–1582.

Thuillier, B., Valentin, D., Marchal, R., Dacremont, C., 2015. Pivot_ profile: a new descriptive method based on free description. Food Quality and Preference 42 (2015), 66–77.

Tolessa, K., Rademaker, M., De Baets, B., Boeckx, P., 2015. Prediction of specialty coffee cup quality based on near infrared spectra of green coffee beans. Talanta 150 (2016), 367–374.

Vaccarini, G., Moriondo, C., Thomas, E., Rodriguez, A., 2015. Nespresso Coffee Codex, the Art of Tasting and its Harmonization, third ed. Besanopoli. November 2015.

Valentin, D., Chollet, S., Abdi, H., 2003. Les mots du vin: experts et novices diffèrent-ils quand ils décrivent des vins? Corpus 2.

Vandenbeuch, A., Kinnamon, S., 2009. Why do taste cells generate action potentials? Journal of Biology 8, 42.

Varela, P., Ares, G., 2012. Sensory profiling, the blurred line between sensory and consumer science. A review of novel methods for product characterization. Food Research International 48 (2012), 893–908.

Varela, P., Beltrán, J., Fiszman, S., , 2014. An alternative way to uncover drivers of coffee liking: Preference mapping based on consumers' preference ranking and open comments. Food Quality and Preference 32, 152–159.

Verhagen, J.V., Engelen, L., 2006. The neurocognitive bases of human multimodal food perception: sensory integration. Neuroscience and Biobehavioral Reviews 30, 613–650.

Vignes, A., Gergaud, O., 2007. Twilight of the idols in the market for champagne: dissonance or consonance in consumer preferences? Journal of Wine Research 18 (3), 147–162.
World Coffee Research, 2016. Sensory Lexicon, Unabridged Sensory Definition and References.
Yadessa, A., Burkhardt, J., Denich, M., Bekele, E., Woldemariam, T., Goldbach, H., 2008. Sensory and Bean Characteristics of Wild Arabica Coffee from Southeast Afromontane Rainforests in Ethiopia. ASIC 2008.

Chapter 19

We Consumers—Tastes, Rituals, and Waves

Jonathan Morris
University of Hertfordshire, Hatfield, United Kingdom

The transformation in coffee drinking tastes and rituals among consumers since the start of the speciality revolution has been profound. Although the industry has passed through a successive set of waves, each broadening and deepening the quality of coffee available, this has only been made possible by the preparedness of consumers to engage with these by adjusting their customs and expectations to embrace new propositions that have redefined the ways they drink, and think about, coffee.

These developments parallel the much broader "quality turn" that has taken place within the food sector in the developed (and developing) world over the last 30 years. This has been identified as originating within an "alternative" food sector that distinguished itself from the dominant, industrial, globalized, food networks by emphasizing the environmental, nutritional and health qualities of its products, highlighting the naturalness, tradition, heritage, and craft elements in their production, their connection to particular geographical origins, and the welfare and sustainability schemes under which they operated (Milne, 2013a).

Notable exemplars of this turn would be the rise of the Slow Food movement, with its emphasis on heritage and traditional practices; the Fair Trade movement seeking to redistribute value and risk along the chain linking producers to consumers, and the Organic movement with its rejection of "agri-tech" forms of food production requiring intensive inputs of chemical fertilizers. All of these movements employ a narrative of "qualities" that, in themselves, do not directly translate into a guarantee or descriptor of flavor, even if it is often implicitly assumed that this is the case. Organic coffee, after all, is indistinguishable from nonorganic in the cup—its "quality" derives from the environmental practices followed during its cultivation and their resonances with the modern consumer.

The Craft and Science of Coffee. http://dx.doi.org/10.1016/B978-0-12-803520-7.00019-0

What the quality turn offers to the consumer therefore is not so much the ability to "taste" the difference, as the opportunity to express their "good taste"—using food products to embody those values held by relevant groups of consumers to be important, and creating a set of signifiers such as labels and certifications, that can assure them that this is, indeed, the case (Milne, 2013b).

The speciality coffee movement embraces many of the alternative values articulated within the wider quality turn. However, it explicitly promotes the notion that "specialty" coffee expresses its quality through flavor, experienced as "a distinctive taste" in cup. This definition was employed to differentiate "specialty" from "commodity" coffee—that is the low cost, mass produced, coffee products that were commonly available in the supermarket grocery chains and popular catering outlets (Rhinehart, 2009).

Subsequently the speciality movement has developed its own internal narrative that depicts a journey from so-called "first wave" commodity coffee, via the gourmet espresso beverages of the second wave coffee shop chains, through to single origin coffees offered in the so-called third wave artisan outlets, sympathetically roasted and "scientifically" prepared in a variety of different ways in order to best express their complex flavors.

This narrative is somewhat reductive and can be challenged on both a geographical and historical basis. Although it may capture the essence of what has happened in what one might call the Anglo-American markets (though perhaps not the Australasian ones), it would be difficult to apply in mature markets with their own established coffee cultures such as those of the Mediterranean—most notably Italy—where espresso (as opposed to espresso beverages)—continues to reign supreme; in Japan, whose original coffee culture established prior to the spread of mass consumption has informed many of the practices now associated with the third wave (White, 2012); and Scandinavia, where the third wave was first identified when Trish Rothgeb Skeie, an American roaster and commentator, coined the phrase to describe the dedication to improving coffee quality that she encountered in Norway in 2003, long before any coffee shop chains (local or foreign) established themselves in the Nordic states (Skeie, 2003).

A more nuanced five wave model has recently been proposed by Maurizio Giuli, who combines the role of director of marketing for the leading Italian coffee machine manufacturer Nuova Simonelli, with a part-time academic career (Giuli and Pascucci, 2014, pp. 99–183). He separates the "first wave" into two phases—with a "pioneering" period preceding an "industrial" period whose origins lie at the end of the 19th century, going some way to placate those historians who find it grating to see the six centuries worth of the recorded history of the beverage crammed into a single developmental phase. Significantly Giuli also argues that the "fourth wave" is actually already upon us, due to the explosion in single portion coffee (that is capsule brewing) during the early 21st century. This serves as a valuable reminder of the need to consider both the at home and away from home sectors when considering the evolution of the coffee industry.

Whatever the limitations of the wave models, they do highlight a fundamental question—namely how does one explain changes in consumer practices and coffee preferences? Food history suggests that such shifts can only be understood by placing them in a broader context that would include reference to macroeconomic and regulatory structures, technological advances, and alterations to consumer lifestyles. This needs to be combined with an anthropological analysis of the circumstances in which coffee is drunk, the functions it is intend to fulfill, and the cultural meanings created around the beverage.

Coffee preparation and consumption, within and outside the home, often takes the form of a routine or ritual. A routine is principally a private practice adopted for convenience so that choices made in the past become fixed: for instance, one always makes coffee the same way in the home, rather than contemplate a new set of choices on each occasion. A ritual is a formalized mode of behavior adopted by the members of a particular group that proceeds in accordance with a set of social rules governing the situation—such as hosting visitors for coffee (Watson, 2013).

Routines and rituals often manufacture meanings around coffee—routines, for example, are often learnt from parents so that their repetition is a form of assertion of identity. At the same time, they generate rigidity—rituals restrict the potential for choice within a practice, relocating it from the individual to the community. The character and taste of the coffee is therefore both historically and socially determined.

The leading anthropologist of globalization, Arjun Appadurai, argues that it is the "social history of things, over large periods of time and at large social levels, that constrains the form, meaning, and structure" by which they are understood in the consumer marketplace (Appadurai, 1986, p. 36). The contemporary routines and rituals that structure how a commodity is conceived and consumed constitute a representation of a collective memory of that commodity and its history.

This chapter therefore proceeds through an analysis of coffee drinking practices over the course of the centuries. It examines the "pioneering" and "industrial" waves and the way that these helped to construct understandings of coffee in different locations and cultures, before turning its attention to the "quality turn," and the changing nature of the relationship between coffee and consumers during the second, third, and fourth waves. By charting the development of consumer practices within early modern, modern, and postmodern societies, the chapter seeks to show how tastes for coffee have evolved among new communities of producers and consumers, yet can still be shaped by meanings created in the past.

1. THE PIONEERING PHASE: USES, MEANINGS, AND STRUCTURES

Appadurai argued that the key to understanding a commodity was to "follow the thing." In the case of coffee, that means returning to the first sites of coffee consumption in Arabia, and understanding the ways in which historical and geographical contexts have structured consumption rituals during the pioneering phase. The first records of coffee consumption date to the mid-15th century when Suri mystics began using a beverage to help keep them awake during their all-night religious vigils (Hattox, 1985). This was an infusion produced by boiling water with the desiccated husks of coffee cherries that a Sufi mufti had observed being consumed by locals when traveling in Ethiopia, and had brought back with him across the Red Sea to the southern regions of the Arabian peninsula which today would be part of Yemen. The Ethiopians called this concoction *buno*, but the Sufis named it *qishr*.

Although the original purpose of *qishr* consumption was strictly functional within the context of the Sufi rituals known as *dhikers*, it soon assumed a social purpose after mainstream religious authorities ruled that the beverage was not intoxicating, and therefore permissible for Muslims to consume. Coffee houses spread throughout Arabia, creating a secular public space in which men could meet and socialize as equals. Coffee cultivation started in the mountainous regions of the Yemeni interior, with the beans subsequently being shipped through Mocha, or transported overland by camel caravan (Tuchsherer, 2003).

By the time the drink reached Constantinople, the Turkish capital of the Ottoman Empire in the 1550s, it had evolved significantly in both form and function. It was brewed solely using beans, rather than husks, presumably as these proved easier to transport. Whereas the Arabian version of the drink was still essentially an infusion, using lightly roasted beans and flavored with cardamom, Turkish coffee preparation involved blackened beans, sweetened with sugar. Coffee houses were favored for their egalitarian customs of seating customers in order of arrival (rather than by rank), and enabling even their relatively poor patrons to demonstrate their hospitality by buying coffee for their companions, creating a set of social conventions and consumption rituals. They were, however, also repudiated to promote sedition and transgression—notably the smoking of tobacco—leading to several Sultans attempting to close them down; in one instance ordering that "anyone who opens a coffee shop should be strung up over its front door" (Saraçgil, 1996, p. 187).

It was these two functions of psychoactive stimulation and social lubricant that accompanied the beverage into Europe. Coffee was brought to Venice by Turkish merchants in the 1570s, and in the first half of the 17th century, it appears to have been made available by apothecaries who supplied it as a medicinal product. Exploiting their guild status, the apothecaries retained their control over the trade (such as it was) to the point that it was difficult to sell coffee as an item for household consumption, let alone to serve it as a beverage to customers.

It was not until the 1650s that the first European coffee houses appeared, and even then not on the mainland, but in London. This distinction can be best explained through historical context. The English Civil War had resulted in the monarchy being overthrown, the King executed, and much of traditional law and custom eroded. London was at the very center of the Parliamentary movement. Its guild structures were much weaker, enabling Pasqua Rosee, an Armenian who had grown up in Smyrna in the Ottoman Empire, to set himself up in business with a native of the city, Christopher Bowman, selling coffee initially from a stall, and subsequently within a coffee house. The trade quickly took off, surviving the restoration of the monarchy in the 1660s and several attempts to close it down, reaching its peak in the 1740s when over 550 coffee houses were to be found within the city (Ellis, 2004).

What accounted for the beverage's success? Certainly the supposed medical benefits that drinking coffee conferred were highlighted in Rosee's marketing pitch, with handbills explaining inter alia that

> *... it's very good to help digestion ... It much quickens the Spirits and makes the Heart Lightsome. It is good against sore Eys ... It suppresseth Fumes exceedingly and therefore good against the Head-ach, and will ... prevent and help Consumptions; and the Cough of the Lungs. It is excellent to prevent and cure the Dropsy, Gout and Scurvy ...*

The key quality claim, however, was the linkage made between the psychoactive qualities of coffee and the economic benefits it could confer:

> *It will prevent Drowsiness and make one fit for business, if one have occasion to Watch.*
>
> The Vertue of the Coffee Drink (c.1656?)

Europe had until this point relied principally upon weak alcoholic drinks for refreshment, due to the dangers in drinking untreated water; this was the revelation—coffee was a beverage that not only maintained sobriety, but actually invigorated the functioning of the brain. Coffee houses soon became meeting places for traders who needed to keep their wits about them while they worked—Lloyds famously began in a coffee house, as to all intents and purposes did the London Stock Exchange (Fig. 19.1).

The democratic aspect of the coffee house was also a key part of its appeal—echoing its Middle Eastern forebears. As well as venues for the exchange of news and information—patrons were traditionally greeted with the cry of "what news?" upon entry—coffee house served as meeting places for the *virtuosi*: a term used principally to describe gentlemen with intellectually curious dispositions who wished to learn about the latest advances and

FIGURE 19.1 A London coffee house of the late 17th or early 18th century. *Copyright: Getty Images.*

discoveries in fledgling fields of study such as biology, geography, and science. Isaac Newton, for example, founded the Royal Society, Britain's premier scientific institution in a coffee house (Cowan, 2005).

Coffee houses soon began to appear elsewhere in Europe, notably in the North Sea ports of the Low Countries and Hanseatic League such as Amsterdam, Bremen, and Hamburg. In Paris, too, the institution of the café began to emerge in the 1690s, following the success of the so-called *limonadiers* guild in extending their remit to serve decoctions such as spirits and sherbets to seated customers to include coffee in the face of fierce opposition from grocers and apothecaries (Spary, 2012, pp. 103–106). In contrast to the English coffee house therefore, the French café was from the beginning a site in which both alcohol and coffee were served—while the smoking of tobacco, the other new psychoactive product par excellence—was also catered for (Perluss, 2015).

Coffee's ability to provide a stimulating yet sober start to the day saw it replace weak beer as the breakfast drink of choice in many bourgeois European households during the 18th century—a breakthrough made possible by the extension of coffee growing into the Dutch territories in East Asia, notably Java and the French colonies in the Caribbean—principally Martinique, and, above all, Saint Domingue (modern-day Haiti) that rapidly became the leading source of coffee in the world.

A quality hierarchy into which these new origins were situated developed, anchored in the expert opinions produced by botanists, most of whom were connected to the companies controlling supply. According to these writers, the coffee from St. Domingue was as "common" as its dominance in the market suggested, that from Martinique was perceived as the best coffee of the islands, but Mocha still reigned supreme described in 1780 as the "richest in volatile particles, the most agreeable to the senses of taste and smell, and the one which, generally speaking, as always been preferred up to now and probably always will be" (Spary, 2012, p. 88).

Unsurprisingly these became linked into a social perspective of how the coffee one drank exhibited "good taste" that corresponded closely to class distinctions. Jacques-Francois Demanchy, the French poet, wrote in 1775

> *It will always be a singular spectacle to see, on the one hand, a woman of high society, comfortably settled in her armchair, who consumes a succulent breakfast to which mocca has added its perfume for a well-varnished tea table, in a ... gilded porcelain cup, with well-refined sugar and good cream; and on the other a vegetable seller soaking a bad penny loaf in a detestable Liquour, which she has been told is Café au lait, in a ghastly earthenware pot, far from being new, on a willow basket.*
>
> Spary (2012, p. 91)

Note that both the high and low class consumers added milk and sugar to their coffee, turning it from a drink into a calorific foodstuff. This encouraged its use by piece workers such as weavers in the Low Countries as not only could they consume coffee without stopping their work, it might actually increase their rate of production due to its stimulant properties.

The international coffee trade flourished during the 19th century with many more origins entering the market, yet it was still left to the final consumer to do most of the work in transforming green beans into the final beverage. William Ukers, the legendary founding Editor of the *Tea and Coffee Trade Journal*, reported that even in the early 20th century it was not unusual on a Sunday to find Europeans roasting their coffee for the following week on the curb outside their homes. This led to some distinctive customer preferences—in Brittany, for example, the preference was above all for peaberry coffee, because the round beans rolled around better in their pans, and so acquired a more uniform roast (Ukers, 1935, p. 562).

Ukers similarly reported that Old Brown Java—coffee that was shipped in the sealed holds of Dutch sailing vessels, and turned brown as a result of sweating during the voyage—enjoyed a premium in New York because

> *Every good housewife in those days knew that green coffee changed its colour in aging and that, of course, aged coffee was best. ... The sweating frequently produced a musty flavour which, if not too pronounced, was highly prized by experts.*
>
> Ukers (1935, p. 344)

Malabar coffee enjoyed the same reputation on the London market, to the point that after shipping times and storage practices improved, Indian producers developed the process of "monsooning"—weathering the beans in open-sided buildings during the high humidity of the monsoon season—to retain customer favor (Thorn and Segal, 2007, pp. 145–146). Further information on the monsoon process can be found in Chapter 3. That these styles of coffee maintained their traction is testimony to the enduring power of consumer practices in shaping the market through shared social understandings of taste.

During the pioneering phase of coffee consumption, therefore, we can already discern the development of uses and structures surrounding coffee, that remain fundamental in influencing the way coffee is perceived and consumed today.

2. THE INDUSTRIAL ERA: THE CREATION OF TASTE COMMUNITIES

The consumer's involvement with coffee was fundamentally changed by the industrialization of the coffee business and the rise of the mass market. The work involved in preparing the coffee prior to the final act of brewing was transferred back to the roasters, along with the knowledge that went with it. Soluble coffee products and ready to drink beverages that became popular in the post Second World War era were the ultimate exemplars of a trajectory that began with the emergence of coffee roasting as a stand-alone business in the late 19th century.

Three qualities were critical for making coffee accessible to the emerging mass market in this period—cheapness, consistency, and convenience. The ways that products fulfilled these expectations varied according to context. The functional and social uses to which coffee was put, and the routines and rituals that surrounded it, all shaped a set of taste preferences among consumer communities and created a set of expectations about the flavor of coffee and how it should be served. The result was the creation of taste communities that were characterized by a significant degree of internal conformity—often resulting in a quasinationalistic meaning becoming attached to coffee—yet displayed a significant diversity between them.

2.1 The Making of the American "Cup of Joe"

The American Civil War proved the critical stimulus for the industrialization of the coffee trade within the United States. Union generals were quick to see the advantage that coffee conferred upon a fighting force, providing the men with both stimulation before battle and solace thereafter: furthermore coffee-making relieved boredom between conflict. Their troops were entitled to a remarkable ration of roughly 36 pounds of coffee each year (current per capita consumption in the United States is around 7 pounds). The word "coffee" appeared more frequently than "rifle," "cannon," or "bullet" in Union soldiers' diaries, whereas Confederate forces frequently lamented their lack of access to the bean because of the blockade of southern ports (Grinspan, 2014).

Even before the conflict was over, entrepreneurs began capitalizing upon the mass demand for coffee it had created. The first ground-coffee package, "Osborn's Celebrated Prepared Java Coffee," appeared in New York, and the Jabez Burns self-emptying roaster was patented in 1864, revolutionizing the potential for industrial coffee roasting. One beneficiary was John Arbuckle of Pittsburgh, who, in 1873, launched Ariosa, the first branded coffee to achieve widespread acceptance across the United States, benefitting from a roasting process that glazed the coffee in a mixture of sugar and egg, before grinding it into powder to be sold in small sealed packages. Rising real wages and falling real coffee prices—the latter being the consequence of the massive expansion of supply from Brazil—gave consumers much greater access to coffee. Between the 1870s and the 1930s, consumption of coffee in the United States rose from around 6 pounds to just under 14 pounds per capita (Ukers, 1935, pp. 403, 451, 521; Biderman, 2013, pp. 140−154).

In the 1870s, the majority of coffee was still sold loose in the green for home roasting and grinding; by the 1930s more than 90% of the coffee sold by retailers was preroasted and supplied in ready-weighed, trade-marked packages. Usually it was also ready-ground, although some grocery chain stores in which the majority of coffee was now purchased made a show of grinding it at that point. The development of mass distribution systems—chain stores, mail order, print, and radio advertising—enabled large scale roasters to establish themselves by developing branded products that took advantage of the latest breakthroughs in packaging technology. Hills Brothers, for example, introduced canned coffee—roast and ground beans sealed in vacuum packed tins—in 1900. Consumers were given little clue as to the contents of these blends—none of the 34 popular branded products illustrated in the 1935 edition of Ukers' classic text *All About Coffee* indicated a geographical origin, even the supposedly premium Yuban (Ukers, 1935, pp. 383, 388, 403, 441, 451, 408).

Consequently, as a J. Walter Thompson survey found in the 1920s, although 87% of housewives cited flavor as the most important factor in their brand choice, "it was extremely difficult for the average person to make clear distinctions where flavor is concerned." One copywriter highlighted how "the housewife experiments with percolators, with drip coffee, with Silex machines, and still most of the time the coffee isn't right. She is battered and bewildered by new packages and new brands, by advertising" (Prendergast, 2010, pp. 157, 189).

Advertisers played upon consumer fears about their lack of coffee knowledge. The major roasters ran campaigns suggesting housewives ran the risk of being hit, having coffee thrown in their face, or even being spanked by their husbands for serving them poorly prepared coffee on their return from work. The solution, of course, was simple—stick faithfully to using their branded products (Morris and Thurston, 2013, pp. 220–222) (Fig. 19.2).

FIGURE 19.2 As simple as child's play. Nescafe reassures US housewives that they can rely on its soluble coffee. *By kind permission of Nestlé Historical Archives.*

Women were targeted in other ways too. Eleanor Roosevelt broadcast to the women of the nation in her radio talks entitled "Over our Coffee Cups," a show sponsored by the Pan-American Coffee Bureau set up to promote coffee consumption. Slots around the talks sought to persuade listeners of additional reasons and occasions to drink coffee. For example, the cover girl and actress Jinx Falkenburg told radio listeners in 1941

> *The best way I know for a girl to look her best at all times is to eat the right things, get plenty of sleep, yes, and drink plenty of coffee. Why do I mention coffee? Because I've found that when I want to look fresh and, well, you might say, "blooming in the evening," coffee is really a wonderful help. After all, you look as well as you feel, and coffee makes me feel cheerful and peppy." ANNOUNCER: Why not try a delicious, flavorful cup of coffee with your evening meal tonight? And see how much more you get out of life with coffee.*
>
> Roosevelt (1941)

Coffee consumption was also rising outside the home as a result of new dietary habits, notably a preference for light midday meals during the working day. These were being taken in places such as lunch counters, soda fountains, and diners, with coffee being served as a beverage to accompany, rather than follow, a snack such as a sandwich. Some factories began offering free coffee as a work incentive, a practice that became widespread after it was demonstrated to improve productivity among defense workers during World War II. By the end of 1952, 80% of American companies had instituted a coffee break, encouraged by a campaign by the Pan-American Coffee Bureau (Prendergast, 2010, p. 221).

These new consumption practices shaped the creation of a distinctive American coffee style. The volume in the cup (or cups) was comparatively large, to last sufficiently long to accompany eating. The body was thin and flavor was weak, reflecting the dominance of bland Brazilian beans in the blends, as well as the parsimonious nature of housewives who used very low brew ratios when preparing coffee, principally by using percolators that overextracted the beans by repeatedly passing boiling water through them.

If anything those characteristics were reinforced in the post-Second World War era. The major roasters responded to a decline in consumption per capita by cutting costs and attempting to steal market share from each other by suggesting their brands could be used even more economically—with Folgers claiming consumers could use "one quarter less" of its "rich" blend. Some regional roasters who supplied the restaurant business switched to selling coffee in 14 oz. packs claiming that these could now be used to brew the same amount of coffee as the old 1 pound bags (Prendergast, 2010, p. 223). This was what was used in the "unlimited refills" offered by diners and fast food outlets, served out of the pour over jugs bulk brewed and left on the hotplate, that came to define the taste of the American "cup of Joe"—a term first coined in the 1930s (Mikkelson, 2009).

By continuing to focus on price, while obscuring the nature of their blends (thereby enabling them to further cheapen them by introducing Robusta), the corporate giants missed the chance to educate the public into paying a premium for superior coffee products. This despite the fact that one campaign that did set out to do this—the Colombian coffee grower's federation's creation of the character Juan Valdez, a peasant farmer growing coffee in the shade at 5000 ft up in the Andes, led to a 300% increase in Americans identifying Colombian coffee as the world's finest within 5 months of its introduction in 1960, pressurizing many companies into introducing "all Colombian" brands (Prendergast, 2010, pp. 259–260).

Similarly the continuing preoccupation of the corporate giants on focusing on housewives and their fears around preparing coffee for their husbands, not only succeeded in patronizing a new generation of women whose ambitions extended much further than becoming "a good little Maxwell Housewife" as a 1965 ad put it, but ignored the fact that a younger generation had become more susceptible to the lure of soft drinks, notably cola, sold to the soundtrack of popular music. "Food goes better with, Fun goes better with, you go better with Coke" the lyrics to one spot of the time put it presciently, foreseeing the way that soft beverages would take over as the meal accompaniment of choice (Prendergast, 2010, pp. 255–258).

2.2 The Construction of Taste Communities in Europe

In Europe these developments started somewhat later. Shop roasting started gaining traction among grocers in the 1890s, but stand-alone businesses largely operated at a local or regional level. The First World War did much to popularize coffee as rations and grinders were issued to the troops, but much of the demand was still frequently met through the use of substitutes. According to Ukers:

> *In Europe chicory is not regarded as an adulterant—it is an addition, or modifier, if you please. And so many people have acquired a coffee-and-chicory taste, that it is doubtful if they would appreciate a real cup of coffee should they ever meet it.*
>
> Ukers (1935, p. 554)

In Germany, which witnessed one of the largest growth in consumption in the early 20th century, two consumption rituals percolated through the classes. The first was the tradition of a family get-together over coffee and cake on a Sunday afternoon. This would often be when the "real" as opposed to surrogate coffee was produced. During the summer this took on the form of a stroll to coffee gardens sited in public parks. In some such gardens, hot water was for sale, enabling patrons to save money by brewing up the coffee they bought with them (Ukers, 1935, p. 556).

The second was what has become referred to as the *kaffeeklatsch*: essentially a gathering of women, either in or out of the home, for a chat over afternoon coffee and cake. Men, of course, were more likely to socialize over beer consumed in the *Kneipe* or pub. When hosted outside the home, the *kaffeeklatsch* would often be held at the *konditorei*, i.e., the pastry shop—a practice whose legacy can today be found in the continuing dominance of the bakery sector as the largest component of the German out-of-home coffee market (Zeitemann, 2013).

The Nordic countries too developed a strong identification of coffee with the female sphere. Terms such as *jordemoderkaffe* (midwives coffee, i.e., very strong coffee), *barselskaffe* (coffee served to visitors to a women who has recently given birth), and *kaffesøster* (a coffee sister—a woman who drinks large quantities of coffee), all entered the Danish language. Similarly *kaffemoster* (coffee aunt) was a Swedish term for a woman with a fondness for coffee, whereas the concept of the *fika* developed during the late 19th century, taking a comforting break from the day's activities to share a coffee and cake with family, friends, or work colleagues (Kjeldgaard et al., 2011).

Already by the 1930s, coffee consumption per capita was greater in Denmark and Sweden than in the United States, whereas that in the former Swedish territories of Norway and Finland was not that far behind. Arguably the main reason was simply functional—coffee was a warming beverage that could be used to combat the cold while raising energy levels—both necessary for the outdoor work in the low temperatures performed by many of the population. Literary studies confirm the spread of coffee throughout the classes, aided by the strong temperance movement, closely linked to the church, which promoted coffee as an alternative to alcohol. In contrast with Britain, it was tea that became a refined beverage in the Nordic countries, with consumption largely restricted to the bourgeoisie (Lundqvist, 2016).

Anthropologists' observations of Sami reindeer herders in Finland during the 1950s are suggestive of the reasons why Scandinavians came to drink so much coffee (Whitaker, 1970). The men would have two cups of coffee in the morning after getting up and before going out to tend to the herd. When the women who remained in the tents heard the barking of the dogs announcing the men's return, they would immediately begin to prepare coffee to be ready for their arrival—another two cups. A meal followed with two cups of coffee to wash it down. A second meal taken later would be accompanied by another four cups—two before, two after—and at some point a snack, alongside another two cups, bringing the total up to 12. In the summer, when conditions allowed, and the men could remain outside longer, they also brewed coffee while at work.

These routine forms of consumption were augmented by the ritual practices of hospitality that were also constructed around coffee drinking. Coffee drinking was used as a way of constituting social ties within herding units that relied on cooperative working and consensus. Visitors to a tent would automatically be offered two cups of coffee, and a meal or snack to follow. Although they might refuse the food, they could not refuse the coffee, and only once it had been prepared, would they be invited to move from the edge to the center of the tent. Guests would always be made fresh coffee, even if a previously brewed pot was still by the fireside. Hospitality rituals could easily push consumption to 20 cups a day. Although coffee was still rationed during the 1950s, and the herding groups might be traveling for 7 months at a time, observers never saw them run short of coffee, such was its importance to the cohesion of the community.

Even in urban communities the working day came to be organized and measured around coffee routines. The Danish incorporated terms for "morning coffee," "prenoon coffee," "afternoon coffee," and "evening coffee"—each taken at a set time (6:00 a.m., 9:00 a.m., 3:00 p.m., 9:00 p.m.). In Finland, the legislation that established two work breaks a day, simply referred to these as "coffee breaks" (Kjeldgaard and Ostberg, 2007).

The style of coffee was shaped by the contexts in which it was drunk. Light roasts predominate in Scandinavia, particularly in Finland, which now has the highest per capita consumption of coffee in the world (Ojaniemi, 2010). One reason for this may be that up until the 1920s most coffee was still bought green and roasted at home, and it may have been believed that light roasting resulted in less wastage of both the bean and energy. Brewed directly in a coffee kettle, it was largely served black, unsurprising given the absence of a dairy industry in the most of the northern regions—indeed among the Sami, if milk was ever added, it took the form of dried reindeer milk, although a more common Swedish addition was cheese. Coffee's more functional properties were hinted at by the ways it was consumed—men often drinking quickly from the saucer, whereas women in private might suck it through a sugar lump held in the lips to add sweetness and succor.

Once industrial roasters emerged they sought to associate themselves with this light coffee style as one that reflected a national distinctiveness. Paulig, the major Finnish roaster, placed an image of the Paula girl—a young woman dressed in national costume pouring coffee from a kettle—on its packaging as early as 1904, and in the 1950s began selecting young women to become the incarnation of the Paula girl promoting the brand through public appearances (Paulig Company History, 2016). This symbolic shift in the identity of coffee from exotic to domestic, even to the point of becoming an icon of national identity, was repeated across Europe as coffee became part of everyday life—even though practices and tastes were significantly different from one region to another.

A significant contributory factor to the development of such distinctions was the replanting of colonial coffee estates in Asia with Robusta coffee (native to Central Africa) following the devastating coffee rust epidemic of the late 19th century. The Dutch authorities began replanting in Java and Sumatra in the 1900s, and already during the First World War the consumption of

Robusta had outstripped that of Arabica in the Netherlands, aided no doubt by the fact that darker roasts were already used for Java and Sumatran Arabicas (Prendergast, 2010, p. 141). Meanwhile French and Belgian administrations were likewise encouraging Robusta cultivation in their African colonies, with most of the output making its way into the blends drunk at the metropole. It was estimated that at least a third of the coffee imported into France in 1938 was Robusta, but it was really in the postwar era that these portions rose dramatically: in 1960, 75% of all the coffee consumed in France was Robusta, requiring a "French roast," i.e., a high roast, to counteract the bitterness through caramelization. Yet the point was that Robusta made the mass consumption of coffee possible—Portugal, for example, notably increased its own per capita consumption on the basis of its imports from Angola, when it was still a Portuguese possession.

The most distinctive southern European coffee culture to emerge, however, was that of Italy, where the evolution of espresso led to the emergence of different cultures of consumption in and out of the home (Morris, 2010). The La Pavoni Ideale of 1905 was the first so-called espresso machine to enter into production, using pressure brewing to speed up coffee delivery, enabling a fresh cup of coffee to be brewed "expressly" for each customer; in contrast to the bulk-brew urns commonly used in catering establishments. Delivery time was still about 50 s and the pressures used were in the region of 1.5−3 bars, so that the resultant beverage tasted like a concentrated version of a drip-brewed coffee (Morris, 2013a).

Coffee in Italy remained very much an elite beverage, and the machines were installed in upmarket establishments, notably the new "American bars" in which cocktails were consumed standing at the counter (Fig. 19.3). The steam

FIGURE 19.3 An "American Bar" in early 20th century Paris. Victoria Arduino espresso machines in the Maison du Café. *By kind permission of the Simonelli group.*

wands attached to the side of the machines were used to warm up alcoholic punches, as well as milk for the coffee if required. As techniques for this developed, so we can see that the cappuccino—initially a term for a small white coffee—was transformed into a beverage assembled at the bar with an espresso base topped with steamed milk. Conversely caffelatte was a popular breakfast drink in the homes of the working classes, but this mixture of milk and bread, warmed in a pan, frequently contained little or no coffee, with branded surrogates made from grains such as orzo used to add flavor (Morris, 2013b).

In 1948 Achille Gaggia introduced the lever machine that utilized a manually operated spring-loaded piston to blast the water through the coffee under around 9 bars of pressure. The resultant extract had a head of essential oils from the coffee on top of it, what we today refer to as crema. Indeed, Gaggia promoted this as entirely new beverage, giving it the name of Crema Caffè—cream coffee (Fig. 19.4). Even more than before, therefore, there was a significant difference between coffee available at the bar and coffee brewed at home.

Modern Italian coffee culture developed during the so-called "economic miracle" experienced in the country during the 1950s and 1960s, when the industrial, manufacturing, economy overtook the agrarian one, leading to massive internal migrations from the countryside to the town. New neighborhoods arose, while high housing densities meant that many migrants had little living space, certainly not enough to socialize in. All of this contributed to a rapid expansion in the number of neighborhood coffee bars, serving a mixture of coffee, snacks, and alcohol, to provide such a facility.

FIGURE 19.4 Crema Caffè. A coffee bar with a Gaggia lever machine in 1950s Rome. *Copyright: Getty Images.*

Bars were a place to grab coffee before work, but equally the site for coffee breaks during the day, as well as a postprandial digestive espresso. That the coffee was both quickly prepared and swiftly drunk facilitated this, as did the service and pricing models. From 1911 onwards, municipal authorities had enjoyed the facility to establish a maximum price for "a cup of coffee without service"—interpreted as meaning drunk while standing at the bar, as in the American bars of the period. The price was deliberately set low to counter inflationary tendencies in the standard of living. The result was the standing culture of consumption still seen today.

Annual consumption per capita doubled between 1955 and 1970 from 1.5 to 3.0 kg as domestic coffee drinking also increased as the aluminum stovetop brewer known as the "moka pot," first introduced by Bialetti, became popular throughout the country. The company's publicity explicitly, if inaccurately, claimed that it produced the same quality of coffee as at the bar, although the pressures generated the steam within the pot were much lower. Again the intention was to capitalize on the distinctiveness of Italian practices, and indeed well over 90% of households in Italy now possess a moka. The emerging large national roasters also sought to turn espresso into an Italian icon—so that from the mid-1970s publicity focused on endorsement by Italian celebrities, and more recently foreign stars paying tribute to the genius of Italian espresso.

One element in that success that is often overlooked is that the espresso process economizes on cost while delivering character. It requires relatively small doses of coffee per cup and its intensification of flavors means that cheaper commodity beans can be used to form significant portions of the blend, assisting in the creation of body and balance. This includes substantial elements of Robusta, which has the added bonus of producing a more developed, visually attractive crema. Robusta heavy blends proved particularly popular in the South, where sugar was often placed in the cup prior to delivery: indeed national roasters were obliged to alter their blends to appeal to southern tastes.

Larger roasters did not seek to enter directly into the out-of-home market, however, due to the low margins forced upon operators by the maximum price regulations (Morris, 2013c, pp. 885–886). Consequently coffee shop chains have not developed within Italy and the sector remains dominated by independent operators who usually rely on low cost, elastic, family labor to make ends meet. They depend heavily on their local small roaster for start-up finance, machines, and so forth, in return to agreeing to an exclusive coffee supply contract. Many argue that this in turn incentivizes those roasters to supply low quality blends in order to recoup their costs. Recent research suggests that despite their very distinctive coffee culture, Italians are largely unaware of the brand of coffee served at their local coffee bar. Most current consumers do not consider themselves capable of making informed value judgments about coffee (Borghi, 2015).

This is perhaps not as surprising as it seems. The common coffee practices that developed among taste communities during the industrial era resulted in coffee becoming a banal everyday good whose form and content were relatively unconsidered provided it conformed to expectation. Coffee acquired meaning as a banal form of national identity reinforced by the performance of the routines and rituals that surrounded it. This spanned the classes rather than dividing them. Consequently although a request for "just a coffee" meant very different things to Americans, Scandinavians, and Italians in the industrial era, it was sufficiently well understood within each community to require no further explanation.

3. POSTMODERN CONSUMERS AND THE QUALITY TURN

Toward the end of the 20th century, academics, marketers, and commentators became increasingly aware that conventional understandings of consumers and consumption no longer applied in what had become a "postmodern era." Industrialization had created a mass consumer society in which access to goods had become open to all. Now with their basic needs easily satisfied as a result of rising affluence, consumers used purchases as props through which to articulate their self-identities to both themselves and others. Branded goods, in particular, developed sets of connotations that extended beyond class distinctions, conveying instead meanings about philosophies and lifestyles, which extend to a sense of membership of consumer communities (Arvidsson, 2006; Bedbury, 2002; Lury, 2004). The advent of the internet has made it ever easier for consumers to seek out niche products and discover groupings of like-minded devotees, transcending national borders, markets, and tastes. The battle for the soul of these new consumers was now all about satisfying their demands for "authenticity" (Lewis and Bridger, 2001).

3.1 The Success of Speciality Coffee in the United States

The 2016 National Coffee Drinking Trends survey carried out by Gallup on behalf of the National Coffee Association reveals just how far the "quality turn" of the last three decades has impacted upon coffee consumption in the United States (National Coffee Association, 2016). For the first time in the 67 years that the poll has been carried out, drip coffee—that is the traditional American "cup of Joe"—constituted only half of all coffee consumed by respondents during the previous day both in and out of home.

The chief cause of this was the rise in the proportion of so-called "gourmet coffee beverages"—a combination of speciality style coffees, espresso-based beverages, and iced or frozen drinks. Although the beverages added to the gourmet category in the past year—flat white, cold brew, nitrogen-infused—are indicative of the importance of the so-called "third wave," the main drivers of this transformation remain the espresso-based beverages most closely associated with the "second wave" coffee shop format: cappuccino, mocha, caffè latte. Daily consumption of such espresso-based beverages has nearly tripled since 2008.

The consumer demographic leading this transformation is the so-called "Older Millennials" aged between 25 and 39 years whose past-day consumption of gourmet beverages has risen from 19% to 41% between 2008 and 2016, whereas 18–24 year olds showed an increase from 13% to 36%. They are, of course, precisely the generations who, it was feared, were being lost to coffee due to the challenge from soft drinks.

So how can we explain this dramatic transformation? There is no space here for a detailed history of the evolution of the speciality movement in the United States, let alone worldwide, but we should remember that initially this began as a movement of small roasters, setting themselves up in opposition to the corporate giants. They were primarily concerned with selling whole beans for domestic consumption, dark-roasting their offerings to highlight the contrast with the insipidness of the mass market blends, and developing a diverse and distinctive set of products to contrast with the narrow standard range offered within the supermarkets. Instead they sold their beans through the fancy goods and gourmet deli stores whose clientele sought alternative and authentic products through which to demonstrate their taste. Not all consumers were affluent: Alfred Peet, seen by many as the godfather of speciality coffee, built his original client base among the students of Berkeley.

One of the first academic analysis of the quality turn in relation to coffee was that of the anthropologist William Roseberry in an essay first published in 1996 (Roseberry, 2002). It began with a sketch of the coffee selection in Zabar's, one of the original speciality roasters and retailers that operated a gourmet food emporium on Manhattan's Upper West Side. Here Roseberry saw beans labeled "Kona style, Colombian Supremo, Gourmet Decaf, Blue Mountain style, Mocha style, French, Italian, Vienna, Decaf Espresso, Water Process Decaf, Kenya AA," whereas in a deli across the street from his New York apartment he encountered 43 coffee products including "Jamaican Blue Mountain, German Chocolate, Swedish Delight, and Swiss Mocha Almond."

He interpreted this selection as proof of a new tactic of market segmentation addressing the two groups for whom mass market coffee products held least appeal: "yuppies" and college students. The products designated with geographical and/or bean descriptors were designed to appeal to those "urban, urbane, professional men and women who ... consumed or hoped to consume variety and quality, as well as quantity" as a way of constituting their individual and peer group identities through practices of taste that distinguished them from the mass of society. Meanwhile the flavored coffees, along with the addition of syrups to the sample beverages served over the counter, were used to recapture the youth market that it was feared had been lost to the lure of sugary, soft, drinks. This combination of "new coffees, more choices, more diversity, less concentration (and) new capitalism" led Roseberry to suggest that we might "now consider coffee to be the beverage of postmodernism" (Roseberry, 2002, p. 162).

Yet, he argued, this postmodern coffee world also relied heavily upon the social meanings constructed around coffee consumption during the early history of the bean. By restoring the notion of variety within the coffee world, highlighting the variety of geographical origins, the grades of beans, and, intriguingly, referencing the varieties of "national" taste styles (French, Italian, German, Swedish, Swiss, and Viennese), the speciality selection at Zabar and elsewhere appeared to reach back to the preindustrial era of coffee as an exotic luxury product. Such an identification was taken further by the displays of old coffee mills, roasters, and brewing apparatus within the stores. Speciality was being sold as by connecting it to "a more genuine past before the concentration and massification of the trades" (Roseberry, 2002, pp. 166–167).

This mobilization of coffee's past as a way of creating symbolic meanings around which new consumption practices could be structured was similarly fundamental to the next phase of the speciality movement—the spread of the so-called "second wave" coffee houses selling espresso beverages. In 1989, the Specialty Coffee Association of America (SCAA), counted 585 speciality coffee outlets operating in the United States; by 1994 that figure had risen to 3600. In 2000 there were 12,600, in 2005, 21,400, and by 2013, the overall number had reached 29,308 outlets (SCAA, 2016).

Gourmet retailers had begun selling beverages in-store in the 1980s as part of a simple marketing strategy of getting consumers to taste the quality in the cup in the hope this would generate sales for home consumption. The SCAA itself, as part of its college coffee house program of the mid-1980s, suggested to operators that "the inclusion of espresso drinks and the attractive appearance of an espresso machine gives the operation an "upscale" quality image. This, in turn, can promote sales of other speciality coffees" (Foodservice Director's College Coffee House Manual, 1986, p. 41). By blending two coffee cultures—that of the 20th century Italian espresso bar, and the 18th century European coffee house—the coffee shop format was able to satisfy the desire for authenticity, by developing an offering that could be located within the "Experience Economy" (Pine and Gilmore, 1999, 2007).

The "authenticity" was supplied through the use of Italian-styled, espresso-based beverages. As in Italy itself, a key element to the coffee shops success was that they were serving beverages that could only be prepared outside the home. In Italy, where most coffee was served black, the distinction was that between the crema-topped espresso served at the bar and the plain black liquor consumed in the home. In America, however, the most striking visual clue to the different nature of the beverages from the traditional cup of Joe, were the steamed and foamed milk toppings, whose preparation provided an additional theatrical experience, that reinforced their value as "hand-crafted, artisan, beverages"—individually made to order by the barista on the machine in front of the customer. This, in turn, appeared to justify the most critical distinction between the cappuccinos served in the coffee shop, and the cups of Joe available in the diners and delis—the premium price.

The range of beverages, and the ability to customize them, further assisted in promoting the spread of the coffee shop. The flexibility of one drink in particular stood out—the caffè latte. In Italy, the term caffè latte was used for a mixture of home-brewed coffee (usually within a stovetop moka pot) and milk warmed in a pan. Occasionally, one might see a latte macchiato—a glass of milk with an espresso shot dropped into it—served in a bar. It seems to have been particularly attractive to German holiday makers, presumably because of the theatrical transformation in color when the coffee is added. In America, however, the caffè latte was effectively developed as an espresso version of café au lait, using textured, steamed milk topped with a flat thin layer of froth, rather than the macrofoamed and domed cappuccino. The caffè latte also proved a great vehicle for the addition of various syrups, further sweetening the beverage, and could be adapted to utilize a wide variety of milks (skim, full, half and half, as well as soy, almond, etc.). Sevety-five percent of the sales from the espresso carts that had sprung up in Seattle during the early 1990s were of caffè latte (Morris, 2013b, p. 272).

By 1994 speciality coffee retailers were reporting that gourmet espresso beverages were outselling those of drip coffee within their stores. A Gallup survey for the SCAA discovered that the typical gourmet coffee drinker was a college educated, 18–34 year olds, with a household income of over $50,000 (Gallup, 1994). As well as the coffee, however, they were consuming the coffee shop itself, its brand image, and the lifestyle it represented.

The college coffeehouses sponsored by the SCAA and the early coffee shop chains such as Starbucks that developed during the 1990s made explicit in their communication materials that they regarded themselves as inheritors of the values promoted by the European coffee houses of previous centuries. This assisted them in their attempts to structure new meanings around coffee consumption, by suggesting that, as in the past, it could be used to promote progressive forms of social integration (Fig. 19.5). This was highly pertinent at a time that the American middle classes were believed to be withdrawing from the outside world and "cocooning" themselves within their homes, going out only to indulge in such solitary behaviors as "bowling alone" (Simon, 2009, pp. 104–106).

The sociologist Ray Oldenburg suggested that a central cause of the problem was the decline of those "great good places" such as "cafés, coffee shops, bookstores, bars, hair salons and other hangouts at the heart of the community" (Oldenburg, 1989). These informal public gathering places—so-called "third places" as distinct from the home and the workplace—allowed, as he put it, unrelated people to relate to each other, nurturing a sense of inclusivity that crossed divisions of class, race, and gender, thereby generating a vitality in the public life of the community.

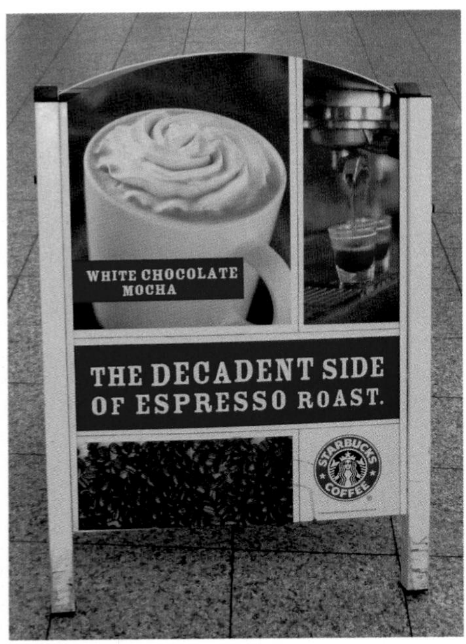

FIGURE 19.5 Starbuck's reinterprets espresso for the Anglo-American market. *Photo: J. Morris.*

Oldenburg identified coffee houses, and, in particular the London coffee houses of the 18th century, as an exemplar of such an institution. He highlighted the democratic features of these institutions, both in terms of their accessibility to all who were prepared to pay the price of a dish of coffee to enter into the "penny university," and the fostering of discussion and interaction through strategies such as seating newcomers at large collective tables. Oldenburg also framed both the French café and the Viennese coffee house as historical models of community institutions in his plea to encourage the development of a new set of "third places" within late 20th century America.

Howard Schultz, the eminence grise behind Starbucks, repeatedly invoked the language of the "third place" in articulating his early vision of the store, and continues to do so today. In his first book, he highlighted how a visit to Starbucks provided an oasis from the stresses of home and work life within which to take a quiet moment to gather one's thoughts, and an opportunity for casual social interaction in the company of fellow human beings. In addition there was a sense of romance connecting customers to products with origins in far-away places (similar to the exoticism surrounding coffee in the 18th century). Above all, he argued, the ambience and the price for admission were democratic—a cup of coffee was an affordable luxury, so that one would see policemen and utility workers standing in line with wealthy surgeons, because although "the blue collar man may not be able to afford the Mercedes the surgeon just drove up in, ... he can order the same $2.00 cappuccino" (Schultz and Yang, 1997, p. 119).

This sense of inclusivity does not just happen, but is inherent in the way that stores operate. The first element of arriving in the coffee shop is the queue to place one's order. This is the modern day version of guests being served in order of arrival, rather than rank, as in the Middle Eastern coffee house. The fact that the surgeon has to stand behind the utility worker is a critical element in the coffee house's success. Arriving at the counter, the smiled greetings to customers, the writing of names onto cups, and the casual chat as the drink is prepared, all serve to reinforce the informal image: staff are trained to deliver this script as part of the service. When customers leave the counter, they enter an ambience that indicates that their transaction is far from over, but extends to the experience of consumption. This is communicated not just through the comfortable seating made available, such as the sofas, but, even more so, the lack of pressure to relinquish it. Add in facilities such as the provision of newspapers (reminiscent of Vienna), the clean and well equipped bathrooms, and the child friendly policies (not least the absence of alcohol), and it is easy to understand how the second wave coffee house worked to position itself as a center for community encounters.

The irony is that relatively few coffee shop customers actually consume the experience in this way. An ad agency that ran a customer focus group for Starbucks in Los Angeles in the early 1990s reported that typical feedback was "Starbucks is so social. We go to Starbucks stores because of a social feeling," yet when it observed the stores themselves, it found that fewer than 10% of customers actually ever talked to anyone (Schultz and Yang, 1997, p. 121). Fieldwork conducted in Starbucks in the mid-2000s found that 65% of all tables were occupied by people sitting on their own. Indeed, the company furnished its stores to accommodate these customers, using round, rather than square, tables, as these feel less formal, have no "empty" seats, and feel less isolated from the rest of the public, even though they discourage people who do not know each other from sitting together and talking (Simon, 2009, pp. 105–108). Instead of escaping the workplace, many customers bring it with them, in the form of the laptops, mobile phones, and other devices whose exploitation of the digital revolution has led to the decentralization of work over the last few decades.

It is easy to decry the coffee shop for simply offering a safe simulacrum of the old-style coffee houses, one in which spectators can observe, rather than participate in, community life. The $2 cappuccino might have appeared to be an affordable luxury, but because of the premium charged over regular diner coffee, this seemed more like the "bourgeois universalism" denounced by Marx—that is a notionally democratic experience whose exclusive cover price restricted it to the middle classes. Similarly the supposed authenticity of the beverages bears little examination once they were translated into more approachable forms for the American consumer—using not just syrups but sizing to sweeten the propositions to the point that a Starbucks "tall" cappuccino, the smallest regularly available, is twice that of the standard Italian version.

Such criticisms, however, miss the fundamental point that by restructuring a consumption experience using a variety of cultural reference points, the second wave chains succeeded in transforming the coffee drinking habits of Americans—not by increasing the overall quantity of consumption per capita, but in persuading them that drinking gourmet coffee justified paying a premium price. And gradually that perception has percolated democratically through society, as the conversion of such nonspecialist operators as Dunkin Donuts and McDonalds to serving cappuccino and latte in the mid-2000s demonstrated (Baker, 2013, p. 317). Italian-style beverages have now become embedded within American mainstream coffee drinking.

3.2 The UK Consumer: From Tea to Speciality

The success of the second wave has been even more acute in those markets where the everyday beverage of choice was not coffee. Significantly, however, even among these, historical meanings have been mobilized to help in structuring contemporary consumption.

The United Kingdom, for example, although the progenitor of the early European coffee house, had long been a tea drinking society. The industrial revolution established cheap tea as the workers' beverage of choice, supplied by plantations from the Empire, including those in Ceylon that were converted from coffee growing after their devastation by coffee rust in the second half of the 19th century. Domestic coffee drinking only really started to spread during the 1950s, with the introduction of "instant," that is soluble, coffee products. Their initial success owed much to the advent of commercial television, as, unlike a pot of tea, they could be prepared during the length of time devoted to an advertising break. This led tea suppliers to retaliate by developing the teabag (Bramah, 1972).

The 1950s also saw a brief flourishing of out of home coffee bars after the first Gaggia machine was installed in a coffee bar in London's Soho entertainment district in 1952. By 1960 it was claimed that there were some 500 in London and 2000 across the country as a whole. The key to their success was that they became venues for teenagers to socialize and, in particular, listen and dance to music, whether it was live or on the juke box, without having to meet their parents in the pub (Clayton, 2003). Although there was little emphasis on coffee quality, this established an important connection between coffee and Italy in the British mentality, reinforced by the large number of Anglo-Italians working in the catering trade. Many establishments simply served "frothy coffee"—essentially any form of coffee base (including soluble) topped by frothy milk, which, once chocolate sprinkles were added, could be presented as a "cappuccino" (Morris, 2005). By the mid-60s, however, coffee bars were dying out, as pubs began embracing younger customers.

The importance of this became clear in the 1990s when local entrepreneurs began to translate the coffee shop format that had developed in the United States for the United Kingdom. They saw branding as a critical component of this, as it would enable them to distinguish themselves from the sandwich bars and cafes already in operation, effectively lending legitimacy to charging the all-important premium for a coffee beverage that underpinned the business proposition. Several, of course, chose to brand themselves American, such as the Seattle Coffee Company, who opened what is generally regarded as the first London coffee shop in 1995, and sold its chain of 65 stores to Starbucks in 1998. So too Coffee Republic, also established in that year, which even sought to recruit its baristas from American students traveling abroad (Morris, 2013c).

However, the two most enduringly successful local chains, Costa and Café Nero, instead positioned themselves as Italian in their proposition, thereby structuring their appeal to British understandings of coffee and the aspiration for a "continental café culture." Costa, in particular, could draw on its heritage of having been established by two members of the Anglo-Italian community in the 1970s, and repeatedly referenced this, rather than the fact it was now owned by the Whitbread leisure conglomerate, to justify its slogans such as being "Italian about Coffee." Café Nero had no such heritage on which to draw, but still had no hesitation in describing itself as "the Italian coffee company," attempting to deliver this in part through the more edgy, dark wood, interiors of its stores, and most notably by allowing smoking up until the introduction of the legal ban in 2007. That said, both Nero and Costa, utilize the coffee shop format of "dwell friendly stores," rather than the upright drink and go of the Italian espresso bar.

Instead they delivered a coffee experience that was structured as purely Italian—any customer requests for "ordinary coffee" were met by serving Americanos. Both used Italian style espresso blends including a proportion of Robusta, prepared on "traditional" Italian-made machines. As Paul Ettinger, the first Food and Beverage Director for Nero explained:

> *People knew if they came to Nero, it was all going to be around espresso .... We never did any drip coffee ... never tried to be smart, we didn't do the big American jugs, ... we didn't do the gimmicks, the syrups, so we were early on very pure ... an Italian coffee bar.*
>
> Interview cited in Morris (2013c, p. 890)

These branding strategies appear to have appealed to customers. As of December 2015 the three main coffee shop chains in the United Kingdom were Costa with 1922 outlets, Starbucks with 849, and Café Nero with 620. This accounts for 53% of the branded coffee shop market, but only 17% of the total number of 20,728 speciality coffee outlets in the United Kingdom.

Allegra Strategies, the market research company that most closely follows the coffee industry, has followed the evolution of the UK market since 1997 and has kindly made available the data upon which the analysis of UK consumer habits that follows is drawn (Allegra, 2016). Since Allegra started following the market, the number of outlets serving speciality coffee—understood as hand prepared, espresso-based beverages—has more than quadrupled.

Speciality Coffee Outlets in the United Kingdom 1997–2015

Year	Independent	%	Branded	%	Nonspecialist	%	Total
1997	3900	82	371	8	485	10	4756
2000	4260	69	1382	22	562	9	6204
2005	4869	55	2766	31	1287	14	8887
2010	5336	37	4693	33	4312	30	14,341
2015	6257	30	6495	31	7976	39	20,728

Allegra Strategies, *Project Café 16*, p. 25.

Different phases of growth can be distinguished within this overall explosion of coffee culture. Although independent operators dominated the limited provision available in 1997, the driving force for growth was supplied by the branded chain operators whose outlet share increased from 8% to 31% by 2005. Thereafter, however, it has been the nonspecialist operators, such as department store cafes, supermarket cafes, pubs, motorway, and forecourt coffee shops and quick service restaurants, who have moved to include a speciality coffee offer in their menu, thereby increasing their proportion of outlets from 14% in 2005 to 39% in 2015. Their dominance within the market is reflective of a broader truth—within the United Kingdom, speciality coffee has become mainstream.

For the British consumer, espresso-based beverages have become the everyday coffee choice outside the home. Latte, cappuccino, and americano are the top three choices of coffee shop customers—with latte now three times more popular than tea. The reason coffee drinking and coffee shops have become banal features of UK coffee culture is perhaps because "functional" reasons such as "need to have a coffee" dominate customer motivations for visiting a coffee shop, and convenience is still cited as the primary selection criteria. The most often-cited convenient location, however, is one close to shopping facilities, suggesting the ways that coffee has become incorporated as a ritual within household routines.

Some interesting gender and generational differences can be observed. Women are notably more inclined to use a visit to a coffee shop to socialize than men (29% as opposed 17%), and, perhaps as a consequence, are more likely to visit in the afternoons, during the later days of the week, and to dwell longer. Men are more likely to visit daily, and in the early morning, suggesting again the incorporation of coffee into their working day. Similarly although customers between 25 and 44 years are most likely to visit in the early mornings, and are the principal purchasers of takeaway, the under 24 year olds are most likely to visit in the afternoons and show the greatest propensity for socializing, studying, and using wi-fi, whereas the over 65 year olds who visit during the mid-mornings have the longest dwell times of all. These differences are suggestive of the ways in which consumer groups have routines that enable them to embed the use of the coffee shop into their habitual practices.

Nonetheless, since the financial crises of the mid-2000s, there have been some notable changes in consumer motives and behaviors. The emphasis on convenience as a driver of coffee shop choice has notably diminished, and that placed on coffee quality increased, alongside loyalty to both operator and coffee supplier brands, and the patronage of one preferred coffee shop. This is even more the case among frequent consumers of speciality coffee. There are strong indications that this is reflective of a rising interest in so-called third wave coffee, reflected in much higher consumer "promoter" scores for independent operations, and within consumer households where 33% now own pod machines, and 7% have some form of artisan brewing equipment, such as an Aeropress of V60.

3.3 New Wave Consumers?

In Britain, the third wave is strongly identified with a new beverage form—the Flat White. Its significance lies as much in its geographical origins as its taste. It traveled alongside a miniwave of Australasian baristas and entrepreneurs who opened coffee shops in London, starting with Flat White itself, again established in Soho in 2005. It required a more technical "data-led" approach to preparation of both the espresso shot and the milk texturing than British baristas were used to and became emblematic of a more skilled and sophisticated approach to coffee coming from the Southern Hemisphere. It was joined by other innovations such as cold brewed coffee, shifting attention away from the increasingly standardized Italian style beverage menu.

The Flat White has now become associated with another area of London—the old East End of the City that today houses many of the digital start-ups and online advertising companies, employing what is widely portrayed as a "hipster" workforce. In his study of the "Flat White Economy," McWilliams notes that those who work in this sector are usually well-educated young people, earning relatively low professional salaries, who economize by living a "backpacker lifestyle" renting small cheap rooms and therefore seeking public spaces and like-minded people with which to socialize and pursue their passions (McWilliams, 2015). Given their circumstances they cannot afford the fine dining and champagne bar lifestyles of London's banking elite; instead speciality coffee provides them with an alternative, affordable product about which they can acquire learning and demonstrate their taste and sophistication.

The barriers between operators and consumers have become blurred so that "third wave coffee" has effectively become an iconic product at the center of its own subculture. Many of the proprietors of third wave establishments treat them primarily as vehicles to support their own passion for coffee, rather than as sources of profitability. Artisan roasters have carved out niche markets by recruiting their customers as online "friends" through social media sites such as Facebook. The relationship between the second wave and third wave coffee providers has become not unlike that between high street brands and the couture sector with developments in the latter often finding their way into the former, not least due to consumer demand. The Flat White, for example, is now a standard offering in most UK coffee shop chains, and regularly consumed by 14% of customers—only one point below tea.

In America the "Third Wave" can be interpreted as a return to the spirit of the original speciality pioneers—indeed when Rothgeb Skeie first coined the phrase, she explained it as a "reaction to those who want to automate or homogenise speciality coffee"—in other words the second wave coffee shop chains (Skeie, 2003). It is important to note that consumers were already being included in debates about speciality coffee, most notably through the internet. *Coffee Review*, established by Ken Davids in 1997, connected consumers and roasters throughout the continent, utilizing independent reviews based on a 100-point scale that echoed that employed by Robert Parker for wine. The community website *Coffee Geek*, established by Mark Prince in 2001, connected up enthusiastic home brewers, notably espresso machine owners who wanted to get the best out of their equipment, to the point of re-engineering it, and reviewing new machines as they came onto the market. These discussions literally took place, at one point, under the banner Alt-Coffee. Subscribers could be said to be "prosumers"—hobbyist consumers using professional equipment, and thereby creating a consumer market for this.

It is misleading, however, to consider Third Wave, through a national lens. Rather it has developed a transnational community of fans, reflecting both the mobile lifestyles of many of its members, and the digital connectivity within a virtual community. Third wave enthusiasts around the world follow blogs such as that of James Hoffman's blog, *jimseven*, using them as a kind of "open source guide" to developing a third wave offering. The digital magazine *Perfect Daily Grind*, for example, founded in 2015, sees its mission as connecting all elements of the supply chain within the third wave community—producer to consumer. It projects 2 million page views for 2016 across the various social media platforms in which it is present with readership roughly equally split between the Americas, Europe, and Asia (Perfect Daily Grind, 2016).

The Third Wave has, however, borrowed from a variety of coffee cultures and historical reference points in order to structure its own consumption

Bedbury, S., 2002. A New Brand World. Penguin, London.

Biderman, B., 2013. A People's History of Coffee and Cafes. Black Apollo, Cambridge.

Borghi, A., 2015. The Role of Coffee in the Out of Home Food and Drink Market. Presentation at HOST Hospitality Exhibition, Milan.

Bramah, E., 1972. Tea and Coffee. Hutchison, London.

Brem, A., Maier, M., Wimschneider, C., 2016. Competitive advantage through innovation: the case of Nespresso. European Journal of Innovation Management 19 (1), 133–148.

Clayton, A., 2003. London's Coffee Houses: A Stimulating Story. Historical Publications, London.

Comunicaffe, February 23, 2016. E l'inventore delle capsule K-Cup Keurig non vuole più usarle. Available from: http://www.comunicaffe.it/.

Cowan, B., 2005. The Social Life of Coffee. The Emergence of the British Coffeehouse. Yale University Press, New Haven.

Ellis, M., 2004. The Coffee House. A Cultural History. Weidenfeld and Nicolson, London.

Foodservice Director's College Coffee House Manual, 1986. Coffee Development Group, Washington.

Gallup, 1994. The 1994 Gallup Study of Awareness and Use of Gourmet and Speciality Coffees Survey. Princeton.

Giuli, M., Pascucci, F., 2014. Il ritorno alla competitività dell'espresso italiano. Angeli, Milan.

Grinspan, J., July 9, 2014. How Coffee Fueled the Civil War. New York Times. Available from: http://opinionator.blogs.nytimes.com/2014/07/09/how-coffee-fueled-the-civil-war/.

Hattox, R., 1985. Coffee and Coffee Houses. The Origins of a Social Beverage in the Medieval Near East. University of Washington Press, Seattle.

Kjeldgaard, D., Ostberg, J., 2007. Coffee grounds and the global cup: glocal consumer culture in Scandinavia. Consumption, Markets and Culture 2, 175–187.

Kjelgaard, D., Hemetsberger, A., Luomala, H., Mastrangelo, D., Pecoraro, M., Ostberg, J., Pilcher, E., 2011. Revisiting the Euroconsumer: transnational history of European coffee consumption. European Advances in Consumer Research 9, 367–372.

Lewis, D., Bridger, D., 2001. The Soul of the New Consumer. Authenticity –What We Buy and Why in the New Economy. Brearley, London.

Lundqvist,, P., 2016. A Question of Taste—The Consumption of Colonial Goods and Alcohol in Three Swedish Nineteenth Century Novels. Paper Presented at ESSHC Congress, Valencia, Spain.

Lury, C., 2004. Brands. The Logos of the Global Economy. Routledge, London.

Matzler, K., Bailom, F., von der Eichen, S., Kohler, T., 2013. Business model innovation: coffee triumphs for Nespresso. Journal of Business Strategy 34 (2), 30–37.

McWilliams, D., 2015. The Flat White Economy. Duckworth Overlook, London.

Mikkelson, B., 2009. Cup of Joe. Available from: http://www.snopes.com/language/eponyms/cupofjoe.asp.

Milne, R., 2013a. Quality. In: Jackson, P., CONANX Group (Eds.), Food Words. Essays in Culinary Culture. Bloomsbury, London, pp. 166–170.

Milne, R., 2013b. Taste. In: Jackson, P., CONANX Group (Eds.), Food Words. Essays in Culinary Culture. Bloomsbury, London, pp. 214–221.

Morris, J., Thurston, R., 2013. Coffee: a condensed history. In: Thurston, R., Morris, J., Steiman, S. (Eds.), Coffee. A Comprehensive Guide to the Bean, the Beverage, and the Industry. Rowman and Littlefield, Lanham, pp. 215–225.

Morris, J., 2005. Imprenditoria italiana in Gran Bretagna. Il consumo del caffè stile italiano. Italia Contemporanea 241, 540–552.

Morris, J., 2010. Making Italian espresso, making espresso Italian. Food and History 8 (2), 155–183.

Morris, J., 2013a. Espresso by design. In: Lees Maffei, G., Fallan, K. (Eds.), Made in Italy: Rethinking a Century of Italian Design. Berg, New York, pp. 225–238.

Morris, J., 2013b. The espresso menu: an international history. In: Thurston, R., Morris, J., Steiman, S. (Eds.), Coffee. A Comprehensive Guide to the Bean, the Beverage, and the Industry. Rowman and Littlefield, Lanham, pp. 262–278.

Morris, J., 2013c. Why espresso? Explaining changes in European coffee preferences from a production of culture perspective. European Review of History: Revue europeenne d'historie 20 (5).

National Coffee Association, 2016. What Are We Drinking? Understanding Coffee Consumption Trends. Available from: http://nationalcoffeeblog.org/2016/03/19/coffee-drinking-trends-2016/.

Ojaniemi, T., 2010. Coffee as a Finnish Institution. Available from: https://www15.uta.fi/FAST/FIN/GEN/to-coffe.html.

Oldenburg, R., 1989. The Great Good Place. Cafés, Coffee Shops, Bookstores, Bars, Hair Salons and Other Hangouts at the Heart of a Community. Paragon, New York.

Paulig Company History, 2016. www.paulig.com/en/company.company-history.

Perfect Daily Grind, 2016. Media Kit. Available from: http://www.perfectdailygrind.com/advertise-with-us.

Perluss, P., 2015. Smoke, Mirrors and Mirth: Café Diversity and Business Expansion in 18th Century Paris. Paper Presented at 7th AISU Conference, Padova, Italy.

Pine, J., Gilmore, J., 1999. The Experience Economy. Work Is Theatre and Every Business a Stage. Harvard Business School Press, Boston.

Pine, J., Gilmore, J., 2007. Authenticity. What Consumers Really Want. Harvard Business School Press, Boston.

Prendergast, M., 2010. Uncommon Grounds. The History of Coffee and How It Transformed Our World. Basic Books, New York.

Rhinehart, R., 2009. What Is Specialty Coffee? Available from: http://www.scaa.org/?page=RicArtp2.

Roosevelt, E., 1941. Shall We Arm Our Merchant Ships? Broadcast October 12—programme transcript. Available from: http://www.americanradioworks.org/shall-we-arm-merchant-ships/.

Roseberry, W., 2002. The rise of yuppie coffees and the reimagination of class in the United States. In: Counihan, C. (Ed.), Food in the USA. A Reader. Routledge, London, pp. 149–168.

Saraçgil, A., 1996. 'Generi voluttuari e ragion di stato: politiche repressive del consumo di vino, caffè e tabacco nell'impero Ottomano nei secc. XVI e XVII. Turcica 28, 163–194.

SCAA, 2016. US Specialty Coffee Shops. Available from: http://www.scaa.org/?page=resources&d=facts-and-figures.

Schultz, H., Yang, D., 1997. Pour Your Heart into It. Hyperion, New York.

Shiau, H.-C., 2016. Guiltless consumption of space as an individualistic pursuit: mapping out the leisure self at Starbucks in Taiwan. Leisure Studies 35 (2), 170–186.

Simon, B., 2009. Everything but the Coffee. University of California Press, Berkeley.

Skeie, T., 2003. Norway and Coffee. The Flamekeeper. Newsletter of the Roasters Guild, Spring issue.

Spary, E., 2012. Eating the Enlightenment. Food and the Sciences in Paris, 1670–1760. University of Chicago Press, Chicago.

The Vertue of the Coffee Drink, c.1656, London—available through Early English Books Online Collection.

Thorn, J., Segal, M., 2007. The Connoisseur's Guide to Coffee. Apple Press, London.

Tuchsherer, M., 2003. Coffee in the red sea area from the sixteenth to the nineteenth century. In: Clarence-Smith, W., Topik, S. (Eds.), The Global Coffee Economy in Africa, Asia and America, 1500–1989. Cambridge University Press, Cambridge.

Ueshima, T., 2013. Japan. In: Thurston, R., Morris, J., Steiman, S. (Eds.), Coffee. A Comprehensive Guide to the Bean, the Beverage, and the Industry. Rowman and Littlefield, Lanham, pp. 197–200.

Ukers, W., 1935. All about coffee, New York. Tea and Coffee Trade Journal.

Watson, M., 2013. Practices. In: Jackson, P., CONANX Group (Eds.), Food Words. Essays in Culinary Culture. Bloomsbury, London, pp. 157–160.

Whitaker, I., 1970. Coffee-drinking and Visiting Ceremonial Among the Karesuando Lapps. Svenska landsmål och svenstkt folkiv: tidskrift, pp. 36–40.

White, M., 2012. Coffee Life in Japan. University of California Press, Berkeley.

Zeitemann, B., 2013. Germany. In: Thurston, R., Morris, J., Steiman, S. (Eds.), Coffee. A Comprehensive Guide to the Bean, the Beverage, and the Industry. Rowman and Littlefield, Lanham, pp. 201–205.

Chapter 20

Human Wellbeing—Sociability, Performance, and Health

Britta Folmer[1], Adriana Farah[2], Lawrence Jones[3], Vincenzo Fogliano[4]
[1]*Nestlé Nespresso SA, Lausanne, Switzerland;* [2]*Federal University of Rio de Janeiro, Rio de Janeiro, Brazil;* [3]*Huntington Medical Research Institute, Pasadena, CA, United States;* [4]*Wageningen University, Wageningen, The Netherlands*

It (coffee) fortifies the members, it cleans the skin, and dries up the humidities that are under it, and it gives an excellent smell to all the body.

Avicenna (980—1037)

1. INTRODUCTION

It is well known that coffee is one of the most widely consumed beverages worldwide. Since its discovery, it has played an important role in the life of many people. With its delicious aroma and intense taste, this is a beverage that consumers savor first thing in the morning, during a small break in the day or to complete a meal. Although the first coffee is often drunk to wake up, a coffee taken as a small break in the day signifies a moment of relaxation or bonding with family or friends. Each cup provides a moment to shape the day.

Coffeehouses were first introduced in Europe in the 17th century and until today, cafés are increasingly becoming a global phenomenon. But in parallel to its social development in the world, physicians and scientists have disputed coffee's role with respect to lifestyle and health. For centuries, coffee has gone through waves of praises and attacks, evoking passions and reactions. But somehow, it has always managed to go back from villain to hero, because of its stimulating properties or because it prevented and healed diseases. Today science is also showing more long term, less directly perceivable benefits on health and well-being, opening a path for coffee to be considered as a functional food in the near future. This chapter will explore these aspects bringing an overview of the evolution of coffees role in the well-being of people through history.

The Craft and Science of Coffee. http://dx.doi.org/10.1016/B978-0-12-803520-7.00020-7

2. A HISTORICAL VIEW ON COFFEE AND WELL-BEING

Throughout history, people have debated the consequences of drinking coffee to the human body and mind. The pleasurable taste and stimulating properties have been worshiped and hated. In this section we will give an insight in its perception and the role it played in society since its first mention till today.

The earliest references to coffee consumption are possibly seen in the Old Testament, where a bean was referred to as "parched pulse," and the first written mention of coffee by Razes, a 10th century Arabian physician says coffee cultivation may have begun as early as AD 575 (Smith, 1987).

There are many stories about the origin of coffee consumption. The oldest one seems to be that coffee was introduced by Mohammed, for when he lay ill and prayed to Allah, the angel Gabriel descended with a beverage "as black as the Kaaba of Mecca" that gave him "enough strength to unseat 40 men from their saddles and make love to the same number of women" (Smith, 1987).

The most well-known story is that of Kaldi, a 9th century legendary goat herder across the Red Sea in Abyssinia (the old name for Ethiopia). According to the tale, Kaldi noticed that certain berries, presumed to be coffee cherries, caused his goats to prance with delight. So he decided to try them for himself, successfully curing his own depression (Ukers, 1935). Another version of Kaldi's story continued saying that a monk from a monastery down the hills where Kaldi tended his flock saw the goats' behavior and took some of the berries back to the monastery, roasted, and brewed them, and tried out the beverage on his brethren. As a result, they were kept more alert during their long prayers at night (Ukers, 1935; Smith, 1987).

The first known written documentation on the medicinal properties and uses of coffee was by the great Middle Eastern physician, Avicenna (980–1037)—also known as the Prince of Physicians and author of the Canon of Medicine (Ukers, 1935). His passion was wine, women, and sensual pleasure. For him, coffee served as a decongestant, muscle relaxant, and diuretic (Encyclopedia Britannica, 1910).

It is said that in 1258, the doctor-priest, Sheikh Omar, from Mocha discovered coffee in Arabia. He was in exile at Ousab and was starving when he came across some coffee cherries. As they were too hard and bitter to eat, he decided to boil them to soften them a bit, which created a brown soupy liquid he drank instead. He claimed that this brown liquid, refreshed and enlivened him, while arousing his drooping spirits (Ukers, 1935). His coffee became known as a cure for many different types of illnesses. When Sheikh Omar returned to Mocha, the governor honored him by building a monastery for him and his disciples.

We may never know which, if any, of these stories were true. But stories and suppositions apart, we can say with certainty that as soon as a group of people started drinking coffee, others tried to ban it, for health, religious, political, or cultural reasons. For example, in 1511, the governor of Mecca,

Kair Bey, related the story of two influential physician brothers named Hakimani, who banned coffee because this exhilarating drink caused people to extravagances prohibited by law. When the Sultan in Cairo heard of this, he was outraged and successfully ordered them to be executed (Smith, 1987). Just two decades later, in 1534, a group of Islamists and some influential physicians took opposite sides in the debate on coffee, its healthful properties and religion. The chief justice took the side of the physicians, thereby laying to rest any concerns about coffee's potential harmful effects. However, people were also concerned that roasted coffee appeared similar to charcoal, a substance which the Koran explicitly banned (Smith, 1987).

In Europe, coffee and its influence on health have also enjoyed a rocky reputation ever since Dr. Leonard Rauwolf, a German physician, introduced it to Western Europe in 1582, following a trip to Aleppo. Dr. Rauwolf wrote: "They have a very good drink by them called chaube (coffee) that is almost as black as ink and very good in illness, chiefly that of the stomach," (Ukers, 1935). Although his advice on the healing properties of coffee helped increase its popularity, some people in the church considered it diabolical. That was before Pope Clement VIII famously exclaimed, "Why, this Satan's drink is so delicious that it would be a pity to let the infidels have exclusive use of it. We shall fool Satan by baptizing it and making it a truly Christian beverage" (Ukers, 1935).

Yet the debate did not end there. A 1679 dissertation before the Marseille College of Physicians criticized coffee's medicinal worth. The verdict of the Faculty of Aix was: "... we must necessarily conclude that coffee is hurtful to the greater part of the inhabitants of Marseille" (Ukers, 1935). However, this verdict was widely ignored as drinking coffee became more and more popular.

The debate on coffee and health then quieted until the late 19th century when Charles (Charlie) William Post, a successful cereal manufacturer, introduced a grain-based beverage, which was intended as a coffee substitute. Charlie was one of the first entrepreneurs to use mass advertising to promote his product, and his ads made a variety of unproven health claims about his beverage, which impacted negatively the popularity of coffee (Pendergast, 1999).

Unsupported health claims both for and against coffee continued through the middle of the 20th century. For example, in the 1930s, Dr. Max Gerson introduced coffee enemas as a treatment for cancer. He claimed that coffee drew toxic products from the blood across the bowel mucosa and/or the bile ducts by dialysis, thus eliminating toxic products from the liver (Gerson, 1978). Despite the lack of any supporting scientific evidence, some holistic practitioners still use this therapy today, especially in Asia, even though recent studies have refuted the value of this procedure (Teekachunhatean et al., 2013).

With the advent of modern scientific technology, our understanding of coffee and its healthful properties has changed dramatically. This, in combination with large and reliable databases, and sophisticated statistics have enabled the separation of confounding factors in epidemiological studies, such as existing medical conditions, smoking, or a poor quality diet. Techniques have included both observational studies that collect large amounts of data on the incidence of diseases, as well as self-reported questionnaires on food consumption frequency.

Over the past decade, dozens of scientific papers have been published, dispelling coffee's bad reputation and highlighting its positive effects on human health. Currently, strong evidence shows that consumption of coffee within the moderate range, approximately two to four cups per day, is not associated with increased long-term health risks among healthy individuals. Therefore, moderate coffee consumption can be incorporated into a healthy diet, along with other healthful behaviors (US Department of Agriculture, 2015). Because of this increased availability of quality health data, we might finally have the hope of getting out of the angel-or-demon cycle that coffee's reputation has had for hundreds of years.

3. COFFEE AND LIFESTYLE

Although the health and well-being aspects of coffee were debated for centuries, its role as a social beverage has been well recognized, even if not always open to everyone. The Ethiopian Coffee Ritual which holds communication and connection as the central themes is described by Ukers as existing from time immemorial (Ukers, 1935). A women first lights incense and then starts to roast the coffee on a charcoal fire. During the roasting, people gather and talk. The discussions continue while the woman pounds the beans and prepares the brew, finally serving it.

When coffee moved to Europe, and more specifically into London, in the 17th century, conversations and gatherings were still central. At first, coffee houses were only open to the elite, but in the 18th century they became places for social gathering and commerce. Over time, these coffee spaces increasingly opened up and became accessible to the community. They remain a place for conversations and social gatherings with family or friends, and the importance of their social role continues to expand and change with society (Fig. 20.1). An example is the fika culture, which is a widespread phenomenon in Sweden and gaining popularity elsewhere in the world. "Fika" basically means "to have a coffee." It is a social event where colleagues, family, or friends catch up on personal or professional matters as a break from daily routine. And even though coffee may be seen as a social beverage, it is often consumed alone, allowing oneself a moment of relaxation or reflection.

Another example illustrates the social role of coffeehouses and how coffee was, and is still used in an act of charity for the less fortunate. In Naples, the tradition was that someone who had experienced good luck would order a caffè sospeso (a pending coffee), paying for two coffees but receiving and

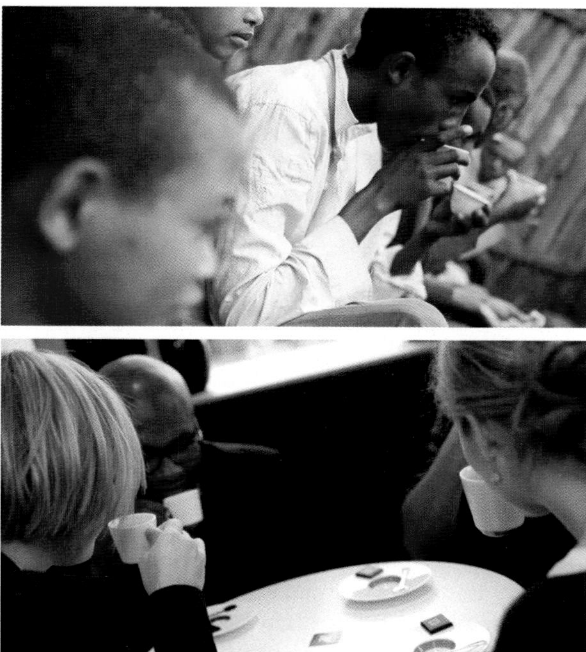

FIGURE 20.1 Social gathering around coffee in Ethiopia (upper) and Europe (lower).

consuming only one. A poor person entering the coffeehouse later on could enquire if there was a "sospeso" and then receiving a coffee for free. This tradition was booming during World War II, but is still common, especially in difficult economic times.

According to the International Coffee Organization, world consumption has more than doubled over the last 50 years with an average growth rate in production of 2% since 1990 (ICO, 2014). This increase in consumption has been driven mainly by producing and emerging coffee markets and not so much by traditional markets. In traditional markets there is, instead, a movement toward increased quality awareness and appreciation, which has been also expanding to other markets. As part of this movement, an increase in small roasting companies and coffee houses offering consumers specialty coffee, marketed by its origin or farmer has been observed. The growing market for specialty coffee in portioned systems is also a sign that consumers have an increasing wish to appreciate high quality coffee at home.

It may be speculated that the perception of the coffee beverage in general is improving with respect to health, and thus also possibly playing a role in the way coffee is consumed, being enjoyed as part of a healthful moment. Today, many people appreciate the potential beneficial effects on health, mostly because of the presence of caffeine and its perceived stimulating properties. Other bioactive compounds such as chlorogenic acids (a group of phenolic compounds, or polyphenols), trigonelline (an alkaloid), and melanoidins (the resulting mixture of compounds formed during roasting of coffee) are generally less well known to consumers, but becoming more and more popular as scientific research knowledge reaches them, especially via Internet.

Today, the debate around the stimulating properties of caffeine has moved from "good versus bad" to the quantities needed for a beneficial impact, and amounts corresponding safe consumption. Although some people can drink a large cup of (caffeinated) coffee just before going to sleep, others feel the arousing and stimulating effects just after consuming one small cup. The fact is that some people are more sensitive to caffeine effects because they metabolize and eliminate caffeine slower than others and the effects may be cumulative for a while (Clark and Landholt, 2016). Effects tend also to be stronger when fatigued, or for elderly people (van Boxtel and Schmitt, 2004). So, there is no magic formula to calculate the amount of caffeine which is just right to give a sense of pleasure and well-being. It is up to each person to pay attention to his or her body response to caffeine intake at different times of the day, and learn from it.

Although coffee is the major source of caffeine intake, other products largely consumed by people from different ages such as black/green/white teas, maté tea, cola beverages, cocoa derivatives, and occasional energy drinks are also important sources and therefore need to be considered (Lima and Farah, 2014; EFSA 4102, 2015a). This has prompted various authorities to publish guidelines on the safety of caffeine consumption. The most recent report is the European Food Safety Authority's 2015 scientific opinion on caffeine safety (EFSA 4102, 2015a). National health authorities have also published reports (US Department of Agriculture, 2015), and the general agreement is that the habitual consumption of up to 400 mg of caffeine per day (two to four cups a day) does not cause safety concerns for nonpregnant adults. It is noteworthy to point that the amount of caffeine in a cup is very cultural and will depend on different factors like the amount of robusta beans in the blend, the amount of roasted and ground coffee used to prepare one cup, the ratio of ground coffee to water, as well as the brewing method. In this chapter we will, where possible, speak of milligram of caffeine. For cups, we generally consider a range between 100 and 200 mg caffeine per cup.

Communications on "safe" limits of consumption, have led to more awareness among certain populations or individuals that may want to limit caffeine intake. Although it is difficult to state exact numbers, the consumption of decaffeinated coffee varies from around 5% in Germany to 20% in the United States (see Chapter 8). In order to respond to this consumer segment that want to limit caffeine intake, decaffeination companies and roasters have collaborated to develop high quality decaffeinated products, offering flavorful coffees but without the caffeine.

In addition to the well-known mentally stimulating effects of caffeine, a number of additional benefits attributed to compounds different from caffeine have been associated to its consumption. For example, in vitro and animal (in vivo) have revealed a number of benefits linked to chlorogenic acids that involve antioxidative, antiinflammatory, antimicrobial, antimutagenic effects. Additionally, these compounds involve glucose and lipid metabolic regulation. Because of the high amount of chlorogenic acids in coffee, they may play an important role in the diet of consumers as a source of antioxidative compounds. In fact, in a number of reports from different countries, based on their official food consumption database or other types of surveys, coffee was the main contributor to total dietary antioxidant capacity in a number of countries, i.e., Brazil (66%), Norway (64%), Italy (38% for women and 27% for men), Spain (45%), Japan (56%), and Czech Republic (54.6% for women and 43.1% for men) (Torres and Farah, 2016).

Trigonelline is another compound that has gained importance over the last years due to its potential involvement in the protective effect of coffee against diseases. Emerging science shows efficacy on glucose metabolism and type 2 diabetes (Yoshinari and Igarashi, 2010). In vitro and animal studies have also reported different mechanisms of neuroprotective effect (Hong et al., 2008; Tohda et al., 2005), antitumor (Hirakawa et al., 2005), and phytoestrogenic effects among others. However, as with chlorogenic acids, this compound undergoes changes and degradation during roasting and overroasted coffees contain low amounts of it.

Additionally, because of the presence of polysaccharides and melanoidins that may act as soluble fibers, a recent study concluded that the consumption of 0.5–2 g per day (present in two to five cups) contributes to up to 20% of the recommended 10 g of daily soluble dietary fiber intake (Fogliano and Morales, 2011).

It is noteworthy to mention that, although traditional and emerging science can link certain compounds to specific mechanisms, for the preventive action against different diseases, it is likely that most preventive effects are caused by synergistic or additive effects with the various compounds present in coffee.

In addition to bioactive compounds, coffee contains vitamin B3 (nicotinic acid, ~10–50 mg per cup) that is formed during the roasting process, as well as a range of minerals (300–1500 mg per cup), from which potassium stands out, contributing to about 40% of total amount of minerals in the beverage (Farah, 2012).

Since the end of the 20th century, major scientific advances have helped clarify some of the mechanisms linked to the effects that coffee components have on the human body. In vitro and in vivo studies using reference compounds have proven that several components in coffee exhibit biological activities. Although some of these active compounds are present in unroasted coffee, others are formed during the roasting process. This is because coffee roasting creates a complex and still incompletely characterized set of chemical reactions. This complexity in the mixture of molecules makes finding the active compounds and the mechanisms responsible for the physiological actions very challenging. As a further complication, only a small percentage of some of the ingested compounds are absorbed into the circulatory system and reach the tissues. It is believed that some bioactive compounds are partly metabolized or fermented by the microbiota present in the gut, and do not retain the original structure when taken up by the body. This complicates our ability to follow these compounds in the human body as well as to understand the physiological impact they have in preventing various diseases.

In the next sections we will go deeper into the relation of coffee and caffeine to well-being and specifically focus on groups of people such as women, children, and athletes. The downsides of caffeine are also discussed. The ongoing scientific research to better understand the impact of coffee consumption on the reduced risk of some diseases are discussed in Section 4 of this chapter.

3.1 Coffee Consumption for Mental Performance

Many coffee drinkers are well aware of the impact coffee has when waking up in the morning, or to increase alertness during the day. It is thus not surprising that the most well researched benefit related to coffee and its caffeine content is its ability to enhance mental performance, which includes improving alertness and perception (see Einöther and Giesbrecht, 2013 for a review). Recent studies on caffeine consumption have also shown that caffeine helps to enhance attention, alertness, and concentration, and that generally a dose of 75 mg is needed to obtain these effects (see EFSA 2054, 2011b for an overview). Other functions such as memory (Nehlig, 2010; Borota et al., 2014) and mood (Smith et al., 2005; Olson et al., 2010) have also been shown to be improved by caffeine.

Although the effects of caffeine are easily perceived and well known, there are very large differences between individuals, explaining the sometimes inconsistent results in human studies. Generally, effects have been shown to be strongest in situations of increased fatigue, e.g., in night shift workers, or in elderly populations as caffeine may help to attenuate this fatigue (van Boxtel and Schmitt, 2004).

Here's how it works: The body absorbs caffeine from the gastrointestinal tract. It efficiently circulates to tissue, traverses cellular membranes and can easily cross the blood–brain barrier and enter the brain. About 1 h after consumption, the maximum amount of caffeine appears in the blood stream (Goldstein et al., 2010; Harland, 2000). It is then metabolized by the liver, and enzymatic reactions create a number of metabolites. Following this process, the kidneys excrete the remaining 3% of caffeine and its metabolites. For most people it takes about 3–6 h to eliminate 50–75% of the caffeine from the body (Goldstein et al., 2010).

Once absorbed, caffeine exhibits numerous and well-studied physiological stimulating effects which are attributed to the fact that caffeine can act as antagonist to the adenosine receptors (O'Connor et al., 2004; Gliottoni and Motl, 2008; Gliottoni et al., 2009). More specifically, the adenosine receptors A_1 and A_2 have a high affinity for caffeine which in turn stimulates the release and turnover of several central neurotransmitter substances, including acetylcholine and nonadrenaline (van Boxtel and Schmitt, 2004). The adenosine receptors are located in all parts of the brain with higher concentrations in some areas that are involved in higher-order processing, e.g., the hippocampus, a brain structure that is critical for memory formation. In addition to blocking adenosine, this mechanism elevates the levels of dopamine and adrenaline (Jackmann et al., 1996; Greer et al., 1998), the reason for the sense of arousal, mood, and cognition.

Coffee components other than caffeine have also been shown to have influenced cognitive performance in an elderly population, even if to a smaller extent compared to caffeine. Decaffeinated coffee enriched in chlorogenic acids has somewhat improved alertness, reduced headaches, and mental fatigue in comparison to the comparative nonenriched decaffeinated coffee. Effects may partly be attributed to the chlorogenic acids, but other compounds naturally present in coffee are also suggested to play a role (Camfield et al., 2013; Cropley et al., 2012).

3.2 Coffee Consumption for Physical Performance

Athletes are interested in the effect of caffeine on endurance and exercise capacity. It is well documented that caffeine can enhance endurance events, stop-go events (e.g., team and racket sports), and sports involving sustained high-intensity activity lasting from 1 min up to an hour (e.g., swimming, rowing, and running races) (Goldstein et al., 2010) (Fig. 20.2). For example, it has been shown that work done during a 15 min time trial on stationary bikes increased by 4% when 3 mg caffeine/kg bodyweight was taken as compared to placebo (Jenkins et al., 2008).

Based on extensive scientific studies the active dose of caffeine has been found to be 3 mg/kg of bodyweight, to be taken 1 h before exercise (EFSA

FIGURE 20.2 Caffeine can enhance sustained high intensity activity such as rowing.

2053, 2011a; Goldstein et al., 2010). For a person weighing about 70 kg, this amount would thus be equivalent to 210 mg.

Coffee and other caffeine vehicles have been used by athletes for a long time, and the first papers discussing the mechanisms date back to 1978 (Costill et al., 1977). Caffeine not only increases attention and alertness, and aids concentration as described in the previous section, but also coordination (Hogervost et al., 2008), and reduces the perception of pain and fatigue (O'Connor, 2004; Gliottoni and Motl, 2008; Gliottoni at al., 2009). Additionally, caffeine may mitigate acute mountain sickness when taken a few hours prior to attaining a high altitude (Kamimori et al., 1995).

Experts have conducted detailed research on caffeine's multiple mechanisms of action for physical performance, so these are well understood. First, caffeine competes with adenosine at its receptor sites (Fredholm et al., 1999; Spriet and Gibala, 2004). As caffeine also elevates adrenaline and dopamine levels (Jackmann et al., 1996; Greer et al., 1998), part of the body's "fight or flight" response, they prepare the body for physical activity by increasing the heart rate and blood pressure. The reduction on pain perception is most likely related to the capacity of caffeine to increase the secretion of β-endorphins. The analgesic properties of these compounds may be responsible for this effect (O'Connor et al., 2004; Gliottoni and Motl, 2008; Gliottoni et al., 2009). Thus, even if the perceived impact of caffeine on sports performance is muscular, the mechanisms responsible are neural (Spriet, 1995).

According to the EFSA caffeine safety report (EFSA 4102, 2015a), athletes can safely consume up to 400 mg of caffeine per day—the same level as generally for adults. In addition, under normal environmental conditions, it is safe to consume single doses of 200 mg of caffeine less than 2 h prior to intense exercise, but the amount and time prior to exercise will vary for different individuals since, as explained before, they may metabolize caffeine at different rates.

Because of caffeine's inherent performance-enhancing capabilities, in 1984, the International Olympic Committee introduced an antidoping program that included caffeine with a serum threshold of 12 μg/mL. This level would only occur with a caffeine intake of around 9 mg/kg. Such a threshold is controversial because it is possible to obtain performance benefits with just 3 mg of caffeine/kg of body weight. In addition, because of the differences in individual's ability to metabolize caffeine over time, it is very difficult to monitor its serum levels (Goldstein et al., 2010). This controversy, along with the inefficiency of the threshold to distinguish between social use and deliberate use to increase performance, is two of the reasons why in 2004, the World Anti-Doping Agency (WADA) removed caffeine from the prohibited list and added it to its monitoring program (WADA, 2009). The monitoring program includes substances that are not prohibited in sport, but which WADA examines in order to detect patterns of misuse.

In addition to caffeine, athletes can benefit from the high content of polyphenols (chlorogenic acids), as well as from its mineral content. It is well known that intense exercise increases oxygen consumption, raises the generation of reactive oxygen species, causes lipid peroxidation of polyunsaturated fatty acids in membranes and DNA damage, and decreases physical performance. Release of inflammatory cytokines may further contribute to exercise-induced oxidative stress as shown for other food products (Chang et al., 2010). A number of studies have shown that the consumption of different beverages containing polyphenols, including chlorogenic acids, before or after intense exercise, decreased plasma levels of oxidative and inflammation markers, thus offering protection against exercised-induced oxidative damage and preventing inflammation (Chang et al., 2010; Panza et al., 2008).

3.3 Coffee, Sleep, and Sleep Quality

As caffeine helps increase alertness and arousal by blocking adenosine, it also controls the sleep–wake cycle. It is, therefore, not surprising that caffeinated coffee can reduce sleep quality by increasing the time required to fall asleep, interfering with the depth of sleep, and reducing the total time spent sleeping. It can also cause more frequent awakening or sleep fragmentation (Huang et al., 2011; Clark and Landholt, 2016).

The speed with which an individual breaks down the caffeine will also help determine the influence on sleeplessness. This variability in the enzymatic breakdown of coffee may account for caffeine's variable effect on sleep induction and arousal (Youngberg et al., 2011). However, habitual caffeine consumers may suffer less from these issues because they develop a tolerance. In this case, caffeine will still disrupt their sleep, but to a lesser extent than for people who are not habitual consumers (Childs and de Wit, 2012; Drapeau et al., 2006).

Travelers' jet lag represents another example of disturbance of the sleep cycle and can often be controlled by using melatonin, a natural hormone which controls sleep cycle. In a recent study consuming caffeine 3 h before habitual bedtime induced a 40-min phase delay of the circadian melatonin rhythm in humans. This magnitude of delay was nearly half of the magnitude of the phase-delaying response induced by exposure to 3 h of evening bright light beginning at habitual bedtime (Burke et al., 2015).

3.4 Caffeine Consumption by Women and Children

In this section we will be discussing the impact of caffeine on women's and children's health. Although the chapter is generally about coffee, here we focus specifically on caffeine which can come from various sources.

There is some evidence to suggest that the normal hormonal changes in pregnancy slow the body's ability to metabolize caffeine. This, in turn, means that a given dose of caffeine can have longer-lasting effects (as long as 15 h in the third trimester) (Kuczkowski, 2009). Even so, the EFSA report on caffeine safety (EFSA 4102, 2015a) concludes that its consumption is safe for pregnant and lactating women, although the report also recommends that these women reduce intake to a maximum of 200 mg throughout the day. Based on scientific findings, there is no risk of adverse birth weights for caffeine consumption below these values. Nevertheless, the risks of very high intakes (more than 600 mg of caffeine per day) include fetal growth retardation and low weight for gestational age (Sengpiel et al., 2013). Although there is no consensus in studies suggesting that caffeine could delay time of conception, it may be prudent for women who are having difficulty conceiving to limit caffeine intake to less than 300 mg per day (Higdon and Frei, 2006).

It is known that caffeine is present in the milk of lactating coffee drinkers with a peak appearing about 1 h after consuming a caffeinated beverage (Stavchansky et al., 1988; Nehlig and Debry, 1994). For this reason, doctors recommend that breastfeeding women keep caffeine intake below 200 mg per day (EFSA 4102, 2015a). At these levels, studies show that sleep time of nursing infants is similar to controls (Santos et al., 2012; Clark and Landholt, 2016).

When it comes to older children and coffee consumption, there are major cultural differences in both overall coffee consumption and consumption guidelines. For example, in most European countries, habitual coffee consumption starts when children become adults, and until the age of 10 chocolate and tea provide the main sources of caffeine intake (EFSA 4102, 2015a). Brazil has implemented an active coffee school program based on findings that 20% coffee added to a glass of whole milk helps children perform better in school (more information can be found from the Brazilian Coffee Industry Association, ABIC). Additionally, there are studies that show that caffeine may attenuate the symptoms of attention deficit syndrome (Garfinkel et al., 1981). European adolescents consume little coffee and their source of caffeine intake is widely distributed among different types of food and beverages (EFSA 4102, 2015a). In the United States this population primarily consumes caffeine from soft drinks (US Department of Agriculture, 2015).

Unfortunately, information on the impact of caffeine on children's and adolescent's health is scarce, and it is therefore difficult to derive general conclusions on safe intakes. Caffeine doses of about 1.4 mg/kg of body weight or more may impact sleep quality in adults, particularly when consumed close to bedtime (EFSA 4102, 2015a). For this reason, and because data on safe habitual caffeine intake for children and adolescents is insufficient, the EFSA suggests the rather conservative level of 3 mg of caffeine/kg of bodyweight per day (EFSA 4102, 2015a). This would equal around 90 mg for a 10 year old.

Canadian authorities are even more conservative and suggest a limit of 2.5 mg/kg of bodyweight per day (Health Canada, 2012). The short-term risk associated with children and caffeine consumption is that caffeine may cause anxiety and nervousness (Nawrot et al., 2003).

3.5 Caffeine Tolerance, Dependence, and Withdrawal

Caffeine is the most widely used psychoactive substance in the world, and experts have debated the issue of possible dependence on caffeine for many years. In fact, different drugs affect different people in different ways, and caffeine is no exception. It is therefore difficult to make general statements on dependence, tolerance, and withdrawal.

The scientific community is, however, in agreement that there is no brain circuit that links caffeine to dependence. Caffeine does not affect areas involved in reinforcing and reward (Nehlig, 2010). According to the standard for measuring any potential drug abuse and dependence (as defined by the *Diagnostic and Statistical Manual of Mental Disorders* IV) (American Psychiatric Association, 2000), there are no criteria that qualify caffeine for potential drug abuse.

It is also important to recognize that habitual users of a variety of drugs tend to develop a tolerance to the effects of that particular substance. This tolerance, in turn, may put users at risk for abuse since they often will increase their dosage in order to achieve the desired effects. As with any drug, regular caffeine users will establish a partial tolerance to caffeine. However, studies have shown that this tolerance only applies to the negative effects of caffeine, such as jitteriness, anxiety, and an increased heart rate. Users do not develop a tolerance to the benefits of caffeine consumption. This means that habitual caffeine consumers still enjoy more positive mental performance, although sometimes slightly higher doses of caffeine are needed (Satel, 2006).

The types of caffeine withdrawal symptoms most often reported are headaches; feelings of weariness, weakness, and drowsiness; impaired concentration; fatigue and work difficulty; depression; anxiety; irritability; increased muscle tension; and occasionally tremors, nausea, or vomiting. Withdrawal symptoms generally peak 20–48 h after the last caffeine consumption, although users can generally avoid these if caffeine consumption is progressively decreased (Nehlig, 2010).

Excessive coffee intake does not cause any significant organic toxicity, but it can generate negative side effects, such as those associated with caffeine withdrawal. At the same time, symptoms related to the toxicity of coffee can occur at levels well below fatal doses. For example, concentrations above 15 mg caffeine/kg of bodyweight may be toxic for the cardiovascular, nervous, and gastrointestinal systems (e.g., ca 1 g of caffeine for a person weighing 70 kg). Although such caffeine levels are not so easily obtained through acute coffee intake, users may easily consume caffeine pills in such quantities. A caffeine overdose may lead to hypertension or hypotension, tachycardia, vomiting, fever, delusion, hallucinations, arrhythmia, cardiac arrest, coma, and death. Fatalities most commonly results from seizures and cardiac arrhythmias at plasma levels of 100–180 μg/mL (Childs and de Wit, 2012). However, virtually all caffeine-related deaths have been associated with an overdose of caffeine tablets, and often associated with the intake of other toxic drugs (Yamamoto et al, 2015). An overdosage of coffee is thus generally considered proportionate to an overdosage of caffeine, since other components generally do not have an acute toxicity (Ludwig et al., 2014). LD 50 is the lethal dose for

50% of subjects: 167–179 mg caffeine/kg bodyweight for mice and 110 mg/kg for rats (Ludwig et al., 2014).

4. COFFEE AND HEALTH

4.1 Cognitive Health

In the mental performance section the acute effects of caffeine were discussed. In this section we will look at the long-term effects of caffeine on the prevention of cognitive degenerative diseases. Cognitive functions such as verbal ability, inductive reasoning, and perceptual speed decrease after 20 years of age. Genetics, life events, and lifestyle factors impact the rate and amplitude of this decline (Hedden and Gabrieli, 2004). A large number of epidemiological studies relate the regular consumption of coffee to a reduced appearance of cognitive decline in the elderly (Arab et al., 2013; Ritchie et al., 2007; Corley et al., 2010). Meta-analysis of these human studies suggests that there is a clear protective effect from caffeine consumption, rather than from coffee itself (Santos et al., 2010; Ryan et al., 2002). However, there is some emerging evidence from animal models that link chlorogenic acids to prevent against neurodegenerative disease and aging (Esposito et al., 2002; Ramassamy, 2006).

Alzheimer's disease is the most frequent cause of dementia, leading to a progressive cognitive decline. Although there is currently no medication against Alzheimer's disease (Waite, 2015), there are studies that show an inverse association between coffee consumption and the development of Alzheimer's disease, with a 27% risk reduction (Waite, 2015).

The mechanism is believed to be related to the antiinflammatory effect of caffeine on the A1 and A2 receptors as well as to the reducing deposits of toxic beta amyloid peptide in the brain, a pathological characteristic in patients with Alzheimer's disease (Rosso et al., 2008; Arendash and Cao, 2010).

In addition to caffeine, polyphenols seems to also play a preventive role against Alzheimer's disease. Although the involvement of coffee polyphenols in human cognitive function has not been well studied, the number of findings on the in vitro neuroprotective effects of polyphenols in general is rapidly increasing (Lakey-Beitia et al., 2015). The mechanisms are not well understood but it seems likely that polyphenols antiinflammatory effects play a role in preventing the development of Alzheimer's disease. Other proposed mechanisms could be (1) the inhibition of the enzymes acetylcholinesterase and butyrylcholinesterase in the brain since such inhibition retards acetylcholine and butyrylcholine breakdown and (2) the prevention of oxidative stress-induced neurodegeneration due to its high antioxidative activity (Oboh et al., 2013).

Similar to Alzheimer's disease, a large number of epidemiological studies have reported an inverse relationship between caffeine consumption and the risk of developing Parkinson's disease. The latter is a neuropathological disorder that slows down motor function, while generating resting tremors, muscular rigidity, and gait disturbances and impairing postural reflex. It involves the degeneration of neurons in the brainstem (Kuwana et al., 1999).

Coffee consumption appears to reduce, or delay, the development of Parkinson's disease. From meta-analysis of 26 studies, a 25% lower risk of Parkinson's disease was found in coffee drinkers as compared to noncoffee drinkers. The mechanism is probably related to the capacity of caffeine to block the A_2 adenosine receptors in the brain (Costa et al., 2010).

Studies recently outlined another possible additional mechanism. A rodent model showed that trigonelline may exert a neuroprotective effect inducing a significant reversal of motor dysfunction (Nathan et al., 2014).

4.2 Type-2 Diabetes

Diabetes mellitus is characterized by a high blood glucose level. This can cause complications such as cardiovascular diseases, stroke, chronic kidney failure, foot ulcers, and damage to the eyes (IDF, 2015).

There are three main types of diabetes: Type 1, in which the pancreas fails to produce enough insulin; Type 2, which begins with insulin resistance (lack of insulin may also develop), and is promoted by obesity and a sedentary lifestyle (Coope et al., 2015); and gestational diabetes, an often transient disease that occurs when pregnant women develop a high blood-sugar level (IDF, 2015).

A recent meta-analysis of large epidemiological studies have shown a link between moderate coffee consumption and a reduced risk of developing type 2 diabetes across different populations (Ding et al., 2014). The findings from these systematic studies demonstrate a clear inverse association between coffee consumption and risk of developing diabetes. Compared with no or infrequent coffee consumption, the risk of developing type 2 diabetes was reduced linearly with 33% for six cups of coffee consumed per day. In a similar comparison, drinking up to four cups per day of decaffeinated coffee was associated with a 20% reduced risk (Ding et al., 2014). This illustrates that the protective effects of coffee on diabetes are caused mainly by other components than caffeine.

In vivo and in vitro studies have suggested several plausible pathways. Animal studies have indicated that the main compounds responsible for the effect are the chlorogenic acids (Kempf et al., 2010) and its derivatives as well as trigonelline (van Dijk et al., 2009; Rios et al., 2015). They appear to target preferentially hepatic glucose metabolism by improving insulin sensitivity (Lecoultre et al., 2014). Other proposed mechanisms are regulation of key enzymes of glucose and lipid metabolism, such as glucokinase, glucose-6-phosphatase, fatty acid synthase, and carnitine palmitoyl transferase (Waite, 2015). In a human study trigonelline generated significantly lower glucose and insulin levels after an oral glucose load compared to a placebo (Rios et al., 2015).

4.3 Cholesterol

Another health aspect to consider is the potential impact of coffee on cholesterol levels, which in turn can lead to various cardiovascular diseases. The diterpene compounds cafestol, and, to a lesser extent, kahweol, naturally found in coffee oil and in unfiltered coffees, can alter lipid enzymes and thus influence cholesterol levels. This relation was found to be linear with each additional 10 mg of cafestol increased serum total cholesterol levels on average by 0.15 mmol/L up to a dose of 100 mg cafestol (Urgert and Katan, 1997). A meta-analysis was performed by Jee et al. (2001) of a set of 18 clinical intervention trials on coffee consumption and cholesterol and serum lipids. The authors corroborated the dose—response relationship between coffee consumption and cholesterol. Additionally, strong increase at a consumption of six or more cups of boiled coffee per day was observed. On the other hand, studies with paper-filtered coffee demonstrated very little increase in serum cholesterol.

An important factor that has to be considered is that diterpene levels in cup vary significantly based on preparation methods (Urgert et al., 1995; Gross et al., 1997), but also on blend due to the presence of Robusta which is practically cafestol free (Urgert and Katan, 1997), and as all natural products there are variations between regions, varieties, and years. Although filtered and soluble coffee are practically diterpene-free (ranging around 0—1 mg cafestol per cup), espresso-based methods contain higher levels of diterpenes (ranging around 1—2 mg cafestol per cup). These levels, on the other hand, are significantly lower than those found in French press or Turkish coffee (2—10 mg per cup).

4.4 Cancer

In its broadest sense, cancer represents the final result of abnormal cell growth and can occur in most human tissues. In the past coffee was found to be associated with an increased cancer risk; however, based on recent epidemiological data, the scientific community agrees that coffee consumption actually is associated with a *lower* overall risk of cancer, especially liver and colorectal cancer (for liver cancer see sidebar on liver health). In the past, a slightly increased risk of bladder cancer at the highest level of coffee intake (more than four cups/day) has been reported (Villanueva et al., 2009; Nkondjock, 2012). Subsequently, other studies failed to show an elevated risk of bladder cancer when the studies were also controlled for smoking (Butt and Sultan, 2011).

There are several compounds in coffee that have been found to play a protective role against cancer. The most well-known are the chlorogenic acids and their derivatives. They are frequently referenced as powerful antioxidative compounds, considering the results of in vitro and animal studies, as well as a few human studies. However, as with all polyphenols, most studies have so far failed to show a significant antioxidative activity in human plasma after coffee intake (Stalmach et al., 2009), which could simply be a matter of analytical limitations and or fast tissue uptake. At the same time, these compounds have been found to act as chemopreventive agents by modulating the expression of the gene-encoding enzymes that are involved in endogenous antioxidant defenses (Feng et al., 2005; Ramos, 2008). Chlorogenic acids may also exert anticarcinogenic activity via other mechanisms, including by inhibiting enzymes involved in DNA replication, cell differentiation, and aging (Jurkowska et al., 2011).

In the colon, chlorogenic acids may inactivate free radicals and help to prevent colon cancer. They may also interfere with the progression of cancer but the mechanisms still needs to be further studied (Ludwig et al., 2014). Melanoidins from coffee behave in vivo as dietary fiber and are largely indigestible and thus fermented in the gut (Borrelli et al., 2004; Gniechwitz et al., 2008). As with chlorogenic acids (Passos et al., 2014) they can enhance immune-stimulating properties and contribute significantly to colon cancer risk reduction (Vitaglione et al., 2012; Moreira et al., 2015; Fogliano and Morales, 2011).

It has been suggested that the contribution of melanoidins to the prevention of colon cancer might happen in three different ways: (1) by increasing the elimination rate of carcinogens through higher colon motility and fecal output; (2) by decreasing colon inflammation through improved microbiota balance (prebiotic effect); and (3) by serving as a "sponge" for free radicals (Garsetti et al., 2000).

Despite all these benefits, some people still consider coffee carcinogenic because it contains various carcinogenic compounds such as acrylamide and polycyclic aromatic hydrocarbons. However, epidemiologic studies have failed to find a link between either of these compounds in coffee and an elevated risk of cancer (Lipworth et al., 2012; Nkondjock, 2012). Acrylamide is a compound in roasted coffee that was associated with cancer in one study of laboratory rodents in which they were exposed to extraordinarily high concentrations (1000−10,000 times physiologic ranges) (Mucci and Adami, 2009). Although the US Food and Drug Administration (2016) announced that coffee is a significant source of acrylamide exposure for adults, EFSA's recent recommendation on acrylamide stated that "for most cancers there is no consistent indication for an association between acrylamide exposure and increased risk" (EFSA 4104, 2015b). The International Agency for Research on Cancer (IARC) classifies acrylamides as a "probable carcinogen" (see www.iarc.fr). Polycyclic aromatic hydrocarbons (PAH), and more specifically benzo[α]pyrene, can be formed in coffee and other foods that are severely roasted or exposed to very high temperatures. This compound is classified by the IARC as "carcinogenic to humans" (Loomis et al., 2016). However, the level of exposure to PAH from coffee is very low. Very important is the change communicated in May 2016 by the IARC where the classification of "overall coffee" moved from "possibly carcinogenic" to "unclassifiable as to its carcinogenicity to humans". This change in classification is based on the increased amount of scientific data, more specifically over 1000 human and animal studies were considered (Loomis et al., 2016).

Epidemiologic studies have shown an increased risk of esophageal cancer from drinking hot beverages such as tea or coffee. The high temperature injures to the esophageal mucosa and consequently causes inflammation or forms reactive nitrogen species, a type of free radical. It has been reported that drinking coffee at 65°C increased the intraesophageal temperature by 6–12°C (Islami et al., 2009).

It is noteworthy to mention that the antiinflammatory and anticarcinogenic activities of coffee compounds could possibly help to counterbalance the risk of developing esophageal cancer. However, further research would be needed to understand the influence of the different mechanisms and to what extent drinking hot coffee is a factor of risk.

4.5 Liver Health

There are a number of diseases that can impact liver health. This includes both liver cancer and cirrhosis, a progressive disease caused by liver steatosis (fatty liver) and alcohol abuse, where healthy tissue is replaced by scar tissue and eventually prevents the liver from functioning correctly (Saab et al., 2014). According to recent meta-analysis of 16 human studies, coffee consumption reduces the risk of developing liver cancer by 40% as compared to no coffee consumption (Larsson and Wolk, 2007; Bravi et al., 2013).

A number of in vitro studies have demonstrated the strong role of coffee chlorogenic acids in protecting liver from damage at various levels, possibly by preventing cell apoptosis and oxidative stress damage due to activation of natural antioxidant and antiinflammatory body systems (Ji et al., 2013). Coffee melanoidins have also been reported to have a protective effect on liver steatosis in obese rats (Vitaglione et al., 2012). This suggests that the melanoidins in coffee may have an influence on liver fat and functionality. Although there are no reports of the absorption of melanoidins in humans, they can function as antioxidant dietary fiber, which quenches radicals and improves the reduced/oxidized glutathione balance in the colon. At the same time, it may act as a prebiotic on colon microbiota, improving the inflammatory pathways in the colon and consequently in the liver.

4.6 Longevity

Early epidemiological studies failed to find associations between coffee consumption and decreased risk in specific and total cause of mortality. In fact, the first studies of this kind tended to conclude that drinking coffee was detrimental do health (LeGrady et al., 1987; Lindsted et al., 1992; Klatsky et al., 1993). The main reason for this was that negative lifestyle factors such as smoking could not be dissociated from heavy coffee drinking (Ding et al., 2015).

One of the most consistent and reliable databases allowing a more objective perspective on the impact of coffee consumption on health is the Harvard Health Professional's Follow-up Study and associated Nurses' Health Study

(Lopez-Garcia et al., 2008). From these databases that cross statistical results on lifestyle factors and health data of more than 130,000 people, it was concluded that regular coffee consumption is not associated with an increased mortality rate in either men or women.

Another prospective US cohort study (Freedman et al., 2012) examined the association of coffee drinking with subsequent cause specific and total mortality in the National Institutes of Health—AARP (American Association of Retired Persons) Diet and Health Study. This study involved more than 400,000 people and is so far the largest human study investigating coffee and health. A significant inverse association between coffee and specific deaths due to heart disease, respiratory disease, stroke, injuries and accidents, diabetes, and infections was found. Total mortality was reduced increasingly up to 16% for both men and women who drank four to five cups of coffee a day. Similar associations were observed whether participants drank predominantly caffeinated or decaffeinated coffee.

Comparable results were found in a more recent study by Ding et al. (2015), based on the Harvard Health Professionals Follow-up Study and Nurses' Health Study (1 and 2). Inverse associations were observed between consumption of regular and decaffeinated coffee and deaths due to cardiovascular and neurological diseases, and suicide. When restricting to never smokers the all-cause mortality risk was increasingly reduced up to 15% for three to five cups. Higher consumption reduced the benefit somewhat.

Various additional studies have found similar protective results (Sugiyama et al., 2010; Gardener et al., 2013; Malerba et al., 2013; Crippa et al., 2014) and studies in vitro and in vivo have been offering support to explain mechanisms behind such epidemiological findings. Although most studies show positive impact of coffee on the prevention of diseases, there are other positive lifestyle factors such as good nutrition, exercise, and a low stress that have an even stronger impact on disease prevention and life expectation.

5. OUTLOOK

Coffee has, since its discovery, gone through many roles. Probably the most important role throughout history and still today, is the social role it plays. Coffee has since long been about connection and in today's society where work and performance are drivers in a globalizing world, drinking a cup of coffee remains a small break in the day to connect with friends, colleagues, or family. It is a drink that enhances sociability and cohesion (Fig. 20.3).

Further to these social needs, coffee remains a great tasting source of caffeine, helping people through the day. Its benefits for mental and physical performance are now scientifically well understood and appreciated by consumers. Information on safe levels of caffeine intake is available, ensuring people can appreciate the effects in a positive manner.

FIGURE 20.3 For many people a coffee means a small break in the day to connect with friends, colleagues or family.

Whereas the immediate effects of coffee are well known to consumers, the more long term benefits of coffee on general health and well-being are less obvious. An occasional report in a newspaper gives consumers some awareness of health aspects beyond stimulation, however, the newspapers have a tendency to speak of single studies, whereas more general meta-analyses generally go beyond the scope of such stories. For consumers, a general perspective of coffee and health is not easy to obtain. Science today has, however, created a very interesting dossier on coffee on health and is clearly pointing to a positive direction where coffee has been found to reduce the risk of several diseases (see Pourshahidi et al., 2016 for a recent comprehensive overview). Even if functional relationships may have been established between coffee and health through risk analysis of epidemiological data, the active compounds and their mechanisms leading to benefit are far from elucidated. Therefore, claims that coffee has specific disease preventing benefits still need further substantiation.

However, looking at ongoing scientific work in the field of coffee and health, we see that again new links are being created between the *Coffea* and *Homo sapiens* species. Chemistry is continuing to unveil the complex and dynamic components of roasted coffee, while systems biology is unveiling the secrets of our genes and cells. Together, these will help us to further understand how coffee components have a physiological impact on human bodies. The question is how this knowledge will be used in the future. As mentioned, coffee is consumed socially and for its stimulating effects. But is there a possibility that coffee will gain importance as a healthful beverage, to find its way into dietary recommendations as a functional food? In this respect, we can envisage designing functional coffee by-products that play a role in personalized nutrition. These by-products could harness the positive lifestyle habits associated with coffee consumption while integrating the pleasure of enjoying a cup of coffee with its intrinsic health benefits.

But our thoughts about the future of coffee as a well-being product do not end there. From the initial coffee cherry, only a small fraction of roasted coffee material ends up in the consumer's cup. On the other hand, by-products including pulp, parchment, silverskin, and spent grounds, are abundant. As the extracted cup has been found to bring such goodness to the consumer, the question remains open to what extent these by-products could help create new value by using them as economically viable resources for drug and food industries.

Our final thoughts go back to the pleasure of the coffee that will remain the main reason for consuming it. At the same time, we need to consider that in today's society, many coffee consumers enjoy their beverages with milk, cream, sugar, and syrups—all of which increase overall caloric intake. This fact in today's society where obesity-related health issues are central, it should be a given that the specialty coffee industry even more actively promotes coffee's myriad of delicious flavors. Specialty coffee is optimally consumed without any additions.

To conclude we would like to highlight a very interesting piece of research based on a hypothesis tested by Williams and Bargh (2008). They showed that the warmth felt from a cup of coffee increases feelings of interpersonal warmth, without the person's awareness of this influence. This hypothesis was based on the fact that warmth is a descriptor often used to describe a person's character but also based on research (on monkeys) that warmth in infancy is needed for normal social development as adults. They could indeed show that holding a cup of hot coffee made participants judge a target person as having a warmer, generous, and caring personality. Possibly this research may touch the foundation for the fact that coffee since its discovery has played a role in social bonding in society.

REFERENCES

American Psychiatric Association, 2000. Diagnostic and Statistical Manual of Mental Disorders, fourth ed. American Psychiatric Association, Arlington, VA.

Arab, L., Khan, F., Lam, H., 2013. Epidemiologic evidence of a relationship between tea, coffee, or caffeine consumption and cognitive decline. Advances in Nutrition 4 (1), 115–122.

Arendash, G.W., Cao, C., 2010. Caffeine and coffee as therapeutics against Alzheimer's disease. Journal of Alzheimer's Disease 20 (S1), 117–126.

Borota, D., Murray, E., Kecell, G., Chang, A., Watabe, J.M., Ly, M., Toscano, J.P., Yassa, M.A., 2014. Post-study caffeine administration enhances memory consolidation in humans. Nature Neuroscience 17, 201–203.

Borrelli, R.C., Esposito, F., Napolitano, A., Ritieni, A., Fogliano, V., 2004. Characterization of a new potential functional ingredient: coffee silverskin. Journal of Agriculture and Food Chemistry 52, 1338–1343.

Bravi, F., Bosetti, C., Tavoni, A., Gallus, S., La Vecchia, C., 2013. Coffee reduces risk for hepatocellular carcinoma: an updated meta-analysis. Clinical Gastroenterology and Hepatology 11, 1413–1421.

Burke, T.M., Markwald, R.R., McHill, A.W., Chinov, E.D., Snider, J.A., Bessman, S.C., Jung, C.M., O'Neill, J.S., Wright Jr., K.P., 2015. Effects of caffeine on the human circadian clock in vivo and in vitro. Science Translational Medicine 7, 146–153.

Butt, M.S., Sultan, M.T., 2011. Coffee and it's consumption: benefits and risks. Critical Reviews in Food Science and Nutrition 51, 363–373.

Camfield, D.A., Silber, B.Y., Scholey, A.B., Nolidin, K., Goh, A., Stough, C., 2013. A randomised placebo-controlled trial to differentiate the acute cognitive and mood effects of chlorogenic acid from decaffeinated coffee. Public Library of Science One 8 (12), e82897.

Chang, W.H., Hu, S.P., Huang, Y.F., Yeh, T.S., Liu, J.F., 2010. Effect of purple sweet potato leaves consumption on exercise-induced oxidative stress and IL-6 and HSP72 levels. Journal of Applied Physiology 109 (6), 1710–1715.

Childs, E., de Wit, H., 2012. Potential mental risks. In: Chu, Y.-F. (Ed.), Coffee, Emerging Health Effects and Disease Prevention. Wiley-Blackwell, Oxford, UK, pp. 293–306.

Clark, I., Landholt, H.P., 2016. Coffee, caffeine and sleep: a systematic review of epidemiological studies and randomized controlled trials. Sleep Medicine Reviews 1–9 (in press).

Coope, A., Torsoni, A.S., Velloso, L., 2015. Mechanisms in Endocrinology: metabolic and inflammatory pathways on the pathogenesis of type 2 diabetes. European Journal of Endocrinology 15, 1065.

Corley, J., Jia, X., Kyle, J.A., Gow, A.J., Brett, C.E., Starr, J.M., McNeill, G., Deary, I.J., 2010. Caffeine consumption and cognitive function at age 70: the Lothian Birth Cohort 1936 study. Psychosomatic Medicine 72, 206–214.

Costa, J., Lunet, N., Santos, C., Santos, J., Vaz-Cameiro, A., 2010. Caffeine exposure and the risk of Parkinson's disease: a systemic review and meta-analysis of observational studies. Journal of Alzheimer's Disease 20 (S1), 221–238.

Costill, D.L., Dalsky, G.P., Fink, W.J., 1977. Effects of caffeine ingestion on metabolism and exercise performance. Medicine and Science in Sports 10 (3), 155–158.

Crippa, A., Discacciati, A., Larsson, S.C., Wolk, A., Orsini, N., 2014. Coffee consumption and mortality from all causes, cardiovascular disease, and cancer: a dose-response meta-analysis. American Journal of Epidemiology 180, 763–775.

Cropley, V., Croft, R., Silber, B., Neale, C., Scholey, A., Stough, C., Schmitt, J., 2012. Does coffee enriched with chlorogenic acids improve mood and cognition after acute administration in healthy elderly? A pilot study. Psychopharmacology 219 (3), 737–749.

Ding, M., Bhupathiraju, S.N., Chen, M., van Dam, R.M., Hu, F.B., 2014. Caffeinated and decaffeinated coffee consumption and risk of type 2 diabetes: a systematic review and a dose-response meta-analysis. Diabetes Care 37 (2), 569–586.

Ding, M., Satija, A., Bhupathiraju, S.N., Hu, Y., Sun, Q., Han, J., Lopez-Garcia, E., Willet, W., van Dam, R.M., Hu, F.A., 2015. Association of coffee consumption with total and cause-specific mortality in three large prospective cohorts. Circulation 132 (24), 2305–2315.

Drapeau, C., Hamel-Hébert, I., Robillard, R., Selmaoui, B., Filipini, D., Carrier, J., 2006. Challenging sleep in aging: the effects of 200 mg of caffeine during the evening in young and middle-aged moderate caffeine consumers. Journal of Sleep Research 15 (2), 133–141.

Einöther, S.J.L., Giesbrecht, T., 2013. Caffeine as an attention enhancer: reviewing existing assumptions. Psychopharmacology 225 (2), 251–274.

Encyclopedia Britannica, eleventh ed. Cambridge, England, Copyright 1910, pp. 62–63.

Esposito, E., Rotilio, D., Di Matteo, V., Di Giulio, C., Cacchio, M., Algeri, S., 2002. A review of specific dietary antioxidants and the effects on biochemical mechanisms related to neurodegenerative processes. Neurobiology of Aging 23, 719–735.

European Food Safety Authority, 2011a. EFSA Panel on Dietetic Products, Nutrition and Allergies (NDA), Scientific Opinion on the substantiation of health claims related to caffeine and increase in physical performance during short-term high-intensity exercise (ID 737, 1486, 1489), increase in endurance performance (ID 737, 1486), increase in endurance capacity (ID 1488) and reduction in the rated perceived exertion/effort during exercise (ID 1488, 1490) pursuant to Article 13(1) of Regulation (EC) No 1924/20061. EFSA Journal 9 (4), 2053.

European Food Safety Authority, 2011b. EFSA Panel on Dietetic Products, Nutrition and Allergies (NDA), Scientific Opinion on the substantiation of health claims related to caffeine and increased fat oxidation leading to a reduction in body fat mass (ID 735, 1484), increased energy expenditure leading to a reduction in body weight (ID 1487), increased alertness (ID 736, 1101, 1187, 1485, 1491, 2063, 2103) and increased attention (ID 736, 1485, 1491, 2375) pursuant to Article 13(1) of Regulation (EC) No 1924/20061. EFSA Journal 9 (4), 2054.

European Food Safety Authority, 2015a. EFSA Panel on Dietetic Products, Nutrition and Allergies (NDA), Scientific Opinion on the safety of caffeine. EFSA Journal 13 (5), 4102.

European Food Safety Authority, 2015b. EFSA Panel on contaminants in the food chain (CONTAM) Scientific Opinion on acrylamides in food. EFSA Journal 13 (6), 4104.

Farah, A., 2012. Coffee constituents. In: Chu, Y.-F. (Ed.), Coffee: Emerging Health Effects and Disease Prevention. IFT Press/Willey-Blackwell, USA, pp. 21–58.

Feng, R., Lu, Y., Bowman, L.L., Qian, Y., Castranova, V., Ding, M., 2005. Inhibition of activator protein-1, NF-κB, and MAPKs and induction of phase 2 detoxifying enzyme activity by chlorogenic acid. Journal of Biological Chemistry 280, 27888–27895.

Fogliano, V., Morales, F.J., 2011. Estimation of dietary intake of melanoidins from coffee and bread. Food and Function 2, 117–123.

Fredholm, B.B., Battig, K., Holmen, J., Nehlig, A., Zvartau, E.E., 1999. Actions of caffeine in the brain with special reference to factors that contribute to its widespread use. Pharmacological Reviews 51 (1), 83–133.

Freedman, N., Park, Y., Abnet, C.C., Hollenbeck, A.R., Sinha, R., 2012. Association of coffee drinking with total and cause-specific mortality. New England Journal of Medicine 366 (20), 1891–1904.

Gardener, H., Rundek, T., Wright, C.B., Elkind, M.S., Sacco, R.L., 2013. Coffee and tea consumption are inversely associated with mortality in a multiethnic urban population. Journal of Nutrition 143, 1299–1308.

Garfinkel, B.D., Webster, C.D., Sloman, L., 1981. Responses to methylphenidate and various does of caffeine in children with attention deficit disorder. The Canadian Journal of Psychiatry/La Revue canadienne de psychiatrie 26 (6), 395–401.

Garsetti, M., Pellegrini, N., Baggio, C., Brighenti, F., 2000. Antioxidant activity in human faeces. British Journal of Nutrition 84 (5), 705–710.

Gerson, M., 1978. The cure of advanced cancer by diet therapy: a summary for 30 years of clinical experimentation. Physiological Chemistry and Physics 10, 449–464.

Gliottoni, R.C., Motl, R.W., 2008. Effect of caffeine on leg-muscle pain during intense cycling exercise: possible role of anxiety sensitivity. International Journal of Sports Nutrition and Exercise Metabolism 18 (2), 103–115.

Gliottoni, R.C., Meyers, J.R., Arrigrimsson, S.A., Boglio, S.P., Motl, R.W., 2009. Effect of caffeine on quadriceps muscle pain during acute cycling exercise in low versus high caffeine consumers. International Journal of Sports Nutrition and Exercise Metabolism 19 (2), 150–161.

Gniechwitz, D., Reichardt, N., Ralph, J., Blaut, M., Steinhart, H., Bunzel, M., 2008. Isolation and characterisation of a coffee melanoidin fraction. Journal of the Science of Food Agriculture 88 (12), 2153–2160.

Goldstein, E.R., Ziegenfuss, T., Kalman, D., Kreider, R., Campbell, B., Wilborn, C., Taylor, L., Willoughby, D., Stout, J., Graves, B.S., Wildman, R., Ivy, J.L., Spano, M., Smith, S.E., Antonio, J., 2010. International society of sports nutrition position stand: caffeine and performance. Journal of the International Society of Sports Nutrition 7, 5.

Greer, F., McLean, C., Graham, T.E., 1998. Caffeine, performance, and metabolism during repeated Wingate exercise tests. Journal of Applied Physiology 85 (4), 1502–1508.

Gross, G., Jaccaud, E., Hugget, A.C., 1997. Analysis of the content of the diterpenes cafestol and kahweol in coffee brews. Food and Chemical Toxicology 35, 547–554.

Harland, B., 2000. Caffeine and nutrition. Nutrition 16, 522–526.

Health Canada, 2012. Information for Parents on Caffeine in Energy Drinks. www.hc-sc.gc.ca/fn-an/securit/addit/caf/faq-eng.php.

Hedden, T., Gabrieli, J.D.E., 2004. Insights into the ageing mind: a view from cognitive neuroscience. Nature Reviews Neuroscience 5, 87–97.

Higdon, J.V., Frei, B., 2006. Coffee and health: a review of recent human research. Critical Reviews in Food Science and Nutrition 46, 101–123.

Hirakawa, N., Okauchi, R., Miura, Y., Yagasaki, K., 2005. Anti-invasive activity of niacin and trigonelline against cancer cells. Bioscience, Biotechnology, and Biochemistry 69 (3), 653–658.

Hogervost, E., Bandelow, S., Schmitt, J., Jentjens, R., Oliveira, M., Allgrove, J., Carter, T., Gleeson, M., 2008. Caffeine improves physical and cognitive performance during exhaustive exercise. Medicine and Science in Sports and Exercise 40, 1841–1851.

Hong, B.N., Yi, T.H., Park, R., Kim, S.Y., Kang, T.H., 2008. Coffee improves auditory neuropathy in diabetic mice. Neuroscience Letters 441 (3), 302–306.

Huang, Z., Urade, Y., Hayaishi, O., 2011. The role of adenosine in the regulation of sleep. Current Topics in Medicinal Chemistry 11, 1047–1057.

International Coffee Organization, 2014. World Coffee Trade (1923–2013): A Review of the Markets, Challenges and Opportunities Facing the Sector, 112th Session. http://www.ico.org/news/icc-111-5-r1e-world-coffee-outlook.pdf.

International Diabetes Federation, 2015. IDF Diabetes Atlas, fifth ed. Brussels, Belgium Available from: http://www.coffeeandhealth.org/abstract/international-diabetes-federation-idf-2009-diabetes-atlas-4th-edition/.

Islami, F., Boffetta, P., Ren, J.S., Pedoeim, L., Khatib, D., Kamangar, F., 2009. High-temperature beverages and foods and esophageal cancer risk – a systematic review. International Journal of Cancer 125 (3), 491–524.

Jackmann, M., Wendling, P., Friors, D., Graham, T.E., 1996. Metabolic, catecholamine, and endurance responses to caffeine during intense exercise. Journal of Applied Physiology 81, 1658–1683.

Jee, S.H., He, J., Appel, L.J., Whelton, P.K., Suh, I., Klag, M.J., 2001. Coffee consumption and serum lipids: a meta-analysis of randomized controlled clinical trials. American Journal of Epidemiology 153, 353–362.

Jenkins, N.T., Trilk, J.L., Singhal, A., O'Connor, P.J., Cureton, K.J., 2008. Ergogenic effects of low doses of caffeine on cycling performance. International Journal of Sport Nutrition and Exercise Metabolism 18 (3), 328–342.

Ji, L., Jiang, P., Lu, B., Sheng, Y., Wang, X., Wang, Z., 2013. Chlorogenic acid, a dietary polyphenol, protects acetaminophen-induced liver injury and its mechanism. Journal of Nutrition and Biochemistry 24 (11), 1911–1919.

Jurkowska, R.Z., Jurkowski, T.P., Jeltsch, A., 2011. Structure and function of mammalian DNA methyltransferases. *Chembiochem* 12, 206–222.

Kamimori, G.H., Brunhart, A.E., Eddington, N.D., Lugo, S., Hoyt, R.W., Fulco, C.S., Durkot, M.J., Cymerman, A., 1995. Effects of altitude (4300 M) on the pharmacokinetics of caffeine and cardio-green in humans. European Journal of Clinical Pharmocology 48, 167–170.

Kempf, K., Herder, C., Erlund, I., Kolb, H., Martin, S., Carstensen, M., Koenig, W., Sandwall, J., Bidel, S., Kuha, S., Tuomilehto, J., 2010. Effects of coffee consumption on subclinical inflammation and other risk factors for type 2 diabetes: a clinical trial. American Journal of Clinical Nutrition 91, 950–957.

Klatsky, A.L., Armstrong, M.A., Friedman, G.D., 1993. Coffee, tea, and mortality. Annals of Epidemiology 3 (3), 375–381.

Kuczkowski, K.M., 2009. Caffeine in pregnancy. Archives of Gynecology and Obstetrics 280, 695–698.

Kuwana, Y., Shiozaki, S., Kanda, T., Kurokawa, M., Koga, K., Ochi, M., Ikeda, K., Kase, H., Jackson, M.J., Smith, L.A., Pearce, R.K., Jenner, P.G., 1999. Antiparkinsonian activity of adenosine A_{2A} antagonists in experimental models. Advances in Neurology 80, 121–123.

Lakey-Beitia, J., Berrocal, R., Rao, K.S., Durant, A.A., 2015. Polyphenols as therapeutic molecules in Alzheimer's disease through modulating amyloid pathways. Molecular Neurobiology 51 (2), 466–479.

Larsson, S.C., Wolk, A., 2007. Coffee consumption and risk of liver cancer: a meta-analysis. Gastroenterology 132, 1740–1745.

Lecoultre, V., Carrel, G., Egli, L., Binnert, C., Boss, A., MacMillan, E.L., Kreis, R., Boesch, C., Darimont, C., Tappy, L., 2014. Coffee consumption attenuates short-term fructose-induced liver insulin resistance in healthy men. American Journal of Clinical Nutrition 99 (2), 268–275.

LeGrady, D., Dyer, A.R., Shekelle, R.B., Stamler, J., Liu, K., Paul, O., Lepper, M., Shryock, A.M., 1987. Coffee consumption and mortality in the Chicago western electric company study. American Journal of Epidemiology 126, 803–812.

Lima, J.P., Farah, A., 2014. Contribution of stimulant foods for habitual daily intake of methylxantines in Rio de Janeiro. In: Proceeding of the International Conference on Coffee Science.

Lindsted, K.D., Kuzma, J.W., Anderson, J.L., 1992. Coffee consumption and cause-specific mortality. Association with age at death and compression of mortality. Journal of Clinical Epidemiology 45, 733–742.

Lipworth, L., Sonderman, J.S., Tarone, R.E., McLaughlin, J., 2012. Review of epidemiologic studies of dietary acrylamide intake and the risk of cancer. European Journal of Cancer Prevention 21, 375–386.

Loomis, D., Guyton, K.Z., Grosse, Y., Lauby-Secretan, B., El Ghissassi, F., Bouvard, V., Benbrahim-Tallaa, L., Guha, N., Mattock, H., Straif, K., 2016. Carcinogenicity of drinking coffee, mate, and very hot beverages. Lancet Oncology 17 (7), 877.

Lopez-Garcia, E., van Dam, R.M., Li, T.Y., Rodriguez-Artalejo, F., Hu, F.B., 2008. The relationship of coffee consumption with mortality. Annals of Internal Medicine 148 (12), 904–914.

Ludwig, I.A., Clifford, M.N., Lean, M.E.J., Ashihara, H., Crozier, A., 2014. Coffee: biochemistry and potential impact on health. Food & Function 5 (8), 1695–1717.

Malerba, S., Turati, F., Galeone, C., Pelucchi, C., Verga, F., La Vecchia, C., Tavani, A., 2013. A meta-analysis of prospective studies of coffee consumption and mortality for all causes, cancers and cardiovascular diseases. European Journal of Epidemiology 28 (7), 527–539.

Moreira, A., Coimbra, M., Nunes, F.M., Passos, C.P., Santos, S.A., Silvestre, A.J., Silva, A., Rangel, M., Domingues, M.R.M., 2015. Chlorogenic acid-arabinose hybrid domains in coffee melanoidins: evidences from a model system. Food Chemistry 185, 135–144.

Mucci, L.A., Adami, H.O., 2009. The plight of the potato: is dietary acrylamide a risk factor for human cancer? Journal of National Cancer Institute 101 (9), 618–621.

Nathan, J., Panjwani, S., Mohan, V., Joshi, V., Thakurdesai, P.A., 2014. Efficacy and safety of standardized extract of *Trigonella foenum-graecum* L seeds in an adjuvant to L-Dopa in the management of patients with Parkinson's disease. Phytotherapy Research 28 (2), 172–178.

Nawrot, P., Jordan, S., Eastwood, J., Rotstein, J., Hugenholtz, A., Feeley, M., 2003. Effects of caffeine on human health. Food Additives and Contaminants 20, 1–30.

Nehlig, A., Debry, G., 1994. Consequences on the newborn of chronic maternal consumption of coffee during gestation and lactation: a review. Journal of the American College of Nutrition 13, 6–21.

Nehlig, A., 2010. Is caffeine a cognitive enhancer? Journal of Alzheimer's Disease 20 (S1), 85–94.

Nkondjock, A., 2012. Coffee and cancers. In: Chu, Y.-F. (Ed.), Coffee: Emerging Health Effects and Disease Prevention. Wiley-Blackwell, Oxford, UK, pp. 293–306.

O'Connor, P.J., Motl, R.W., Broglio, S.P., Ely, M.R., 2004. Dose-dependent effect of caffeine on reducing leg muscle pain during cycling exercise is unrelated to systolic blood pressure. Pain 109, 291–298.

Oboh, G., Agunloye, O.M., Akinyemi, A.J., Ademiluyi, A.O., Adefegha, S.A., 2013. Comparative study on the inhibitory effect of caffeic and chlorogenic acids on key enzymes linked to Alzheimer's disease and some pro-oxidant induced oxidative stress in rats' brain-in vitro. Neurochemistry Research 38 (2), 413–419.

Olson, C.A., Thornton, J.A., Adam, G.E., Lieberman, H.R., 2010. Effects of 2 adenosine antagonists, quercetin and cafeïne, on vigilance and mood. Journal of Clinical Psychopharmacology 30 (5), 573–578.

Panza, V.S., Wazlawik, E., Schütz, G.R., Comin, L., Hecht, K.C., Luiz da Silva, E., 2008. Consumption of green tea favorably affects oxidative stress markers in weight-trained men. Nutrition 24 (5), 433–442.

Passos, C.P., Cepeda, M.R., Ferreira, S.S., Nunes, F.M., Evtuguin, D.V., Madureira, P., Vilanova, M., Coimbra, M.A., 2014. Influence of molecular weight on in vitro immunostimulatory properties of instant coffee. Food Chemistry 161, 60–66.

Pendergast, M., 1999. Uncommon Grounds: The History of Coffee and How It Transformed Our World, first ed. Basic Books, New York.

Pourshahidi, L.K., Navarini, L., Petracco, M., Strain, J.J., 2016. A Comprehensive overview of the risks and benefits of coffee consumption. Comprehensive Reviews in Food Science and Food Safety 15, 671–684.

Ramassamy, C., 2006. Emerging role of polyphenolic compounds in the treatment of eurodegenerative diseases: a review of their intracellular targets. European Journal of Pharmacology 545, 51–64.

Ramos, S., 2008. Cancer chemoprevention and chemotherapy: dietary polyphenols and signaling pathways. Molecular Nutrition and Food Research 52, 507–526.

Rios, J.L., Francini, F., Schinella, G.R., 2015. Natural products for the treatment of type 2 diabetes mellitus. Planta Medica 81 (12–13), 975–994.

Ritchie, K., Carriere, I., de Mendonca, A., Portet, F., Dartigues, J.F., Rouaud, O., Barberger-Gateau, P., Ancelin, M.L., 2007. The neuroprotective effects of caffeine. A prospective population study. Neurology 69, 536–545.

Rosso, A., Mossey, J., Lippa, C.F., 2008. Review: caffeine: neuroprotective functions in cognition and Alzheimer's disease. American Journal of Alzheimer's Disease and Other Dementias 23 (5), 417–422.

Ryan, L., Hatfield, C., Hofstetter, M., 2002. Caffeine reduces time-of-day effects on memory performance in older adults. Psychological Science 13 (1), 68–71.

Saab, S., Mallam, D., Cox, G.A., Tong, M.J., 2014. Impact of coffee on liver diseases: a systematic review. Liver International 34 (4), 495–504.

Santos, C., Costa, J., Santos, J., Vaz-Carneiro, A., Lunet, N., 2010. Caffeine intake and dementia: systematic review and meta-analysis. Journal of Alzheimer's Disease 20 (S1), 187–204.

Santos, I.S., Matijasevich, A., Domingues, M.R., 2012. Maternal caffeine consumption and infant nighttime waking: prospective cohort study. Pediatrics 129, 860–868.

Satel, S., 2006. Is caffeine addictive? A review of the literature. American Journal of Drug and Alcohol Abuse 32 (4), 493–502.

Sengpiel, V., Elind, E., Bacelis, J., Nilsson, S., Grove, J., Myhre, R., Haugen, M., Meltzer, H.M., Alexander, T., Jacobsson, B., Brantsaeter, A.L., 2013. Maternal caffeine intake during pregnancy is associated with birth weight but not with gestational length: results from a large prospective observational cohort study. BioMed Central Medicine 11, 42–60.

Smith, A., Sutherland, D., Christopher, G., 2005. Effects of repeated doses of caffeïne and mood and performance of alert and fatigued volunteers. Journal of Psychopharmacology 19 (6), 620–626.

Smith, R.F., 1987. A history of coffee. In: Clifford, M.N., Wilson, K.C. (Eds.), Coffee: Botany, Biochemistry and Production of Beans and Beverage. Croom Helm, Ney York, pp. 1–12.

Spriet, L.L., Gibala, M.J., 2004. Nutritional strategies to influence adaptations to training. Journal of Sports Sciences 22 (1), 127–141.

Spriet, L.L., 1995. Caffeine and performance. International Journal of Sports Nutrition 5, S84–S99.

Stalmach, A., Mullen, W., Barron, D., Uchida, K., Yokota, T., Cavin, C., Stelling, H., Williamson, G., Crozier, A., 2009. Metabolite profiling of hydroxycinnamate derivatives in plasma and urine after ingestion of coffee by humans: identification of biomarkers of coffee consumption. Drug Metabolism Disposition 37, 1749–1758.

Stavchansky, S., Combs, A., Sagraves, R., Delgado, M., Joshi, A., 1988. Pharmacokinetics of caffeine in breast milk and plasma after single oral administration of caffeine to lactating mothers. Biopharmaceutics and Drug Disposition 9, 285–299.

Sugiyama, K., Kuriyama, S., Akhter, M., Kakizaki, M., Nakaya, N., Ohmori-Matsuda, K., Shimazu, T., Nagai, M., Sugawara, Y., Hozawa, A., Fukao, A., Tsuji, I., 2010. Coffee consumption and mortality due to all causes, cardiovascular disease, and cancer in Japanese women. Journal of Nutrition 140, 1007–1013.

Teekachunhatean, S., Tosri, N., Rojanasthien, N., Srichairatanakool, S., Sangdee, C., 2013. Pharmacokinetics of caffeine following a single administration of coffee enema versus oral coffee consumption in healthy male subjects. Pharmacology 1–7, 147238.

Tohda, C., Kuboyama, T., Komatsu, K., 2005. Search for natural products related to regeneration of the neuronal network. Neurosignals 14 (1–2), 34–45.

Torres, T., Farah, A., March 2016. Coffee, maté, açaí and beans are the main contributors to the antioxidant capacity of Brazilian's diet. European Journal of Nutrition 14. http://dx.doi.org/10.1007/s00394-016-1198-9.

Ukers, W.H., 1935. All About Coffee, second ed. Tea and Coffee Trade Journal Co., New York.

Urgert, R., Katan, M.B., 1997. The cholesterol-raising factor from coffee beans. Annual Review of Nutrition 17, 305–324.

Urgert, R., van der Weg, G., Kosmeijer-Schuil, T.G., van de Bovenkamp, P., Hovenier, R., Katan, M.B., 1995. Levels of the cholesterol-elevating diterpenes cafestol and kahweol in various coffee brews. Journal of Agricultural and Food Chemistry 43 (8), 2167–2172.

US Food and Drug Administration, 2016. FDA Guidance for Industry Acrylamide in Foods. FDA Office of Food Safety.

United States Department of Agriculture (USDA), 2015. USDA Scientific Report of the Dietary Guidelines Advisory Committee.

van Boxtel, M.P.J., Schmitt, J.A.J., 2004. Age related changes in the effects of caffeine on memory and cognitive performance. In: Nehlig, A. (Ed.), Coffee, Tea, Chocolate and the Brain. CRC Press, Boca Raton, Florida, pp. 85–97.

van Dijk, A.E., Olthof, M.R., Meeuse, J.C., Seebus, E., Heine, R.J., van Dam, R.M., 2009. Acute effects of decaffeinated coffee and the major coffee components chlorogenic acid and trigonelline on glucose tolerance. Diabetes Care 32, 1023–1025.

Villanueva, C.M., Silverman, D.T., Murta-Nascimento, C., Malats, N., Garcia-Closas, M., Castro, F., Tardon, A., Gracia-Closas, R., Serra, C., Carrato, A., Rothman, N., Real, F.X., Dosemeci, M., Kogevinas, M., 2009. Coffee consumption, genetic susceptibility and bladder cancer risk. Cancer Causes and Control 20 (1), 121–127.

Vitaglione, P., Fogliano, V., Pellegrini, N., 2012. Coffee, colon function and colorectal cancer. Food and Function 3, 916–922.

World Anti-doping Agency (WADA). The 2009 monitoring program. www.wada-ama.org.

Waite, M., 2015. Treatment for Alzheimer's disease: has anything changed? Australian Prescriber 38, 60–63.

Williams, L.E., Bargh, J.A., 2008. Experiencing physical warmth promotes interpersonal warmth. Science 322 (5901), 606–607.

Yamamoto, T., Yoshizawa, K., Kubo, S.I., Emoto, Y., Hara, K., Waters, B., Umehara, T., Murase, T., Ikematsu, K., 2015. Autopsy report for a caffeine intoxication case and review of the current literature. Journal of Toxicologic Pathology 28 (1), 33–36.

Yoshinari, O., Igarashi, K., 2010. Anti-diabetic effect of trigonelline and nicotinic acid, on KK-Ay mice. Current Medicinal Chemistry 17 (20), 2196–2202.

Youngberg, M.R., Karpov, I.O., Begley, A., Pollock, B.G., Buysse, D.J., 2011. Clinical and physiological correlates of caffeine and caffeine metabolites in primary insomnia. Journal of Clinical Sleep Medicine 7, 196–203.

Index

'*Note*: Page numbers followed by "f" indicate figures, "t" indicate tables and "b" indicate boxes.'

D

E

F

G

Q

R

S